Energy Crops

RSC Energy and Environment Series

Editor-in-Chief:
Professor Laurence Peter, University of Bath, UK

Series Editors:
Professor Heinz Frei, Lawrence Berkeley National Laboratory, USA
Professor Ferdi Schüth, Max Planck Institute for Coal Research, Germany
Professor Tim S. Zhao, The Hong Kong University of Science and Technology, Hong Kong

Titles in the Series:
1: Thermochemical Conversion of Biomass to Liquid Fuels and Chemicals
2: Innovations in Fuel Cell Technologies
3: Energy Crops

How to obtain future titles on publication:
A standing order plan is available for this series. A standing order will bring delivery of each new volume immediately on publication.

For further information please contact:
Book Sales Department, Royal Society of Chemistry, Thomas Graham House, Science Park, Milton Road, Cambridge, CB4 0WF, UK
Telephone: +44 (0)1223 420066, Fax: +44 (0)1223 420247,
Email: books@rsc.org
Visit our website at http://www.rsc.org/Shop/Books/

Energy Crops

Edited by

Nigel G Halford
Plant Science Department, Rothamsted Research, Harpenden, Herts, UK

Angela Karp
Centre for Biofuels and Climate Change, Rothamsted Research, Harpenden, Herts, UK

RSCPublishing

RSC Energy and Environment Series No. 3

ISBN: 978-1-84973-032-7
ISSN: 2044-0774

A catalogue record for this book is available from the British Library

© Royal Society of Chemistry 2011

All rights reserved

Apart from fair dealing for the purposes of research for non-commercial purposes or for private study, criticism or review, as permitted under the Copyright, Designs and Patents Act 1988 and the Copyright and Related Rights Regulations 2003, this publication may not be reproduced, stored or transmitted, in any form or by any means, without the prior permission in writing of The Royal Society of Chemistry, or in the case of reproduction in accordance with the terms of licences issued by the Copyright Licensing Agency in the UK, or in accordance with the terms of the licences issued by the appropriate Reproduction Rights Organization outside the UK. Enquiries concerning reproduction outside the terms stated here should be sent to The Royal Society of Chemistry at the address printed on this page.

The RSC is not responsible for individual opinions expressed in this work.

Published by The Royal Society of Chemistry,
Thomas Graham House, Science Park, Milton Road,
Cambridge CB4 0WF, UK

Registered Charity Number 207890

For further information see our web site at www.rsc.org

Preface

The last few years have seen the concept of bioenergy and biofuels, that is energy and fuel from plants, come of age. Rising oil prices have led to more food crops being grown for energy as well as food, and not just for niche or local markets but on a large scale, as significant contributors to global energy supply. This has created controversy as it has added to the upward pressure on crop commodity prices that was already being created by the increasing demand for food of an expanding and to some extent more prosperous world population. Increasing attention has therefore been focussed on meeting the rising demand for bioenergy and biofuels in more sustainable ways. A wider range of crops is being explored, including non-food crops, as well as the use of crop residues rather than grain or seed.

Energy Crops is a comprehensive reference of this topic from the plant and agricultural science perspective. It covers energy crops that are already in use and those that are being developed or researched, and includes species that have been cultivated by humankind for millennia and some that have never been considered as crops before. The introductory chapter defines energy crops, reviews the development and state of the technology, gives a historical perspective and introduces the ethical issues. Each of the subsequent chapters is dedicated to a single crop and describes the current usage of that crop for energy, its potential for future development, the economics of its use for energy production and the research that is being undertaken to tailor it for use as an energy crop. Each chapter is written by a specialist author or authors of international standing and we are extremely grateful to all of the contributors who made this book possible.

Nigel Halford and Angela Karp

RSC Energy and Environment Series No. 3
Energy Crops
Edited by Nigel G Halford and Angela Karp
© Royal Society of Chemistry 2011
Published by the Royal Society of Chemistry, www.rsc.org

Preface

Contents

Chapter 1	**Energy Crops: Introduction**		**1**
	Angela Karp and Nigel G. Halford		
	1.1	Introduction	1
	1.2	Biorenewables, Bioenergy and Biofuel Definitions	3
	1.3	Converting Crops to Energy	4
		1.3.1 Direct Combustion	4
		1.3.2 Thermal Conversion	5
		1.3.3 Biological Conversion	5
		1.3.4 Chemical Conversion	6
		1.3.5 First-, Second- and Advanced-Generation Technologies	7
	1.4	Energy-Crop Types	8
		1.4.1 Grain and Seed Crops	8
		1.4.2 Sugar Crops	8
		1.4.3 Oil Crops	9
		1.4.4 Dedicated Biomass Crops	9
		1.4.5 Algae	10
	1.5	The Energy-Crop Debate	10
	Acknowledgements		11
	References		11
Chapter 2	**Challenges and Opportunities for Using Wheat for Biofuel Production**		**13**
	Peter R Shewry, Jackie Freeman, Mark Wilkinson, Till Pellny and Rowan A C Mitchell		
	2.1	Introduction	13
	2.2	Cell Wall Composition of Wheat Grain and Straw	14

	2.3 Effect of Cell Wall Structure on Digestibility	18
	2.4 Biosynthesis of Arabinoxylan in Wheat	18
	2.5 Sustainability of Wheat Production for Biofuel	23
	2.6 Conclusions	23
	Acknowledgements	23
	References	24
Chapter 3	**Maize**	**27**
	Stephen H. Howell	
	3.1 Corn Ethanol	27
	3.2 Yield Improvements	29
	3.2.1 Critical Growth Stages	31
	3.2.2 Grain-Yield Traits	31
	3.2.3 Heterosis	32
	3.2.4 Nitrogen-Use Efficiency	32
	3.3 Cornstarch	34
	3.3.1 Starch Biosynthesis	35
	3.4 Grain Processing	36
	3.5 Corn Stover and Cellulosic Ethanol	37
	3.5.1 Biomass Traits	38
	3.5.2 Maize Cell Walls and Lignocellulose	39
	3.5.3 Cellulose Synthesis	40
	3.5.4 Hemicellulose Synthesis	43
	3.5.5 Lignin Synthesis	44
	3.5.6 Corn-Stover Composition	46
	3.5.7 Corn-Stover Processing	47
	3.5.8 Corn-Stover Removal	49
	Acknowledgements	50
	References	50
Chapter 4	**Sweet Sorghum as a Biofuel Crop**	**56**
	Gene Stevens and Roland A.Y. Holou	
	4.1 Introduction	56
	4.2 Biofuel from Sweet Sorghum	57
	4.3 Sweet Sorghum Cultivars	60
	4.4 Sorghum Diseases	62
	4.5 Weed and Insect Control	64
	4.6 Sorghum Planting Density	64
	4.7 Soil Fertility for Sweet Sorghum	65
	4.8 Measuring Sugar Content in Juice	66
	4.9 Removing Sorghum Seed Heads	67
	4.10 Fermentation	68
	4.11 Utilising Byproducts – Bagasse and Vinasse	73
	4.12 Challenges for the Future	74

| | Acknowledgements | 74 |
| | References | 74 |

Chapter 5 **Sugarcane** — 77
A. D. Santiago, R. Rossetto, W. de Mello Ivo and S. Urquiaga

	5.1	Introduction	77
	5.2	Production System	81
		5.2.1 Varieties	81
		5.2.2 Pests and Diseases	84
		5.2.3 Nutrient Requirements	86
		5.2.4 Mechanical Harvesting and Residue Management	88
		5.2.5 Water Use	89
	5.3	Sugarcane Products and Coproducts	90
	5.4	Energy Balance of the Production of Sugarcane Ethanol in Brazil	92
	5.5	Land-Use Changes in Sugarcane Areas: The Brazilian Case	95
	References		95

Chapter 6 **Sugar Beet** — 104
Mike May

	6.1	Development of Sugar Beet Industries	104
	6.2	Products from Sugar Beet	105
	6.3	Growth of Beet	106
	6.4	Cane or Beet?	107
	6.5	Factory Efficiency	107
	6.6	Biofuels from Sugar Beet	109
	6.7	Ways Forward and Research	111
	6.8	Why Use Sugar Beet – Pros and Cons	113
	6.9	What Could Be Used in the Future?	114
	References		114

Chapter 7 **Oilseed Rape** — 116
Richard Weightman, Peter Gladders and Pete Berry

	7.1	Introduction	116
	7.2	Current Crop and Use as Biofuel Feedstock	118
		7.2.1 Area Yield and Production	118
	7.3	Utilisation and Economics	120
		7.3.1 Current Use for Food, Feed and Fuel	121
		7.3.2 Global Trade and Assurance	123
		7.3.3 Prices and Crop Gross Margins	125

		7.3.4	Revenue from Support Measures	126
		7.3.5	Coproducts: Rapeseed Meal, Glycerol and Straw	127
	7.4	Quality of Oilseed Rape Oil and Other Methyl Esters		128
		7.4.1	Composition	129
		7.4.2	Genetic and Environmental Effects on Fatty Acid Composition	130
	7.5	Yield Improvement		131
		7.5.1	Estimating the Yield Potential of Oilseed Rape	131
			Seeds/m^2	131
			Seed Growth	133
		7.5.2	Achieving the Yield Potential of Oilseed Rape	134
			7.5.2.1 Rotation Length	135
			7.5.2.2 Disease, Pest and Weed Control	135
			7.5.2.3 Establishment Methods	136
			7.5.2.4 Timing and Rate of Sowing	137
			7.5.2.5 Fertiliser Use	137
	7.6	Carbon Footprint		138
	7.7	Future Prospects		141
	Acknowledgements			142
	References			143

Chapter 8 Soybeans 148
Anthony J. Kinney and Tom E. Clemente

8.1	Introduction: A Brief History of Soybean Production	148
8.2	Soybean-Seed Composition and Uses	149
8.3	Properties of Soybean Biodiesel	151
8.4	Breeding Approaches to Improving Soybean-Oil Quality	153
8.5	Genetic Transformation of Soybean	155
8.6	Biotechnology Approaches to Improving Soybean-Oil Quality	156
8.7	Properties of High-Oleic Soybean Oil in Biodiesel Applications	157
8.8	Future Possibilities: Increasing Soybean Oil Content	159
References		161

Chapter 9 Perspectives on Sunflower as an Energy Crop 165
Zina Flagella and Massimo Monteleone

9.1	Worldwide Scenario of Sunflower Cultivation and Production	165
9.2	Survey of Biodiesel Production and Trade	169

9.3	Biodiesel and Oil Quality in Relation to Genetic and Environmental Influence		171
	9.3.1	Genetic, Environmental and Agronomic Influences on Sunflower-Oil Quality	173
		9.3.1.1 Effect of Genotype	173
		9.3.1.2 Effect of Temperature, Sowing Date and Water Regime	174
		9.3.1.3 Simulation Models Predicting Sunflower-Oil Quality in Relation to Environment and Genotype	175
9.4	Energetic Uses of Sunflower Other than Biodiesel		176
	9.4.1	Anaerobic Digestion and Biogas Production	176
	9.4.2	Thermochemical Conversion Processes	178
9.5	Energy Balance of Sunflower Cultivation and Processing		179
9.6	Criticism and Perspectives on the Use of Sunflower as an Energy Crop		182
Acknowledgements			184
References			184

Chapter 10 Palm Oil as an Energy Crop 187
Keat Teong Lee and Kok Tat Tan

10.1	Introduction	187
10.2	Potential of Palm Oil as Energy Crop	189
10.3	Economic Consideration	192
10.4	Palm-Oil Issues and Challenges	192
10.5	Current Research and Development	194
Acknowledgements		195
References		195

Chapter 11 *Jatropha curcas*: A Source of Energy and Other Applications 196
Satyawati Sharma and Ashwani Kumar

11.1	Introduction	196
11.2	Taxonomy and Botanical Description	198
11.3	Distribution and Ecological Requirement	200
11.4	*Jatropha* Cultivation	201
11.5	Composition of *Jatropha*	202
11.6	Toxicity of Jatropha Seeds	203
11.7	Potential Applications of *Jatropha*	207
	11.7.1 Fuel from *Jatropha*	207
	11.7.2 Other Uses	211
11.8	Potential of *Jatropha* in Wasteland Reclamation	214
11.9	Pests and Diseases	214

11.10	Food *vs.* Fuel	214
11.11	Crop Improvements	216
11.12	Energy, Environment and *Jatropha*	218
11.13	Energy Flow	221
11.14	*Jatropha curcas* and Clean Development Mechanism	223
11.15	Economics	223
11.16	Costs and Returns	224
11.17	Conclusions	225
References		227

Chapter 12 *Pongamia pinnata*, a Sustainable Feedstock for Biodiesel Production 233
Stephen H. Kazakoff, Peter M. Gresshoff and Paul T. Scott

12.1	Introduction	233
12.2	Legume Nodulation and Symbiotic Nitrogen Fixation	234
12.3	*Pongamia pinnata*, an Emerging Biodiesel Feedstock	236
12.4	*Pongamia* Seed Oil for Biodiesel Production	241
12.5	Manipulation of *Pongamia* Seed Oil	250
12.6	Propagation and Genetic Modification of *Pongamia*	250
12.7	Future Challenges and Concluding Remarks	252
Acknowledgements		254
References		254

Chapter 13 Willow 259
S. J. Hanley

13.1	Introduction		259
	13.1.1	The Genus *Salix*	260
	13.1.2	Classification	260
	13.1.3	Willow Biology	261
13.2	Willow as a Bioenergy Crop		261
	13.2.1	Fast Growth and Coppicing Ability	261
	13.2.2	Low-N Input Requirements	262
	13.2.3	Natural Adaption to Marginal Land	262
	13.2.4	Genetic Diversity for Breeding	263
13.3	Cultivation and Agronomy		263
13.4	Willow Genetic Improvement		263
	13.4.1	Germplasm Collections	264
	13.4.2	Breeding Willows for Bioenergy End Uses	265
	13.4.3	Practical Breeding and Selection	265

		13.4.4	Targets for Crop Improvement	266
			13.4.4.1 Increased Biomass Yield	266
			13.4.4.2 Increased Resource-Use Efficiency	267
			13.4.4.3 Increased Resistance to Biotic Stresses	267
			13.4.4.4 Composition Traits	268
	13.5	Genetic and Genomic Research to Underpin Breeding		269
	13.6	Environmental Benefits		270
	13.7	Conclusions		271
	Acknowledgements			272
	References			272

Chapter 14 Poplar — 275

S. Y. Dillen, O. El Kasmioui, N. Marron,
C. Calfapietra and R. Ceulemans

	14.1	The Poplar Genus	275
		14.1.1 Uses and Interests of the Poplar Genus	275
		14.1.2 Subdivision of the Poplar Genus	276
	14.2	Poplar SRC Systems	280
		14.2.1 Synoptic History of Poplar SRC	280
		14.2.2 SRC Principles	281
		14.2.3 Case Study	283
	14.3	Breeding	284
		14.3.1 Quantitative Traits: Yield and Yield Components	284
		14.3.2 Qualitative Traits: Wood Quality	285
		14.3.3 Molecular Genetics and Molecular Biological Tools	286
	14.4	Current and Future Limitations to Large-Scale Poplar SRC	286
		14.4.1 Environmental Impacts	287
		14.4.2 Water and Nutrient Requirements	287
		14.4.3 Pest and Disease Sensitivity	288
		14.4.4 Effects of Global Change on Poplar SRC	288
		14.4.5 Life-Cycle Assessment (LCA)	290
	14.5	Economic and Energetic Analysis of SRC Poplar	290
		14.5.1 Costs of Poplar for Bioenergy	291
		14.5.2 Energy Balance of Poplar SRC	292
	14.6	Conclusions	293
	Acknowledgements		294
	References		294

Chapter 15 Developing *Miscanthus* for Bioenergy   301
John Clifton Brown, Steve Renvoize, Yu-Chung Chiang,
Yasushi Ibaragi, Richard Flavell, Joerg Greef, Lin Huang,
Tsai Wen Hsu, Do-Soon Kim, Astley Hastings, Kai Schwarz,
Paul Stampfl, John Valentine, Toshihiko Yamada,
Qingguo Xi and Iain Donnison

15.1	Introduction	302
	15.1.1 How Was *Miscanthus* Identified as a Key Candidate Bioenergy Feedstock?	302
15.2	*Miscanthus*, Taxonomy and Origins	303
15.3	Physiological Traits	306
	15.3.1 C4 Photosynthesis in *Miscanthus*	306
	15.3.2 Nutrient-Use Efficiency	306
	15.3.3 Water-Use Efficiency	306
	15.3.4 Overwintering Frost Tolerance	307
	15.3.5 Radiation Capture	308
	15.3.6 Pests, Diseases and Weed Control	308
	15.3.7 Crop Modeling	309
15.4	Towards Breeding Better *Miscanthus* for Bioenergy	309
	15.4.1 Setting Breeding Targets	309
	15.4.2 *Miscanthus* Germplasm Collection and Characterisation	311
	15.4.3 Crossing	312
	15.4.4 Selection	312
15.5	Development of the Whole Biomass Chain	313
	15.5.1 Propagation and Establishment	313
	15.5.2 Harvest	314
	15.5.3 Utilisation/Fuel Matching with End Uses	314
15.6	Life-Cycle Analysis and Energy Balance	315
15.7	Biodiversity and Visual Impact	316
15.8	Conclusions	317
Acknowledgements		317
References		317

Chapter 16 Subtropical and Tropical Reeds for Biomass   322
Mihály Czakó and László Márton

16.1	Introduction	322
16.2	Giant Reed – *Arundo*: Botany and Nonbiomass Uses	323
	16.2.1 Giant Reed – *Arundo* for Biomass	325
	16.2.1.1 Energy Input and Output Considerations	325
	16.2.1.2 Natural Biomass Accumulation Potential	326
	16.2.1.3 Sustainable Harvested Yields	326

		16.2.1.4	Water-Use Efficiency	331
		16.2.1.5	Harvesting	331
	16.2.2	Other Subtropical and Tropical Reeds		331
References				334

Chapter 17 Switchgrass 341
Kenneth P. Vogel, Gautam Sarath, Aaron J. Saathoff and Robert B. Mitchell

17.1	Introduction		341
17.2	Adaptation and Distribution		342
17.3	Morphology and Taxonomy		344
17.4	Physiology and Growth		344
	17.4.1	Abiotic Factors	344
	17.4.2	Biotic Factors	346
17.5	Genetics and Breeding		347
	17.5.1	Genetics	347
	17.5.2	Breeding	349
17.6	Management as a Biomass Energy Crop		353
	17.6.1	Establishment	353
	17.6.2	Fertility Management	355
	17.6.3	Harvest Management	356
	17.6.4	Seed-Production Management	357
17.7	Composition and Conversion		357
	17.7.1	Composition	357
	17.7.2	Switchgrass Pretreatment	361
17.8	Economics and Net Energy		367
17.9	Sustainability and Ecological Services		368
References			369

Chapter 18 Algae 380
Ira A. Levine

18.1	Introduction		380
18.2	Macroalgae		383
18.3	Microalgae		395
18.4	Microalgae Biofuel Value Chain (Table 18.4, Figure 18.10)		396
	18.4.1	Stage 1: Project Planning	396
	18.4.2	Stage 2: Cultivar Selection and Enhancement	396
	18.4.3	Stage 3: Cultivation System	401
	18.4.4	Stage 4: Harvesting	407
	18.4.5	Stage 5: Extraction	408

	18.4.6 Stage 6: Conversion	409
	18.4.7 Stage 7: Policy and Regulatory Review	411
18.5	Conclusion	413
References		413

Subject Index   **416**

CHAPTER 1
Energy Crops: Introduction

ANGELA KARP AND NIGEL G. HALFORD

Departments of Plant and Invertebrate Ecology and Plant Science, Centre for Bioenergy and Climate Change, Rothamsted Research, Harpenden, Herts AL5 2JQ, UK

1.1 Introduction

Two major events that impacted significantly on the development of humankind involved the use of plants: our ability to make fire and our change from being hunter-gatherers to food-producers. The exact dates in which these advances first occurred are subject to debate but it is certain that they occurred in this order. Estimates suggest that agriculture arose some 10 000 years ago, whilst the control of fire may date back some 790 000 years.[1] As food production became more efficient, it became possible for larger numbers of people to live together. Human populations expanded and civilisations were born.[2] During this expansion, the requirement for plants to provide fuel was not in conflict with food production. Rather, this requirement diminished as alternative sources of energy were developed. As a result, twenty-five years ago, although plants were still being used for fuel in underdeveloped regions of the world, it was oil, coal, natural gas and nuclear power that together fulfilled most of the world's energy needs.[3]

Within a remarkably short timeframe, a major change then took place. The first steps of this process can be traced to the first oil crisis of the 1970s, when a sudden rise in the price of oil led to the first push for the development of renewable energies. In addition to food production, many governments supported the development of novel nonfood crops (see Chapters 11–18).

These crops were harvested and combusted to produce heat and power, a contribution that continues today. However, a drop in oil prices in the 1990s stemmed enthusiasm and, other than the production of bioethanol in Brazil (see Chapter 5), bioenergy markets remained small.

A culmination of events then precipitated a "green energy" boom. Oil prices spiked and there was increasing concern over energy security. Moreover, numerous reports, particularly from the Intergovernmental Panel on Climate Change (IPCC)[4] and from Nicholas Stern,[5] focused attention on the substantial cost to humankind of *not* acting to reduce the current rate of increase in greenhouse gas (GHG) emissions. As the use of fossil fuels is a major contributor to climate change, the need for alternative sources of energy, which save carbon and are renewable, was placed quickly and firmly back at the top of global agendas.

There are many possible alternatives to fossil fuels, particularly for heat and power generation, *e.g.* wind, hydro, solar, as well as plant biomass, all of which are expected to play a role. However, there are few alternatives to replace transport fuels (*e.g.* electric, hydrogen). As the number of vehicles on the roads is continually rising, it is clear that, unless emissions from the transport sector can be curbed, they will counter any reductions achieved by other sectors. Combined with the desire of some nations to reduce their reliance on the fuel supplies of a few major producers, this resulted in a swift, strong push to increase the production of liquid transport fuels from crops and many food crops were exploited for this purpose (see Chapter 2–10). Simultaneously, the need for fossil-fuel substitutes for a whole range of products that currently rely on the refining of oil became apparent to the chemical industries.

As a consequence of these environmental and political drivers, three new markets have emerged for plants, all of which are potentially huge: bioenergy, biofuels and biorenewable materials (Figure 1.1). However, securing sufficient food for future populations remains a major challenge, particularly in the face of climate change. To balance all of these demands on plants will require little short of another revolution. One decade into this millennium, major challenges face humankind. "A Perfect Storm" were the words adopted by Professor Beddington, the UK government's chief scientific adviser, to draw attention to this recently.[6] He pointed specifically to research that indicated that by 2030 "a whole series of events come together": The world's population will rise from 6bn to 8bn (33%); demand for food will increase by 50%; demand for water will increase by 30% and; demand for energy will increase by 50%.[6]

Whilst crops certainly hold solutions to these challenges, they are also part of the problem. Agriculture is a major user of resource, including energy from fossil fuels, and is also another major contributor to GHG emissions. Research has drawn attention to the fact that producing energy from crops could have the opposite effect to what is intended, if the energy inputs exceed the carbon benefits and if the GHG emissions associated with any new cultivation for displaced food production are not taken into account.[7]

In this book, the potential contributions of different energy crops will be reviewed within the context of the different and sometimes conflicting

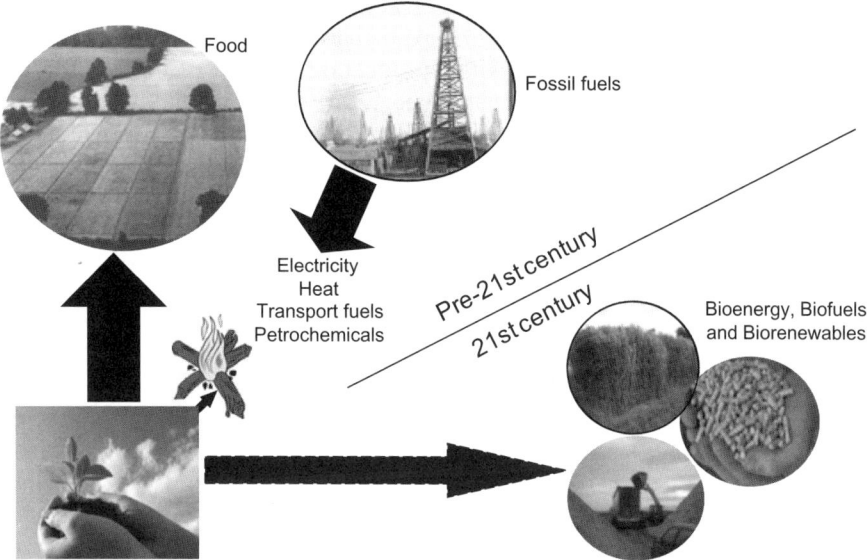

Figure 1.1 Major challenges lie ahead. Until now, plants have predominantly provided food and although they have been used for fuel since early mankind, fossil fuels and nuclear power have increasingly supplied the world's energy needs. In the 21st century, crops will also be needed to supply raw materials for expanding bioenergy, biofuel and biorenewable markets.

challenges and drivers that are currently acting. By way of introduction, this chapter will introduce various bioenergy terms, outline how feedstock from energy crops can be converted into energy, briefly define the different types of energy crops and finally draw attention to the main components of the debate surrounding energy crop production. The principle aim will be to introduce the topics and not to comprehensively review them, as the subsequent chapters will provide the details.

1.2 Biorenewables, Bioenergy and Biofuel Definitions

"Biorenewables" is an all-embracing term that covers the production of heat, power, transport fuel and other products from organic matter of recent (as distinct from fossil) origin. These converge in the "biorefinery" concept, which can be defined as the sustainable coproduction of a spectrum of biobased products (food, feed, materials, chemicals) and energy (fuels, power, heat) from biomass. However, not all biorenewable products are derived from biorefining *per se*. The need for more sustainable products with reduced carbon footprints has led to the development of many new supply chains from nonfood crops and wastes.

In this book we will confine our discussions to the production of renewable energy from biological materials, or "bioenergy". In practice, this term is sometimes used with reference only to renewable heat and power and sometimes also to include renewable transport fuels. Here, we adopt an earlier definition[8] and take it to mean the production of any renewable energy, reserving the use of the term "biofuels" for renewable liquid transport fuels only. Biofuels can be subdivided into "bioethanol" made from grain and sugar crops, which provides a substitute for petrol, and "biodiesel", made from vegetable oil, which is a substitute for diesel (see Sections 1.3.3 and 1.3.4).

Energy crops can be grouped on the basis of the main market their production is targeted at (*e.g.* "biofuel crops") and on the type of feedstock that they provide for conversion (*e.g.* "biomass crops"). This is then further qualified by taking into account the different conversion processes that can be applied and by virtue of the fact that more than one type of energy source can be obtained from the same crop.[8,9] Thus, the term "biomass crops" is most often used with reference to growing crops for the purpose of producing total biological mass as a feedstock, whilst "lignocellulosic crops" refers more specifically to use of the cell wall components of plants. "Bioethanol crops" and "biodiesel crops" refer to crops grown for production of bioethanol and biodiesel, respectively, and so on. There are also secondary classifications of "first" and "second" generation, depending upon whether conversion is based on simple sugars or more complex lignocelluloses as the energy source (see below). Whilst this book is not concerned with detailed descriptions of the conversion processes (a topic to be covered by another volume in this series), it is important to outline what they are in order to have clarity over the specific energy-crop system being referred to at any one time.

1.3 Converting Crops to Energy

There are four main routes to energy conversion that can be summarised as: direct combustion, thermal conversion (pyrolysis, gasification), biological conversion (anaerobic digestion, fermentation) and chemical conversion (transesterification).

1.3.1 Direct Combustion

Direct combustion is the oldest and still the most commonly used route for converting biomass to heat, power and combined heat and power (CHP). It is applicable at all scales from domestic to industrial, using boilers that range from small domestic stoves (1 to 10 kW) to the largest devices used in power and CHP plants (>5 MW). Biomass of many different forms can be used, including wood chip, pellets and different straws. The most efficient combustion is achieved when dedicated biomass boilers are used. However, solid biomass particles can also be mixed with coal in cofiring or cogeneration, enabling the large power-generation industries to reduce their carbon footprint by incorporating biomass as a percentage of their feedstock.

1.3.2 Thermal Conversion

In thermal conversion processes, high temperatures are used but energy is not produced directly from biomass. Instead, the biomass is converted to energy carriers such as synthetic gases, oil or methanol that have higher energy densities (and thus lower transport costs) and/or more predictable and improved combustion characteristics.[10] The two main thermochemical processes, pyrolysis and gasification, differ with respect to the presence/absence of oxygen.

Pyrolysis occurs in the complete absence of oxygen at temperatures in the range of 400–800 °C.[11,12] During this process, most of the cellulose and hemicellulose and part of the lignin, which together make up the lignocellulosic fraction, disintegrate and form gases. As these gases cool, some of the vapours condense to form bio-oil, which has potential as a substitute for fuel oil and as a feedstock for synthetic gasoline or diesel production. The remaining biomass, mostly comprised of lignin, forms charcoal.

Gasification requires partial oxidation and temperatures of around 800 °C. The biomass is partially burned to form "producer gas" and charcoal. The carbon dioxide (CO_2) and water (H_2O) in the producer gas are chemically reduced by the charcoal to form carbon monoxide (CO) and hydrogen (H_2). Producer gas contains 18–20% H_2, 18–20% CO, 8–10% CO_2, 2–3% methane (CH_4), trace amounts of higher hydrocarbons (*e.g.* methane and ethane), water, nitrogen (if air is used as the oxidising agent) and various contaminants such as small char particles, ash, tars and oils. The partial oxidation in the gasification process can be carried out using air, oxygen, steam or a mixture of these. When air is used, a low heating value gas is produced that is suitable for use in boilers, engines and turbines but not for transporting in pipelines, because of the low energy density. Gasification with oxygen or steam produces a medium heating value gas, which is suitable for limited pipeline distribution, as well as a synthesis gas or "syngas" (typically 40% CO, 40% H_2, 3% CH_4 and 17% CO_2, dry basis). Syngas can be used to make methanol, ammonia and diesel using Fischer–Tropsch synthesis,[10] first developed by F. Fischer and H. Tropsch in 1923.

1.3.3 Biological Conversion

There are two main forms of biological conversion: anaerobic digestion and fermentation. Both are well-developed technologies.

Anaerobic digestion (AD) is the breakdown of organic materials by bacteria in the absence of oxygen. Almost any organic material can be processed, including waste paper, grass cuttings, food waste, industrial effluents, sewage and animal waste. The result is a biogas made up of around 60% methane and 40% CO_2. Biogas can be burnt to generate heat or (once it is scrubbed) electricity. It can also be used as a biofuel. A solid and liquid residue called digestate is also produced, which can be used as a soil conditioner. The amount of biogas and the quality of digestates obtained vary according to the feedstock used. More biogas will be produced if the feedstock is putrescible, which means it is more liable to decompose. The use of sewage and manure produces less biogas as the animal that produced it has already removed some of the energy content.

Fermentation, followed by distillation, is the biological conversion process used for converting sugars to ethanol or, depending on the microbial strain, other low molecular weight alcohols. Most ethanol fermentation is based on Baker's yeast (*Saccharomyces cerevisiae*), which requires simple (monomeric) sugars as raw material. Conventional yeast fermentation produces 0.51 kg of ethanol from 1 kg of any of the C6 sugars, such as glucose and mannose, or sucrose. However, not all feedstocks contain simple sugars. Starch, for example, is a polymer and when starch is used a hydrolysis step is first required to break it down into simple sugars for fermentation. Bioethanol production from starch is now well established in the USA, with maize (Chapter 3) being the major raw material.

Polymeric carbohydrates are also present in all cell walls of the plant and potentially provide the most abundant source of carbon for bioenergy and biofuel production available on earth. However, in this form the sugars are not readily accessible, existing in the form of fibres, sometimes interlinked with lignin, and additional steps of pretreatment and hydrolysis are required to release the sugars for fermentation. The cell wall polymers of wheat are described in Chapter 2.

Fermentation and distillation to produce ethanol from sugar have been at the heart of the brewing and of the wine and spirit industries for a long time in human history. The same well-developed and efficient procedures have been used effectively to produce alcohols that can be used as fuels for internal combustion engines. This is well established in Brazil using sugar from sugarcane (Chapter 5), and is being developed in the USA using sweet sorghum (Chapter 4). When used as a biofuel, ethanol is blended with gasoline. The percentage blend depends on the engine type: a 10% blend (E10) can be used by most engines, whereas an 85% blend (E85) requires specialised flexifuel vehicles.

1.3.4 Chemical Conversion

The main chemical conversion process is transesterification which results in the production of biodiesel; effectively, biodiesel is the fatty acid methyl esters (FAME) produced from different oil-containing crops. In fact, lipids (*e.g.* from algae (Chapter 18)) and oils, either directly produced by crops or derived from processed vegetable oil from the food industry, can all be used as feedstock for conversion into biodiesel.

Biodiesel production begins with pressing the crop to produce a liquid oil fraction and an oil cake byproduct, which can be used as cattle feed. The liquid vegetable oils can be used directly as engine fuels, but this requires engine modification because of their very high viscosity, poor thermal and hydrolytic stability and less favourable ignition qualities. As a result, transesterification is used to transform the large, branched molecular structure of the oils into smaller, straight-chained molecules similar to those of standard diesel. There are three basic routes: (i) Base-catalysed transesterification; (ii) Direct, acid-catalysed transesterification; (iii) Conversion of the oil to its fatty acids and then to biodiesel. The base-catalysed method is the most commonly used because it uses low temperature (50–66 °C) and pressure (around 1.4 bar), has a high yield (98%) with minimal side reactions and reaction time, and it is a direct conversion to biodiesel with no intermediate compounds. The base catalyst is methoxide (CH_3O^-), which is generated by dissolving sodium hydroxide or potassium hydroxide in methanol. The process results in the formation of glycerol, a second byproduct, as well as the biodiesel that can be used as a substitute for, or additive to, petroleum-based diesel.

1.3.5 First-, Second- and Advanced-Generation Technologies

Biofuel production from crops in which the sugars/starches and oils are stored in forms that are easily accessible is often referred to as "first-generation" biofuel. This is because the fermentation of sugars and starches and transesterification of lipids and oils for biofuel production involves well-developed conversion technologies, the only real change being the use of the product for transport rather than human consumption.

In contrast, although cell wall polysaccharides represent an abundant potential source of sugars, their recalcitrance to breakdown presents a major challenge. The additional pretreatment and hydrolysis steps are currently inefficient, energy intensive and expensive to perform. Considerable optimisation of the enzymatic and physicochemical processes is needed to improve the efficiency of the conversion chain, or modification of the polysaccharides themselves through plant breeding or genetic modification, before biofuels can be cost-effectively produced in this way. As a result, the conversion of cell wall polysaccharides has become known as "second-generation" biofuel. The same is true for production of liquid fuel from gasification and pyrolysis. Although possible, large biomass volumes are required and considerable improvements in efficiency are needed for these to be truly commercially viable at the industrial scale.

In many ways, the terms first and second generation are misleading, since in practice there is a continuum of technological processes and numerous ways in which feedstocks and processes can be combined. This is also a very fast-changing field, and one into which considerable investment has been placed. Similarly, the distinction between second generation and more advanced generation technologies is often difficult to make.

One alternative that is also being explored is the use of fuel cells, which are devices that convert chemical energy directly into electrical energy. Ethanol or methanol produced from biomass by fermentation or chemical catalysis can be fed into the anode of a fuel cell and air or oxygen into the cathode. Electrolysis oxidises the ethanol or methanol to CO_2 and H_2O, generating an electrical current in the process that can be used to power electric motors. A challenge to this technology is that the membrane separating the anode and the cathode in the fuel cell is somewhat permeable to the alcohol fuels, which reduces efficiency. The ultimate advancement would be the use of hydrogen, instead of alcohols, to generate electrons in the fuel cell. Using hydrogen as a fuel in vehicles has attractions as it would generate only water as an emission. However, much further development is required and there are safety issues to be dealt with concerning hydrogen storage. At present, production of hydrogen from water is not yet feasible in practice but possible ways of using lignocelluloses for hydrogen generation are being developed.

1.4 Energy-Crop Types

In theory, any crop could be used as an energy crop. Over 80 are referred to in a recent handbook, for example.[13] In practice, however, issues relating to the availability of feedstock and the efficiency, cost effectiveness and sustainability of the whole chain, from field to fuel (see Section 1.5) restrict the choice. In this book, the major energy-crop types are covered by separate chapters and will simply be introduced here.

1.4.1 Grain and Seed Crops

The presence of wheat and barley seed in archaeological sites dating back to 6750 BC is testimony to the importance that cereals have played in the development of human societies. Today, maize, rice and wheat dominate world agricultural production and, together with a whole range of other grain and seed crops, provide the staple food of populations worldwide.[2]

Grain and seed crops have also traditionally been used in fermentation to produce beer, wine and spirits, because the stored carbohydrates (sucrose and starch) can be readily broken down by enzymatic systems. The adoption of this process to produce bioethanol for vehicles, however, has resulted in the use of food grain for nonfood purposes on an unprecedented scale. Maize (Chapter 3) and wheat (Chapter 2) currently make the largest contribution to biofuel production, matched only by that from sugarcane (Chapter 5) and oil palm (Chapter 10).[8] Maize is also used for biogas. In addition, the parts of cereal crops that are not used for food (*e.g.* wheat straw and corn stover) can be used as a source of biomass for thermal conversion and lignocellulose for second-generation biological conversion. Similarly, grain from sorghum as well as stalks of sweet sorghum (Chapter 4) can both be used for biofuels, whilst sorghum stovers could be used as a source of lignocellulose.

1.4.2 Sugar Crops

Sugars are transported in plant stems in normal development but some species can also store high concentrations. The main sugar crops used for bioenergy are sugarcane (Chapter 5), sugar beet (Chapter 6) and, as mentioned above, sweet sorghum (Chapter 4).

Bioethanol production from sugarcane is an extremely efficient and well-developed industry in Brazil (Chapter 5). Indeed, it is interesting to contemplate whether or not other biofuel crops would really be competitive if sugarcane could be cultivated throughout the world in the way that it can in the tropics and subtropics. Most certainly, it is among the most productive plants known and it is also able to store high concentrations of sucrose in the stem. In addition, sugarcane bagasse (the fibrous residue) is a primary fuel source, making most sugarcane mills extremely efficient. It could also be a source of lignocelluloses and the feasibility of producing ethanol from bagasse is currently under investigation. Sweet sorghum is adapted to both humid and tropical climates but can be grown in colder climates than sugarcane (Chapter 4). In cooler temperate climates, sugar beet (Chapter 6) can be used as a source of sugar for bioethanol.

1.4.3 Oil Crops

Oil palm (Chapter 10) is by far the largest producer of oil for biodiesel. However, a large range of alternative oil crops are grown in areas where the climate does not favour oil-palm production.[14] This includes soybean (particularly in the Americas) (Chapter 8) and oilseed rape (particularly in Europe and cooler temperate areas) (Chapter 7). More recently, Jatropha (Chapter 11) has been heralded as a promising biofuel crop for drought-prone environments, as well as other species, such as Pongamia (Chapter 12). Other oil crops, such as sunflower (Chapter 9), babassu palm, peanut and even olive are also used for biodiesel.

1.4.4 Dedicated Biomass Crops

As pointed out previously, biomass can be obtained from any crop. Indeed, practices such as using the whole wheat crop (grain included) for combustion are known to have been carried out (this probably exacerbates global warming through the production of nitrogen oxides, which are much more potent GHGs than CO_2). However, the term "dedicated biomass crops" refers to nonfood crops that are solely grown for biomass production. These comprise mostly perennial grasses and fast-growing trees. Dedicated biomass crops were first developed for combustion and thermal conversion technologies but, due to their potential to supply high yields of lignocelluloses, have become of interest for second-generation biofuels. Perennial grasses are also widely used for biogas, but wood chip is not suitable for this process.

An impressive number of perennial grasses are used as energy crops but, in this volume, coverage is restricted to the major ones; *Miscanthus*, (Chapter 15); switchgrass (Chapter 17) and reeds (Chapter 16).[12] Similarly, this volume covers two main fast growing trees: willows (Chapter 13) and poplars (Chapter 14).

1.4.5 Algae

Algae fall into two main types: microalgae (phytoplankton, microphytes or planktonic algae) and macroalgae (seaweed). Both are used for biofuel production, although microalgae have received most attention due to their ability to be grown in ponds and bioreactors. Macroalgae can be grown on ropes. As the photosynthetic efficiency of algae (6–8% on average) is higher than terrestrial plants (1.8–2.2% on average) they are able to accumulate biomass at faster rates. Other advantages are that they do not require the use of high-grade productive land and can utilise a wide range of water sources (fresh, brackish, saline and waste water) (Chapter 18).

1.5 The Energy-Crop Debate

Back in the 1970s, when oil prices first rose, the principle of growing crops for energy was encouraged without challenge. When biofuels first came along, they were heralded as "green gold" but, all too quickly, they became "a crime against humanity" and "food versus fuel" is a commonly heard phrase these days.

Two points remain clear: The challenges ahead are formidable (Figure 1.1) and energy crops have the potential to provide a source of renewable energy that can reduce GHG emissions and help combat climate change. However, put in very simple terms (which it most certainly is not), growing energy crops requires resources (land, water, energy) and using these resources for energy means that they are not available for food. Converting land use to energy crops may result in direct environmental impacts, for example on biodiversity, or water availability. Even if there is enough land to manage all these aspects favourably (a matter of much debate), the justification for energy crops becomes hard to defend if the energy inputs required from crop to fuel result in little or no carbon savings and GHG reductions.

The reason why little (if any) of these concerns were raised back in the 1970s and 80s is because the bioenergy system that was being encouraged was the use of dedicated biomass crops and crop wastes and the scale of land conversion was still relatively small. Out of a large potential number of species, a few perennial grasses and fast-growing trees were promoted because of their advantages as nonfood crops with the potential to produce high biomass yields with relatively low fertiliser inputs. Life-cycle analyses of biomass to heat and power produced in this way shows high carbon savings and GHG reductions. The use of first-generation food crops for biofuels is a different situation. Crops like maize and wheat require high nitrogen fertiliser inputs and the energy savings are therefore much lower, and (depending on the management) may even be negative. When only seed/grain is used there is also wastage and lower yield. Finally, there is direct competition with food production as the grain is diverted to an alternative use, although high-protein animal feed is a significant coproduct.

Much effort has been placed on improving the entire chain from crop to fuel, for all energy crops, and the calculation of energy balances is very much affected by the boundaries set on the system and the coproducts that are included. Nonetheless, there is general recognition that efficient second- and advanced-generation systems are needed, which access the cell wall polysaccharides in the nonedible parts of the plant. However, the biofuels debate took a more complicated turn with the publication of a paper in Science by Searchinger and colleagues.[7] They claimed that the increase in the use of maize for fuel would result in new plantings of maize around the world to make up the shortfall in food supply and that the GHG emissions resulting from this new land conversion would not only offset any savings but even result in a carbon debt. Although the assumptions made in the paper have been rigorously challenged, the outcome has been to slow down the pace of energy crops expansion and the message "proceed with caution" seems to sum up the situation now. These issues, only briefly touched on here, are covered in many of the chapters in relation to the crop being focused on.

Acknowledgements

Rothamsted Research receives grant-aided support from the Biotechnology and Biological Sciences Research Council (BBSRC) of the UK.

References

1. N. Goren-Inbar, N. Alperson, E. Kislev, O. Simchoni, Y. Melamed, A. Ben-Nun and E. Werker, *Science*, 2004, **304**, 725.
2. C. B. Heiser Jr., *Seed to civilization. The Story of Man's Food*, W.H. Freeman and Company, San Francisco, 1973.
3. R. E. H. Sims, A. Hastings, B. Schlamadinger, G. Taylor and P. Smith, *Glob. Change Biol.*, 2006, **12**, 2054.
4. IPCC, *Assessment Report of the Intergovernmental Panel on Climate Change*, ed. S. Solomon, D. Qin, M. Manning, Z. Chen, M. Marquis, K.B. Averyt, M. Tignor, H.L. Miller, Cambridge University Press, Cambridge, UK and New York, NY, USA, 2006, pp. 996.
5. N. Stern, *The Economics of Climate Change. The Stern Review*. Cambridge University Press, New York, 2006.
6. Professor Sir John Beddington's Speech at SDUK 09. http://www.govnet.co.uk/news/govnet/professor-sir-john-beddingtons-speech-at-sduk-09.
7. T. Searchinger, R. Heimlich, R. A. Houghton, F. X. Dong, A. Elobeid, J. Fabiosa, S. Tokgoz, D. Hayes and T. H. Yu, *Science*, 2008, **319**, 1238.
8. A. Karp and I. Shield, *New Phytol*, 2008, **179**, 15.
9. R. E. Sims, *Renew. Energy*, 2001, **22**, 31.

10. M. M. Wright and R. C. Brown, *Biofuels, Bioprod. Biorefin.*, 2008, **2**, 229.
11. B. V. Babu, *Biofuels, Bioprod. Biorefin.*, 2008, **2**, 393.
12. D. A. Laird, R. C. Brown, J. E. Amonette and J. Lehmann, *Biofuels, Bioprod. Biorefin.*, 2009, **3**, 547.
13. N. El Bassam, *Handbook of Bioenergy Crops; A Complete Reference to Species, Development and Applications*, Earthscan Ltd, London, 2009, pp. 640.
14. S. Al-Zuhair, *Biofuels, Bioprod. Biorefin.*, 2007, **1**, 57.

CHAPTER 2
Challenges and Opportunities for Using Wheat for Biofuel Production

PETER R SHEWRY, JACKIE FREEMAN, MARK WILKINSON, TILL PELLNY AND ROWAN A C MITCHELL

Department of Plant Science, Rothamsted Research, Harpenden, Hertfordshire, UK, AL5 2JQ

2.1 Introduction

Over 600 million tonnes of wheat are harvested annually in the world, with about 15–16 million tonnes being grown in the UK. It is the major crop grown in the UK and Europe and, together with maize and rice, one of the three major cereal crops that dominate world agricultural production. Intensive plant breeding and, in particular, the exploitation of dwarfing genes [reviewed by refs. 1,2], has resulted in a harvest index of about 50%, making wheat one of the most efficient conversion systems for solar energy into harvested organs. Furthermore, about 70% of the dry grain is starch, which is readily fermented to give ethanol or butanol. It is therefore an attractive "primary" biofuel crop and two major new facilities in the north of England (operated by Ensus at Wilton and Vivergo Fuels at Saltend) will process up to 2 million tonnes of wheat a year when fully operational. However, two other factors must also be taken into account.

First, although substantial amounts of wheat are already used for livestock feed and as industrial feedstock, it is still viewed in most parts of the world as a food crop, and is deeply embedded in many cultures and religions. Hence, it is inevitable that the use of wheat for biofuel will be a focus of the food *vs.* fuel debate.

Secondly, in many countries the production of economically viable yields of wheat requires energy, in terms of the production, distribution and application of fertilisers and agrochemicals, and other uses of farm machinery. Hence, it is essential that a detailed life-cycle analysis is carried out.

The utilisation of wheat as a source of starch for biofuel production means that currently only about 40% of the harvested dry weight is used, with the remaining 60% (in the straw and other grain components) not being fermentable. However, if the barriers to fermentation can be overcome, the use of byproducts of wheat and other food crops for biofuel production has effectively no input costs for production (and no land-use implications) and is therefore a far more desirable option than use of starch.[3] The development of modified varieties and/or fermentation systems to utilise these crop and grain residues should therefore be a key target if wheat is to have a long-term future as a biofuel crop.

2.2 Cell Wall Composition of Wheat Grain and Straw

The composition of the wheat grain is affected by genetic and environmental factors, including nitrogen availability that largely determines grain protein content and grain size, which affects the overall composition (due to changes in the proportions of the major tissues). Hence, it is only possible to give broad figures for composition. The major component is carbohydrates, principally starch ($\approx 70\%$ grain dry weight), and cell wall polysaccharides ($\approx 10\%$), with about 10–14% protein. Other components include oligosaccharides and monosaccharides, lipids, minerals (measured as ash) and a range of phytochemicals and vitamins.

However, the grain comprises a number of tissues (Table 2.1) which differ in their compositions, including the amounts and proportions of cell wall polysaccharides (Table 2.2).

The starchy endosperm is the major storage tissue of the grain, accounting for about 83% of the dry weight, and forms the white flour fraction on milling. It comprises 70–80% starch, with about 8–12% protein and 2–3% cell wall polysaccharides. The major cell wall components in wheat are arabinoxylan (AX), accounting for about 70% of the cell walls, with about 20% (1,3:1,4) β-glucan, 7% glucomannan and 2% cellulose.

Arabinoxylan has a linear backbone of (1→4) linked β-D-xylopyranosyl units that are either unsubstituted, monosubstituted with arabinose on position O-3 or disubstituted with arabinose on the O-2 and O-3 positions (Figure 2.1). The degree to which this substitution occurs varies, for example, 60.6% to 70.3% of the units were unsubstituted, 16.1% to 27.3% were monosubstituted

Table 2.1 Histological comparison of mature wheat grains from two wheat cultivars (Caphorn and Crousty) (% of dry weight).

	Caphorn	Crousty
Germ	3.0	3.2
Embryonic axis	1.5	1.7
Scutellum	1.5	1.5
Starchy endosperm	82.7	83.7
Peripheral layers	14.3	13.1
Aleurone layer	6.5	6.4
Intermediate layer[a]	3.8	3.2
Outer pericarp	4.0	3.5

[a]composed of the hyaline layers + testa + inner pericarp
Taken from Barron et al.[45]

and 7.7% to 19.3% were disubstituted in 90 lines derived from a cross between wheat varieties differing in total AX content.[4] They occur in water-extractable (WE-AX) and water-unextractable (WU-AX) forms, with WE-AX generally forming about 25% of the total.[5]

The degree of substitution with arabinose has been suggested to affect the solubility of the polymer, with highly substituted AX being more soluble due to reduced aggregation of the xylan backbone.[6] However, Saulnier et al.[4] reported an opposite trend in wheat, with water-extractable WE-AX having a lower ratio of A:X than WU-AX. This decreased solubility probably results from a further structural feature: the feruloylation of arabinose units. A proportion of the arabinose units is esterified with ferulic acid (or more rarely, related other phenolic acids such as p-coumaric acid) at the O-5 position (Figure 2.1). Ferulic acid has been estimated to account for 0.2–0.4% (w/w) of WE-AX and 0.6–0.9% (w/w) of WU-AX of wheat flour.[7] The feruloylated arabinose appears to be exclusively attached to the O-3 of monosubstituted xylose in AX.[4,8] Oxidative crosslinking of ferulic acid attached to different AX chains to give forms of diferulic acid occurs, leading to insolubility of the AX. Consequently, WU-AX is generally high in arabinose and ferulic acid and it is necessary to break this association to give increased soluble fibre.

While the degree of substitution and crosslinking are the dominant factors determining solubility, the chain length of the xylan polymer may also have an influence. WU-AX has a higher molecular weight than WE-AX,[4] so decreasing xylan extension could increase solubility.

The second major polymer present in wheat endosperm cell walls is (1,3;1,4)-β-D-glucan (β-glucan). This is similar to cellulose (1,4)-β-D-glucan in that it comprises β-D-glucopyranosyl residues linked mainly by 1,4 glucosidic linkages, but has (1,3)-β-linkage for every three or four (1,4)-β-linkages. Longer stretches of 5 to 20 (1,4)-β-linkages also occur, resulting in "cellulose-like" domains. Like AX it occurs in soluble and insoluble forms, with about 10–15% of the β-glucan in wholemeal wheat flour being extractable with hot water.[9]

Table 2.2 Compositions of cell wall types in wheat grain tissues (% dry weight).

	Cell walls (% dry weight)	Cell wall components (% total polysaccharide)						
		cellulose	lignin	pectin	xylan	β-glucan	glucomannan	AGP[a]	References
Starchy endosperm	2–3	2	0	–	70	20	7	(10)	A
Total bran (pericarp, testa, aleurone)		29	8	–	64	6	–	–	B
Aleurone	40	2–4	0	–	62–65	29–34	–	–	C
Outer pericarp (beeswing)		30	12	–	60	–	–	–	D
Straw	85	37–40	14–17	0.5	39	–	–	–	E

A,[46,13]; B,[47]; C,[14,48,49]; D,[50]; E,[51,52]
[a]AGP is not usually reported in balance sheets of cell wall composition but can be estimated as accounting for 10% or more of the total cell wall polysaccharide in the starchy endosperm cell walls.

Challenges and Opportunities for Using Wheat for Biofuel Production

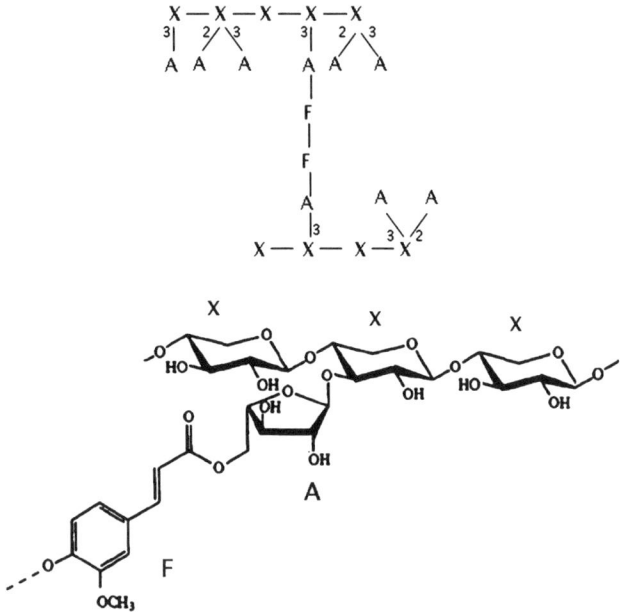

Figure 2.1 Structure of starchy endosperm arabinoxylan. X: (1→4) linked β-D-xylopyranosyl units; A: arabinose substitution; F: ferulic acid substitution.

The starchy endosperm cell also contain an arabinogalactan peptide (AGP) comprising 15 amino acids including three hydroxyproline residues that are *O*-glycosylated with arabinogalactan chains (the latter accounting for about 92% of the total dry weight) and is assumed to be present in the cell wall.[10,11] This is predominantly water soluble and hence is rarely included in balance sheets of cell wall polysaccharides. However, estimates of 0.24% to 0.33%[12] and 0.22% to 0.33%[13] grain dry weight indicate that it accounts for a significant proportion (certainly 10% or greater) of the total cell wall polysaccharides.

The outer layer of endosperm cells, called the aleurone, lacks starch and is rich in oil and protein. The aleurone cells also have thick cell walls, which account for about 40% of the dry weight.[14] These walls have a broadly similar composition to those of the starchy endosperm cells, although they apparently lack AGP. However, most of the aleurone AX is not water-soluble, with an A:X ratio of 0.38–0.39.[14] The aleurone cell AX is also highly esterified and crosslinked, with about 3.2% of the AX dry weight being ferulic acid and 0.45% diferulic acid,[14,15] and additional esterification with *p*-coumaric acid and acetyl groups.[16,17]

The outer layers of the grain have been characterised in less detail, with the exception of the outer pericarp that is readily prepared and often called beeswing bran on account of its filmy appearance. The composition of this fraction is more similar to that of the straw, with about 30% cellulose, 12% lignin and about 40% AX (Table 2.2) that has a highly complex branched structure with galactose and glucuronic acid residues (and is hence often termed glucuroarabinoxylan, GAX) and high contents of ferulic acid and dehydroferulic acids[14–15,18] and acetylation.[19]

Cell walls account for about 85% of the dry weight of straw, comprising 37% to 40% cellulose, 14% to 17% lignin and 60% GAX (Table 2.2), the latter being similar in complexity and composition to the GAX present in the outer layers of the grain.

2.3 Effect of Cell Wall Structure on Digestibility

At present, the only plant polysaccharide that can be readily digested to give fermentable sugars on an industrial scale is starch, with long-established technologies for the production of alcoholic beverages as well as more recent processes for bioethanol production. However, a range of enzymes are known to be capable of hydrolysing cell wall polysaccharides (xylanases, glucanases, cellulases, mannanases, *etc.*), providing a basis for the development of modified recombinant enzymes for the industrial-scale saccharification of plant tissues. These will in most cases result in the release of a mixture of hexose and pentose sugars that will in turn require specific enzymes for their fermentation into ethanol, butanol or other fuel compounds. Consequently, the main limitation to the exploitation of plant cell walls is not the basic polysaccharide structure but the extent to which this is modified and crosslinked, and in particular associated with lignin. The presence of lignin and other phenolics is a particular problem as lignin itself cannot readily be degraded and the linkage of lignin to cell wall polysaccharides also limits the digestion of these polymers. In the case of wheat the high degree of substitution of GAX (including acetylation) and lignification in the outer bran layers and straw are particularly important in this respect as they confer resistance to digestion, compared with the simple starchy endosperm cell walls that are readily digested during seed germination.

In wheat the major limitation to digestion is the presence of ferulate and diferulate crosslinks, both between arabinosyl residues of GAX and between GAX and lignin. These crosslinks also limit the digestion of cellulose.[20] Feruloyl esterases appear to be ineffective in decreasing crosslinking of lignified cell walls, so the most promising approach is to develop new cereal varieties with less ferulate crosslinking;[20] thus decreasing feruloylation of GAX is a major target to improve the utilisation of wheat whole grain and straw as biofuel. Other targets are to decrease or alter lignification and to decrease the substitution of GAX.

2.4 Biosynthesis of Arabinoxylan in Wheat

The pathway of AX biosynthesis is summarised in Figure 2.2. The enzymes catalysing the early steps, shown in light gray, have been established in wheat or other species and genes encoding the enzymes have been identified. In contrast, the enzymes catalysing xylan backbone synthesis and the transfer of arabinose

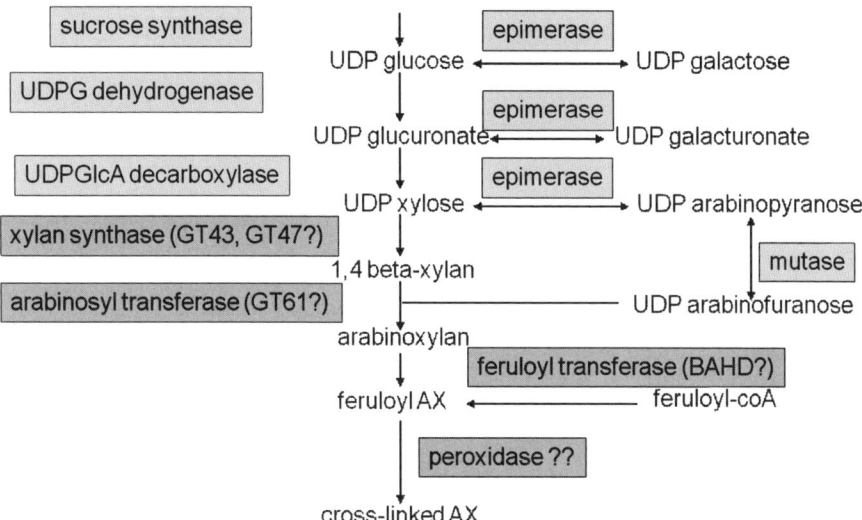

Figure 2.2 Proposed pathway of arabinoxylan synthesis. Light gray boxes show steps where enzymes and encoding genes are known, dark gray boxes where they are unknown. The gene families that have been recently implicated in the unknown steps (see text) are indicated in parentheses.

residues onto this (shown in dark gray) have not been characterised and their encoding genes remain unknown, although there have recently been promising developments, reviewed in ref. 21. Feruloylation may be catalysed by a currently unidentified feruloyl transferase using feruloyl-CoA as a substrate,[22] while oxidative crosslinking may occur in the wall catalysed by one or more of a range of peroxidases[23] or other oxidases.

Due to the significant impact of AX on human and animal nutrition and health, and its relevance to deriving biofuels from grasses, there has been great interest worldwide in identifying the genes that are responsible for AX synthesis. These have focused on bioinformatics approaches[24,25] and high-throughput screening of mutants,[26] since standard methods have proved difficult to apply to glycosyltransferases (GTs).[21] We have used a novel bioinformatics approach that exploits the differences between grass and dicot cell walls to identify candidate genes from EST distribution for xylan synthesis in the CAZy GT43, GT47 and GT61 families and for feruloylation of xylan in the BAHD family of acyl-coA transferases.[27]

When this work was carried out there was some evidence for involvment of the GT43 gene *IRX9* in xylan synthesis from analysis of the knock-out mutant in Arabidopsis.[28] Since then, it has been shown that *irx9* mutants have shorter xylan chains,[29,30] as do *irx14* mutants (another GT43 gene).[30] Furthermore, knock-out mutants in Arabidopsis for some of the GT47 genes identified by Mitchell *et al.*[27] (*irx10* and *irx10-like*) also have shorter xylan chains, exhibiting a similar xylan phenotype to the GT43 mutants.[31] At the time of writing, it

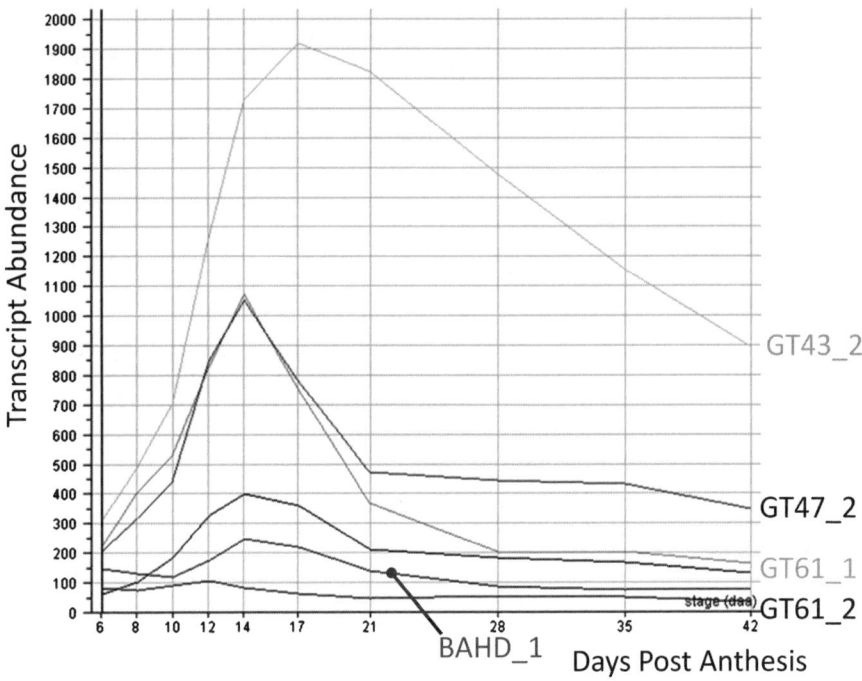

Figure 2.3 Expression of candidate genes during grain development. Average of two replicates at each time point measured with Affymetrix wheat array.[34]

therefore appears that both GT43 and GT47 gene products may be involved in xylan synthase activity with the mechanism being unknown. To date, nothing has been published on the GT61 candidate genes, nor on the BAHD candidate genes. The BAHD family comprises a large number of acyl-coA transferases including known hydroxycinnamoyl transferases.[32] A spermidine feruloyl transferase was also recently identified in this family.[33]

Indirect evidence for the roles of the candidate genes in wheat can be derived from analysing transcriptome profiles determined by Affymetrix array analysis of developing grain.[34] Figure 2.3 shows the expression of the most highly expressed forms of the candidate genes in developing wheat grain. The expression profiles of the GT43, 47 and 61 genes match the period of endosperm cell expansion and correspond to the pattern of xylan synthase activity that has been determined biochemically in developing barley endosperms, peaking at about 17 days postanthesis.[35]

Mitchell et al.[27] suggested that GT61 genes encode enzymes that transfer xylosyl side chains to feruloylated arabinosyl residues. However, these side chains do not exist in endosperm AX[4] and the high expression of these particular GT61 genes in the array analysis is not consistent with this. A possible explanation for this inconsistency is that the arrays shown in Figure 2.3 used

Table 2.3 Comparison of expression of candidate genes in whole grain and endosperm-enriched wheat samples (average of similar ratios at 14 and 21 days postanthesis). Unpublished data from M. Wilkinson et al.

Gene family	Closest rice locus	Ratio expression Endosperm-enriched/whole grain
GT43_2	Os03g17850	1.8
GT47_2	Os01g70200	1.3
GT61_2	Os03g37010	0.5
GT61_1	Os02g22650	1.8
BAHD_1	Os01g09010	0.4

mRNA from whole developing caryopses and the outer grain tissues contain xylosyl side chains. We therefore investigated this further by measuring gene expression using mRNA from preparations that were highly enriched in endosperm tissue, with the outer pericarp and embryo having been removed. Table 2.3 compares the expression of the candidate genes in this endosperm-enriched fraction with that in the whole caryopsis sample. This shows that the GT61-1 gene is actually more highly expressed in the endosperm-enriched sample (Table 2.3), which has led us to hypothesise that these particular GT61 genes encode arabinosyl transferases. This identity is also consistent with the GT61 genes being far more highly expressed in cereals than dicots, to a much greater degree that any of the other GT candidates,[27] since Ara substitution of xylan is much more prevalent in grasses than in dicots.

Feruloylation of AX is much more widespread in the AX of the aleurone and pericarp than it is in the starchy endosperm.[4] Consistent with this, the two acyl transferase-like candidate genes from the BAHD family that we have identified are both moderately expressed in the mRNA from whole developing caryopses (Figure 2.3) and are less highly expressed in the mRNA from the endosperm-enriched fraction (Table 2.3).

The coexpression of genes may also give important clues to function. All of the candidate genes show some degree of coexpression with each other and with the precursor steps in xylan synthesis (Figure 2.2) in cereal EST libraries.[27] This coexpression is also seen in our wheat array data[34] and in the public rice and barley array data, summarised in Table 2.4.

Given our current state of knowledge of the candidate genes discussed above we have ascribed putative encoding genes to each of the unknown steps in Figure 2.2. These candidates are currently under experimental investigation in our laboratory and others around the world and it is likely that our knowledge of genes responsible for determining GAX structure in cereals will advance rapidly in the next few years. Once the genes are identified, they will become major targets for manipulation in order to facilitate the saccharification and fermentation processes to derive biofuels from cereal cell walls in feedstocks such as straw and bran.

Table 2.4 Coexpression of candidate genes. Pearson correlation coefficient between expression of candidate genes in wheat grain[34] and in rice and barley public array data, analysed with the GeneVestigator tool.

		GT43_1 (IRX14)	GT43_2 (IRX9)	GT47_2 (IRX10)	GT61_1	GT61_2	GT61_7	BAHD_1
		Os06g47340	Os03g17850	Os01g70200	Os02g22650	Os02g22480	Os02g04250	Os01g09010
GT43_1	rice	i.d.	0.51	0.69	0.17	0.43	**0.74***	**0.60***
	barley		0.29	**0.57***	0.17	0.23	0.55	0.46
	wheat		0.85	0.69	0.80	0.39	−0.36	0.22
GT43_2	rice	0.51	i.d.	0.59	0.19	0.57	0.43	0.32
	barley	0.29		**0.43***	0.40	**0.56***	−0.14	0.32
	wheat	0.85		0.66	0.79	0.39	−0.40	0.27
GT47_2	rice	**0.69***	**0.59***	i.d.	0.32	0.67	**0.76***	**0.61***
	barley	**0.57***	0.43		0.15	0.22	0.35	0.40
	wheat	0.69	0.66		**0.88***	0.61	−0.23	0.34
GT61_1	rice	0.17	0.19	0.32	i.d.	0.46	0.35	0.30
	barley	0.17	0.40	0.15		**0.54***	0.19	0.57
	wheat	0.80	0.79	**0.88***		0.58	−0.29	0.32
GT61_2	rice	0.43	0.57	0.67	**0.46***	i.d.	**0.65***	0.40
	barley	0.23	0.56	0.22	0.54		0.15	0.56
	wheat	0.39	0.39	0.61	0.58		0.05	**0.82***
GT61_7	rice	**0.74***	0.43	**0.76***	0.35	0.65	i.d.	**0.68***
	barley	**0.55***	−0.14	**0.35***	0.19	0.15		0.51
	wheat	−0.36	−0.40	−0.23	−0.29	0.05		0.11
BAHD_1	rice	**0.60***	0.32	0.61	0.30	0.40	**0.68***	i.d.
	barley	**0.46***	0.32	**0.40***	0.57	**0.56***	0.51	
	wheat	0.22	0.27	0.34	0.32	**0.82***	0.11	

* Values which are in the top 1% of all genes on the array for the gene indicated in the column header.

2.5 Sustainability of Wheat Production for Biofuel

The high yields of wheat grown in the UK and many other countries are dependent on high inputs of agrochemicals, particularly nitrogen fertiliser. This is illustrated by the classical Broadbalk experiment at Rothamsted, which was established in 1843 and is still yielding important information. Yields on this site only increased above 3 tonnes/ha in the 1960s with the introduction of semidwarf wheat varieties and the application of herbicides, fungicides and high levels of nitrogen fertiliser (above 144 kg/ha). Current yields for first wheat crops (cv Hereward in an oats, maize, wheat, wheat, wheat rotation) range from 3.8 tonnes/ha with 48 kg N/ha to 8.9 tonnes/ha with 288 kg N/ha (means of 10 harvests, 1998–2007).[36] Such high levels of inputs may not be sustainable if wheat is produced only for biofuel and low-input management systems may be required.[37–39]

Kindred et al.[40,41] have reported detailed studies of the most suitable varieties and optimum nitrogen applications for biofuel production. Kindred et al.[40] showed that the economic optimum nitrogen application rate for alcohol production was about 12% lower than that for grain production, but that a significantly lower application rate was required to optimise saving in greenhouse-gas emissions. However, a more detailed study of two cultivars (Riband and Option) showed that the economic optimum rates for grain yield and alcohol production were similar.[41]

Although the precise energy balance for wheat is still a matter of debate and will vary with different crop production and bioethanol-production systems, it is clear that it will be economically more attractive if more of the grain than just the starch can be utilised. Thus, Murphy and Power[42] reported improved efficiency by combining the production of ethanol from starch with digestion of stillage (the grain residue after distillation) to give biomethane and the use of straw for combustion, the net energy production increasing from 25 GJ/ha/a to 72 GJ/ha/a. Another approach is to combine the production of ethanol from starch with the production of high-value coproducts such as arabinoxylan for food and pharmaceutical use.[43,44]

2.6 Conclusions

The use of wheat as a biofuel with current technology is dependent on the fermentation of starch that may be unattractive in terms of energy balance and have undesirable implications for land use and food prices. Technological advances are therefore required, particularly in the derivation of biofuel from the cell wall polysaccharides of wheat. Such advances would enable the use of byproducts of wheat food crops, straw and bran, as well as whole grain as feedstock for biofuel production.

Acknowledgements

Rothamsted Research receives grant-aided support from the Biotechnology and Biological Sciences Research Council (BBSRC) of the UK. The authors are grateful to Dr. Daniel Kindred (ADAS UK Ltd.) for discussions.

References

1. J. R. Lenton in *Advances in Botanical Research – Biotechnology of Cereals*, ed. P. R. Shewry, P. A. Lazzeri and K. J. Edwards, Academic Press, London, 2001, pp. 127–164.
2. T. Worland and J. W. Snape, in *The World Wheat Book. A History of Wheat Breeding*, ed. A. P. Bonjean and W. J. Angus, Lavoisier Publishing, Paris, 2001, p 59–100.
3. A. E. Farrell, R. J. Plevin, B. T. Turner, A. D. Jones, M. O'Hare and D. M. Kammen, *Science*, 2006, **311**, 506–508.
4. L. Saulnier, P.-E. Sado, G. Branlard, G. Charmet and F. Guillon, *J. Cereal Sci.*, 2007, **46**, 261–281.
5. J. J. Ordaz-Ortiz and L. Saulnier, *J. Cereal Sci.*, 2005, **42**, 119–125.
6. B. Andrewartha, D. R. Phillips and B. A. Stone, *Carbohydrate Res.*, 1979, **77**, 191–204.
7. E. Bonnin, A. Le Goff, L. Saulnier, M. Chaurand and J. F. Thibault, *J. Cereal Sci.*, 1998, **28**, 53–62.
8. T. Ishii, *Plant Sci.*, 1997, **127**, 111–127.
9. C. Nemeth, J. Freeman, H. D. Jones, C. Sparks, M. D. Wilkinson, T. Pellny, J. Dunwell, A. A. M. Andersson, P. Åman, F. Guillon, L. Saulnier, R. A. C. Mitchell and P. R. Shewry, *Plant Phys.*, 2010, **152**, 1209–1218.
10. G. B. Fincher, W. H. Sawyer and B. A. Stone, *Biochem. J.*, 1974, **139**, 535–545.
11. K. Van den Bulck, K. Swennen, A.-M. A. Loosveld, C. M. Courtin, K. Brijs, P. Proost, J. Van Damme, S. Van Campenhout, A. Mort and J. A. Delcour, *J. Cereal Sci.*, 2005, **41**, 59–67.
12. R. Andersson, E. Westerlund and P. Åman, *J. Cereal Sci.*, 1994, **19**, 77–82.
13. A. Loosveld, C. Maes, W. H. M. van Casteren, H. A. Schols, P. J. Grobet and J. A. Delcour, *Cereal Chem.*, 1998, **75**, 815–819.
14. C. Antoine, S. Peyron, F. Mabille, C. Lapierre, B. Bouchet, J. Abecassis and X. Rouau, *J. Agr. Food Chem.*, 2003, **51**, 2026–2033.
15. M. L. Parker, A. Ng and K. W. Waldron, *J. Sci. Food Agric.*, 2005, **85**, 2539–2547.
16. C. Antoine, S. Peyron, V. Lullien-Pellerin, J. Abecassis and 2. Rouau, *J. Cereal Sci.*, 2004, **39**, 387–393.
17. D. I. Rhodes, M. Sadek and B. A. Stone, *J. Cereal Sci.*, 2002, **36**, 67–81.
18. L. Saulnier and J. F. Thibault, *J. Sci. Food Agric.*, 1999, **79**, 396–402.
19. G. Mandalari, C. B. Faulds, A. I. Sancho, A. Saija, G. Bisignano, R. L. LoCurto and K. W. Waldron, *J. Cereal Sci.*, 2005, **42**, 205–212.
20. J. H. Grabber, *Crop Sci.*, 2005, **45**, 820–831.
21. G. B. Fincher, *Plant Phys.*, 2009, **149**, 27–37.
22. T. Yoshida-Shimokawa, S. Yoshida, K. Kakegawa and T. Ishii, *Planta*, 2001, **212**, 470–474.
23. A. Encina and S. C. Fry, *Planta*, 2005, **223**, 77–89.
24. T. Girke, J. Lauricha, H. Tran, K. Keegstra and N. Raikhel, *Plant Physiol.*, 2004, **136**, 3003–3008.

25. N. Farrokhi, R. A. Burton, L. Brownfield, M. Hrmova, S. M. Wilson, A. Bacic and G. B. Fincher, *Plant Biotech. J.*, 2006, **4**, 145–167.
26. W. D. Yong, B. Link, R. O'Malley, J. Tewari, C. T. Hunter, C. A. Lu, 2. M. Li, A. B. Bleecker, K. E. Koch, M. C. McCann, D. R. McCarty, S. E. Patterson, W. D. Reiter, C. Staiger, S. R. Thomas, W. Vermerris and N. C. Carpita, *Planta*, 2005, **221**, 747–751.
27. R. A. C. Mitchell, P. Dupree and P. R. Shewry, *Plant Physiol.*, 2007, **144**, 43–53.
28. D. M. Brown, L. A. H. Zeef, J. Ellis, R. Goodacre and S. R. Turner, *Plant Cell*, 2005, **17**, 2281–2295.
29. M. J. Peña, R. Zhong, G.-K. Zhou, E. A. Richardson, M. A. O'Neill, A. G. Darvill, W. S. York and Z.-H. Ye, *Plant Cell*, 2007, **19**, 549–563.
30. D. M. Brown, F. Goubet, V. W. Wong, R. Goodacre, E. Stephens, P. Dupree and S. R. Turner, *Plant J.*, 2007, **52**, 1154–1168.
31. D. M. Brown, Z. Zhang, E. Stephens, P. Dupree and S. R. Turner, *Plant J.*, 2009, **57**, 732–746.
32. J. C. D'Auria, *Curr. Opin. Plant Biol.*, 2006, **9**, 331–340.
33. E. Grienenberger, S. Besseau, P. Geoffroy, D. Debayle, D. Heintz, C. Lapierre, B. Pollet, T. Heitz and M. Legrand, *Plant J.*, 2009, **58**, 246–259.
34. Y. Wan, R. L. Poole, A. K. Huttly, C. Underwood, K. Feeney, S. Welham, M. J. Gooding, E. N. C. Mills, K. J. Edwards, P. R. Shewry, R. A. C. Mitchell, *BMC Genomics*, 2008, **9**, Article 121.
35. T. Urahara, K. Tsuchiya, T. Kotake, T. Tohno-oka, K. Komae, N. Kawada and U. Tsumuraya, *Physiol. Plantarum*, 2004, **122**, 169–180.
36. Anonymous, *Rothamsted Long-Term Experiments*, Lawes Agricultural Trust Co. Ltd., Harpenden, Hertfordshire, 2006.
37. C. Loyce and J. M. Meynard, *Ind. Crops Prods.*, 1997, **6**, 271–283.
38. C. Loyce, J. P. Rellier and J. M. Meynard, *Agric. Systems*, 2002a, **72**, 9–31.
39. C. Loyce, J. P. Rellier and J. M. Meynard, *Agric. Systems*, 2002b, **72**, 33–57.
40. D. R. Kindred, T. C. Smith, R. Sylvester-Bradley, D. Ginsberg and C. J. Dyer, *HGCA Project Report 417*, HGCA, London, 2007, p 44.
41. D. R. Kindred, T. M. O. Verhoeven, R. M. Weightman, J. S. Swanston, R. C. Agu, J. M. Brosnan and R. Sylvester-Bradley, *J. Cereal Sci.*, 2008, **48**, 46–57.
42. J. D. Murphy and N. M. Power, *Fuel*, 2008, **87**, 1799–1806.
43. R. M. Weightman, H. Davis-Knight, R. H. Wang, N. Misailidis, A. L. Villaneuva and G. M. Campbell, *Aspects Appl. Biol.*, 2008, **90**, 153–160.
44. R. M. Weightman, H. R. Davis-Knight, G. M. Campbell, N. Misailidis, R. Wang and A. L. Villaneuva, *HGCA Project Report 448*, HGCA, London, 2009, p 54.
45. C. Barron, A. Surget and X. Rouau, *J. Cereal Sci.*, 2007, **45**, 88–96.
46. D. J. Mares and B. A. Stone, *Aust. J. Biol. Sci.*, 1973, **26**, 793–812.
47. R. R. Selvendran, S. G. Ring, M. A. O'Neill and M. S. Du Pont, *Chem. Ind.* (London), 1980, **22**, 885–888.
48. A. Bacic and B. A. Stone, *Aust. J. Plant. Physiol.*, 1981, **8**, 475–495.

49. D. I. Rhodes and B. A. Stone, *J. Cereal Sci.*, 2002, **36**, 83–101.
50. M. S. Du Pont and R. R. Selvendran, *Carbohydr. Res.*, 1987, **163**, 99–113.
51. J. M. Lawther, R. Sun and W. B. Banks, *J. Agr. Food Chem.*, 1995, **43**, 667–675.
52. X.-F. Sun, R. Sun, P. Fowler and M. S. Baird, *J. Agr. Food Chem.*, 2005, **53**, 860–870.

CHAPTER 3
Maize

STEPHEN H. HOWELL

Plant Sciences Institute, 1035A Roy J. Carver Co-Laboratory, Iowa State University, Ames, Iowa, USA

3.1 Corn Ethanol

Probably no subject in the area of bioenergy has been more controversial than corn ethanol. Corn ethanol launched the biofuels industry in the US some thirty years ago with the Energy Policy Act of 1978.[1] Most recently, the debate surrounding the production of corn ethanol has focused on several issues, such as net energy gain/loss, sustainability and the economics of corn ethanol production. The advantages or disadvantages of using corn as a feedstock for biofuel production are most frequently weighed against other feedstocks, and against those odds, other lignocellulosic sources usually win. However, the history, knowhow and the existence of an infrastructure for corn and ethanol production is a compelling argument favouring corn as the principal feedstock for ethanol production in the near future.[2]

It has been argued by many that the corn-ethanol industry in the US owes its existence to government mandates and subsidies. The Energy Policy Act of 1978 started off with a $.40/gallon subsidy, and since that time subsidies have ranged from $0.40–$0.60/gallon and are currently pegged at $0.51/gallon. In addition, a tariff of $0.54/gallon was established as an import barrier in support of domestic production. It appears, as well, that mandates have assured the future for the corn-ethanol industry in the US at least until 2016. The Energy Independence and Security Act of 2007 established a renewable fuel standard (RFS), which calls for an increase in biofuels production from its present (2008) level of 9B gallons per year to 36B gallons by 2022. Of that, 21B gallons will be cellulosic ethanol and advanced biofuels and the remaining 15B gallons, a level to be reached by 2016, will be from conventional sources, such as grain ethanol.[3]

Whatever the motivation, the US corn-ethanol industry has, indeed, ramped up in the past few years. The annualised growth rate in ethanol production from 1980 to 2007 was 14.0%. Around 2000, ethanol gained traction as it surfaced as the substitute for methyl tertiary butyl ether (MTBE), an oxygenate that reduced air pollution, but contaminated water supplies. That resulted in greater demand for ethanol as an oxygenate and as gasoline prices started ratcheting higher, ethanol production increased at an annual rate of $\sim 25\%$ from 2003 to 2007.[4] The USDA projects a 20 per cent increase in corn usage by the ethanol industry in 2009/10 over the previous year, to 4.2 billion bushels, or 33 per cent of the corn crop.

Most recently, however, the ethanol industry has taken a hit from falling ethanol prices and rising corn prices. The industry enjoyed its best returns (ethanol operating margins) in mid-2006 when the price of corn was low and the demand for ethanol was rising as many of the oil companies phased out MTBE. The return over operating costs at that time were a bit over $3 per gallon of ethanol. Those returns have diminished and as of April, 2009, are less than $0.25 per gallon with the price of ethanol at $1.57 gallon and corn prices at $3.46/bushel.[5]

Nonetheless, meeting the mandated goals for corn ethanol will require a monumental effort in improving corn production. Present US yields are 154 bushels per acre,[6] and historical trends for corn yield increase during the high-input era (1950–2005) for corn production were about 1.8% per year.[7] At that rate, corn yields will only be expected to increase by 13% by 2016 – not enough to support the production of 15B gallons of corn ethanol per year plus meeting all the other needs for corn. However, others predict that new technologies will boost corn yield considerably in the next few years.[2]

Biofuels gained in popularity a few years ago as a way to free nations from their dependence on foreign oil and as a "green alternative" to fossil fuels. More recently biofuels, particularly corn ethanol, have come under fire as possibly doing more environmental harm than good. In early 2008, two articles in the journal Science[8,9] argued that forests or grasslands converted for the production of biofuels release soil organic carbon and incur a "biofuel carbon debt" with a long payback period. The Searchinger *et al.*[8] study demonstrated that biofuel production often displaces crops, moving them to new areas where further land-use conversions are required. The argument, termed "indirect land-use change" (ILUC) represents a proposition that increased biofuels production in the United States, causes crop prices to rise, encouraging the conversion of land elsewhere in the world from grasslands or forests to crops. The Science articles were highly publicised in the lay press, and Time magazine ran a cover on April 7, 2008, with an ear of corn under the title of the "Clean Energy Myth".

ILUC has been controversial, but has worked its way into US energy policy. The Energy Independence and Security Act of 2007 amended the Clean Air Act by requiring that renewable fuels produced in new facilities whose construction commenced after the date of enactment achieve at least a 20 per cent reduction in life-cycle greenhouse-gas emissions, compared to baseline life-cycle greenhouse-gas emissions. Life-cycle greenhouse-gas emissions are defined as the aggregate quantity of greenhouse-gas emissions, including direct emissions and significant indirect emissions (such as emissions for land-use changes as determined by the EPA Administrator) related to the full fuel life cycle.[10] Critics of ILAC and the EPA policy argue that indirect emissions cannot be accurately measured and that they are products of market forces and not direct action of utilising land for growing crops for biofuels.

Corn ethanol took another hit in 2008, when a wet spring in the Midwest delayed corn planting and predictions for the corn harvest were poor. That, and fuel costs, drove up food prices inciting food riots in several countries and sparked a food versus fuel debate.[11] A UN expert, Jean Ziegler, called the practice of converting food crops into biofuel "a crime against humanity", saying it creates food shortages and raises food price that cause millions of poor people to go hungry. However, the debate in the US soon cooled as the growing season moved on unexpectedly to produce another bumper corn crop. With the prospect of a near record crop, the price of corn and ethanol fell toward the end of 2008.

Corn ethanol was again the target of study by Chiu *et al.*[12] who reported on irrigated water consumption associated with growing corn for ethanol production. They reported that corn requires from 5 to 2000 l of water per gallon of ethanol depending on where the corn is grown. In Iowa, where most corn production is rain fed, irrigated water usage is the lowest, but moving further west to drier climates, irrigated water usage increases to 2138 l of water per gallon of ethanol in California. In the western Midwest states, such as Nebraska and Kansas, irrigated water usage is intermediate (~ 500 l of water per gallon of ethanol), however, most of the water used is fossil water from the Ogallala aquifer. Thus, the expansion of land use in western states for biofuels production could tap heavily into water reserves.

3.2 Yield Improvements

Clearly, one of the major challenges for the corn industry in meeting government mandates for biofuels is to increase corn production. Corn production in the US has risen six-fold over the past 70 years since the introduction of hybrid corn, occupying fewer acres of land (Figure 3.1), owing to improvements in technology (seed varieties, fertilisers, pesticides, and machinery) and in production practices (reduced tillage, irrigation, crop rotations, and pest-management systems).[13] It has been estimated that plant breeding contributed about 50 per cent of the yield increases and management practices to the rest.[14]

Duvick[15] analysed the increases in US corn yields from 1930 to 2000 through ERA hybrids developed by Pioneer Hi-Bred International.[16,17] All of the hybrids used in the study were widely grown hybrids, representing the elite

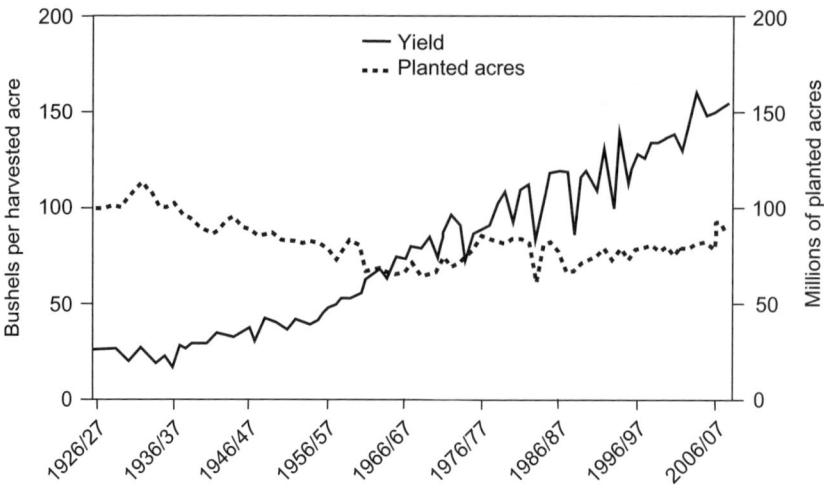

Figure 3.1 US corn yield and acreage. Source: USDA, World Agricultural Outlook Board, World Agricultural Supply and Demands Estimates.

germplasm of each decade. Despite the yield gains over the years, it was surprising to find that "yield potential" did not increase in hybrids introduced between 1930 and 1990. Yield potential is considered to be the yield of a hybrid under nonlimiting conditions for sunlight, nutrients, and water.[18] Yield potential of modern corn hybrids did not differ when plants were grown under widely spaced, noncompetitive conditions.[15] When plant density was increased, grain yield per unit area increased for 1990s hybrids, but not for 1930s hybrids. Hence, 1990 hybrids were better able to compete under conditions when resources were limiting.[19] From 1930 to 2000, plant population densities gradually increased from 30 000 plants/ha to 80 000 plants/ha.[20] So, much of the yield gain in modern hybrids has come from planting much denser crops with hybrids that are better able to compete under crowded conditions.

The increase in yield of ERA hybrids grown under denser conditions has been attributed to enhancement in light interception due to increased leaf area index and to changes in light utilisation by more erect upper leaves.[21] Leaf area per plant has not changed much during the hybrid era,[22] but increased plant population density from 1930 to 2000 has doubled the leaf area index (the ratio of total upper leaf surface of vegetation divided by the surface area of the land on which the vegetation grows). Other yield gains can be attributed to delay in leaf senescence (or maintenance of green-leaf area and leaf photosynthesis) during the grain-filling period. Normally, leaf photosynthesis peaks out at silking and declines during the grain-filling period. In the newer hybrids, leaf photosynthesis declines less during the grain-filling period.[23]

3.2.1 Critical Growth Stages

There are several growth stages that are critical for a corn crop to reach maximum grain yield potential according to the Iowa State University Extension Service.[24] The first stage is emergence. A uniform, optimum population stand is important in producing a good crop, because unlike soybeans, corn is unable to make up for poor stands. Since modern hybrids perform better at higher plant density, planting population have been adjusted upward to achieve optimum stands. Flooding and poor drainage are responsible for some of the most significant stand losses.

The second critical growth stage occurs during ear development, when the number of kernel rows and the number of potential kernels (ovules) per row are determined.[24] The number of kernel rows per ear is heavily influenced by genetics and is generally determined by growth stage V12 (twelve vegetative leaves). The number of potential kernels per row is strongly influenced by field conditions and is usually determined between V12 and V17. Practices to reduce environmental, pest, moisture and nutrient stress during this time will maximize the potential number of harvestable kernels.

The next most critical stage is pollination.[24] Pollen shed typically lasts from 1 to 2 weeks, and good pollination is highly dependent upon the weather. Drought and heat conditions can desynchronise pollen shed and silk emergence and also can desiccate silks and pollen grains. The final critical growth stage is the grain-fill stage. Grain fill begins at pollination and ends at kernel black-layer formation. The "black layer" is an abscission layer located at the base of a kernel, which normally forms about 60 days after silking. The abscission layer cuts off water and dry matter transfer into the kernel. Stress on the corn plant during the grain-fill period can affect final yield by reducing the size and/or weight of harvested kernels.

3.2.2 Grain-Yield Traits

Optimal growth and grain yield require a balance between source and sink – a balance between the accumulation of dry matter in biomass and its redistribution to grain. Excess source capacity relative to sink results in the development of alternative sinks and development of typical symptoms such as the purpling of leaves, sheath tissues, and stalks during the grain-filling period. On the other hand, excess sink capacity relative to source capacity, causes premature senescence of leaves and stalks during the grain-filling period.[21] Sink size in maize is related to kernel number and weight, however, genetic improvement in corn to improve sink size has been correlated with increase in kernel number, and not kernel size or weight.[25]

Lee and Tollenaar[21] have cited several traits that represent opportunities for improvement in grain yield. They emphasize one – improvement in performance of plants under high planting density stress. It is not entirely clear what form that stress might take – likely a compilation of stresses arising from a competition for various resources. However, it is also instructive to understand what traits were not improved during the hybrid era as planting densities increased. The traits that did not change were:

1. maximum potential photosynthesis[25] and plant growth rate at silking;[22]
2. dry mass accumulation up to silking;[26]
3. harvest index or the proportion of total dry matter allocated to grain;[14]
4. magnitude of heterosis effects, *i.e.* the per cent yield improvement in hybrids versus inbreds.[16]

3.2.3 Heterosis

Although heterosis (or hybrid vigour) has not improved during the hybrid era, it is still a major component of yield in corn. Despite its importance to the hybrid corn industry, the genetic or molecular mechanism underlying heterosis is still not known. However, the opportunities for understanding heterosis have improved as more maize genomic information is made available. The first draft sequence of the *Zea mays* B73 genome with over 32 000 genes was announced at the 50th Annual Maize meetings in Washington D.C. and recently submitted for publication.[27]

In one molecular study on heterosis by a group at Pioneer Hi-Bred, Guo *et al.*[28] profiled gene expression in analysing a group of maize hybrids that shared a common female parent and that varied in the degree of yield heterosis. They produced hybrids by crossing a common female inbred (S1) with various male inbreds that differed in the level of their pedigree relationship to the female parent. In profiling the RNA from nonpollinated immature ear tissues, they observed that most gene expression in the hybrids approximated the midparental level, some was similar to either the maternal or paternal level and ~20% fell outside the parental range.

Swanson-Wagner *et al.*[29] conducted another molecular study of heterosis in corn and found all possible modes of gene expression changes in comparing a F1 hybrid to the two parental inbred lines, B73 and Mo17. They observed gene action that could be interpreted as additivity, high- and low-parent dominance, underdominance, and overdominance. Although most of the expression differences were additive effects, 22% of the differentially regulated genes exhibited nonadditive modes of gene expression. Thus, the patterns of gene expression in heterosis appear to be highly complex, however, there may be fewer genes that exercise master control over the phenomenon.

3.2.4 Nitrogen-Use Efficiency

Nitrogen fertilisation, a crop-management practice that escalated in the US after World War II, has had profound effects on corn production during the hybrid era. Therefore, nitrogen-use efficiency (NUE) is an important attribute for corn since N is a costly input and can have a significant environmental impact. NUE is defined as "the proportion of all N inputs that are removed in harvested crop biomass, contained in recycled crop residues, and incorporated into soil organic matter and inorganic N pools."[30] Another measure of N uptake and utilisation is N fertiliser recovery efficiency (RE_N, defined as the percentage of fertiliser N recovered in above-ground plant biomass during the growing season). In the US cornbelt, mean RE_N for corn is 37% (SD = ±30), with a huge variation.[30] Although corn has a better RE_N than rice, for example, the corn RE_N means that there is, on average, about 63% loss (non recovery) on N fertiliser on the corn crop within a growing season.

Corn has a larger increase in grain yield per unit N uptake compared to some other crops because corn is a C_4 plant with efficient photosynthesis. C_4 plants have a higher photosynthetic rate per unit leaf N content, resulting in greater biomass production per unit of plant N accumulation.[31] For example, Cassman et al.[30] reported that a corn crop, which produces a grain yield of 10 metric tons/ha, needs to take up about 190 kg ha^{-1} of N. Certain indigenous soils supply about 130 kg ha^{-1} of N, which leaves 60 kg ha^{-1} of N required for the crop. Given that RE_N is 37% for corn, then a N fertiliser rate of 162 kg ha^{-1} must be applied to meet crop N demands. Because of the variability, but larger contribution of indigenous soil N to crop demands, the prediction of local soil N is a key in optimising N fertiliser application.

Since 1980, the ratio of crop yield per unit of applied N fertiliser (partial factor productivity for N fertiliser or PFP_N) has risen in the US (Figure 3.2).[30] Several factors are thought to contribute to the rise: (a) greater yields and increased stress tolerance of modern hybrids; (b) improved crop-management practices, such as conservation tillage and higher plant densities; and (c) shift to N fertiliser application in spring or at planting; or (d) greater use of split N fertiliser applications during the growing season. In addition, increased planting of corn varieties with lower corn kernel protein content may also have contributed to higher PFP_N. Despite the improvement in efficiency since 1980,

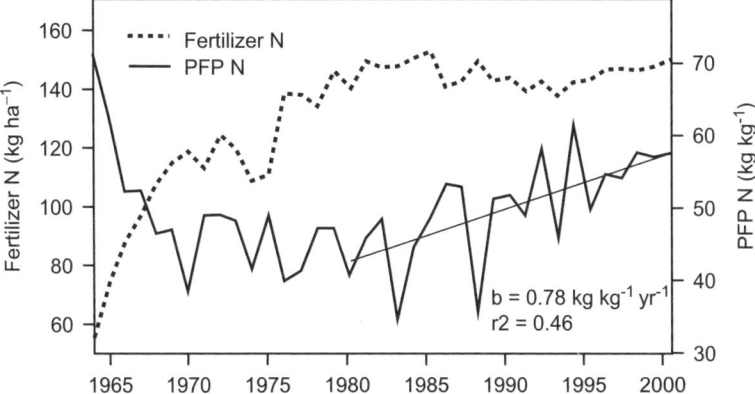

Figure 3.2 Trends in maize partial factor productivity from applied N fertiliser (PFP_N, kg grain yield/kg N applied) in the US.[30]

the best estimate by Cassman *et al.*[30] of average RE_N in farmer's fields is less than 40% of applied N. However, RE_N values may continue to improve in the future as corn varieties with even lower protein content are further developed for biofuels use.

Grain NUE is defined somewhat differently as the grain yield per unit N available from the soil, including N fertiliser. Genetic variance in NUE has been observed both at low and high N fertilisation levels. At high N levels, genetic variation in NUE has been explained by variation in N uptake, whereas at low N levels, NUE variability has been largely attributed to differences in total NUE, as defined above.[32] This suggests that the limiting steps in N assimilation may be different when plants are grown under high or low levels of N fertilisation. In general, about 25% of the total genotypic variation in grain NUE represents variance in genotype × N interaction. The yield component that contributes most to this interaction is kernel number per ear.[32] Genotypes with low grain NUE react more to N fertilisation. Thus, genotype × N interaction seems to be largely due to variation in the adaptation of the plant to low-N input rather than to variation in the adaptation to high-N input. The absence of such interactions for traits relative to vegetative growth means that they are due to grain development.[33]

3.3 Cornstarch

Corn grain has been the ideal feedstock for ethanol production because the endosperm is rich in starch. Corn (at 15% moisture content) in one compositional study was reported to contain 72% starch, 10.5% hemicellulose/cellulose, 9.5% protein, 4.5% oil, 2.0% sugars and 1.5% ash.[34] Since cornstarch is presently the major raw material for the production of biofuels in the US, much work has been done in studying the biosynthesis of starch and optimising its production. Several extensive reviews have been written on this subject,[35–42] and this review will deal with more current studies and focus on starch synthesis in amyloplasts in corn endosperm.

Cornstarch is chemically composed of two glucan fractions – amylose and amylopectin. Amylose is made up of long chains of glucose residues joined together by α1–4 glycoside bonds and sparsely branched owing to occasional α1–6 linkages. Amylopectin, on the other hand, is made up of short α1–4 linked glucose chains (ranging in size from 6–30 residues, averaging about 12) with frequent branch points (α1–6 linkages). The polymers of glucose are radially arranged in a starch granule in concentric crystalline and amorphous growth rings. The crystalline rings are themselves composed of crystalline and amorphous lamellae. Crystallinity within the crystalline lamellae arises from the formation of hydrogen-bonded, double-helical glucose chains in the packed amylopectin branch clusters. The crystalline nature of the amylopectin is thought to be important in developing a sink – drawing the metabolic equilibrium toward carbohydrate accumulation.[43] However, reduction in the crytallinity of the crystalline lamellae, *e.g.* by producing shorter amylose chains, is a target for biofuels production, because the less crystalline the structure, the less time and energy are needed to disrupt starch structure for processing.

3.3.1 Starch Biosynthesis

Amylose and amylopectin are synthesised from ADP-glucose, which in turn is made from glucose-1-phosphate and ATP in a reaction catalysed by ADP-glucose pyrophosphorylase (AGPase). The formation of crystalline starch involves the coordinated activities of several enzymes – starch synthases (SSs), starch-branching enzymes (SBEs) and starch-debranching enzymes (DBEs). SSs catalyse the addition of glucose residues from ADP-glucose onto the nonreducing ends of growing glucan chains. SBE catalyses the formation of α1–6 linkages by cleaving the α1–4 linkages within a glucan chain and catalysing the formation of the α1–6 linkage between the reducing end of the cleaved chain and the C6 on a residue on a neighbouring glucan chain. DBEs are thought to hydrolyse branch linkages, thus rearranging glucan chains to form crystalline structures creating a sink for starch.[44]

A recent finding on starch biosynthesis in corn is that the enzymes are thought to function in large multisubunited complexes.[37,45] This is important because the organisation of the enzymes may be responsible for the complex structure of starch and the starch granule. The investigators found that two isoforms of sucrose synthase (SSIIa and SSIII) and two isoforms of the branching enzyme (SBEIIa and SBEIIb) all comigrated on gel permeation chromatography in a 670-kD complex.[43] Mutations eliminating any one of the subunits prevent the formation of the 670-kD complex indicating the interdependence of the subunits on the assembly of the complexes. A smaller 300-kD complex consisting of SSIIa, SBEIIa and SBEIIb was also found that may represent a subassembly of the larger complex. The involvement of SSIII in the larger complex is important because of SSIII is the product of the dull1 (*du1*) gene and implicated as a regulator of other starch biosynthesis genes in maize. For example, *du1* mutations not only eliminated SSIII activity but also reduce SBEIIa activity.[46]

In addition, other activities are associated with the complex including AGPase, the sucrose synthase isoform SUS-SH1 and pyruvate orthophosphate dikinase (PPDK).[43] The association of AGPase with the complex is highly significant because the enzyme is allosterically regulated and is thought to be the rate-limiting step in starch biosynthesis.[47] The inclusion of PPDK in the complex is relevant in that it might be responsible for partitioning carbon in corn amyloplasts by channeling pyrophosphate to AGPase diverting ADP-glucose away from starch synthesis toward breakdown and synthesis of amino acids and lipids.[43,48] The regulation of PPDK could have implications for starch production by diverting more carbon into starch, rather than into protein (assuming that other yield factors are not compromised in doing so). The association of the complex with SUS-SH1, a product of the shrunken1 (*shl*) gene, is also intriguing, but still somewhat controversial because the enzyme has not been thought to be an amyloplast protein.

Starch biosynthesis and starch structure are important issues in ethanol production. There is significant variability (23%) in corn hybrids for ethanol production from dry milling operations. It is estimated that 75% of the variability is due to genetics and 25% is due to environment.[49] A number of seed companies are producing high fermentable corn hybrids that allow dry mill operations to obtain higher ethanol yields.[50] However, it is not clear what physical or chemical properties underlie these modest, but important gains in ethanol yield. What has become apparent is that high total fermentables trait is a more accurate indicator of dry grind ethanol production than total starch or extractable starch.[51] There is only a negligible or weak correlation between starch content or extractability and starch fermentability.[49] On the commodity market, premiums are not being paid at the elevator for high fermentable corn. However, Pioneer has developed a point-of-sale assay using near-infrared (NIR) technology that allows ethanol plants to predict the value of corn for ethanol production by identifying total fermentables.[52]

3.4 Grain Processing

Most corn used for ethanol production is processed by dry milling, a process in which corn is ground with hammer mills into a flour. Ethanol can also be produced from wet milling, a process in which corn kernels are soaked in water (often containing sulfur dioxide) to separate out germ, fibre and gluten before converting the remaining starch to ethanol. Wet milling is associated with larger corn-processing operations producing a variety of food and feed products. Dry milling is the method of choice because it can be carried out by smaller operations and produces higher yields of ethanol (2.8 gallons of ethanol per bushel of corn for dry milling, 2.5 gallons per bushel for wet milling).[53]

Starch is found in partially crystalline granules in the corn endosperm. Corn starch contains about 27% amylose; the rest being amylopectin. Starch needs to be broken down into soluble dextrins for fermentation. To do so, the dry mill product is made into a mash to which is added a thermostable, starch-hydrolysing enzyme, alpha-amylase. Alpha-amylase breaks down starch to produce dextrins by hydrolysing α1–4 bonds. The mash is then heated to more than 100 °C using a jet cooker in which high temperature and mechanical shear breaks down large starch molecules. The mash is then pumped through a holding tube and into a flash tank where more alpha-amylase is added and the mash is allowed to liquify. The dextrinised mash is cooled and glucoamylase (gamma-amylase) is added. Glucoamylase converts the dextrins into simple sugars (glucose) and also hydrolyses 1–6 linkages. After this, the mash is further cooled, and yeast is added to begin the fermentation process, which lasts for 2–4 days, at which time the ethanol concentration reaches 10–12%.[50] Some ethanol plants have converted to the use of simultaneous saccharification and fermentation (SSF) to increase ethanol production, save energy, reduce microbial contamination and enhance fermentation activity by preventing the exposure of yeast to osmotic conditions (high glucose) at the beginning of the fermentation process.[50] The fermentation product or beer is subjected to distillation, yielding 95% ethanol and from which residual water is removed by molecular sieves. Currently, the maximum amount of pure ethanol that can be made from a bushel of corn is 2.74 gallons (98 gallons per ton at 15% moisture or 115 gallons per dry ton).

An advance on the horizon in dry mill processing is the production of transgenic corn expressing endogenous liquefaction enzymes, such as alpha-amylase.[49] Singh et al.[54] developed corn that produces an endogenous, thermostable alpha-amylase enzyme that is activated in the presence of water at elevated temperatures. They compared the performance of the amylase producing corn in the dry-grind process to nontransgenic corn to which exogenous alpha-amylase was added. The transgenic corn (1–10% by weight) was added to normal dent corn (of the same genetic background as the amylase corn) as treatments and resulting samples were evaluated for ethanol production. Following fermentation, ethanol concentrations in batches in which 1% transgenic corn was added to normal dent corn were comparable to controls in which a conventional amount of exogenous alpha-amylase enzymes was added to nontransgenic dent corn. Hence, the use of transgenic corn expressing alpha-amylase could cut down on the costs of enzymes used in processing, although this is not as much of an issue as it is in cellulosic ethanol production (see sections below).

The solid and liquid fraction remaining after distillation is referred to as "whole stillage". Whole stillage includes the fibre, oil, and protein components of the grain, as well as the nonfermented starch. This coproduct of ethanol manufacture is a valuable feed ingredient for livestock, poultry, and fish. Although it is possible to feed whole stillage, it is usually processed further to create a feed product known as wet distillers grains with solubles (WDGS).[50] WDGS can be used directly as a feed product. In fact, it is often preferred for dairy and beef feed, however, WDGS has a short shelf life. To prevent spoilage and reduce transportation costs, WDGS is usually dried to 10–12% moisture, giving rise to dry distillers grains (DDGs). Drying WDGS is energy intensive, consuming about one-third of the energy requirements of the entire dry grind plant. However, producing a uniform, stable, high-quality feed coproduct is essential to the profitability of the plant, resulting in most plants producing DDGs rather than WDGS.[50] Another advance on the horizon is the expression of various high-valued proteins in transgenic corn to increase the value of the whole stillage or WDGS fractions.[55] However, considerable regulatory hurdles presently stand in the way of doing so.

3.5 Corn Stover and Cellulosic Ethanol

If one asks how much of US energy supply can be met by converting corn grain to ethanol, the answer is: not much. If the entire US corn crop was converted to

Table 3.1 US Dept. of Energy EERE theoretical ethanol yield calculator based on C6 and C5 polymer weight percentages, http://www1.eere.energy.gov/biomass/ethanol_yield_calculator.html.

Feedstock	Theoretical yield ($1\,dry\,ton^{-1}$)
Corn grain	470.9
Mixed paper	439.8
Corn Stover	427.7
Bagasse	422.0
Rice straw	416.0
Hardwood sawdust	381.5
Forest thinnings	308.5
Cotton gin trash	215.0

ethanol, it is estimated that it would supply about 12% of the transportation fuels Americans use each year.[56] Given the limitations of grain ethanol, energy planners have turned to more plentiful feedstocks for biofuels. A compelling argument was made in the billion-ton study to use cellulosic biomass for biofuels.[57] The major thesis of that study was that US land resources are capable of supplying enough biomass to displace 30% of the country's petroleum usage by the year 2030. A number of cellulosic resources were cited in that study including corn stover, which could account for 20% of the billion-ton total (actually 1.3 billion ton total). Corn stover is, indeed, attractive because it is the residue from the harvest of the corn crop and it fares pretty well as a biofuels feedstock when compared to other resources (Table 3.1). However, stover comes with a whole set of issues, some of which are discussed below.

3.5.1 Biomass Traits

Corn stover traits as they relate to biofuels have not been studied extensively, because corn has been bred traditionally for grain production and quality and not for biomass, with the exception of corn grown for forage purposes.[58,59] However, one breeding study that assessed corn stover traits in relation to biofuels is that of Kirkpatrick,[60] who evaluated 50 maize genotypes for their cellulosic ethanol production. She analysed germplasm ranging from population crosses to commercial hybrids in an effort to determine the amount of genetic variation in these traits. Stover (stalks and leaves), cobs, and husk fractions were evaluated for yield and chemical composition mostly by near-infrared reflectance spectroscopy (NIRS). Pentose and hexose sugar content predicted by this method was used to calculate theoretical ethanol potential (TEP).[61] Other fractions were analysed by detergent fibre methods[62] from which cellulose and hemicellulose composition was estimated and also used to predict TEP. Theoretical ethanol yields were then calculated on a land-area basis by multiplying TEP and dry matter yield for each fraction. Significant variation was found among hybrids for agronomic and compositional traits in stover, cobs and husks.

Maize

Among the genotypes in the Kirkpatrick study, corn stover yield correlated positively with grain yield and with plant height. Therefore, taller and higher grain yielding genotypes generally produced higher stover yields. Stover dry matter yield averaged 6.6 t ha^{-1} for the two growing seasons and accounted for 38% of total plant above ground dry matter (TDM). The harvest index (HI), the ratio of grain to total dry matter, averaged about 0.5 for the two growing seasons. HI varied across genotypes, however, when averaged among environments HI was consistent. TEP (liters/ton or lt^{-1}) estimated on hexose and pentose sugars averaged about 460 lt^{-1}. Theoretical ethanol yields (TEY) were calculated from TEP and stover yield and were found to average about 3000 l ha^{-1} over two seasons. TEY was highly correlated with stover yield and in turn stover yield showed significant correlations with grain yield. Therefore, Kirpatrick[60] concluded that the best way to improve genotypes for cellulosic ethanol production may be by increasing stover yields, which can be estimated from grain yields.

As for cobs, they yielded on average 1.3 t ha^{-1} and accounted for 7.6% of total plant above-ground dry matter. Cob yields also differed significantly among genotypes and correlated positively with grain yields, indicating that cob yields too can be estimated from grain yields. Cob compositional characteristics also significantly differed among genotypes, suggesting that it may be possible to select for higher cob cellulose and hemicellulose content. Significant differences were also found among genotypes for TEP, which was most greatly influenced by cob yields, not by cob composition. From this, it was concluded that selection for higher cob yields as opposed to compositional traits would have the greatest effect on cob ethanol yields. Furthermore, given the positive correlation between grain and cob yields, the most practical approach to selecting for higher cob yields would, again, be by selecting genotypes with higher grain yields.[60]

Husks accounted for 3.7% of the total above ground dry matter. Average husk TEP was about 615 lt^{-1}, and theoretical ethanol yields calculated by multiplying husk yields (t ha^{-1}) and husk TEP (lt^{-1}) averaged 438 l ha^{-1} over all hybrids. TEY was highly correlated with husk yield, indicating that the best approach to improving ethanol yield from husks may be by selecting for higher husk dry matter yields.[60]

3.5.2 Maize Cell Walls and Lignocellulose

Corn stover has value as a feedstock for biofuels production because stover is a source of lignocellulose, the material that makes up the structure of plant cell walls. The amount of ligncellulose produced by corn obviously is related to the amount of biomass (vegetative material or stover) produced, as described above. To improve lignocellulose production, therefore, one would expect to apply breeding strategies to select for increased biomass.

The composition of cell walls (particularly with respect to the relative amount of cellulose, hemicellulose and lignin) is also an important trait for

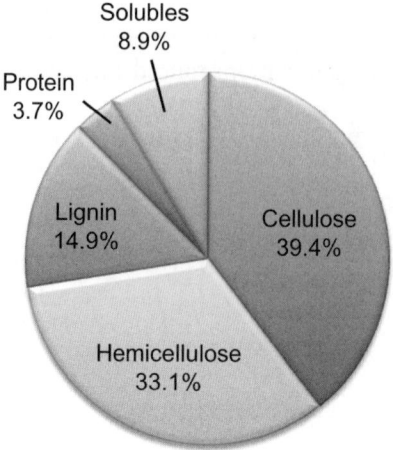

Figure 3.3 Example of a compositional analysis of corn stover.[115]

biofuels production, which, too, can be subject to selection by breeding (Figure 3.3). However, the importance of compositional characteristics of stover depends on the technology by which stover is converted to biofuels.[63] Compositional traits may be more important for biochemical, rather than for thermochemical conversion technologies, since the recalcitrance of biomass to enzymatic breakdown relates, in part, to polymer composition.[64] In general, high cellulose and low lignin appear to be compositional characteristics best suited for biochemical conversion.

Underlying biomass production and composition issues are the genetics and biochemistry relating to the synthesis of the precursors and polymers of which plant cell walls are composed. In the future it is likely that these features will be targeted for improvement as we understand more about lignocellulose production. In general, corn plants produce cell walls that change in character and composition during cell development. Primary cell walls, which are classified as type-II primary cell walls in maize, are laid down early during and after cell division and they contain cellulose and a hydrated matrix consisting mostly of hemicelluloses.[65] Primary cell walls are pliable and accommodate the growth of cells during cell-elongation phases. During subsequent development secondary cell walls are laid down interior to the primary wall. Secondary walls are generally much thicker, particularly in structural tissues or the xylem vasculature, and the hydrated space in the wall matrix is filled with lignin, which makes the walls more impermeable to water.[65]

3.5.3 Cellulose Synthesis

A major component of primary and secondary cell walls, and therefore of vegetative biomass in corn, is cellulose. Hence, the synthesis of cellulose is an

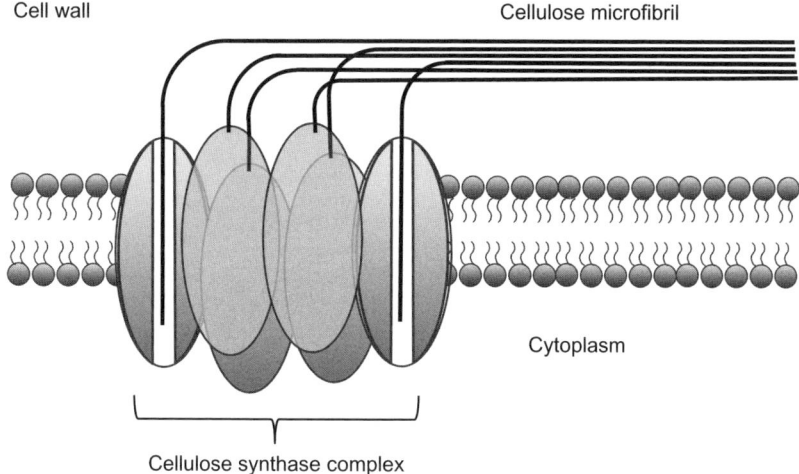

Figure 3.4 Diagram of cellulose synthase rosette in the plant cell plasma membrane. Redrawn from ref. 116.

important issue in biomass production. Much has been learned in just the past few years about cellulose synthesis in plants since the genes and the proteins involved in cellulose synthesis have been identified. A number of excellent reviews are available on this subject.[66–71] In brief, the cellulose synthase complex in the plant cell membrane is the molecular machine responsible for cellulose biosynthesis. The cellulose synthase complex catalyses the incorporation of UDP-glucose into growing 1,4 α-linked glucose chains.

The complex is composed of multiple glycosyl transferase enzyme subunits embedded in the plant cell membrane (Figure 3.4). The glycosyl-transferase subunits are encoded by 12 CesA genes identified so far in maize.[72] The cellulose synthase complex appears by freeze fracture/electron microscopy to be a rosette structure in the plant cell plasma membrane with six lobes arranged with a 6-fold axis of symmetry.[73] It is thought that each lobe is made up of six glycosyl-transferase subunits, each of which grows a glucan chain. The complex is highly processive and glides in the plane of the membrane along paths of cortical microtubules.[74] The glucan chains are spun off into a cellulose microfibril in which the glucan chains hydrogen bond into a tight crystalline structure. Cellulose microfibrils, in corn are about 3.5 nm in cross section, consistent with the crystal structure for a microfibril composed of about 36 glucan chains.[75] In current models of the cellulose synthase complex, each lobe is a heterohexamer of three different CesA subunits.[68] The heterohexamer concept is based on the work of Taylor *et al.*[76] who showed in *Arabidopsis* that three CesA mutants belong to a single complementation group called irregular xylem5 (irx5). The genes in this group are all expressed in the same tissues (involved in secondary wall synthesis), and the proteins encoded by these CesA genes coimmunoprecipitate in cell extracts.[76] Furthermore, the proteins do not associate, and presumably do not assemble the cellulose synthase complex, when one of the subunits is missing in a knockout mutant. Thus, the authors concluded that the complex must be composed of at least three different gene products.

It is thought that different members of the CesA gene family function in the synthesis of either primary or secondary cell walls. In *Arabidopsis*, CESA1, CESA3, and CESA6 are required for cellulose biosynthesis in primary cell walls,[77] whereas CESA4, CESA7, and CESA8 are required for cellulose biosynthesis during secondary wall deposition.[78] The same functionality has been ascribed to the maize CesA genes primarily based on their sequence relatedness to the *Arabidopsis*.[79] Most of the genes identified so far appear to group by sequence relatedness with *Arabidopsis* genes associated with primary cell wall synthesis. However, one maize gene, ZmCesA-8, has been shown to be expressed in regions of developing vascular bundles and epidermis, and likely is associated with secondary cell wall formation.[79]

It is clearly of interest to determine what limits the rate of cellulose synthesis and what can be done to enhance cellulose production in biomass crops. One possibility is that the initiation of glucan chains may be rate limiting. Peng *et al.*[80] proposed that a plant sterol conjugate, sitosterol-beta-glucoside (SG) serves as a primer to initiate glucan polymerisation. They concluded this from an *in vitro* system involving cotton fibre membranes that produces sitosterol-cellodextrins (SCDs) from SG and uridine 5′-diphosphate-glucose (UDP-Glc). The authors speculate that cellulase encoded by the Korrigan (Kor) gene, which is also required for cellulose synthesis in plants, may function to cleave SG from the growing glucan chain.[80] However, since the *in vitro* system is derived from a highly specialised tissue for abundant cellulose fibre production in cotton and since the system does not produce long glucan chains, it is not yet resolved whether SGs are physiologically important primers in other plants, such as maize.

Another possibility for enhancing cellulose synthesis would be to modify or overexpress CesA genes. However, as yet, there are no reports of successful overexpression of CesA genes leading to increased cellulose levels in plants. The fact that cellulose synthase is a heteromeric complex associated with other proteins may mean that several genes will have to be introduced into transgenic plants to make more functional complexes.

It is also possible that precursor levels limit cellulose production. Cellulose synthesis can be a major sink for carbon in plants, and it may be possible to manipulate patterns of carbon flux to enhance cellulose synthesis.[81] For example, sucrose synthase (SuSy) has been implicated in channeling substrate UDP-Glc to cellulose synthase in cotton fibre production.[82] Sucrose synthase plays an important role in providing UDP-glucose to cellulose synthesis in sink tissues, because the synthesis reaction is reversible in the presence of high levels of sucrose driving the degradation of sucrose to UDP-glucose and fructose. A form of sucrose synthase has been identified in cotton that is plasma membrane bound and is thought to be responsible for channeling substrates to the cellulose synthase complex. Ruan *et al.*[83] developed antisense constructs to the cotton sucrose synthase and demonstrated that sucrose synthase suppression inhibited fibre initiation and elongation. It is not known whether these mechanisms operate in corn, but sink strength and UDP-glucose production *via* sucrose synthase or through alternative pathways, such as UDP-glucose pyrophosphorylase, may be important factors in efforts to enhance cellulose synthesis in corn.

3.5.4 Hemicellulose Synthesis

Other polymers in plant cell walls important to biofuels are hemicellulose and lignin. The hemicelluloses are noncrystalline glycans that are tightly bound to and tether the cellulosic microfibrils in the cell wall. In maize, the major hemicelluloses in mature vegetative tissue are the glucuronoarabinoxylans (GAXs, Figure 3.5) while mixed linkage (1-3),(1-4)-β-D-glucans are produced during cell expansion.[84] The cell walls of grasses in general are pectin-poor and have very little structural protein compared to dicot plants. Instead, they have networks of polyphenolics that form primarily when cells stop expanding.

The backbone chains of hemicellulose and cellulose are structurally similar, which suggested that cellulose synthase-like (*Csl*) genes with sequence similarity to the *CesA* genes might be involved in the hemicellulose biosynthesis.[85] The *Csls* are a family of genes composed of nine subfamilies encoding various glycan synthases. Certain *Csl* subfamilies are common to all plants, whereas other subfamilies, CslFs, *CslHs*, and *CslJs*, are present only in the Poaceae.[86] The *CslFs* appear to be involved in the synthesis of mixed-linked glucans, a backbone that is specific to members of the grass family. Burton *et al.*[87] introduced rice *CslF* genes into *Arabidopsis*, which lacks mixed-linked glucans. They demonstrated with a specific monoclonal antibody that epidermal cells of the transformed *Arabidopsis* plants produced mixed-linked glucans. The Fincher laboratory has also reported preliminary data that members of the *CslHs*, and *CslJs* subfamilies may also be involved in mixed-glucan synthesis.[86]

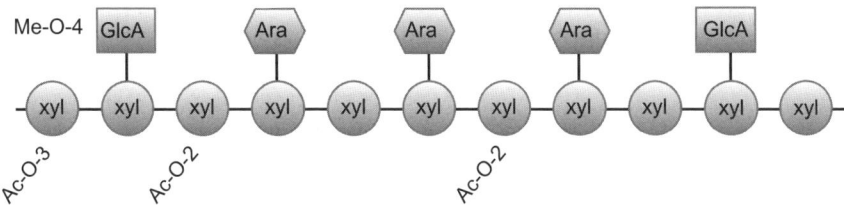

Figure 3.5 Structure of glucuronoarabinoxylan, the most abundant hemicellulose found in corn cell walls. This hemicellulose component has a xylopyranose backbone joined by β1–4 glycosidic bonds and glucuronic acid and arabinofuranose side chains linked to the backbone by α1–2 bonds. Redrawn from ref. 65.

To date, no one has been able to identify *Csl* genes involved in xylan backbone synthesis. Liepman *et al.*[88] tested members of the *Arabidopsis* and/or rice *CslA*, *CslC*, *CslD*, *CslE*, and *CslH* gene subfamilies for heterologous xylan synthase expression in insect cells, but failed to find activity. It may be that *Csl* genes do not catalyse xylan synthase activity or that conditions for expression of the active enzyme are not met in heterologous systems expressing single recombinant *Csl* genes.[89] Nonetheless, xylan synthase and glucuronyl-transferase activities leading to the production a GAX-like product have been demonstrated in *in vitro* reactions using crude microsomal fractions from wheat.[90] Thus, the fundamentals of the *in vitro* assay appear to work, but the right combination of recombinant genes to drive the reaction have not yet been identified.

GAX, but not mixed glucans have side chains (Figure 3.5) presumably appended onto the backbones by glycosyltransferases in the membranes of the secretory pathway. However, the topology of the enzymes in the membranes and the mechanism of synthesis of hemicelluloses is not yet known.[89] One model is that catalytic domains of the glycosyltransferases face the Golgi lumen and the enzyme uses sugar nucleotides from the lumen as substrate. Another model is that the Csl proteins function like CesA proteins in which the sugar nucleotides are derived from the cytosolic side of the Golgi but the polymer is transported across the membrane into the Golgi lumen.[89] From there, the hemicellulose would be deposited in the cell wall, perhaps in coordination with cellulose synthesis.

3.5.5 Lignin Synthesis

Lignins have received considerable attention because they are regarded as a major obstacle in converting corn biomass in biofuels by biochemical conversion technologies. In corn, lignins are mainly found in secondary cell walls in which they occupy the normally hydrated space between fibrillar components adding rigidity and water impermeability to cell walls. Lignin polymers are mainly composed of three monolignols, *p*-coumaryl, coniferyl and sinapyl alcohols, which, when polymerised, form *p*-hydroxyphenyl (H), guaiacyl (G) and syringyl (S) units in lignin.[91] Monolignols are synthesised *via* the phenylpropanoid pathway beginning with the deamination of phenylalanine to form cinnamic acid[92] (Figure 3.6). Maize stalk lignin is characterised by an abundance of S units and little H units (60.3% S, 37.4% G and 2.3% H).[93] After lignin monomers are biosynthesised, they are translocated to the cell wall, where they are polymerised by peroxidases or laccases. Other products of the phenylpropanoid pathway, ferulic and *p*-coumaric acids, link lignin to hemicelluloses by ester or ether linkages.

The role of the phenylpropanoid pathway in determining the composition and the digestability of lignin has been extensively explored in maize through the analysis of brown-midrib (*bm*) mutants at four independent loci (*bm1*, *bm2*, *bm3*, and *bm4*). These mutants have a reddish brown pigmentation of the leaf

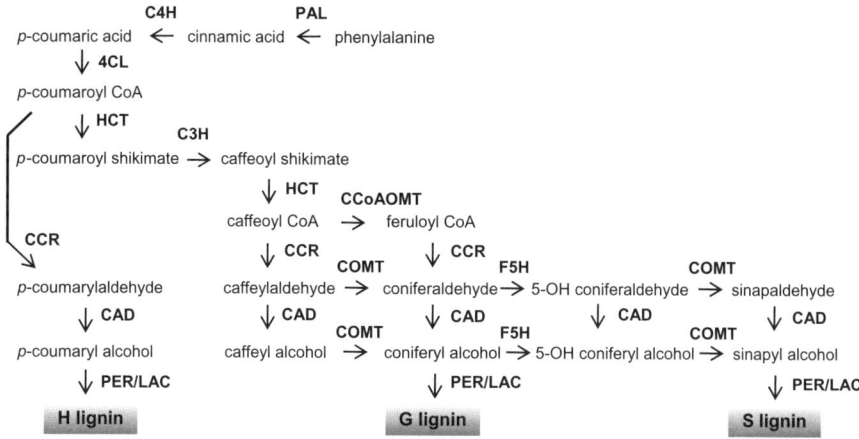

Figure 3.6 Lignin biosynthesis pathway leading to H, G and S lignin production. Redrawn from ref. 117.

midrib and stalk pith associated with lignified tissues.[94] While considerable work has been carried out on these mutants, the mutations are not fully understood at a gene level, with the exception of *bm3*. Nonetheless, they have been extensively characterised with respect to their chemical defects.

For example, maize bm1 mutants have a somewhat reduced lignin content (10–20%), a slight decrease in ferulic acid (FA) esters and a substantially reduced content of *p*-coumaric (pCA) esters and FA ethers.[95,96] The relative amounts of H/G/S monolignols units are comparable to normal plants. The lignins in *bm1* also contain substantial amounts of other phenylpropanoid pathway intermediates, coniferaldehyde and, to a lesser extent, sinapaldehyde (Figure 3.6).[95,97] For that reason, it is thought that the *bm1* mutation could be in a gene that regulates the expression of the CAD gene family.[98]

Maize *bm2* mutants also have a reduced lignin content, a slight reduction in p-courmaric and ferulic acid ester levels and a substantial decrease in ferulic acid ether levels (about 60%).[95] The *bm2* mutant also has a lower content of guaiacyl (G) units in lignin, but in contrast to *bm3* and *bm1*, does not show the incorporation of unusual monolignols. Maize *bm3* mutants have a lignin content that is reduced by 25–40%. In these plants, the content of p-coumaric esters is reduced by about 50%, but *bm3* plants are not altered in their content of alkali-releasable ferulic acid.[95] A similar trend was observed in *bm4* as in *bm2* plants, with lignin content reduced by about 15% and a slight change in ferulic and *p*-coumaric ester levels, but no change in etherified ferulic acid content.[95]

The only *bm* mutant for which the gene defect has been found is bm3. Vignols *et al.*[99] showed that two independent *bm3* mutations in the (COMT) gene, encode caffeic acid/5-hydroxyferulic acid *O*-methyltransferase (Figure 3.6). The *bm3-1* allele has a B5 element retrotransposon inserted near the junction of the 3′ coding region of the COMT gene intron and the bm3-2 allele resulted from a deletion of part of the COMT gene.

The content and composition of maize lignin is correlated with differential expression of genes on the phenylpropanoid pathway leading to the synthesis of monolignols,[100] and it may be that lignins can be modified by engineering changes in expression of those genes. Various studies reviewed by Vanholme et al.[101] demonstrated that lignin content can be reduced by downregulating genes such as those encoding phenylalanine ammonia-lyase (PAL); cinnamate 4-hydroxylase (C4H); 4-hydroxycinnamoyl CoA ligase (4CL); p-hydroxycinnamoyl-CoA: quinate shikimate p-hydroxycinnamoyltransferase (HCT), p-coumarate 3-hydroxylase (C3H); caffeoyl-CoA O-methyltransferase (CCoAOMT); cinnamoyl-CoA reductase (CCR), and, to a lesser extent, cinnamyl alcohol dehydrogenase (CAD) (Figure 3.6). Lignin composition with respect to the balance between H/G/S lignins can be engineered as well. For example, downregulation of the ferulate 5-hydroxylase (F5H) gene results in G-enriched lignin, while overexpression of the same gene can produce plants enriched in S-lignins. HCT-and C3H-downregulated plants are enriched in H lignin, which is usually a minor component of normal lignin. Downregulation of caffeic acid/5-hydroxyconiferaldehyde O-methyltransferase (COMT) reduces the production of S-lignin and promotes 5-hydroxyconiferyl alcohol incorporation into lignin.

Grabber et al.[102] conducted an interesting study in which they incorporated ester interunit linkages into cell wall lignins in an effort to enhance the alkaline solubility of lignin and the enzymatic hydrolysis of corn fibre. They substituted coniferyl alcohol with various amounts of coniferyl ferulate (an ester conjugate from secondary metabolism) into artificially lignified maize cell walls. The coniferyl ferulate was extensively copolymerised into lignin, but led to a reduction in cell wall lignin content and the ferulate copolymerisation into lignin. The modification increased the extractability of cell wall lignin by up to two-fold in mild aqueous alkaline conditions. Thus, the conjugate provides a milder and more economical means for delignifying cell walls.[102]

3.5.6 Corn-Stover Composition

Stover chemical composition varies during corn crop growth and maturity. Most of studies on the composition of corn stover have been conducted regarding its use as a forage, however, a study by Pordesimo et al.[103] looked at the changes relevant to the use of the crop as a feedstock for bioenergy production. They found in the two commercial corn hybrids under study that dry matter in the stover fractions peaked at the time of physiological maturity, which in their case was 118 days after planting. The distribution in dry matter of plants, which were cut at 15 cm above the ground, was in rounded terms: 46% grain, 28% stalk, 11% leaf, 8% cob and 7% husk (or 46% grain, 54% stover). At ∼110 days of the Pordesimo et al.[103] study, during maturation, senescence and weathering, they found that that the chemical composition, but not the energy content of the stover varied. (The energy content averaged about 17–21 kJ g^{-1}.) These findings, therefore, are likely to be of consequence to the use of stover as a feedstock for biochemical conversion where composition counts, but less of an issue for use of the stover for thermochemical conversion to biofuels (Section 3.5.7).

Pordesimo et al.[103] reported that the chemical composition (as determined by NIR) of the different stover fractions (stalk, leaf, and husk) differed from each other at all stages of development. In particular, stalks had a higher lignin content than husks, and husks had a higher xylan content than stalks. Leaves showed the greatest change in chemical composition with time. Of all the stover fractions, leaves had the highest level of soluble solids at grain maturity, but those levels dropped precipitously with further leaf senescence. Lignin and xylan content of leaves increased proportionately at that time. Thus, Pordesimo et al.[103] concluded that the timing for the collection of corn stover is relevant to its use as a fermentation feedstock.

3.5.7 Corn-Stover Processing

Stover is a tough lignocellulosic material that is recalcitrant to processing involved in the biochemical conversion to biofuels.[64] Stover is largely composed of crystalline cellulose, a beta-linked glucose polymer, which is more difficult to degrade than the alpha-linked polymer in starch. Stover also contains hemicellulose, a more complex polymer of several sugars but rich in xylose. Hence, the principal monosaccharides produced in the hydrolysis of corn-stover polymers are glucose and xylose. In a composition study by McAloon et al.,[34] corn stover in their samples was composed of 37.3% cellulose, 20.6% xylan, 17.5% lignin, 6.1% ash, 2.1% arabinans, 2.0% acetate, 1.4% galactan/mannan and 13% extractives.

Methods to convert lignocellulose to biofuels are being hotly pursued, and in that context corn stover does not present any special issues relative to other lignocelluosic resources. In general, two platforms have been proposed for the conversion of lignocellulosic material into biofuels: the biochemical and the thermochemical platform.[104] The biochemical platform involves pretreatment to loosen the lignocellulose structure, followed by cellulase and other hydrolytic enzymes to release soluble sugars, which are then be fermented to biofuels. In the thermochemical platform, biomass is heated to a gaseous or liquid form, which can then be converted into other products.

A major step in the biochemical conversion of lignocellulose to biofuels is the breakdown of cellulose by cellulase enzymes. Two approaches are envisioned for doing so, either by incubating a lignocellulose-containing slurry with microorganisms producing cellulases or by adding cellulase enzymes. Cellulases (endo-glucanases) such as those derived from filamentous fungus, *Trichoderma reesei*, are notoriously slow acting, and high levels of the enzymes are required for efficient breakdown. It is estimated that cellulases at 2.5% (w/w) of input cellulose must be added for industrial processing, which becomes a major cost in biofuels production. An alternative approach is to produce cellulases in plants used as feedstocks for cellulosic ethanol production. A major hurdle in doing so is to produce an enzyme that does not interfere with the normal growth and development of the plant and that retains activity through pretreatment processes. For these reasons Ziegler et al.[105] and Ziegelhoffer et al.[106] used an approach similar to that described in a previous section on the transgenic expression of starch liquifaction enzymes in kernels. They expressed a transgene encoding the catalytic domain of endo-1,4-beta-D-glucanase E1 from *Acidothermus cellulolyticus*, a thermophilic cellulose-degrading eubacterium in *Arabidopsis* and tobacco. This enzyme has a high-temperature optimum and low activity at ambient temperatures. Therefore, the enzyme would be inactive during normal plant growth, but active at high temperature during processing.

Ransom et al.[107] used a similar approach expressing the catalytic domain of the *A. cellulolyticus* enzyme in corn. They targeted the enzyme to the apoplast by fusing the signal peptide of tobacco pathogenesis-related protein onto the enzyme and reported expression level of the cellulase up to 1.1% of transgenic plant total soluble proteins. However, the retention of the cellulase activity during pretreatment is still a concern. In a companion study with the enzyme expressed in tobacco, they found only 35% retention of activity when the plant material was subjected to ammonia fibre explosion (AFEX) treatment, one of the possible pretreatment regimes.[108] In lieu of activating the enzyme after pretreatment, an approach that has been suggested is to add extracts from cellulase-expressing plants to pretreated materials, to activate the enzymes prior to pretreatment or to develop milder pretreatment methods.

Recently, Donohoe et al.[109] determined whether pretreatment truly increases the accessibility of cellulases to corn stover lignocellulose. They examined the ability of cellulase enzymes to penetrate into corn stover that had been subjected to three dilute acid pretreatments: a low severity pretreatment (20 min at 100 °C), moderate severity pretreatment (20 min at 120 °C) and a high severity (20 min pretreatment at 150 °C). They used *T. reesei* enzymes, Cel7A (CBH I) and Cel7B (EG 1) and detected the presence of the enzymes by immuno electron microscopy. They observed very little penetration of the cellulases into stover (>1% of the thickness of secondary cell walls) that had been subjected to low severity pretreatment. On the other hand they observed total penetration (100% of the thickness of secondary cell walls) in stover subjected to high severity treatment. Of interest was stover subjected to the moderate severity pretreatment. The cellulase penetrated 20% into the thickness of the cell wall, but the enzyme appeared to move in a planar front through the cell wall rather than drilling through in limited places – a situation seen in the enzymatic degradation of starch granules. The study is a dramatic demonstration of the effectiveness of pretreatment.[109]

Acid pretreatment not only loosens cell wall structure but also hydrolyses hemicellulose.[110] The major hydrolytic product of corn-stover hemicelluloses is the pentose sugar, xylose. Xylose has presented some challenges as a raw material for fermentation because most industrial micro-organisms, such as *Saccharomyces cerevisiae*, do not ferment this sugar, although another yeast, *Pichia stipitis*, can. However, when *P. stipitis* genes encoding xylose reductase and xylitol dehydrogenase were introduced in *S. cerevisiae* in combination with the endogenous gene xylulokinase, the resulting yeast strains were able to utilise xylose for growth and ethanol production.[111]

A major goal in cellulosic biofuels production is consolidated bioprocessing or simultaneous saccharification and fermentation (SSF) – producing a single micro-organism with the capacity to breakdown cellulose and to ferment hexose and pentose sugars. According to Lynd et al.[112] "Progress in developing micro-organisms capable of consolidated bioprocessing is being made through two strategies: engineering naturally occurring cellulolytic micro-organisms to improve product-related properties, such as yield and titer, and engineering noncellulolytic organisms that exhibit high product yields and titers to express a heterologous cellulase system enabling cellulose utilisation".

3.5.8 Corn-Stover Removal

A major question concerning the harvesting of corn stover for biofuels is how much crop residue can be collected. Leaving stover in the field helps in preventing soil erosion caused by wind or rain. Leaving stover also replaces lost nutrients and sequesters carbon in the soil, lessening CO_2 accumulation in the atmosphere as a greenhouse gas. There are no generalities for how much stover can be removed. An argument justifying collecting stover is that in one study about 85% of the stover rotted on the ground releasing CO_2, while the rest contributed to soil organic material.[113] Thus, the collection of stover may actually reduce the GHG emission from a corn crop.

Other studies have examined the amount of carbon lost from soils at a given rate of stover removal.[114] In this example (Table 3.2) corn-stover carbon remaining and lost from soil over time was estimated at two corn production levels and with and without 70 per cent stover harvest. In the first year, 67 per cent

Table 3.2 Corn-stover carbon remaining in soil over time and with stover harvested at 70%. From ref. 114.

Time period	240 bu/acre grain crop (5.3 ton/acre stover)		178 bu/acre grain crop (3.3 ton/acre stover)	
	No stover harvest	70% stover harvested	No stover harvest	70% stover harvested
	--------------lb carbon (C)/acre--------------			
Starting corn stover	4240	1270	2600	780
Remaining after 1 yr	1400	420	860	260
Remaining after 4 yr	850	250	520	160
Remaining after 8 yr	640	190	390	120
Total lost from soil	3600	4050	2210	2480

of stover C was estimated to be lost as CO_2 from microbial processing, 80 perc ent lost after four years, and 85 per cent lost after eight years. While the effect of stover harvest on the amount of C remaining in soil is not large on the short term, it will affect soil organic matter over a long time period.

Acknowledgements

I am grateful for the input and suggestions from Kendall Lamkey and Alan Myers, both at Iowa State University.

References

1. W. E. Tyner, *Bioscience*, 2008, **58**, 646.
2. T. M. Crosbie, in *Biobased Outlook Conference*, Iowa State University, 2008.
3. F. Sissine, *Congressional Research Service*, 2007.
4. K. C. Dhuyvetter, T. L. Kastens and T. C. Schroeder, http://www.agmanager.info/energy/TriStateDairyShortcourse(Jan2008)-2.pdf.
5. http://www.card.iastate.edu/research/bio/tools/hist_eth_gm.aspx.
6. *World Agricultural Supply and Demand Estimates*. http://www.usda.gov/oce/commodity/wasde/latest.pdf.
7. D. B. Egli, *Agron. J.*, 2008, **100**, S79.
8. T. Searchinger, R. Heimlich, R. A. Houghton, F. Dong, A. Elobeid, J. Fabiosa, S. Tokgoz, D. Hayes and T. H. Yu, *Science*, 2008, **319**, 1238.
9. J. Fargione, J. Hill, D. Tilman, S. Polasky and P. Hawthorne, *Science*, 2008, **319**, 1235.
10. K. Cole, B. Dale, B. Davison, B. Erickson and M. Parr, *Industrial Biotech.*, 2008, **4**, 322.
11. L. Brown. http://cleantech.com/news/2360/why-ethanol-production-will-drive-world-food-prices-even-higher-in-2008.
12. Y.-W. Chiu, B. Walseth and S. Suh, *Environ Sci Technol.*, 2009, **43**, 2688.
13. Briefing Room, Corn: Background. http://www.ers.usda.gov/Briefing/Corn/background.htm.
14. D. N. Duvick, *Adv. Agron.*, 2005, **86**, 83.
15. D. N. Duvick, in *Developing Drought and Low N-tolerant Maize*, ed. G. O. Edmeades, B. Bänziger, H. R. Mickelson and C. B. Pena-Valdivia, CIMMYT, El Batan, Mexico, 1997, pp. 332.
16. D. N. Duvick, J. C. S. Smith and M. Cooper, *Plant Breed. Rev.*, 2004, **24**, 109.
17. D. N. Duvick, *Maydica*, 1977, **22**, 187.
18. L. T. Evans and R. A. Fischer, *Crop Sci.*, 1999, **39**, 1544.
19. M. Tollenaar and E. A. Lee, *Field Crop. Res.*, 2002, **75**, 161.
20. T. M. Crosbie, S. R. Eathington, G. R. Johnson, M. Edwards, R. Reiter, S. Stark, R. G. Mohanty, M. Oyervides, R. E. Buehler, A. K. Walker, R. Delannay, J. C. Pershing, M. A. Hall and K. R. Lamkey, in *Plant*

breeding: *The Arnel R. Hallauer International Symposium*, ed. K. R. Lamkey and M. Lee, Blackwell Publishing, Oxford, UK, 2006, pp. 3.
21. E. A. Lee and M. Tollenaar, *Crop Sci.*, 2007, **47**, S202.
22. T. M. Crosbie, in *Changes in physiological traits associated with long-term breeding efforts to improve grain yield of maize, Proc. Annu. Corn and Sorghum Ind. Res. Conf, Washington, D.C.*, ed. H. D. Loden and D. Wilkinson, Am. Seed Trade Assoc., Washington, D.C., 1982.
23. J. Ying, E. A. Lee and M. Tollenaar, *Field Crops Res.*, 2000, **68**, 87.
24. Corn Yield Potential. http://www.extension.org/pages/Corn_Yield_ Potential.
25. M. Tollenaar, L. M. Dwyer and D. W. Stewart, *Crop Sci.*, 1992, **32**, 432.
26. M. Tollenaar, D. E. McCullough and L. M. Dwyer, in *Genetic improvement of field crops*, ed. G. A. Slafer, Marcel Dekker, Inc., New York, 1994, pp. 183.
27. P. S. Schnable, D. Ware, R. S. Fulton, J. C. Stein, F. Wei, S. Pasternak, C. Liang, J. Zhang, L. Fulton, T. A. Graves, P. Minx, A. D. Reily, L. Courtney, S. S. Kruchowski, C. Tomlinson, C. Strong, K. Delehaunty, C. Fronick, B. Courtney, S. M. Rock, E. Belter, F. Du, K. Kim, R. M. Abbott, M. Cotton, A. Levy, P. Marchetto, K. Ochoa, S. M. Jackson, B. Gillam, W. Chen, L. Yan, J. Higginbotham, M. Cardenas, J. Waligorski, E. Applebaum, L. Phelps, J. Falcone, K. Kanchi, T. Thane, A. Scimone, N. Thane, J. Henke, T. Wang, J. Ruppert, N. Shah, K. Rotter, J. Hodges, E. Ingenthron, M. Cordes, S. Kohlberg, J. Sgro, B. Delgado, K. Mead, A. Chinwalla, S. Leonard, K. Crouse, K. Collura, D. Kudrna, J. Currie, R. He, A. Angelova, S. Rajasekar, T. Mueller, R. Lomeli, G. Scara, A. Ko, K. Delaney, M. Wissotski, G. Lopez, D. Campos, M. Braidotti, E. Ashley, W. Golser, H. Kim, S. Lee, J. Lin, Z. Dujmic, W. Kim, J. Talag, A. Zuccolo, C. Fan, A. Sebastian, M. Kramer, L. Spiegel, L. Nascimento, T. Zutavern, B. Miller, C. Ambroise, S. Muller, W. Spooner, A. Narechania, L. Ren, S. Wei, S. Kumari, B. Faga, M. J. Levy, L. McMahan, P. Van Buren, M. W. Vaughn, K. Ying, C. T. Yeh, S. J. Emrich, Y. Jia, A. Kalyanaraman, A. P. Hsia, W. B. Barbazuk, R. S. Baucom, T. P. Brutnell, N. C. Carpita, C. Chaparro, J. M. Chia, J. M. Deragon, J. C. Estill, Y. Fu, J. A. Jeddeloh, Y. Han, H. Lee, P. Li, D. R. Lisch, S. Liu, Z. Liu, D. H. Nagel, M. C. McCann, P. SanMiguel, A. M. Myers, D. Nettleton, J. Nguyen, B. W. Penning, L. Ponnala, K. L. Schneider, D. C. Schwartz, A. Sharma, C. Soderlund, N. M. Springer, Q. Sun, H. Wang, M. Waterman, R. Westerman, T. K. Wolfgruber, L. Yang, Y. Yu, L. Zhang, S. Zhou, Q. Zhu, J. L. Bennetzen, R. K. Dawe, J. Jiang, N. Jiang, G. G. Presting, S. R. Wessler, S. Aluru, R. A. Martienssen, S. W. Clifton, W. R. McCombie, R. A. Wing and R. K. Wilson, *Science*, 2009, **326**, 1112.
28. M. Guo, M. A. Rupe, X. Yang, O. Crasta, C. Zinselmeier, O. S. Smith and B. Bowen, *Theor. Appl. Genet.*, 2006, **113**, 831.
29. R. A. Swanson-Wagner, Y. Jia, R. DeCook, L. A. Borsuk, D. Nettleton and P. S. Schnable, *Proc. Natl. Acad. Sci.USA*, 2006, **103**, 6805.

30. K. G. Cassman, A. Dobermann and D. T. Walters, *AMBIO*, 2002, **31**, 132.
31. D. J. Greenwood, G. Lemaire, G. Gosse, P. Cruz, A. Draycott and J. T. Neeteson, *Ann. Bot.*, 1990, **66**, 425.
32. P. Bertin and A. Gallais, *Maydica*, 2001, **45**, 53.
33. A. Gallais and B. Hirel, *J. Exp. Bot.*, 2004, **55**, 295.
34. A. McAloon, F. Taylor, W. Yee, K. Ibsen and R. Wooley, *Determining the cost of producing ethanol from corn starch and lignocellulosic feedstocks*, National Renewable Energy Laboratory: Golden, CO, 2000.
35. A. M. Smith, K. Denyer and C. Martin, *Annu. Rev. Plant. Phys. Biol.*, 1997, **48**, 67.
36. S. G. Ball and M. K. Morell, *Annu. Rev. Plant Biol.*, 2003, **54**, 207.
37. I. J. Tetlow, M. K. Morell and M. J. Emes, *J. Exp. Bot.*, 2004, **55**, 2131.
38. S. Ball, H. P. Guan, M. James, A. Myers, P. Keeling, G. Mouille, A. Buleon, P. Colonna and J. Preiss, *Cell*, 1996, **86**, 349.
39. J. L. Prioul, V. Mechin and C. Damerval, *C. R. Biol.*, 2008, **331**, 772.
40. L. C. Hannah and M. James, *Curr. Opin. Biotechnol*, 2008, **19**, 160.
41. M. G. James, K. Denyer and A. M. Myers, *Curr. Opin. Plant Biol*, 2003, **6**, 215.
42. A. M. Smith, *Biomacromolecules*, 2001, **2**, 335.
43. T. A. Hennen-Bierwagen, Q. Lin, F. Grimaud, V. Planchot, P. L. Keeling, M. G. James and A. M. Myers, *Plant Physiol.*, 2009, **149**, 1541.
44. A. M. Myers, M. K. Morell, M. G. James and S. G. Ball, *Plant Physiol.*, 2000, **122**, 989.
45. T. A. Hennen-Bierwagen, F. Liu, R. S. Marsh, S. Kim, Q. Gan, I. J. Tetlow, M. J. Emes, M. G. James and A. M. Myers, *Plant Physiol.*, 2008, **146**, 1892.
46. C. D. Boyer and J. Preiss, *Plant Physiol.*, 1981, **67**, 1141.
47. D. M. Stark, K. P. Timmerman, G. F. Barry, J. Preiss and G. M. Kishore, *Science*, 1992, **258**, 287.
48. V. Mechin, C. Thevenot, M. Le Guilloux, J. L. Prioul and C. Damerval, *Plant Physiol.*, 2007, **143**, 1203.
49. V. Singh, K. Rausch and C. Miller and J. Graeber, in *Bioenergy – I Conference*, Tomar, Portugal, 2006.
50. R. J. Bothast and M. A. Schlicher, *Appl Microbiol. Biotechnol.*, 2005, **67**, 19.
51. Driving to better ethanol hybrids. http://www.pioneer.com/CMRoot/Pioneer/media_room/biofuels/documents/driving.pdf.
52. T. Bryan, *Ethanol Prod. Mag.*, 2003, **2003**, 36.
53. S. Butzen and D. Haefele, *Crop Insights*, 2008, **18**, 1.
54. V. Singh, C. J. Batie, G. W. Aux, K. D. Rausch and C. Miller, *Cereal Chem.*, 2006, **83**, 317.
55. E. E. Hood and J. M. Jilka, *Curr. Opin. Biotechnol.*, 1999, **10**, 382.
56. C. F. Runge and B. Senauer, *How Biofuels Could Starve the Poor*. New York Times, May 7, 2007.
57. R. D. Perlack, L. L. Wright, A. F. Turhollow, R. L. Graham, B. L. Stokes and D. C. Erbach, *Biomass as Feedstock for A Bioenergy and Bioproducts*

Industry: The Technical Feasibility of a Billion-Ton Annual Supply; A357634; Oak Ridge National Lab, Oak Ridge TN, 2005.
58. K. J. Shinners and B. N. Binversie, *Biomass Bioenerg*, 2007, **31**, 576.
59. J. G. Lauer, J. G. Coors and P. J. Flannery, *Crop Sci.*, 2001, **41**, 1449.
60. K. M. Kirkpatrick, *The evaluation of maize genotypes for potential use in cellulosic ethanol production*, Iowa State University, Ames, Iowa, 2008.
61. Theoretical Ethanol Yield Calculator, http://www1.eere.energy.gov/biomass/ethanol_yield_calculator.html.
62. H. M. Goering and P. J. Van Soest, in *U.S. Dept. Agri. Sci. Handbook*, 1970, **379**.
63. A. Carroll and C. Somerville, *Annu. Rev. Plant. Biol.*, 2008, doi:10.1146/annurev.arplant.043008.092125.
64. M. E. Himmel, S. Y. Ding, D. K. Johnson, W. S. Adney, M. R. Nimlos, J. W. Brady and T. D. Foust, *Science*, 2007, **315**, 804.
65. M. Pauly and K. Keegstra, *Plant J.*, 2008, **54**, 559.
66. D. P. Delmer, *Annu Rev Plant Physiol Plant.*, 1999, **50**, 245.
67. K. S. Dhugga, *Curr. Opin. Plant Biol.*, 2001, **4**, 488.
68. M. S. Doblin, I. Kurek, D. Jacob-Wilk and D. P. Delmer, *Plant Cell. Physiol.*, 2002, **43**, 1407.
69. C. Somerville, S. Bauer, G. Brininstool, M. Facette, T. Hamann, J. Milne, E. Osborne, A. Paredez, S. Persson, T. Raab, S. Vorwerk and H. Youngs, *Science*, 2004, **306**, 2206.
70. C. Somerville, *Annu. Rev. Cell. Dev. Biol.*, 2006, **22**, 53.
71. A. R. Paredez, S. Persson, D. W. Ehrhardt and C. R. Somerville, *Plant Physiol.*, 2008, **147**, 1723.
72. L. Appenzeller, M. Doblin, R. Barreiro, H. Wang, 3. Niu, K. Kollipara, L. Carrigan, D. Tomes, M. Chapman and K. S. Dhugga, *Cellulose*, 2004, **11**, 287.
73. S. C. Mueller and R. M. Brown, Jr,, *J. Cell Biol.*, 1980, **84**, 315.
74. A. R. Paredez, C. R. Somerville and D. W. Ehrhardt, *Science*, 2006, **312**, 1491.
75. R. M. Brown, *Pure Appl. Chem.*, 1999, **71**, 201.
76. N. G. Taylor, R. M. Howells, A. K. Huttly, K. Vickers and S. R. Turner, *Proc. Natl. Acad. Sci. USA*, 2003, **100**, 1450.
77. S. Robert, G. Mouille and H. Höfte, *Cellulose*, 2004, **11**, 351.
78. N. G. Taylor, J. C. Gardiner, R. Whiteman and S. R. Turner, *Cellulose*, 2004, **11**, 329.
79. N. Holland, D. Holland, T. Helentjaris, K. S. Dhugga, B. Xoconostle-Cazares and D. P. Delmer, *Plant Physiol.*, 2000, **123**, 1313.
80. L. Peng, Y. Kawagoe, P. Hogan and D. Delmer, *Science*, 2002, **295**, 147.
81. D. P. Delmer and C. H. Haigler, *Metab. Eng.*, 2002, **4**, 22.
82. C. H. Haigler, M. Ivanova-Datcheva, P. S. Hogan, V. V. Salnikov, S. Hwang, K. Martin and D. P. Delmer, *Plant Mol. Biol.*, 2001, **47**, 29.
83. Y. L. Ruan, D. J. Llewellyn and R. T. Furbank, *Plant Cell*, 2003, **15**, 952.
84. W. Yong, B. Link, R. O'Malley, J. Tewari, C. T. Hunter, C. A. Lu, 3. Li, A. B. Bleecker, K. E. Koch, M. C. McCann, D. R. McCarty, S. E.

Patterson, W. D. Reiter, C. Staiger, S. R. Thomas, W. Vermerris and N. C. Carpita, *Planta*, 2005, **221**, 747.
85. K. S. Dhugga, R. Barreiro, B. Whitten, K. Stecca, J. Hazebroek, G. S. Randhawa, M. Dolan, A. J. Kinney, D. Tomes, S. Nichols and P. Anderson, *Science*, 2004, **303**, 363.
86. G. B. Fincher, *Plant Physiol.*, 2009, **149**, 27.
87. R. A. Burton, S. M. Wilson, M. Hrmova, A. J. Harvey, N. J. Shirley, A. Medhurst, B. A. Stone, E. J. Newbigin, A. Bacic and G. B. Fincher, *Science*, 2006, **311**, 1940.
88. A. H. Liepman, C. G. Wilkerson and K. Keegstra, *Proc. Natl. Acad. Sci USA*, 2005, **102**, 2221.
89. O. Lerouxel, D. M. Cavalier, A. H. Liepman and K. Keegstra, *Curr. Opin. Plant Biol.*, 2006, **9**, 621.
90. W. Zeng, M. Chatterjee and A. Faik, *Plant Physiol.*, 2008, **147**, 78.
91. W. Boerjan, J. Ralph and M. Baucher, *Annu Rev. Plant Biol.*, 2003, **54**, 519.
92. J. M. Humphreys and C. Chapple, *Curr. Opin. Plant Biol.*, 2002, **5**, 224.
93. Y. Barrière, C. Riboulet, V. Méchin, S. Maltese, M. Pichon, A. Cardinal, C. Lapierre, T. Lubberstedt and J.-P. Martinant, *Gene, Genom. Genom.*, 2007, **133**.
94. Y. Barriere and O. Argillier, *Agronomie*, 1993, **13**, 865.
95. Y. Barriere, J. Ralph, V. Mechin, S. Guillaumie, J. H. Grabber, O. Argillier, B. Chabbert and C. Lapierre, *C. R. Biol.*, 2004, **327**, 847.
96. G. J. Provan, L. Scobbie and A. Chesson, *Agric. Food*, 1997, **734**, 133.
97. H. Kim, J. Ralph, F. Lu, G. Pilate, J. C. Leple, B. Pollet and C. Lapierre, *J. Biol. Chem.*, 2002, **277**, 47412.
98. S. Guillaumie, M. Pichon, J. P. Martinant, M. Bosio, D. Goffner and Y. Barriere, *Planta*, 2007, **226**, 235.
99. F. Vignols, J. Rigau, M. A. Torres, M. Capellades and P. Puigdomenech, *Plant Cell*, 1995, **7**, 407.
100. S. Guillaumie, H. San-Clemente, C. Deswarte, Y. Martinez, C. Lapierre, A. Murigneux, Y. Barriere, M. Pichon and D. Goffner, *Plant Physiol.*, 2007, **143**, 339.
101. R. Vanholme, K. Morreel, J. Ralph and W. Boerjan, *Curr. Opin. Plant Biol.*, 2008, **11**, 278.
102. J. Grabber, R. Hatfield, J. Ralph and F. Lu, B*iomacromolecules*, 2008, **9**, 2510.
103. L. O. Pordesimo, B. R. Hames, S. Sokhansanj and W. C. Edens, *Biomass Bioenerg.*, 2005, **28**, 366.
104. J. Houghton, S. Weatherwax and J. Ferrell, in *Breaking the biological barriers to cellulosic ethanol: A joint research agenda, A Research Roadmap Resulting from the Biomass to Biofuels Workshop*, Rockville, Maryland, Rockville, Maryland, 2005.
105. M. T. Ziegler, S. R. Thomas and K. J. Danna, *Mol. Breed.*, 2000, **6**, 37.
106. T. Ziegelhoffer, J. A. Raasch and S. Austin-Phillips, *Mol. Breed.*, 2001, **8**, 147.

107. C. Ransom, V. Balan, G. Biswas, B. Dale, E. Crockett and M. Sticklen, *Appl. Biochem. Biotechnol.*, 2007, **137–140**, 207.
108. F. Teymouri, H. Alizadeh, L. Laureano-Perez, B. Dale and M. Sticklen, *Appl. Biochem. Biotechnol.*, 2004, **113–116**, 1183.
109. B. S. Donohoe, S. R. Decker, M. P. Tucker, M. E. Himmel and T. B. Vinzant, *Biotechnol. Bioeng.*, 2008, **101**, 913–925.
110. R. Torget, P. Walter, M. Himmel and K. Grohmann, *Appl. Biochem. Biotechnol.*, 2008, **28–29**, 75.
111. K. Ohgren, O. Bengtsson, M. F. Gorwa-Grauslund, M. Galbe, B. Hahn-Hagerdal and G. Zacchi, *J. Biotechnol.*, 2006, **126**, 488.
112. L. R. Lynd, W. H. van Zyl, J. E. McBride and M. Laser, *Curr. Opin. Biotechnol.*, 2005, **16**, 577.
113. What is a life-cycle assessment? 2002. http://www.nrel.gov/docs/gen/fy02/31792.pdf.
114. J. Sawyer and A. Mallarino, in *Integrated Crop Management Newsletter*, I. S. University, ed. Ames, Iowa, 2007.
115. S. P. Chundawat, B. Venkatesh and B. E. Dale, *Biotechnol. Bioeng.*, 2007, **96**, 219.
116. N. G. Taylor, *New Phytol.*, 2008, **178**, 239.
117. X. Li, J. K. Weng and C. Chapple, *Plant J.*, 2008, **54**, 569.

CHAPTER 4
Sweet Sorghum as a Biofuel Crop

GENE STEVENS AND ROLAND A.Y. HOLOU

University of Missouri Delta Research Center, P.O. Box 160, Portageville, Missouri 63873, USA

4.1 Introduction

Sorghum is a high biomass grass in the botanical tribe, Andropogoneae. Sorghum plants use the C_4 pathway to fix carbon dioxide for photosynthesis. Other grasses in the Andropogoneae tribe include sugarcane [*Saccharum officinarum* L.], corn or maize [*Zea mays* L.subsp. *mays*], *and Miscanthus* [*Miscanthus* spp.]. The genus, *Sorghum,* is made up of perennial wild species, such as johnsongrass (*S. halepense*), and annual cultivated sorghums, which are all the same species [*S. bicolour* subsp. *bicolour* (L.)]. *Sorghum bicolour* is grown for grain, forage, hay, and syrup. Grain sorghum, sometimes called "milo", is produced primarily for poultry and swine feed.[21,26] Approximately 29% of the grain sorghum produced in the United States in 2008 was used for starch-based ethanol production (J. Dahlberg, National Sorghum Producers, personal communication). Sorghum syrup is produced from sweet sorghum, also called "sorgo". The stalks of sweet sorghum are squeezed to extract the sugars in the juice that are then processed into syrup or ethanol. Most forage sorghums used for silage, grazing, and hay are sorghum hybrids, while most sweet sorghums currently grown for syrup production are open pollinated cultivars.

Grain and sweet sorghums are grown on every continent of the world except Antarctica. It was first domesticated in Sudan and Ethiopia, in North-eastern

Africa. The first introduction of sweet sorghums into the US was the cultivar "Chinese Amber", which was brought from France in 1853.[30] The USDA distributed large amounts of sweet sorghum seeds to American farmers in 1857. One of the goals was to extend sugar production farther north than sugarcane is adapted. In 1863, Isaac Hedges published a book entitled *Sorgo or the Northern Sugar Plant*. Several groups attempted to produce granulated sugar from sweet sorghum juice but discovered that the fructose and glucose in the juice interferes with the crystallisation of the sucrose. Farmers found that concentrating the sugar into syrup is the easiest way to make a stable product for their own consumption and to sell. In the southern United States, a popular method of eating sorghum syrup is to pour it over a hot, buttered biscuit.

Sorghum is normally cultivated as an annual. In warm climates such as the Texas Gulf Coast, grain sorghum can be managed to produce new tillers after the first harvest and grow a second grain crop.[18] In the dry lowlands of eastern and central Africa, grain sorghum can be ratooned in 3 to 4 yr cycles from one seed planting (B. Folk, personal communication). Ratoon cropping in sorghum occurs when a farmer does not destroy the lower stalks and roots after harvest to allow regrowth for additional harvests. In semiarid climates where other cereal crops frequently fail from drought, farmers grind sorghum grain from their fields into porridge and pat the meal into bread cakes.

Sorghum seed panicles are located on the top of the stalks and come in a wide range of shapes and sizes, with the seed clusters on secondary and tertiary branches. Generally, sweet sorghum plants contain more sugar in their stalks than grain sorghum and typically have poor grain yield. In the United States, sweet sorghum cultivars are taller than grain sorghum because breeders have selected for reduced plant height in grain sorghum to make mechanical harvesting easier and to reduce plant lodging. In favourable environments, sweet sorghum cultivars can grow 4.5 m tall and produce 45 to 110 Mg of fresh weight biomass per hectare with less N fertiliser than maize (Figure 4.1). Since nitrogen fertiliser is produced from natural gas, nitrogen efficiency is important in the net energy budget of a biofuel crop.[29] Bitzer[4] reported that maize has an energy efficiency of 1:1.8 while sweet sorghum has an efficiency of 1:8.

4.2 Biofuel from Sweet Sorghum

Sweet sorghum was promoted as a source of motor fuel during the energy crisis of the 1970s.[6,19] Over the next two decades, interest in alternative fuels and energy conservation declined when gasoline and diesel prices receded. Through the early 1990s, some research with sweet sorghum for ethanol continued in Virginia and Louisiana.[11,12,27] Today, in many countries, research funding has increased for "green" alternative fuels due to public concern for possible climate change from carbon dioxide produced by burning fossil fuels and depletion of the world supply of petroleum. India, Brazil, and China have programs developing sweet sorghum for ethanol fuel production[26,38] (M. Bitzer, personal communication). In the United States, most of the research in sweet sorghum is

Figure 4.1 "Dale" cultivar sweet sorghum in Southeast Missouri. Photo: Gene Stevens.

being conducted in the central and southern regions. In Texas, the economic feasibility of large-scale production of ethanol from sweet sorghum was studied.[8,22] Also, a new technology for processing sweet sorghum biomass was developed at Texas A&M University.[17] In Oklahoma, research is being conducted to evaluate onfarm sweet sorghum ethanol-production systems.[3] In the Mississippi River Delta, a regional strategy report, sponsored by Memphis Biodimensions, Inc., identified sweet sorghum as "the near-term sugar/ dedicated energy crop for the region".[33]

The traditional process of extracting sugar from the stalks for syrup is very labour intensive involving hand stripping leaves and feeding stalks through a mill. Many farmers still use this method today. Before gasoline engines were available, horses and mules were used to power mills to extract the sugary juice (Figure 4.2). The process of making syrup from juice has not changed much over the years. Most of the water is slowly boiled off to concentrate the sugars into sorghum syrup. The syrup has a similar consistency to honey but with a unique flavor. During Prohibition in the United States, farmers making illegal whiskey (called "moonshine") found that they could add inexpensive sorghum

Figure 4.2 Antique horse-powered sweet sorghum mill used to make sorghum syrup. Photo reprinted by permission of Morris Bitzer.

syrup, produced on their own farms, to corn mash to speed up the fermentation process and increase ethanol production.

Sweet sorghum is adapted to be both humid and tropical climates. Sorghum can be grown in colder climates than sugarcane and will produce a sugar crop in 4 months versus 12 months for sugarcane.[4] Today, sweet sorghum can be grown as far north and south as 45° latitude. Farmers have grown sweet sorghum in southern Canada in the Northern Hemisphere and Uruguay in the Southern Hemisphere.

Sweet sorghum stalks contain both structural carbohydrates (cellulose fibre) and nonstructural carbohydrates (sucrose, glucose, and fructose). Several options are available for converting structural carbohydrates to useable energy. Although technically possible today in the laboratory, the expense of producing cellulosic enzymes to break fibres into sugars for fermentation is not commercially feasible at this time. Other methods of converting biomass to useable energy are (1) burning to produce steam for generating electricity, (2) reacting with oxygen at high temperatures to make synthetic gas, (3) liquefying (several methods are available depending on desired endproduct), and (4) anaerobic digestion. Lau et al.[17] reported on technology called MixAlco that converts sweet sorghum biomass "into alcohol and carboxylate acids using microorganisms, water, steam, lime and hydrogen through an anaerobic process".

Some studies suggest that a tradeoff exists between producing high sugar content in sugarcane stalks and maximum biomass yields. The same may be true with sweet sorghum. Sugarcane cultivars/hybrids bred for biomass are sometimes called "energy cane". In Puerto Rico, Alexander[1] reported "that growth-oriented management can increase whole cane yield in the order of 250–300 per cent, at cost increases in the order of 35–40 per cent". To provide a 365-day supply feedstock to a biofuel processing facility, Alexander proposed harvesting and stockpiling biomass from high-yielding tropical grasses. He speculated that the relatively small cost increases of growing grasses for biomass, relative to sugarcane, would be overshadowed by the higher yields after lignocellulose technologies are established.

Farmers in Texas, Kansas, Oklahoma, Nebraska and the Mississippi River Delta region often plant grain sorghum because it is more drought tolerant than maize. In Texas, Morris et al.[22] developed economic models for farm-level sweet sorghum sugar production from juice and ethanol facilities. "Monte Carlo simulations" were used to compare returns from three feedstock scenarios in different regions of Texas. The added machinery costs of harvesting and milling equipment to extract the juice were large capital input items when sweet sorghum was used as a primary feedstock. Although ethanol plants using sweet sorghum generated returns on assets of 11% in their models, from the perspective of an investor considering construction of a new ethanol facility, corn alone was the most profitable feedstock material.

The amount of juice that is recovered from the sweet sorghum stalks is dependent on the harvest and sugar extraction methods. The three main ways to extract sugar are roller mills, screw presses, and diffusers. Cundiff[11] reviewed results from earlier research that showed juice expression ratios ranged from 0.47 to 0.55 (juice mass divided by input mass) for roller mills and a 0.35 ratio from a screw press. The performance of a screw press extraction of sugars in juices was increased by 67% when the rind-leaf was mechanically removed prior to processing. Diffusers use hot water to extract sugars by pumping the water countercurrent to moving shredded cane on a system of belts or conveyors. Goven Investments[15] reported that diffusers extract a higher ratio of sugars from stalks than roller mills with a lower investment cost to build new facilities.

The short sweet sorghum harvest period is a management obstacle for farmers. New types of harvest equipment are needed to gather stalks and extract sugar from large fields in a timely manner while the sugar contents in stalks are still at the optimum levels. Also, there must be liquid tanks available to hold 18 000 to 55 000 l of juice per hectare. Less capacity would be needed if the juices could be easily concentrated by removing some of the water. Kundiyana et al.[16] suggested storing juice in large bladders. In Oklahoma, Dr. Danielle Bellmer has been developing a process of small-scale conversion of sweet sorghum juice into ethanol. Bellmer[3] reported that infield fermentation is possible without adding nutrients and with no temperature controls. When inoculation of the juice with yeast was delayed, significant losses in ethanol occurred. She found that after the fermentation occurred the diluted ethanol liquid is stable for long-term storage.

4.3 Sweet Sorghum Cultivars

Most of the sweet sorghum cultivars grown in the United States today were developed from germplasm maintained at the US Sugar Crops Field Station at

Meriden, Mississippi. Four of the most widely grown cultivars for ethanol developed at the station are "Theis", "Keller", "Dale" and "M81E". After the closing of the facility in 1983, about 1200 sweet sorghum seed samples were transferred to the Plant Genetics Resources Conservation Unit in Griffin, Georgia. Gary Pederson and Merrelyn Spinks USDA-ARS,[24] organised data containing Brix and sugar content information from old field and laboratory notebooks. Brix is the per cent sugar solids in a liquid measured with a refractometer. Many of the sweet sorghum plant introductions available in the USDA Germplasm Resources Network (GRIN) came from the Meriden station.

"M81E" was the last cultivar released by Dr. Kelly Freeman before the field station closed. The cultivar still retains its experimental line name from his breeding work. M81E and Dale are the most popular sweet sorghums for syrup and ethanol. "Keller" is a major cultivar grown in China and "URJA" grown in India is a selection from Keller (M. Bitzer, personal communication). At Virginia Polytechnic Institute, Dr. Howard Harrison developed the cultivar "Della" by backcrossing plants from "Dale". Della is one week earlier in maturity than Dale.

Sweet sorghum grown for juice has a short harvest period with a significant risk of waste if rainy fall weather or equipment failure causes harvest delays. Farmers can expand the harvest period by planting multiple cultivars to utilise the 20-day maturity spread that exists between the major sweet sorghum lines. Della has the shortest and M81E the longest time from planting to maturity. Maturities of several cultivars are Della 114 days, Dale 120 days, Keller 125 days, Theis 135 days, and M81E 135 days. Another method of expanding the harvest time is to stagger the planting dates of different fields. If planting continues into mid-June, scout closely for armyworms. In the northern and southern extremes of sweet sorghum production, fields planted late in the season may be lost from freezing temperatures at the end of the season. Dr. Ismail Dweikat at University of Nebraska is developing sweet sorghum cultivars with increased cold tolerance. He is also breeding for chinch bug resistance and selecting lines that do not produce flowers.[13]

Sugar production using hydrid sweet sorghum production is a relatively new development. The Nimbkar Agricultural Research Institute in Maharashtra, India has released a multipurpose sweet-stalked hybrid-"Madhura".[23] It is used for making syrup, using a jaggery production scheme, and fermented for ethanol. Sweet sorghum hybrids developed in China produced higher total soluble sugar and cellulose yields than inbred cultivars.[38] In 2007, Dr. Todd Pfeiffer released, jointly between University of Kentucky and University of Nebraska, the first commercial male-sterile sweet sorghum hybrid in the United States called "KN Morris". The female parent is male sterile and crossed with the male parent Dale. This hybrid was named after Dr. Morris Bitzer, emeritus University of Kentucky extension professor, who has spent most of his career helping farmers produce sweet sorghum syrup. The "K" is for Kentucky and "N" for Nebraska. When hybrid seed from KN Morris are planted the resulting plants are also male sterile. Dr. Bill Rooney at Texas A&M University has released female inbred lines with higher sugar content for use in the production of sweet sorghum hybrids.

4.4 Sorghum Diseases

Sweet sorghum is susceptible to most of the same fungal and bacterial diseases as all other types of sorghum. In wet, cold soil conditions, sorghum seeds germinate slowly and produce low-vigour seedlings. Necrotic roots on sorghum seedlings can result from *Fusarium* spp., *Pythium* spp., or *Rhizoctonia* spp. (Figure 4.3). The chance of this occurring in sweet sorghum, can be minimised if seeds are pretreated with a labeled seed protectant such as captan and planting is delayed until average soil temperatures are 16 to 19 °C. Planting sweet sorghum in a field the following year after grain sorghum or maize should also be avoided if possible, sweet sorghum should be rotated with a nongrass species such as soybean or cotton. Gray leaf spot (*Cercospora sorghi* Ellis & Everh.) is an example of a disease common in both maize and sorghum.

Other sorghum fungal diseases include leaf blight (*Exserohilum turcicum* (Pass.) K. J. Leonard & E. G. Suggs), zonate leaf spot (*Gloeocercospora sorghi* D. Bain & Edgerton ex Deighton), rough leaf spot (*Ascochyta sorghina* Sacc.), sooty stripe (*Ramulispora sorghi* (Ellis & Everh.) L. S. Olive & Lefebvre in Olive *et al.*), rust (*Puccinia purpurea* Cooke), charcoal rot (*Macrophomina phaseolina* (Tassi) G. Goidanich), downy mildew (*Peronosclerospora sorghi* (W. Weston & Uppal) C. G. Shaw), stalk rots, various smuts, and head moulds.[36] Zonate leaf

Figure 4.3 Black-coloured necrotic roots are an indication of seedling disease (*Fusarium, Pythium*, or *Rhizoctonia*) on sorghum plants. Photo was reprinted by permission of Allen Wrather.

Figure 4.4 Target-like lesion on leaf due to zonate leaf spot (*Bipolaris sorghicola*). Photo reprinted by permission of Allen Wrather.

spot is shown in Figure 4.4. Sorghum diseases caused by bacteria include bacteria leaf stripe (*Burkholderia andropogonis* (Smith) Gillis *et al.*), bacteria leaf streak (*Xanthomonas campestris* pv. *holcicola* (Elliot) Dye), and bacterial leaf spot (*Pseudomonas syringae* pv. *syringae* van Hall Smith). Azoxystrobin (Quadris®) is a broad spectrum fungicide labeled for control of anthracnose stalk rot (*Colletotrichum graminicola* (Ces.) G. W. Wilson), charcoal rot, and damping-off seedling diseases. Local extension personnel should be consulted for application rates, timings, and use restrictions.

During harvest, if a red colour is observed in the juice, it is probably caused by stalk infection with *Fusarium* spp. or *Anthracnose*. *Fusarium* spp. root and stalk rot tends to be more common in sorghum fields with high plant population and drought stress. *Anthracnose* occurs frequently in very humid conditions and causes a brick red colour in the inside of stalks. Severe cases can cause sorghum stalk lodging especially during high winds. Sweet sorghum is also susceptible to maize dwarf mosaic virus (MDM). For MDM and *Anthracnose* control, cultivars with resistance to the diseases should be chosen. Ergot, caused by the fungus *Claviceps afriana* (Frederickson, Mantle and de Milliano), can be a problem in warm climates. The sorghum seed embryos are infected at pollination. Heads must be removed or the disease will reduce sugar content and cause a sticky deposit on harvest machinery. Bacterial disease infections on sorghum vary year to year depending on weather conditions. Crop rotation is one the most effective control strategies for fungal and bacteria diseases.

Each cultivar has its strengths and weaknesses for disease and lodging resistance. Dale and Della are resistant to most diseases. Keller and M81E are resistant to red stalk rot (*Anthracnose*) but are susceptible to MDM. Theis is resistant to both red stalk rot and MDM and has superior lodging resistance.

4.5 Weed and Insect Control

Several herbicides and insecticides are labeled for grain and forage sorghums to control both plant and insect pests. Before applying pesticides, the company that manufactures a specific chemical should be consulted to determine the correct application for sweet sorghum.

In sweet sorghum research plots at Missouri, good grass and broadleaf control occurred without crop injury by broadcast spraying a pre-emergence application of atrazine and metolachlor. The sweet sorghum seed were treated with a fluxofenin herbicide seed safener before planting.

The best control strategy for sorghum insects is integrated pest management (IPM) that combines "cultural, mechanical, biological or chemical options".[5] Insects often cause the most damage in late-planted sorghum fields. Pheromone traps can be used to provide early warning that moths may be laying eggs for worms on sorghum leaves. Aphid pests in sorghum include corn leaf aphids (*Rhopalsiphum maidis* (Fitch)), greenbugs (*Schizaphis graminum* (Rondani)), and yellow sugarcane aphids (*Sipha flava* (Forbers)). Ladybugs and lacewings are important beneficial insects that eat aphids. When ladybug larvae are present in high numbers, spraying chemicals for aphid control may not be needed. Chinch bugs (*Blissus leucopterus leucopterus* (Say)) typically are worse in drought conditions. They sometimes migrate to sorghum from neighbouring grass pastures, prairies and wheat fields that are "drying down" at maturity. Effective grasshopper control can usually be accomplished by spot treating hatching sites in field borders and grass waterways. Sorghum midge and sorghum webworms should be protected against, if the sweet sorghum heads are to be harvested for seed production.

4.6 Sorghum Planting Density

Mechanical cultivation is an important weed control tool in sweet sorghum and requires rows to be spaced wide enough apart (~ 75 cm) to avoid ploughing up plants. Dr. Morris Bitzer recommends a sweet sorghum plant spacing of 13 plants per metre of row for ethanol production (Bitzer, personal communication). Farmers should plant enough seed to achieve a final plant population of 150 000 to 250 000 plants per hectare. From 1967–1972, Broadhead and Freeman[6] evaluated the effects of sweet sorghum row spacing (50 and 100 cm rows) and in-row population (plants spaced 15, 30, 45, and 60 cm apart) on juice Brix, sucrose, purity, and yield of sugar per ton of stalks. Stalk and sugar yield per unit area was significantly higher in 50 cm rows compared to 100 cm rows. However, the sugar per ton of stalks was less in 50-cm rows and plants lodged more frequently in the narrow-row plots because the individual stalks were smaller.

Most plant-density information for sorghum is from grain and forage research. This information should be applicable to sweet sorghum. Sorghum plants respond to low plant density by producing additional tillers per plant.[14] Conley et al.[10] reported that grain sorghum plants partially compensated for plant densities less than 150 000 plants per hectare by making additional heads per plant. One-meter plant skips in rows planted on 75-cm row spacing had no significant effect on grain yields. However, longer plant skips of 1.8 m and 2.8 m in the rows reduced yields when total plant populations declined to less than 165 000 plants per hectare. The highest yields were found in high and low populations with 100 kg N ha^{-1}.

4.7 Soil Fertility for Sweet Sorghum

The optimum nitrogen (N), phosphorus (P), and potassium (K) fertiliser rates for sweet sorghum in fields vary depending on soil test levels, soil pH, yield potential, rainfall, organic matter, and crop-rotation history. Sorghum grows best when soil pH is 6.0 to 7.0. In Alabama, Soileau and Bradford[32] found that liming an acid soil (pH 4.8) significantly increased fresh and dry sweet sorghum biomass and sugar in the juice. The test was conducted on a Wynnville silt loam soil with 0.9% organic matter. Because of the confounding effect of soil pH in the study, consistent biomass or sugar yield response to N, P, K were not found. In limed plots, the maximum biomass was 54 Mg ha^{-1} fresh weight with 16.7% Brix. In Louisiana, Ricaud and Arceneaux[27] measured the effect of N and P on sweet sorghum biomass, and sugar from "Wray", "M81E", and "Cowley" cultivars. The highest mean biomass yields occurred with M81E. When no K was applied, averaged across cultivars, 100 kg N ha^{-1} produced the highest yields with 47 Mg stalks ha^{-1} and 15.1% Brix. When K was applied with N, the highest sugar yields occurred with Wray receiving 180–0–80 fertiliser. Field research in Kentucky and Oklahoma showed that 67 to 84 kg N ha^{-1} produced maximum juice yields (M. Bitzer, personal communication). In Iowa, Anderson et al.[2] reported that total N content in sweet sorghum stalks (M81E) was 92 to 180 kg N ha^{-1}. The amount depended on the rate of N fertiliser applied. Fertiliser N-use efficiency ranged from 30 to 80% with the highest efficiencies occurring at the lowest N rate treatments.

Rotating sweet sorghum with soybeans can help reduce N-fertiliser requirements. Generally legume residues have lower C:N ratios than grasses, which promotes N mineralisation by soil micro-organisms. In Nebraska, Clegg[9] estimated that a soybean crop contributed 95 ± 23 kg N ha^{-1} to grain sorghum planted the following year. Under nonirrigated conditions, grain sorghum did not respond to supplemental N after soybeans.[35] On a silt loam soil in Missouri, Stevens et al.[31] reported that highest sweet sorghum stalk and sugar yield occurred with 67 kg N ha^{-1} in a field that had been rotated with soybeans. However, this rate was not statistically significantly different (0.05 level) from the zero N check.

To determine how much K and P fertiliser to apply on sweet sorghum, soil samples should be collected from 0 to 15-cm depth. Since sorghum is deep rooted, soil sampling 1 to 1.5 m in the soil profile might be helpful, but few farmers have the labour and financial resources for this amount of sampling and lab testing. The Missouri Extension Service recommends maintaining soil test levels in the plough layer for all types of sorghum above 50 kg P ha^{-1} using Bray 1 extraction solution and K per hectare above 246×5.6 CEC (cation exchange capacity) using ammonium acetate extraction solution (Brown *et al.*, 2004). Consult local extension specialists to determine critical soil test levels for sorghum in your region.

To maintain soil P and K at current levels in fields, it is necessary to apply enough P and K fertiliser to offset crop removal.[31] The amount of nutrients that are removed by growing sweet sorghum on a field will depend on how the biomass is processed. The University of Missouri Soil Test Laboratory[7] calculates P and K removal for harvested sorghum silage using the following equations:

$$\text{P removal as kg } P_2O_5 = 2.3 \times \text{Mg ha}^{-1} \text{ of fresh weight silage}$$
$$\text{K removal as kg } K_2O = 5.0 \times \text{Mg ha}^{-1} \text{ of fresh weight silage}$$

To convert kg of P_2O_5 to P multiply by 0.437. To convert kg of K_2O to K multiply by 0.830. In recent sweet sorghum fertility tests in Missouri, large year-to-year variations in the amounts of P and K removal in stalks and leaves were observed. If the harvest equipment strips the leaves from the stalks and deposits the leaves back in the field, the amount of nutrients removed will be less than whole-plant sorghum chopped for silage. Mills[20] reported that the K content in healthy sorghum leaves at heading growth stage is 1.4 to 1.7% K on a dry weight basis. The least amount of PK removal occurs when juice only is collected and all the bagasse including leaves are redeposited to the soil.

4.8 Measuring Sugar Content in Juice

For onfarm testing of sweet sorghum juice, a Brix (%) measurement from a refractometer provides satisfactory results for estimating sugar content. If Brix readings are taken in the field on bright days, the handheld light refractometer is usually easy to read (Figure 4.5). Two or three drops of juice are placed under the hinged glass plate located at the front of the metre. By looking into the eyepiece the boundaries between light blue and dark blue areas can be found, as illustrated in Figure 4.5. The tick marks on the centre line indicate the Brix reading. If the juice sample is collected and taken back to the office or lab, it may be necessary to stand under a bright light to see the blue shade change with this refractometer. If possible, a model with auto temperature compensation should be purchased to avoid having to adjust Brix readings for current temperature.

Sweet Sorghum as a Biofuel Crop

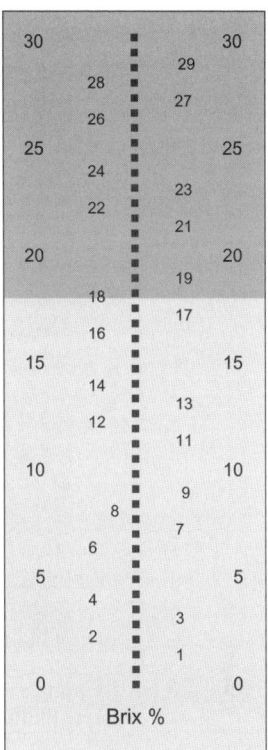

Figure 4.5 Estimating sugar content in sorghum juice (Brix) with light refractometer. Illustration on right is view through eye piece with Brix reading of 18 denoted by edge of dark blue area. Photo: Gene Stevens.

Pocket digital refractometers are easier to use (Figure 4.6) but are about three times more expensive to buy than nonelectronic light refractometers. To use digital meters, the meter has to be calibrated first to zero by placing water on the sensor well and pushing "Start". The screen should read "0". Juice samples can be tested by squeezing two to three drops on the sensor. In bright field conditions, sometimes this meter may display an error message "nnnn" on the screen. In most cases, this can be rectified by simply turning the body to create a shadow on the meter and pushing "Start" again. If that does not work, the meter and sample need to be taken indoors to conduct do the test. For both types of refractometer, sensor areas need to be cleaned with water and dried with a soft cloth before storing.

4.9 Removing Sorghum Seed Heads

To produce maximum sugar content in sorghum stalks, the reproductive heads on sweet sorghum stalks should be removed before the grain matures. To do this, sorghum syrup growers sometimes modify high-clearance sprayers by

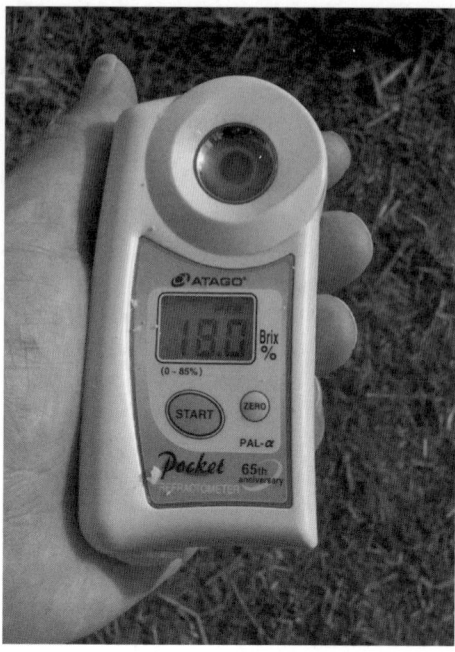

Figure 4.6 Estimating sugar content in sorghum juice (Brix) with digital refractometer. Photo: Gene Stevens.

adding rotating cutters on the boom. Morris Bitzer[4] recommends removing seed heads "when the seed is in the late milk stage". Check Brix readings from juice samples in 3 to 4 day intervals after deheading. Usually in about 2.5 weeks sugar contents will increase to optimum levels for harvest. Figure 4.7 shows the Brix levels of sweet sorghum in Kentucky deheaded at different stages of maturity.

4.10 Fermentation

Fermenting sweet sorghum juice is a relatively simple process compared to grains. Grains such as sorghum and maize are ground, hydrated to make mash, and enzymes added to break the starches into sugars. In fermenting sweet sorghum, these steps are skipped since the carbohydrates are already available to brewing yeast as sucrose, fructose, and glucose. Adding nutrients to the juice is usually not necessary. To minimise bacteria growth in the juice on the day of harvest, the harvest equipment needs to be thoroughly washed when harvesting is finished on the previous day. In Kansas, Wun *et al.*[37] reported that contaminating bacteria can consume 20% of the fermentable sugars within 3 days. All juice fermentation and storage tanks should be disinfected and airlock systems used to maintain aerobic conditions and prevent vinegar production from bacteria (Figure 4.8).

Sweet Sorghum as a Biofuel Crop

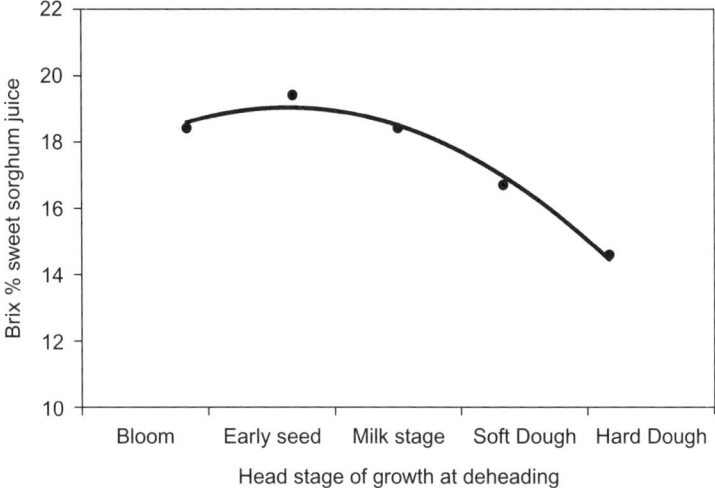

Figure 4.7 Brix sugar content in Kentucky sweet sorghum stalks 2.5 weeks after removing heads at different growth stages. Source: Graphed from data provided by M. Bitzer.

The amount of fermentable sugars recovered is dependent on the sugar content in the stalk and efficiency of extraction equipment used (roller mill, screw press, or diffuser). In the Missouri case study below, an equation reported in Cundiff[12] was used to estimate weight of sugar extracted in juice for fermentation [$TSS_j = 0.87\ W_{ws}\ E_j\ B_r/100$]. TSS_j is total soluble sugars in juice. W_{ws} is weight of water in whole stalk; E_j is juice expression ratio based on the processing equipment. B_r is Brix measurement of the juice. Eighty-seven per cent of the solid in the juice was assumed to be soluble. For juice-extraction efficiency, we used an expression ratio for a commercial roller mill ($E_j = 0.75$ kg juice per kg input to press). Fulton Iron Works (St. Louis, MO) manufactures a three-roller mill that has a 75–80% efficiency from a single pass (Figure 4.9). They also produce a four-roller mill which the company reports to have greater extraction efficiency. Other values in the equation (W_{ws} and B_r) were collected in sweet sorghum field test plots.

M81E sweet sorghum was planted at the University of Missouri-Delta Research Center in Portageville, Missouri. In 2007, the plots averaged 46 Mg fresh weight per hectare of stripped stalks (no leaves). The Brix readings for the juice averaged 20.6% using a digital refractometer and the stalk moisture content was 67%. The calculated total weight of the juice in the stalk ($W_{ws} = 46\ \text{Mg} \times 67\%/100$) was 30.8 Mg ha^{-1}. Using the Cundiff's formula below,[11] the total potential soluble sugar expressed in juice (TSS_j) was 4.1 Mg by extraction from a roller mill.

$TSS_j = 0.87\ W_{ws}\ E_j\ B_r/100$
4.1 Mg soluble sugar = 0.87×30.8 Mg total juice $\times\ 0.75 \times 20.6\%/100$

Figure 4.8 Airlock on fermentation tank to relieve carbon dioxide pressure. After the lid is closed the end of the flexible hoses are submerged in bleach water to prevent bacteria in the air from contaminating the juice. Photo: Gene Stevens.

To get a more precise measurement of total sugars, frozen fresh stalk samples from the plots were sent to the analytical laboratory at FutureFuel™ Chemical Company located in Batesville, Arkansas. Dr. Becky Edwards used a Soxhlet apparatus to separate the soluble and insoluble components in the sorghum stalks and high-performance liquid chromatography (HPLC) to determine the identity and amounts of the chemicals. Most of the sugar was in the sucrose form (Table 4.1). Combining the weights of sucrose, glucose, and fructose per hectare [(7228 kg sucrose + 386 kg fructose + 464 kg glucose)/1000] equaled 8.08 Mg. The 2-fold discrepancy in sugar amounts (Cundiff's equation versus actual lab measurement) is partly due to the juice-expression value used for a roller mill compared to the Soxhlet apparatus. The Soxhlet apparatus probably extracted close to 100% percentage of the sugars. Praj Industries Limited reported 96 to 97% sugar-extraction efficiencies with sweet sorghum using thermopermeation and low-pressure milling.[25]

To estimate ethanol yield from sugars, Ricaud and Arcenaux[27] used a theoretical conversion of 584 l of ethanol per Mg of fermentable sugar (0.07 gallons of

Figure 4.9 Fulton Iron Works© sugarcane mill (St. Louis, MO) used by Memphis Bio-dimensions to process extract sweet sorghum juice at Whiteville, Tennessee.

ethanol per pound of sugar). Most farmers do not have expensive laboratory equipment for testing juice, but if the amount of sucrose, fructose, and glucose in juice is known, more precise estimates of ethanol yield can be predicted. During fermentation, all fermentable sugars are converted to glucose before yeast make ethanol. Sucrose is a disaccharide and breaks into two glucose compounds. 360 g formula weight of glucose has more mass than 342 g of sucrose. Therefore, using ratios ($2\times 180/342 = 1.053$) and the FutureFuel lab results from 2007 Missouri sweet sorghum, 7228 kg sucrose ha^{-1} is equivalent to 7608 kg glucose ha^{-1}.

$$\text{Sucrose} + H_2O \rightarrow 2\,\text{Glucose}$$
$$342 \quad\quad 18 \quad\quad 2\times 180 = 360 (\text{formula wt})$$
$$7228\text{ kg sucrose} = 7611\text{ glucose}$$

Fructose is an isomer of glucose with a slightly lower formula weight. Likewise using ratios ($180/164 = 0.56$), 386 kg sucrose ha^{-1} is equivalent to 424 kg glucose ha^{-1}.

$$\text{Fructose} + \tfrac{1}{2}O_2 \rightarrow \text{Glucose}$$
$$164 \quad\quad 16 \quad\quad 180$$
$$386\text{ kg sucrose} = 424\text{ kg glucose}$$

Adding the glucose measure in the stalks (464 kg glucose ha^{-1}) with the equivalent amounts from sucrose and fructose equals 8499 kg glucose per hectare (7611 + 424 + 464). For each glucose molecule, yeast produce two ethanol molecules.

$$\text{Glucose} \rightarrow 2\,\text{Ethanol} + 2\text{CO}_2$$
$$180 \qquad 2 \times 46 = 92$$

Using ratios again, (2×46/180 = 0.511), 8499 kg glucose is theoretically equal to 4343 kg ethanol per hectare. Then, divide the density of ethanol (0.79 kg per liter) to convert kg per hectare to liters per hectare. In summary, from the 2007 Missouri study, 54 971 per hectare (588 gallons per acre) was the calculated potential yield from sucrose, fructose, and glucose in 46 Mg fresh weight per hectare of stripped stalks. Efficiency losses during distillation were not included in the calculations. Hydrometers can be used to estimate the ethanol content after distillation (Figure 4.10).

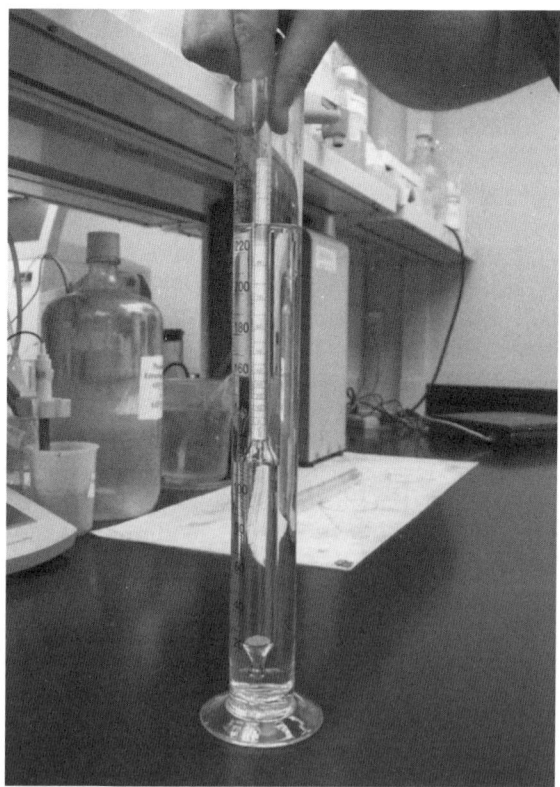

Figure 4.10 Hydrometer used to measure alcohol content in distilled liquid. Photo: Gene Stevens.

4.11 Utilising Byproducts – Bagasse and Vinasse

The crushed sweet sorghum stalks remaining after the juice extraction is called bagasse. The residual slurry after the distillation of the fermented juice is called vinasse. Finding valuable uses for bagasse and vinasse may be the difference between economic success and failure of the whole system. Currently, the two primary uses for bagasse are livestock feed and solid fuel for combustion. Long-term options for converting structural carbohydrates to useable energy include enzymatic cellulose breakdown, burning, gasification, liquefaction, and anaerobic digestion. Large research and development programs are dedicated to developing organisms to ferment 5-carbon sugars. Dr. Randy Powell, organic chemist, with Memphis Biodimensions, Inc. said, "The Department of Energy and National Renewable Energy Laboratory projections of around 80 gallons/dry ton of lignocellulose are based upon conversion of the cellulose (glucan 6-carbon sugars) and hemicellulose (xylan 5-carbon sugars) to ethanol, which is in turn dependent upon the composition of cellulose, hemicellulose, and lignin in the bagasse" (R. Powell, personal communication, http://www.biodimensions.net/).

Ensiled bagasse from sweet sorghum is a satisfactory feed for livestock. But, the lignin and hemicellulose in the bagasse are marginally digestible. The nutritional value of forages is often measured by testing acid detergent fibre (ADF) and neutral detergent fibre (NDF).[28] These indexes are relative measurements used for calculating feed ration formulations. ADF refers to the cell-wall components made up of cellulose and lignin. NDF includes the ADF fraction plus hemicelluloses. Generally, as the numbers increase, feed value decrease. M81E sweet sorghum bagasse tested from Missouri plots in 2007 averaged 39 ADF and 58 NDF. The composition of the bagasse from 2007 tested by the laboratory at FutureFuel was 21% acid insoluble lignin and 9% acid soluble lignin (dry weight basis). In 2008 at Portageville, Missouri, M81E plots produced 87 fresh Mg per hectare compared to $46\,\text{Mg}\,\text{ha}^{-1}$ in 2007. However, the feed value of the bagasse was lower (52 ADF and 72 NDF) in 2008. The main difference between years was temperature and solar radiation.

Bagasse can also be burned to create heat for distillation or produce steam for generating electricity. Overestimation of the weight of dry bagasse that will remain after the juice is extracted should be avoided. Using a simple gravimetric measurement of unprocessed stalks based on fresh and oven dry weight is a good method to determine moisture content and total dry matter of sweet sorghum. However, estimating bagasse yield by this method only subtracts the water weight from the fresh stalks but ignores the sugars and other plant material extracted in the juice.

Vinasse is a mixture of the dead yeast and residual plant material and minerals from the sweet sorghum juice after the fermentation process is complete. The two most common uses for vinasse are fertiliser and methane production. To make fertiliser, the solids in the vinasse are concentrated and composted. The compost can be bagged and sold as a high-value material for fertilising lawns and gardens. The vinasse is also used to produce methane in a thermophilic digester. Vandenburgh[34] reported on the construction and operation of a thermophilic digester by the city of St. Joseph, Missouri.

4.12 Challenges for the Future

In the near future, fermentation of sugars extracted from juice is the best option for making affordable motor fuel from sweet sorghum. However, as sweet sorghum biofuel research continues, processes such as gasification and anaerobic digestion may provide more useable energy per hectare than sugar fermentation. Continued genetic research is needed to develop cultivars and hybrids such as "KN Morris" with high biomass and sugar yields. For fermentation to be competitive with other systems, equipment is needed to extract juice with improved expression ratios.

Acknowledgements

The authors greatly appreciate the information and suggestions provided for this chapter by Dr. Morris Bitzer, professor emeritus at University of Kentucky; Dr. Jeff Dahlberg, United States Sorghum Checkoff Program; Hal Debor, President Sweet Sorghum Ethanol Association; and Dr. Randall Powell, Memphis Biodimensions, Inc.; Dr. Kelly Tindall, University of Missouri entomologist; and Dr. Allen Wrather, University of Missouri, plant pathologist.

References

1. G. Alexander, in *Sugarcane as feed*, Paper 72 in FAO Animal Production and Health Paper. FAO Corporate Document Repository. ISBN: 9251029695, www.fao.org/docrep/003/s8850e/S8850E04.html, 1988, p. 46.
2. I. D. Anderson, *Agron. J.*, 1997, **89**, 881.
3. D. Bellmer, *Small-scale conversion of ethanol from sweet sorghum*. Annual Sweet Sorghum Ethanol Association General Meeting. Orlando, FL, PDF, www.sseassociation.org/, 2009.
4. M. Bitzer, *Early deheading of sweet sorghum-research paper*, National Sweet Sorghum Producers and Processing Association. University of Kentucky. www.ca.ky.edu/nssppa/production. html. 2009.
5. K. L. Bradley, *Extension Guide M171*, University of Missouri, Columbia, MO, 2009.
6. D. Broadhead, *Agron. J.*, 1980, **72**, 523.
7. J. D. Brown, *Soil Test Interpretations and Recommendations Handbook*, University of Missouri-Plant Science Division, Columbia, MO, 2004.
8. G. Buchanon, *Research, education, and economics*, International Conference on Sorghum for Biofuel, Houston, TX, 2008.
9. M. D. Clegg, *Field Crops Res.*, 1982, **5**, 233.
10. S. Conley, *Crop Manage.*, doi:10.1094/CM-2005-0718-01-RS, 2005.
11. J. Cundiff, *Biomass Bioenergy*, 1992, **3**, 403.

12. J. Cundiff, *1990 annual report: sweet sorghum for a Piedmont ethanol industry*, Virginia Polytechnic Institute, Blacksburg, VA, 1991.
13. I. Dweikat, *Sweet sorghum: a dual bioenergy crop*. Annual Sweet Sorghum Ethanol Association General Meeting. Orlando, FL, PDF, www.sseassociation.org/, 2009.
14. R. Escalada, *Agron. J.*, 1975, **67**, 473.
15. Goven Investments, Inc. *Integrated solutions for the sweet sorghum to ethanol global market*. Annual Sweet Sorghum Ethanol Association General Meeting. Orlando, FL, PDF, www.sseassociation.org/, 2009.
16. D. Kundiyana, *Paper 066070*, Am. Soc. Agr. Eng. Biol. Eng., St. Joesph, MI, 2006.
17. M. Lau, *Report 06-2*. Agric. Food Policy Center, Texas A&M University, College Station, TX, 2006.
18. S. Livingston, *Extension Bull L-5168*, Texas A&M University, College Station, TX, 1996.
19. F. Miller, "*Sorghum-A New Fuel.*" Paper presented at the American Seed Trade Association Annual Corn Sorghum Res. Conference, Chicago, IL, 1980.
20. H. Mills, *Plant Analysis Handbook II- A Practical Sampling, Preparation, Analysis, and Interpretation Guide*, MicroMacro Publishing, Inc. Athens, GA, 1996.
21. R. Myer, *Extension Bull. AS-SS*, University of Florida Extension, Gainsville, FL, 2006.
22. B. Morris, *Economic feasibility of ethanol production from sweet sorghum juice in Texas*, Southern Agricultural Economics Association Annual Meetings, Atlanta, GA, 2009.
23. Nimbkar Agricultural Research Institute (NARI), *Multipurpose sweet-stalked sorghum hybrid- "Madhura"*. Manharashtra, India, http://nariphaltan.virtualave.net, 2006.
24. G. Pederson, *Utilizing old data to improve germplasm documentation: sweet sorghum collection*. ASA-CSSA-SSSA Annual Meeting Abstracts. 2006 International Meetings, Indianapolis, IN CDROM, 2006.
25. Praj Industries, *Sweet sorghum to bioethanol*. Conference on "Global consultation on pro-poor sweet sorghum development for bioethanol production", Rome, Italy, 2007.
26. B. Reddy, *Extension Bull. 627-2006*, International Crops Research Institute for the Semi-Arid Tropics. Andhra Pradesh, India, 2006.
27. R. Ricaud, *Report from Department of Agronomy and St. Gabriel Research Station*, Louisiana State University, Baton Rouge, LA., 1990.
28. J. Schroeder, *Extension Bull. AS-1080*, North Dakota State University, Fargo, ND, 1994.
29. H. Shapouri, Office of Chief Economist, Office of Energy Policy and New Uses. Agric. Econ. Report No. 814, United States Department of Agriculture, Washington, DC, 2002.
30. W. Smith, in *Sorghum: origin, history, and production* John Wiley & Sons, New York, NY, ISBN: 0471242373, 2000.

31. G. Stevens, *Better Crops with Plant Food. International Plant Nutrition Inst.*, 2007, **91**(Issue 4), 20.
32. J. Soileau, *Agron. J.*, 1985, **77**, 471.
33. S. Tripp. *Regional strategy for biobased products in the Mississippi Delta.* Memphis Bioworks Foundation. Memphis, TN. Battelle Technology Partnership Practice, Cleveland, Ohio, 2009.
34. S. Vanderburgh, *Proc. Water Environment Federation, Residuals and Biosolids*, Water Environment Federation, 2008, **9**, 851.
35. C. Yamoah, *Agr. Ecosyst. Environ.*, 1998, **68**, 233.
36. A. Wrather, *Extension Guide G 4356*, University of Missouri, Columbia, MO, 2009.
37. X. Wu, *Paper 080037*, Proc. Am. Soc. Agric. Engineers and Biol. Engineers. Providence, RI, 2008.
38. Y. Zhao, *Field Crops Res.*, 2009, **111**, 55.

CHAPTER 5
Sugarcane

A. D. SANTIAGO,[1] R. ROSSETTO,[2] W. DE MELLO IVO[1] AND S. URQUIAGA[3]

[1] Brazilian Agricultural Research Corporation, Embrapa Tabuleiros Costeiros, Caixa Postal 2013, Maceió, CEP 57061-970, Alagoas, Brazil
[2] Agência Paulista de Tecnologia do Agronegócio, Instituto Agronômico de Campinas, Piracicaba, São Paulo, Brazil [3] Brazilian Agricultural Research Corporation, Embrapa Agrobiology, Caixa Postal 74.505, Seropédica, Rio de Janeiro, CEP 23890-000, Brazil

5.1 Introduction

Sugarcane, one of the main cultivated crops used in tropical areas, is originally from New Guinea, from where it was diffused to several regions of the world. That diffusion occurred by three routes and at different times. The first was in 8000 BC, towards the Salomon Islands and New Caledonia. The second, approximately 6000 BC, was towards the Philippines, Java, Malaya, Burma and India; and the third, between the years 500 and 1100 AD was from Fiji to Tonga, Tahiti and Hawaii.[1]

The initial description by Linnaeus in 1753 stated that sugarcane belongs to the genus *Saccharum,* L., presenting the following species: *S. officinarum* L., *S. spontaneum* L., *S. sinensis* Roxb. and *S. barben* Jew. There are two additional species: *S. robustum* Jew and *S. edule* Hask.

The *officinarum* species encompasses plants referred to as noble canes, which are referred to as such due to their high sugar contents and low percentage of fibres. The canes of genus *spontaneum* are known as wild sugarcane, with short slender stalks and rich in fibres. The *sinensis* includes the canes referred to as

Chinese or Japanese, which are adapted to poor and dry soils, as they have a well-developed root system. The *barben* species is composed of varieties that present a medium content of sucrose and high percentage of fibres. They are resistant to low temperatures and are referred to as Indian canes due to their origin. The varieties of the *robustum* have low sucrose contents and high fibre percentage.[2]

The interest in sugarcane cultivation was mainly based on the idea of obtaining food and energy. Besides these options, sugarcane was seen as an opportunity for being used as the raw material of hundreds of byproducts of different manufacture, which include waxes, thermal insulators, fine alcohol, paper, MDF panels, vegetable hormones and plastics, among many others.[3]

The main sugarcane producers in the world are: Brazil, India, China, Thailand, and Pakistan. Over the last years there has been little change in terms of the main producers, but significant changes have, however, occurred in production volume (Table 5.1).[4,5] Overall, there has been an increase over the past 50 years in the worldwide crop area used for this culture. The exception occurred in Cuba, as a result of the crisis of the former Soviet Union, which severely affected the country's balance of trade.

Since the late 1970s, Brazil has been the main sugarcane producer, as a result of the National Alcohol Program (Proálcool). At that time, the country, which was a large petroleum importer, suffered with the well-known world oil crisis, which increased the price four-fold in the international market, forcing local

Table 5.1 Sugarcane production (million tons) of the major producers (reproduced from ref 4 and ref 5).

	2007	*2001–07*	*1991–2000*	*1981–90*	*1971–80*	*1961–70*
Brazil	549.7	424.5	302.8	230.8	108.5	71.2
India	355.5	284.0	267.6	185.8	140.1	110.4
China	113.7	92.6	74.6	51.2	26.7	17.6
Thailand	64.4	58.6	48.2	26.9	14.8	–
Pakistan	54.7	49.2	44.6	34.5	24.8	19.1
Mexico	52.1	49.1	43.8	38.9	33.3	28.4
Colombia	38.5	38.5	32.3	25.0	19.7	13.5
Australia	36.4	35.0	33.3	25.2	21.3	15.1
Philippines	32.5	30.6	26.3	26.1	31.8	21.1
Argentina	30.0	23.1	17.4	14.7	14.9	10.8
USA	27.8	28.5	29.1	26.1	24.2	21.3
Guatemala	25.4	19.9	14.6	7.2	6.4	–
Indonesia	25.3	26.5	28.8	24.8	14.4	10.1
South Africa	20.3	20.8	18.8	18.9	17.2	11.9
Vietnam	17.4	16.2	12.6	6.1	–	–
Egypt	17.0	16.3	13.7	9.7	8.0	5.5
Cuba	11.9	21.0	45.0	73.0	58.2	47.3
Venezuela	9.7	9.3	7.5	7.3	5.7	–
Ecuador	8.4	8.4	6.9	6.3	6.7	7.3
Peru	8.2	8.2	7.1	7.0	8.7	8.1
Total	1498.9	1260.2	1074.9	845.6	585.5	418.5

authorities to search for other, alternative sources to fossil fuel. Ethanol produced from sugarcane was the alternative of choice. The strategy of the Brazilian Government to include ethanol in the national energy matrix significantly changed the importance of sugarcane cultivation, making it strategic not only as a source of human food, but also as the raw material for generating renewable energy, through the production of biofuel and the energy produced from sugarcane biomass. Besides the political innovation, the technical, scientific and institutional innovations implemented by Brazilian public and private sectors were important for the program's success.[6]

The effects of these investments in the Proálcool program can be observed in the average 80% increase in ground sugarcane from the 1975–1976 harvest (68 million tons) to the 1980–1981 harvest (124 million tons). The evolution of agroindustrial productivity is illustrated in Figure 5.1. The increase in agriculture productivity (t ha^{-1}) reveals the large implementation of technologies, where improved varieties allied to management techniques, such as mechanisation, soil conservation and residue recycling, using fertilisers, controlling pests and weeds were responsible for that rise. Also in Figure 5.1, it is possible to observe the evolution of industrial productivity, achieved by means of innovations such as the use of selected micro-organisms and adjustments in the plants. In the last 30 years, there has been an increase of 3.1% per year, which resulted in doubling the volume of ethanol produced in the same hectare.

In the agricultural as in the industrial area, there is still room for productivity increments. In this respect, genetic improvement is, unquestionably, one of the most important techniques to increase sugarcane productivity, certainly allied to an efficient production system.

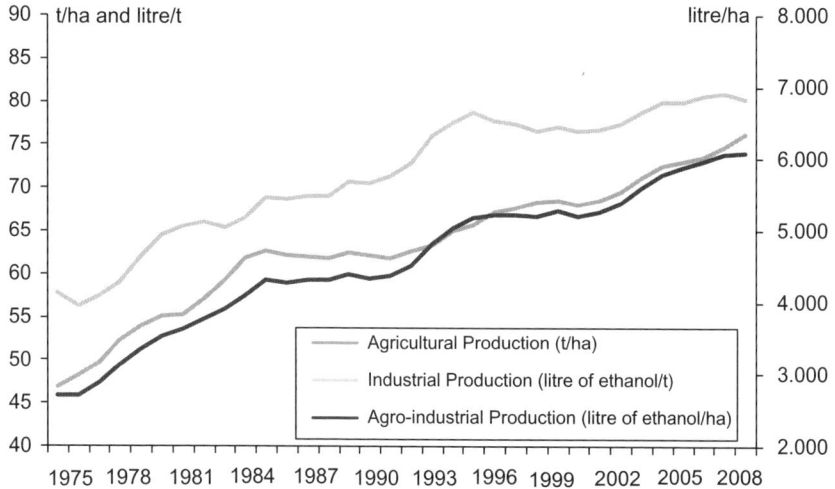

Figure 5.1 Evolution of sugarcane and ethanol production in Brazil (reproduced from ref. 69).

Table 5.2 Internal energy sources in Brazil (reproduced from ref. 8)

	mil tep			(%)	
	2007	2008	08/07 (%)	2007	2008
NOT RENEWABLE	**129 103**	**137 333**	**32**	**54.1**	**54.7**
Oil and derivatives	89 239	93 711	5.0	37.4	37.3
Natural gas	22 199	25 625	15.4	9.3	10.2
Coal and derivatives	14 356	14 294	–0.4	6.0	5.7
Uranium (U3O8) and derivatives	3309	3703	11.9	1.4	1.5
RENEWABLE	**109 656**	**114 193**	**22.0**	**46.0**	**45.3**
Hydraulic and electricity	35 505	35 013	–1.4	14.9	13.9
Firewood and charcoal	28 628	28 717	0.3	12.0	11.4
Derivatives of sugarcane	37 847	41 820	10.5	15.9	16.6
Other renewables	7676	8643	12.6	3.2	3.4
	238 759	251 526	53.9	100.1	100.0

Globally speaking, Brazil gathers a number of comparative advantages to lead in energy agriculture. The country has the perspective of incorporating new areas in agriculture for energy, without any competition with the agriculture for food;[7] a problem that some countries have been facing, such as the United States. In Brazil there is a considerable amount of land with degraded pasture, which benefits from the cultivation of sugarcane and other crops, such as soy and peanut, necessary for the rotation system, and that contribute not only to increasing the production of alcohol but of sugar and vegetable protein as well. It is likely that the expansion of sugarcane over pastures also benefits cattle farmers, who may observe an increase in the profitability of their rural property, besides improving soil fertility compared with its original conditions.

Data from the Brazilian Ministry of Mines and Energy (Table 5.2) show that, in 2008, the Internal Energy Supply was equivalent to 251.5 million tons of petroleum, 45.4% of which (114.2 million tons) originated from renewable sources.

Figure 5.2 shows that sugarcane byproducts account for 16.6% of the internal energy supply, against 13.9% from hydroelectric sources.

A comparison between the internal energy supply of Brazil and the world (Figure 5.3), even in different years, shows that there are significant differences in terms of using renewable energy. In the world, 10.1% of the total originates from biomass, whereas in Brazil that contribution is 31.5%. Of that total, 16.6% are from sugarcane products, thus generating more energy than hydroelectric sources.[8] These data reveal the importance of sugarcane to energy generation. The main part used from sugarcane is the bagasse, the fibrous residue remaining after the stalks are crushed and juice extracted, with a possibility, in a near future, of also using the sugarcane straw for this purpose. Because of the great potential of using sugarcane to generate energy, the main programmes for the improvement of sugarcane cultivation have directed their efforts to developing genotypes rich in fibres (energy cane).

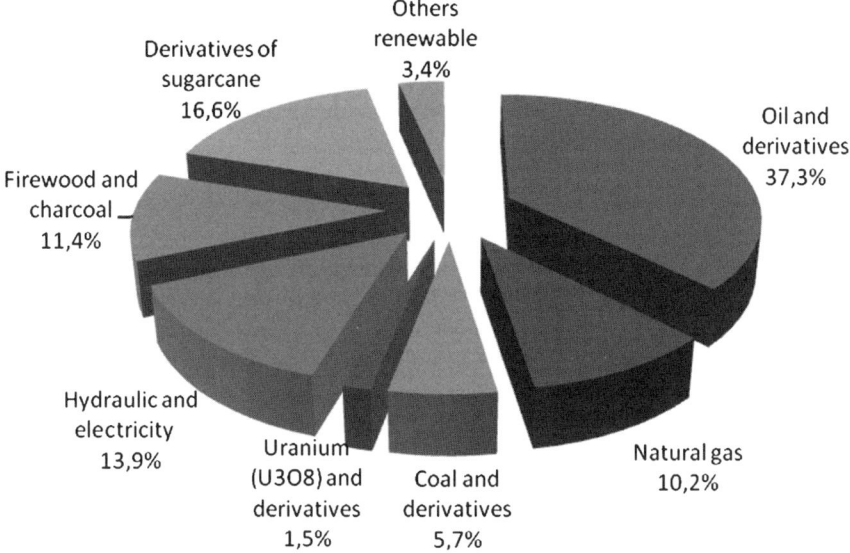

Figure 5.2 Internal energy sources in Brazil (reproduced from ref. 8).

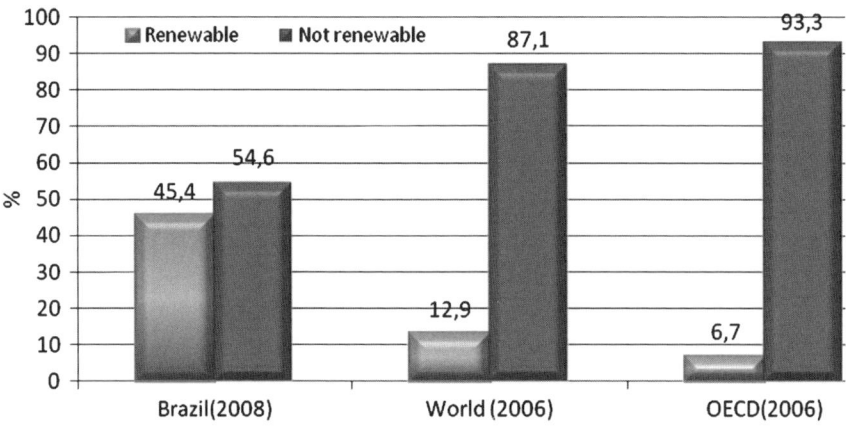

Figure 5.3 Brazil's internal energy sources in comparison with the world (reproduced from ref. 8).

5.2 Production System

5.2.1 Varieties

Sugarcane is a crop with high potential for producing biomass. It is a C_4 plant, and therefore its great effectiveness in photosynthesis results in a high production of biomass, particularly in tropical regions, where there is an abundance of sunlight and water. It is possible to increase biomass production through genetic improvement and soil-management practices. Therefore, improvement programmes have a preponderant role in meeting the new demands of raw material to obtain ethanol.

In Brazil, the agricultural productivity of sugarcane has increased an average 2% per year. Genetic improvement is responsible for half that gain, and the other half is the result of better management practices implemented in sugarcane-production systems.[9]

Until a few years ago, the sugarcane varieties obtained in genetic improvement programmes were focused on the high production of sucrose, allied to the production of biomass and resistance to pests and diseases. To do this, crossbreeding was used, as follows: *Saccharum officinarum* (noble cane) with a rustic cane, *Saccharum spontaneum*, obtaining a hybrid that was very productive and with high sucrose content.

Current improvement faces new challenges, where the objective is to use all parts of the plant that are not normally used to produce energy, such as straw, leaves, and the residues and not only the juice extracted in grinding. The new varieties should, therefore, meet the characteristics of greater productivity and high sugar content, which are needed for producing sugar and ethanol using the conventional system – first-generation ethanol; and should also contain new characteristics, aimed at producing energy from biomass. That energy is transformed into electrical energy and may be used in the production of second-generation ethanol, which is the ethanol produced from cellulosic materials, such as the straw and bagasse. Within this perspective, the concept of "energy cane"[10] was proposed to differentiate materials with high fibre content, improved to better meet energy-production purposes. In this case it refers to materials with about 25% fibres.

With the new search focused on varieties with higher fibre contents, it is important that the sugarcane germplasm used by breeders contain gene sources with specific characteristics for high biomass production and high fibre contents. Other characteristics, however, such as resistance to diseases and pests, as well as the adaptability to mechanical harvesting cannot be forgotten. One of the main problems is, exactly, that the genetic basis used in improvement programmes has always considered high sucrose content as the main focus for selecting clones. Sucrose content and fibre content are antagonist characteristics (when one increases, the other reduces), therefore the genes that contribute to increasing fibre content are usually eliminated by improvement aimed at increasing sucrose levels (Creste, personal communication).

The current varieties are interspecific hybrids whose crossbreeding occurred through genetic improvement programmes or naturally among different species of the *Sachharum* genus or among species of the genera *Erianthus*, *Miscanthus*, *Narenga* and *Sclerostachya*. The number of chromosomes of these current varieties ranges between 100 and 120, with about 80% originating from *S. officinarum*, 15% from *S. spontaneum* and 5% from recombinants. It has been proven that these materials present characteristics of high productivity, high sucrose content, resistance to diseases and environmental stress, but they were, however, improved from very narrow genetics. The variability present in the gene pool usually managed by breeders does not represent the full genetic variability that exists in the genus *Saccharum* and even less in the whole *Saccharum* complex.

Genetic bases for fibre content and also vigour, which are characteristics that contribute to high productivity, may be found in *Saccharum spontaneum* and *Miscanthus,* which is being used in Europe as an energy crop.

Some improvement programmes have prioritised crossbreeding between commercial hybrids and *S. spontaneum* for the introgression of genes that are able to increase fibre content, while maintaining average sucrose contents and other agronomic characteristics. Although the fibre and sucrose content are antagonist characteristics, it is possible to select families in which this correlation is positive, thus providing the opportunity to select FS1 clones with the desired characteristics. The Australian improvement programme seeks gene introgression, precisely to explore the variability existing in *S. spontaneum* in the search for "energy cane" varieties.

The molecular-markers technique has also been used in genetic improvement programmes in Brazil. This technique involves searching for differences between the genomes of different species or different clones that could point at specific characteristics of a certain individual, such as resistance to stress, high fibre content, *etc.* Using this technique begins by sequencing the genomes of the plants involved in the breeding. Using bioinformatics software, the DNA sequences produced are evaluated and the differences between the genomes are identified. These differences are referred to as markers. After this step, a search is performed for correlations between the markers and the agronomic characteristics of each individual, such as high fibre content. Once the markers are identified, they can be used in laboratory tests to identify specific characteristic in the germplasm bank. Individuals that have the characteristics wanted by the breeder are chosen for breeding.

Molecular markers make genetic introgression feasible by following whole genetic blocks, thus providing gains in the time required for traditional genetic improvement. It is expected that by using this technique it will be possible to make further reductions in the time necessary to obtain improved varieties, compared with the traditional system, which takes 10 to 12 years.

Genetic engineering has also been used for the genetic improvement of sugarcane. There are two main techniques currently in use: transformation-mediated *Agrobacterium tumefaciens* and biolistics. The main strategies sought through transgenics are resistance to diseases, such as the mosaic virus and *Xanthomonas albilineans* bacterium, which causes ratoon stunting disease, and pests such as the sugarcane borer. Another objective is to introduce genes resistant to herbicides through genetic transformation. There are also promising studies on genetic transformation to introduce genes that aim at improving the technological features of the sugarcane, increasing the production of sucrose through enzymatic action.[11]

As a further perspective, improvement programmes should work to gather germplasm collections representative of the existing genetic variability in the

"*Saccharum* Complex". Moreover, it is also necessary for collections to have a programme to characterize the efficiently deposited accessions, integrating molecular, phenotypic and morphologic data. It will be necessary to prioritize financial and human resources for the *in vitro* conservation of sugarcane germplasm, associated with *in vivo* conservation, ensuring the reposition of any material lost in the collections that are kept in the field. Finally, the molecular-markers technique has a strong potential to make germplasm characterisation feasible and enable a program for genetic introgression, but there must be investments in high-throughput genotyping systems.

Therefore, improvement programmes have a central role in establishing the new energy matrix coming from sugarcane. New varieties will represent a great technological advance in generating biomass and clean renewable energy.

5.2.2 Pests and Diseases

The sugarcane sector is a model in the employment of technologies that reduce the use of chemical protectants; for example the biological control of pests such as the sugarcane borer. Infestations by this pest once caused considerable economic losses, and biologic control effectively reduced infestations to levels close to 1%. In Brazil, a smaller amount of pesticide is used per hectare of sugarcane (0.36 kg/ha) than for soybean (1.17 kg/ha). Pesticide use in sugarcane crops is limited to controlling soil pests, ants and, more recently, spittlebugs. The greatest use of chemical protectants on sugarcane is in the chemical control of invasive plants.

Such increases in the environmental quality of sugarcane fields in Brazil occurred as a result of programmes for the genetic improvement of sugarcane. These programmes were successful in introducing materials resistant to or tolerant of the main pests and diseases that affect sugarcane. In addition, it is expected that the use of transgenic cultivars will result in a greater reduction in the use of chemical protectants, though the environmental impacts of transgenic varieties should be better defined.

The most important of these pests is known as the sugarcane borer, *Diatraea* spp. This pest occurs in the main productive regions, harming the production of stalks, and the quality of the juice, as it permits the appearance of pathogens that cause sucrose inversion. In several sugarcane-producing regions of the world, this pest is biologically controlled through the use of the parasitoid *Cotesia flavipes* (Cameron, 1891), bred in laboratories and then released on the sugarcane fields.

Another important pest is *Mahanarva posticata*, which, like the sugarcane borer, harms production and raw material quality. This pest occurs mainly in the states of Northern Brazil. In Southeast Brazil, on the other hand, there has lately been a significant increase in the occurrence of *Mahanarva fimbriolata* (Stål) or spittlebug. This occurred due to an increase in the area subjected to mechanical harvesting, leaving sugarcane residue on the soil surface. These important pests are controlled using the *Metarhizium anisopliae* fungus, at the dosage of 1 to 2 kilograms per hectare. In addition to not harming the environment, using this fungus is economical for the producers, since its cost is only 10% that of agricultural protectants.

In spite of the success achieved by genetic improvement and some methods of biological control, new pests and diseases have appeared, posing new challenges. In relation to new diseases, *Puccinia kuehnii* is a rust fungus that causes the orange rust disease of sugarcane in Asia, Africa, Australia, Brazil and in Florida, where it can cause extensive losses. One of those pests is the giant sugarcane borer (*Telchin licus licus,* Drury 1773), with no efficient control currently. This pest is mainly present in Northern Brazil, but there have been reports of its occurrence since the year 2007 in the major national producing state, São Paulo. As reported for the spittlebug, the increase in mechanical harvesting has established better conditions for the development of this pest in the sugarcane field. There are losses in both the agricultural and industrial output. Penetration occurs through the rhizome and, as the larva develops, the pest moves to the stalk, causing severe damage (Figure 5.4). By the time of harvesting the borer has migrated to the base of the rhizome, making control more difficult. The current control method is manual removal. Several studies are being performed on the control of this pest, mainly through biological control and pheromone use, because the currently used methods are inefficient and costly.

The technology of biological pest control in sugarcane permits significant reductions in the insect population without harming the environment, and with lower costs. Therefore, worldwide studies in the sugar-alcohol sector should have the same focus, as well as addressing the use of new technologies, such as

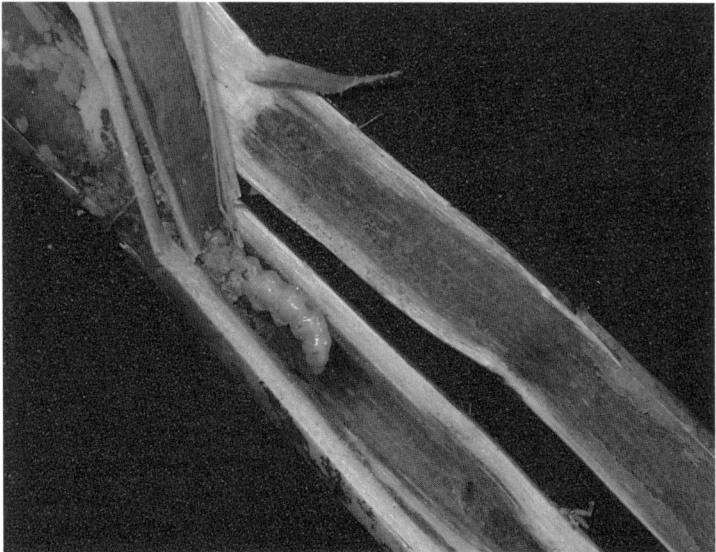

Figure 5.4 *Telchin licus licus* damage in a sugarcane stalk.

the identification of chemical substances that could work as pheromones for sugarcane pests. Furthermore, transgenics could be an important tool in reducing the use of chemical protectants, provided their impacts are exhaustively studied.

5.2.3 Nutrient Requirements

Sugarcane is a crop that is able to remove significant amounts of soil nutrients due to the great amount of biomass that is produced. In Brazil, sugarcane produces, in general, from 25 to 40 t ha^{-1} of dry material, 60 to 70% of which is removed from the field in the form of stalks. However, as sugar and ethanol, the final products of the process are composed exclusively of carbon, hydrogen and oxygen, all of which are elements that are removed from air and water during photosynthesis. Every nutrient removed from the soil may therefore be recycled by means of residues from the sugar-ethanol industry, such as vinasse, filter cake, straw, ash or bagasse. Vinasse is a liquid waste from ethanol production. Brazil produces more than 300 bilion litres of vinasse every year.

Several authors have reported that the aerial part of the sugarcane plant contains the following amounts of nutrients for each 100 t of stalks produced: 100 to 154 kg of N, 15 to 25 kg of P_2O_5, 77 to 232 kg of K_2O, and 14 to 49 kg of S.[12–15]

In global terms, sugarcrops, sugarcane and beet, account for 4.2% of fertiliser consumption. In Brazil, sugarcane is the third crop in fertiliser consumption, after soybean and corn. In 2007/2008, to cultivate approximately 8.3×10^6 hectares of sugarcane, about 3.4×10^6 t of NPK fertiliser was used, corresponding to 13.8% of Brazilian consumption.[16] This crop is considered relatively efficient in using nutrients: the mean dose of mineral fertiliser used is 408 kg/ha, similar to that of several other crops with much smaller productions of biomass per area. The nutrient doses recommended for sugarcane cultivation in Brazil, especially of nitrogen, are, in general, smaller than those used in several other countries with similar productivity.[17] One important aspect for this rational use of fertilisers is nutrient recycling, common in the sugarcane agroindustry, by returning solid and liquid residues, such as filter cakes, ashes, straw, and, especially, vinasse.

Much of the Brazilian soil where sugarcane is cultivated is acid with low fertility. Nevertheless, sugarcane is a plant that is relatively tolerant of soil acidity and the presence of aluminium.[18,19] Studies published in Brazil[20] show that this crop does not respond much to limestone application on soils with base saturation above 25% of the cation exchange capacity, corresponding to a pH of 4.5 in calcium chloride. An aspect that should be taken into consideration is that sensitivity to soil acidity depends on the variety. On the other hand, lime increases sugarcane longevity, as it is also a source of calcium and magnesium.

Gypsum is a material that has been largely used in sugarcane cultivation, with the objective to increase Ca in the subsoil (20–60 cm) and reduce the activity of aluminium on soil solution, thus improving the chemical environment for root development. This element is also a source of sulfur. A combination of limestone and gypsum produces better results for the crop than using either alone.[21,22]

The mineral nutrient that most accumulates in sugarcane after K, is N. Considering the whole plant, the requirement for N reaches 2.1 to 2.4 kg N per ton of stalks.[23,24] These numbers indicate that sugarcane extracts over 200 kg ha^{-1} of N for the production of 100 t ha^{-1} of stalks, of which 90 to 110 kg ha^{-1} are exported with harvesting.

Despite the considerable need for N, sugarcane cultivation in Brazil is performed with relatively small doses of nitrogenous fertilisers: 30 to 60 kg ha^{-1} of N in each sugarcane plant and 60 to 120 kg ha^{-1} of N in ratoons.[25-27] In other sugarcane-producing countries, with productivity similar to the aforementioned, N doses are usually above 120 kg ha^{-1} and, in some cases, reach 200 kg ha^{-1}.[28-30] In addition to the application of relatively low doses, the recovery by sugarcane plants of the N applied as fertiliser is usually low, ranging between 20 and 40%, according to most studies conducted with ^{15}N.[31-36] These values are smaller than those observed for grain cultures.[37,38] One of the reasons for this is that sugarcane has a long cycle, which allows for exploiting the soil for a longer period of time, and fertilisers are applied in the beginning of the growing cycles, thus they become subject to losses and/or immobilisation by soil organic matter (SOM). In fact, the soils eventually become the final deposit of a considerable part of the fertiliser N (20 to 40%) where it is added to the N stock of the SOM.[25]

Another point to be considered is that sugarcane has been continuously cultivated for several decades, sometimes more than a century, in some regions, with no reduction in productivity or apparent degradation of soil fertility, or significant decrease in the amount of SOM, despite exporting N quantities equal to or above those applied by fertiliser.[39] This has been considered to be indirect evidence that biological nitrogen fixation (BNF) has an important role in the nitrogenous nutrition of sugarcane in Brazil,[40,41] and in the world.

Most Brazilian soil is poor in P. Therefore, sugarcane responses to the application of this element are high, especially for the first year of sugarcane cultivation. The plant requirement ranges from 10 to 40 kg ha^{-1} of P_2O_5 for every 100 tons of sugarcane produced.[15] P is a nutrient used in higher amounts in Brazil than in most sugarcane producers. The median dose is approximately 120 and 30 kg ha^{-1} of P_2O_5 in the sugarcane plant and ratoon cane, respectively.[15,42] The corresponding values for sugarcane plant and ratoon in Australia are 58 and 57 kg ha^{-1} P_2O_5,[17,42] 60 and 60 in Mexico, and, in Costa Rica, 150 and 75 kg ha^{-1} of P_2O_5, respectively.[15]

The nutrient extracted in greatest amounts by the sugarcane cultivation is K, and, therefore, high responses to potassium are expected, especially in soils deficient of this nutrient, as evidenced in recent review articles on this topic.[43,44]

Potassium fertilisation management is relatively easy compared with that of nitrogenous or phosphate fertilisation, because of the smaller interaction of K with the mineral and organic soil components. The exchangeable contents of K in the soil are considered good indicators of the need to apply K and are employed as parameters in the recommendations of fertilisation, together with the expected productivity, which is an important consideration for nutrients that are accumulated in great amounts in the plants.[12,19,23,44–46] For soils with very low K contents ($< 0.8\,\text{mmol}\,L^{-1}$), the doses recommended for the sugarcane cycle in the state of São Paulo range from 100 to 200 kg ha^{-1} of K_2O, depending on the range of expected productivity. The corresponding doses for soils with intermediate contents (1.6 to 3.0 mmol L^{-1}) are 40 to 80 kg ha^{-1}; for soils with contents above 6 mmol L^{-1} potassium fertilisation is not recommended. In the ratoon cycle, the dose ranges between 30 to 150 kg ha^{-1} of K_2O due to the availability of the nutrient in the soil and the goal for productivity.[25] The mean K doses applied to sugarcane – plant cane and ratoons, respectively – in some countries are 120 and 145 kg ha^{-1} K_2O in Australia, 175 and 150 kg ha^{-1} K_2O in Costa Rica.[44] In Florida, USA, the doses range from 0 to 280 kg ha^{-1} K_2O in the cane plant and from 0 to 170 kg ha^{-1} K_2O in rations,[30] but in Louisiana, USA, smaller doses are used – up to 55 and 67 kg ha^{-1} K_2O for cane plant and ratoon, respectively,[47] showing the capacity of K soil supply at different locations.

5.2.4 Mechanical Harvesting and Residue Management

The sugar-alcohol sector has two major impacts regarding the quality of atmospheric air. One impact is highly positive, represented by the use of alcohol as fuel, either mixed with gasoline or not. The other impact is represented by the emission of gases during sugarcane burning. Although the carbon balance is highly positive, that is, during photosynthesis sugarcane fixates much more carbon than is emitted during the burning, the eradication of this practice could mean further increases in atmospheric air quality. Considering this improvement in environment quality, mechanical harvesting is a certain target in sugarcane cultivation. In some countries, such as Australia, this is the current practice. In Brazil, it has been implemented to comply with the legislation that determines the end of field burns.[6] In São Paulo and in other states where sugarcane is grown, mechanical harvesting already represents a considerable area and, compared with hand harvesting, has greater efficiency and reduced costs.[48]

In addition to the cost factor, mechanical harvesting leaves a great amount of residue on the soil, which represents a significant storage of nutrients that are recycled throughout the years. This material interferes with the recommendation for fertilisation that used to be given when the cane was burned. In relation to nitrogen, the mineralisation of the organic matter is quite slow, but storage is gradually formed in the soil. To compensate for this high C/N ratio, and the possible immobilisation of N by micro-organisms, the recommendation is to add 20% more N, giving a mean recommended dose of about 120 kg ha^{-1} of N, or about 1.2 or 1.3 kg of N per ton of sugarcane to be produced.

Every year, about 10 to 12 t ha^{-1} of residue is left on the soil. The nutrients from the residue are not always quickly transferred to the cane. This depends

Table 5.3 Nutrient quantities of sugarcane trash (reproduced from ref. 49)

Nutrient recycling	N	P	K	Ca	Mg	S
Trash (kg ha^{-1} yr^{-1})	54.7	4.4	76	54.9	25.5	15
Mineralisation rate (% yr^{-1})	20	60	85	50	50	60
Total per year (kg ha^{-1} yr^{-1})	10.9	2.6	64.6	27.5	12.8	9

on the rate of mineralisation. In terms of recycling, it is estimated that the residue adds, yearly, small amounts of N, P, Mg and S nutrients, intermediate amounts of Ca and very significant quantities of K (Table 5.3).[49]

There are many advantages in sugarcane residue, including: it increases soil protection against erosion, reducing the direct impact from rain drops on the soil surface and surface draining; it reduces the range of temperature on the surface layers of the soil; it increases biological activity; it increases water-infiltration rates on the soil; it reduces evaporation; it partially controls weed growth; it recycles nutrients (N, P, K, Ca, Mg, S, Si); it increases SOM (soil organic matter) contents it improves water retention capacity; it improves surface rooting and reduces compaction due to mechanisation.

On the other hand, there are negative impacts from the residue that stays on the soil surface: it increases the risk for fires, makes cultivation operations more difficult, causes delay or failures in budding, especially in cooler regions, and it increases some pests. Such technological hindrances should, in a near future, be reduced or eliminated by scientific research.

5.2.5 Water Use

The rational use of water has been implemented by the sugarcane-producing sector in Brazil, both in the field and in industry. The region with the largest production of this crop, the state of São Paulo, does not use irrigation except in special circumstances, such as in seedling nurseries and in sugarcane fields close to water springs.[6] Irrigation is a practice mostly used in North and Central-West areas due to the water deficiency in some periods of the year. Besides the water deficit, another important factor that motivates using irrigation is the need to use the vinasse that is produced in the distilleries. This residue has moved from being an environmental problem in the 1970s and 1980s to being an important source of nutrients, especially potassium, with its use being strongly associated with the practice of irrigation. On average, two or three irrigations with water (40 mm or 60 mm) are applied in the driest period, diluting the vinasse or washing water. To achieve this result, production units have made high investments over recent years in storing pluvial water and in new irrigation systems that are more efficient in saving water.

In the processing industry, rational water use has been adopted in recent years. Depending on the technology used, the industry used to require great amounts of water, in the range of $5\,m^3$ to $13\,m^3$ per ton of processed sugarcane.[6] These values, however, have dropped from $5\,m^3$, between 1990 and 1997, to 1.83 m³ per ton of sugarcane in 2004, in São Paulo.[50] Also regarding this information, if we consider 1.83 m³ of water per ton of sugarcane, and exclude mills with the highest specific consumption, the mean rate for the remaining mills that account for 92% of the total crushed sugarcane is 1.23 m³ of water per ton of sugarcane. In addition, the recycling rate has increased since 1990. Mills with the best management practices replace only 500 l of water in the industrial system, with a recycling rate of 96.67%.[50]

5.3 Sugarcane Products and Coproducts

The sugarcane-production system and the processing of its main products, sugar and alcohol, generate a significant number of coproducts that have high added value (Figure 5.5). It has certainly not always been this way. Until recently, some products were considered a problem, as is the case of vinasse, which, if well used, becomes an important source of nutrients for the plants.

Because of the great production of vinasse during the processing of alcohol, more specifically 13 l of vinasse for every liter of alcohol produced, this residue, which has great biochemical oxygen demand (BOD), causes severe harm to the environment if poured into water streams. This residue was once an environmental issue, and in Brazil its disposal in rivers is against the law (according to Decree 303 of 1967), and the country has other juridical norms that regulate its

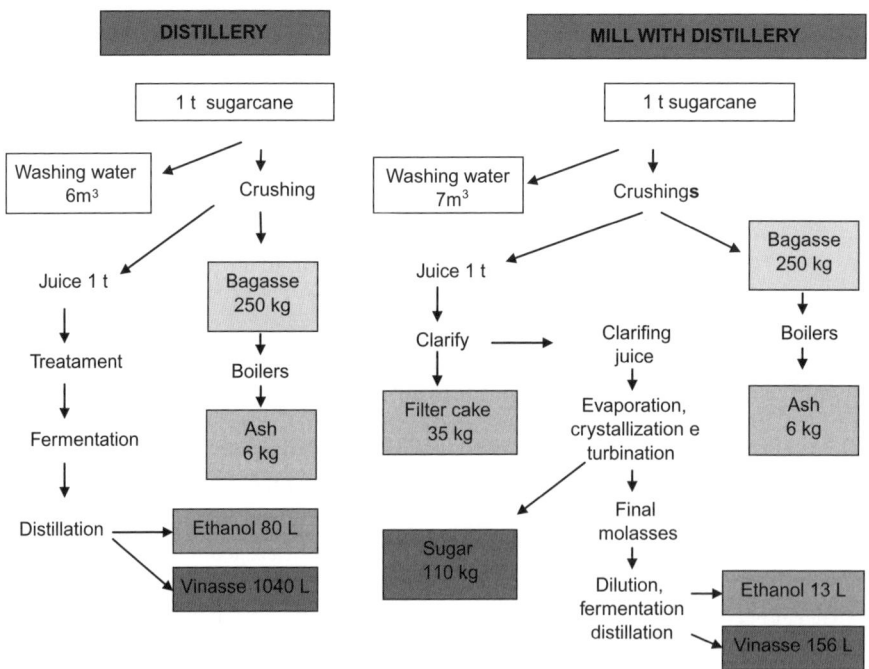

Figure 5.5 Production phases of sugar and ethanol, and residues generated.

use on the soil. Currently, the application technique that is used most often is the fertirrigation of sugarcane fields, with distribution by ducts and application to the soil by aspersion.

Another coproduct, filter cake, is defined as the residue eliminated in the process of sugarcane juice decantation in the production of sugar or alcohol. It is used as a fertiliser due to the great amount of nitrogen, phosphorus, calcium and organic matter that it provides to the soil.[51]

It is mainly used in the form of an organic compound and can be employed at doses ranging from 80 t ha^{-1} to 100 t ha^{-1}, in the preplantation period at doses between 20 t ha^{-1} and 35 t ha^{-1}; and between plantations, at doses ranging from 40 t ha^{-1} to 50 t ha^{-1}.[6]

In addition to vinasse and filter cake, the bagasse is the coproduct most used in the agroindustrial system of sugarcane. The amount of bagasse produced in the industrial plants can be used in several ways. However, it is mainly used to produce energy, used in the industry and on the field or sold directly to authorised dealers. This use is expected to increase, because of the development of new cultivars with greater fibre content, referred to as energy canes, which have the objective to increase the energy efficiency of the bagasse.

Each ton of sugarcane (stalks) produces 140 kg of bagasse, 90% of which is used to produce energy (thermal and electric) in the mill,[52] in addition to 150 kg of sugar (used for sugar, ethanol and, now, plastics) and 140 kg of straw.

Basically, all of the thermal energy and about 95% of the electric energy, is produced at the mill, with cogeneration systems using the bagasse. Another important aspect is the renewable energy produced by the mills for external use, today mainly ethanol, which is about nine times greater than the fossil fuel input used in its production. Hence, this process is considered the most attractive among the commercial uses of alternative energy in the world, from the perspective of sustainability, as it reduces the emission of greenhouse-effect gases in about 12.7 million tons of equivalent carbon.[53]

The most anticipated technology in the business and scientific worlds is the production of cellulosic ethanol, also referred to as second-generation alcohol. To produce this fuel, cellulosic materials will be used, such as the bagasse, and trash. Studies[54] state that the fact that the bagasse is processed in the mill and that there is no need to transport the raw material make bagasse hydrolysis competitively advantageous in producing alcohol.

Several research groups are working to identify enzymes capable of promoting the hydrolysis of cellulosic material to obtain alcohol, thus tripling the world production of ethanol, without the need for increasing the cultivated area. This technology will promote extra gains for corn and sugarcane-ethanol producers, and of other products such as grass for energy purposes and energy forests. The United States has invested in studies in this area, as has Brazil, through the newly established Center for Bioethanol Technology, with Embrapa and several universities.

Another source of cellulosic materials is sugarcane straw. The increase in mechanical harvesting has raised the supply of this raw material, which can be used to produce energy by burning it in boilers, or as material for hydrolysis.

An efficient hydrolysis of lignocellulosic materials and further fermentation of the resulting sugar has been a great challenge, with two acknowledged routes: the acid and the enzymatic route.[55] The former has the advantage of being efficient and of lower costs; however, it produces polluting residue and a product that inhibits further fermentation. The second route, the enzymatic process, has been more studied, as it has fewer disadvantages in terms of production and is less polluting. It does, however, have a high cost due to the enzymes used.

In addition to the traditional sugarcane products, sugar and alcohol, there are new products with a high potential of usage. These products can be grouped by process families: biotechnological products (agricultural protectants, nitrogen fixators and silage inoculums), chemical products (chemical industry, paper, cellulose and concentrated vinasse), veterinary drugs, foods (derivatives of yeast, fructose and glucose, inverted syrups, bagasse derivatives and honey) and structural components (MDF plywood, bagasse plywood and cement).[56]

5.4 Energy Balance of the Production of Sugarcane Ethanol in Brazil

It is not difficult to produce biofuels from different biomass sources of animal (fats, oil) or plants biomass (oils, carbohydrates). The true difficulty is to obtain significant energy gains compared with the fossil-fuel energy invested in its production. This has been a major concern among Brazilian scientists worried about the energy sustainability of sugarcane-ethanol production in the country. In this sense, several Brazilian specialists have estimated the amount of fossil energy necessary to produce ethanol (energy per liter, in $MJ\,L^{-1}$, or energy per ton, $GJ\,t^{-1}$) of sugarcane in Brazilian conditions.[52,57–60] Nowadays, in Brazil there is a consensus that the energy balance (ratio between renewable energy contained in the produced biofuel and the total amount of energy, fossil or not, invested in its production) is approximately 8 or 9, in the conditions of São Paulo and of Brazil in general.[52,57,58,60]

However, several international teams have also calculated the energy balance for sugarcane ethanol in Brazilian conditions[61–64] and the estimates were considerably smaller (between 3.7 and 1.1:1). There are various reasons for this. It was considered[61] that the energy used in the production of ethanol was mostly of fossil origin, as it was in the past, but, currently, all of the energy used in the mills to process sugarcane to produce ethanol is renewable, provided by burning the bagasse produced from crushing the sugarcane. In addition, it should be noted that the low energy balance values mentioned above are associated with the use of outdated information regarding the use of fossil energy on field operations. In addition, the estimates used for the energy cost of transporting sugarcane from the field to the mill were highly exaggerated, and did not represent either the agricultural or industrial production system currently in use in the country. All of the differences were discussed[59] and adjusted in a more updated balance.[60]

To determine the sustainability of sugarcane-ethanol production, in addition to the energy balance, the greenhouse-gas emission (GHG) balance should be evaluated, including the emission of those gases caused by the changes made to the soil to produce ethanol, such as the indirect emissions of GHG associated with the expansion of the area cultivated with sugarcane to produce biofuel.

To estimate or calculate the energy balance of a biofuel it is necessary to know in detail the biomass agricultural production system, as well as the industrial part of the process. In terms of the production of sugarcane ethanol, important studies that clearly show the current situation in Brazil have recently been reported.[57,59,60] These studies surveyed all the data regarding the consumption of agricultural input (fertilisers, machinery, fuels, *etc*.), harvesting and transportation, as well as the inputs associated with the industrial part of the process of ethanol production, including the fraction corresponding to the facilities and machinery involved. With all this information made available and after the weighting analysis so that the study would reflect the country's mean production system for ethanol, all the data was transformed into energy-equivalent units: Joules.[65] Based on this information, the energy balance of the production of sugarcane ethanol for Brazilian conditions was calculated, and these results are summarised and shown in Table 5.4.

The estimate of the total fossil energy used in field operations, including the transportation of sugarcane to the mill and the provision of inputs, is 12 329.7 MJ ha^{-1} year^{-1}. Considering that one liter of ethanol produces 21.45 MJ of energy in the combustion, one hectare of sugarcane currently producing an average 6510 L of ethanol could generate 139 639 MJ of energy, approximately 11 times the fossil-fuel energy invested in the agricultural operations.

Hence, the overall energy balance resulting from the production of ethanol is 9.35:1, which means that for every 1.0 MJ of fossil energy consumed, 9.35 MJ of renewable energy is produced, in the form of sugarcane-derived ethanol.

It should be highlighted that, in Brazil, there is still a surplus of bagasse that could potentially be used in the coproduction of energy. Another residue of sugarcane cultivation, the straw, which is currently burnt in over 60% of the national sugarcane field area, without generating any energy benefit, thereby contributing to the GHG emission also offers a high potential for use in the cogeneration of energy; or alternatively as the raw material in the production of second-generation ethanol, which would likely increase the overall energy gain of this crop.

The knowledge about the energy balance of sugarcane-ethanol production allowed determination of the limiting factors for this parameter to be even more positive. The rational use of nitrogenous fertilisers is particularly important, as they are an input that requires large amounts of fossil energy (54 MJ kgN^{-1}). Table 5.1 shows that the 57 kgNha^{-1} annually applied to this crop accounts for 25% of the agricultural energy cost and 20% of the total energy cost. It should be highlighted that Brazil is practically the only country in the world in which it is possible to obtain good sugarcane yields in N-poor soils and with low doses of N fertiliser. This phenomenon is related to the significant contribution of biological nitrogen fixation (BNF) in sugarcane cultivation, which has contributed

Table 5.4 Fossil energy input, total energy yield and energy balance of bioethanol produced from sugarcane under Brazilian conditions (6510 l ha^{-1}). Energy values expressed on a per ha per year basis, with a production of 79.5 t ha^{-1} (reproduced and adapted from ref. 60)

Energy input mj ha^{-1} year^{-1}	
Field operations	
Labour	501.8
Machinery + Diesel	2972.5
Nitrogen	3061.8
Phosphorous + Potassium + Lime	1018.8
Seeds^a + Inseticides	339.5
Herbicides	1445.3
Vinasse disposal	656.0
Transport of consumables^b + Cane transport^c	2334.8
Total field operations	**12 329.7**
Factory inputs	
Chemicals used in factory^d	487.6
Structural mild steel, equipment and cement	2123.5
Total factory inputs	**2611.1**
Total all fossil energy input	14 940.8
Output	
Total ethanol yield (6510 l ha^{-1})	139 639.5
Final Energy Balance^e	**9.35**

^{a}This calculated from 2.6% of all field operation inputs.
^{b}Transport of machinery, fuels, *etc*. to plantation/factory.
^{c}Transport of cane from field to mill.
^{d}Taken from ref. 57.
^{e}Total energy field/fossil energy invested.

with an average 35% of the N requirements of the crop. In efficient varieties the BNF contribution is over 60%.[40,66,67,68] In this respect, it is important to state that, in 2008, Embrapa launched the first inoculants of diazotrophic bacteria for this crop, which is about to become commercially available, and is expected to reduce the consumption of N fertiliser, thus helping to increase the energy gain in ethanol production.

Another input that should be used rationally is the herbicide applied for the maturation of the sugarcane plant. In Brazil this input is applied at the average dose of 1.5 kg ha^{-1} yr^{-1} of the principal active constituent, and therefore consumes 1.450 MJ ha^{-1} (Table 5.4), which accounts for about 15% of the total fossil-origin energy invested in the agricultural sector of sugarcane production. In this respect, there is a need for genetic improvement studies that lead to the production of varieties that have better yields with fewer inputs that require fossil-fuel energy. The recommendation, in the meantime, is to rationalize the use of these inputs in the whole ethanol-production system.

In conclusion for the aforementioned statements, it should be noted that the Brazilian bioenergy programme based on the production of sugarcane ethanol is energetically positive and at this moment there is no liquid biofuel in the world with a similar energy gain.

5.5 Land-Use Changes in Sugarcane Areas: The Brazilian Case

Brazil is the largest sugarcane producer in the world. Therefore, a brief analysis about the possible tendencies of land-use changes will be presented, focused on the Brazilian case.

The land-use changes made to cultivate sugarcane in Brazil and in other countries occurred, initially, due to colonisation by European countries that mastered the art of navigation in the 14th century. In Brazil, the first vegetation that was replaced was that of the tropical Atlantic forest (Mata Atlântica). That replacement occurred mainly in the Northern and Southern regions, with most of the occupation occurring in the flood plain and other areas with more fertile soils. In the 1970s, this native forest was once again devastated with a new emphasis, now further into the interior with soils that were not so fertile, because the use of fertilisers permitted their occupation. That increase in the area cultivated with sugarcane occurred as a result of the high price of sugar on the international market and, in the early 1980s, due to an increase in the price of petroleum.[6] Until that period, sugarcane cultivation continued to be more concentrated in the North and Southeast regions.

More recently, with the introduction of flexifuel vehicles and the need to reduce the emission of GHG, there has been an increase in the external and internal demands for Brazilian ethanol. Therefore, there has been a new wave of expansion for sugarcane in Brazil, outlining a new spatial distribution for this crop. In the 2000/2001 harvest, the country had a production of 255 million tons of crushed sugarcane and a cultivated area of 4.8 million hectares, which increased to 564 Mt and an area of 9 Mha in the 2008/2009 harvest.[69] A great part of that increase (95.4%) occurred in the Central-South region of Brazil, which presented a growth rate of 16% in the period from 2005 to 2008.[70] The most productive states in the Southeast are São Paulo and Minas Gerais; in the Central-West, the states of Goiás, Mato Grosso and Mato Grosso do Sul; as well as the state of Paraná, in the South. This expansion occurred in the last decade, by the replacement of pastures, other crops and *cerrado* vegetation by sugarcane fields.

An important analysis of the process of land-use change has recently been conducted for the areas that have seen sugarcane expansion in Brazil.[70] The objectives of the study were to measure and evaluate direct changes of land use caused by sugarcane expansion over recent years, as well as the consequences of future expected expansion of sugarcane in the country, evaluating two possibilities: direct land-use change (LUC) and indirect land-use change (ILUC). The authors concluded that sugarcane is localised and expanding in areas that have been designated for agricultural production since long ago. The projections indicated that sugarcane expansion will continue in those areas. This means that there will be no sugarcane expansion in the agricultural frontier, which is the place where agriculture has changed the natural landscapes. The results showed that sugarcane is not putting any direct pressure on natural vegetation in any region in Brazil. One point that supports these conclusions is that sugarcane is established only in regions where investments have already made in mills, at a maximum distance of 60 km from the mill.[71] Therefore, the crop will only expand in the regions that provide the infrastructure for that activity. As for indirect land-use change, the authors stated that the expansion of pastures and other cultures is occurring regardless of the expansion of sugarcane. This fact supports the argument that there is no clear evidence that deforestation caused by the indirect effect of land use is a consequence of sugarcane expansion, though recognising that the expansion of this crop contributes with the displacement of other crops and pastures.

In the future, more value should be given to the form that occupation takes in the frontier region, referred to as MAPITO, which comprises the states of Maranhão, Piauí and Tocantins. This area still has a considerable area of native *cerrado* vegetation, and is considered to be a region with potential for the expansion of sugarcane. Therefore, depending on infrastructure conditions, unplanned land-use changes may occur and consequently replace natural landscapes. Reports from the World Wildlife Foundation (WWF)[71] state that if the situation continues without incentives to increase the conversion of pastures and other areas degraded for agricultural production (soybean, corn, cotton, sorghum, sugarcane, and other crops) it is expected that there will be a deforestation of the *Cerrado* of approximately 10 million hectares in the next ten years. According to the report, that would be an unacceptable scenario, considering the availability of degraded areas, thus making sustainable expansion planning essential for the main crops on the increase in Brazil, *i.e.* soy and sugarcane.

Mechanisms to reduce land-use changes involving natural vegetation areas should be encouraged. Such mechanisms include particularly: (1) increasing production per area, either by genetic improvement or by using more efficient technologies for culture management; (2) recovering degraded areas to use for agriculture; (3) replacing agricultural cultivation with other crops, but not natural vegetation with crops; and (4) implementing public policies (conservation units, zoning, *etc*).

In this sense, in 2009 the Brazilian government established an Agroecological Zoning for Sugarcane,[72] with the objective of providing technical support for the formulation of public policies, aimed at organising the expansion and sustainable production of sugarcane in Brazil. This tool is presented as a decree, with direct influence on the possible land-use changes that may occur with the expansion of this crop. Some of the general guidelines show the concern with environmental preservation, such as the exclusion of: (1) areas in the Amazon, Pantanal and High Paraguay Hydrographic Basin biomes; (2) land with native vegetation; (3) areas with slopes greater than 12%; (4) indigenous lands;

(5) remaining areas from forests, dunes, mangroves, escarpments and outcrops, and reforestations; (6) urban and mining areas. The conclusion of this Agroecological Zoning showed that the country has about 64.7 million hectares of land that are suitable for the expansion of sugarcane cultivation (Figure 5.6), of which 19.3 million hectares were considered as having potential for high production, 41.2 million of intermediate potential, and 4.3 million of low potential for cultivation. The areas suitable for expansion, currently cultivated with pastures, account for about 37.2 million hectares. Supporting this analysis, a document by the WWF[71] emphasizes that, in Brazil, it would be possible to double the national agriculture area by simply recovering degraded pastures for agricultural use.

These data refer to what we have available. But how much will be necessary to meet the needs of the markets? In this sense, the analysis of two scenarios,[73] one concerning ethanol and the other from the perspective of energy

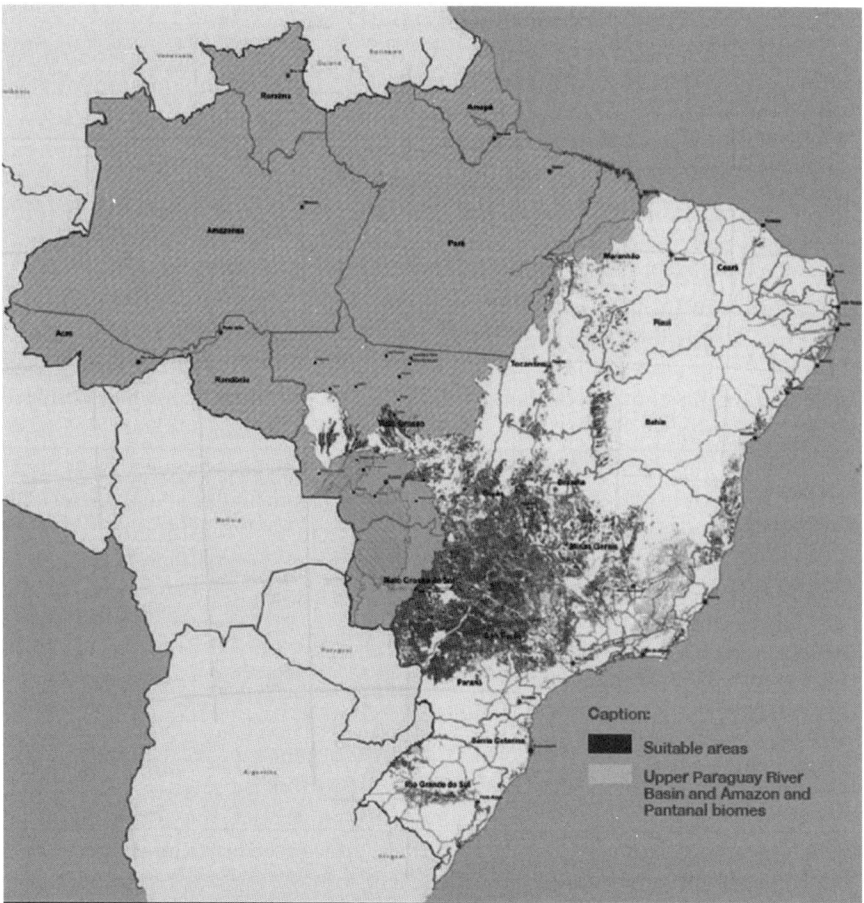

Figure 5.6 Agroecological zoning of sugarcane in Brazil (reproduced from ref. 72).

production, showed that, for the conditions in the year 2020, the additional area needed for the production of sugarcane will be of approximately 5.1 Mha. This new area necessary for sugarcane until 2020 is only 8% of the total area with cultures today, or 2.5% of the area currently cultivated with pastures (Table 5.5).

Combining the information from two important studies,[71,72] it is observed that sugarcane expansion in Brazil has a great chance of being conducted in a way that is less harmful to the environment, with land-use changes that allow for a smaller emission of greenhouse-effect gases and greater sustainability of the adopted production systems. It is expected for the year 2020 that there will be a net reduction of approximately $100\,kg\,CO_2\,eq\,m^{-3}$ of ethanol,[73] in the emissions associated with the land-use change in Brazil, for a production system that does not include the burning of sugarcane trash (Table 5.6). This is

Table 5.5 Land use in Brazil: selected uses (2006) (reproduced from ref. 73)

Land use	Area, M ha	% arable land	% cultivated land
Total land	850		
Forests	410		
Arable land	340 (40%)	100.0	
Pasture land	200	58.8	
Cultivated land (all crops)	63	18.5	100
Soybean	22	6.5	34.9
Maize	13	3.8	20.6
Sugarcane (total)	7	2.1	11.1
Sugarcane for ethanol	3.5	1.0	5.6
Available land	77	22.6	122.2

Table 5.6 Emissions associated with LUC to unburned cane (reproduced from ref. 73)

| Reference crop | Carbon stock change[a] ($tC\,ha^{-1}$) | Emission ($kg\,CO_2\,eq.\,m^{-3}$) | | |
		2006	2020 electricity	2020 ethanol
Degraded pasturelands	10	−302	−259	−185
Natural pasturelands	−5	157	134	96
Cultivated pasturelands	−1	29	25	18
Soybean croplands	−2	61	52	37
Maize croplands	11	−317	−272	−195
Cotton cropland	13	−384	−329	−236
Cerrado	−21	601	515	369
Campo Limpo	−29	859	737	527
Cerradão	−36	1040	891	638
LUC emissions[b]		−118	−109	−78

[a]based on measured values for below and above ground (only for perennials carbon stocks).
[b]Considering the following LUC distribution – 2006: 50% pasturelands (70% degraded pasturelands; 30% natural pasturelands), 50% croplands (65% soybeans, 35% other croplands); 2020: 60% pasturelands (70% degraded pasturelands; 30% natural pasturelands); 40% croplands (65% soybeans, 35% other croplands). Cerrado were always less than 1%.

expected considering that the expansion of sugarcane areas includes a very small fraction of native lands with high carbon storages, and some degraded land.[72]

More details of these estimates are expected to be provided in time, by means of the intensification of the process of generating information on issues such as the emission of GHGs, particularly N_2O, in intact Brazilian ecosystems, allowing for safer comparisons with background emissions[74] emissions of nitrous oxide in the various production systems using vinasse,[60] and others, which should improve the understanding of the effects caused by land-use change due to the expansion of sugarcane and other crops in Brazil.

References

1. F. Blackburn, *Sugar-cane*. Tropical agriculture series, 1984, p. 414.
2. R. Cesnik and J. Miocque, *Melhoramento da cana-de-açúcar*, Brasília, DF: Embrapa Informação Tecnológica, 2004, p. 307.
3. ICIDCA/ABIPTI. *Manual dos Derivados da cana-de-açúcar*. Brasília: ABIPTI, 1999, p. 14.
4. FAOSTAT, 2010. Available at: http://faostat.fao.org/site/339/default.aspx.April, 2010.
5. G. Fisher, E. Teixeira, E. T. Hizsnyik and H. Van Velthuizen, in *Sugarcane Ethanol*, ed. P. Zuurbier, J. Van Vooren, Wageningen Academic Publishers, Wageningen, the Netherlands, 2009, p. 29.
6. W. Mello Ivo, R. Rossetto, A. D. Santiago, G. V. S. Barbosa and J. N. Vasconcelos, in: *Agricultura Tropical – quatro décadas de inovações tecnológicas, institucionais e políticas*, ed. A. C. S. Albuquerque and A. G. da Silva, Embrapa, Brasília, 2008, p. 678.
7. J. Goldemberg, F. E. B. Nigro and S. T. Coelho, *Bioenergia no estado de São Paulo : situação atual, perspectivas, barreiras e propostas*, Ed Imprensa Oficial do Estado de São Paulo, São Paulo, 2008, 152 p.
8. MME, Resenha Enertética Brasileira, exercício 2008. Available at www.mme.gov.br/Publicações/Balanço Energetico Nacional/2.BEN/2.8.Matrizes Energéticas-Exercício 2008 - NOVO/01. Resenha Energética Brasileira.
9. W. Maccheroni, *Opiniões*, 2008, **42**.
10. J. R. Albert-Thenet, *Proceedings of the Barbados Society of Technologists in Agriculture*, 2003.
11. S. Creste, V. E. Rosa, L. R. Pinto, J. C. Albino and A. V. O. Figueira, in: *Cana-de-açúcar*, ed. L. Dinardo Miranda, A.C.M. Vasconcelos and M. G. A. Landell, Instituto Agronômico, Campinas, 2008, p. 157.
12. B. van Raij, H. Cantarella, J. A. Quaggio and A. M. C. Furlani, *Recomendações de adubação e calagem para o Estado de São Paulo*, Instituto Agronômico, Campinas. 1997. 285 p.
13. H. C. J. Franco, H. Cantarella, P. C. O. Trivelin, A. C. Vitti, R. Otto, C. E. Faroni, R. H. Sartori and M. O. Trivelin, *STAB Açúcar, Álcool e Subprodutos*, 2008, **26**(5), 41.

14. G. Moura Filho, H. L. Soriano, L. C. da Silva, G. V. S. Barbosa, I. A. Lyra Neto and V.T. Silva, Acúmulo e eficiência de macronutrientes em cana-planta, sob condição de sequeiro. In: *28 Reunião Brasileira de Fertilidade do Solo e Nutrição de Plantas*, Londrina, 2008. Anais. Londrina : Embrapa Solos/SBCS, p. 1.
15. R. Rossetto, F. L. F. Dias and A. C. Vitti, in: *Cana-de-açúcar*, ed. L. L. Dinardo-Miranda, A. C. M. Vasconcelos and M.G.A. Landell, Instituto Agronômico, Campinas, 2008a, p. 221.
16. ANDA – Associação Nacional para a Difusão de Adubos. Anuário Estatístico Setor de Fertilizantes. São Paulo, ANDA, 2007, p. 160.
17. A. E. Hartemink, *Adv. Agron.*, 2008, **99**, 125.
18. M. L. Marinho and G. A. C. Albuquerque, in: *Nutrição e adubação da cana-de-açúcar no Brasil*, ed. J. Orlando Filho, Planalsucar, Piracicaba, 1983, p. 179.
19. R. Rossetto, A. Spironello, H. Cantarella and J. A. Quaggio, *Bragantia*, 2004, **63**, 105–119.
20. J. A. Quaggio and B. van Raij, in: *Cana-de-açúcar*, ed. L.L. Dinardo-Miranda, A. C. M. Vasconcelos and M. G. A. Landell, Instituto Agronômico, Campinas, 2008a, p. 313.
21. J. L. Morelli, J. L. I. Demattê, E. J. Nelli and A. E. Dalben, *STAB Açúcar, Álcool e Subprodutos*, 1987, **6**, 24.
22. C. P. Penatti and J. A. Forti, Calcário e gesso em cana-de-açúcar. Relatório de projeto. Piracicaba: Copersucar, 1993, p. 80.
23. J. Orlando Filho, E. J. R. Zambello and A. A. Rodela, *Brasil Açucareiro*, 1980, **97**, 18.
24. P. C. O. Trivelin, M. W. Oliveira, A. C. Vitti, G. J. C. Gava and J. A. Bendassolli, *Pesquisa Agropecuária Brasileira*, 2002a, **37**, 193.
25. A. Spironello, B. van Raij, C. P. Penatti, H. Cantarella, J. L. Morelli, J. Orlando Filho, M. G. A. Landell and R. Rossetto, in: *Recomendações de adubação e calagem para o Estado de São Paulo*, ed. B. van Raij, H. Cantarella, J. A. Quaggio and A. M. C. Furlani, Instituto Agronômico de Campinas, Campinas, 1997, 285 p.
26. C. P. Penatti, J. L. Donzelli and J. A. Forti, Doses de nitrogênio em cana-planta. In: 7. Seminário de Tecnologia Agronômica. Piracicaba, 1997, Anais, Piracicaba: Centro de Tecnologia da Copersucar, 1997, p. 340.
27. H. Cantarella, P. C. O. Trivelin and A. C. Vitti, in: *Nitrogênio e Enxofre na Agricultura Brasileira*, ed. T. Yamada, S. R. S. Abdalla, G. C. Vitti, International Plant Nutrition Institute – IPN, Piracicaba, 2007, p. 722.
28. J. L. Donzelli, in: *A energia da cana-de-açúcar. Doze estudos sobre a agroindústria da cana-de-açúcar no Brasil e a sua sustentabilidade*, ed. I.C. Macedo, Berlendis Editores Ltda., UNICA, 2005, p. 160.
29. S. S. Garcia, D. J. P. Lopez, J. Z. Cruz, L. C. L. Espinoza, M. C. Estrada, C. F. O. Garcia, J. F. J. Lopez and J. A. R. Ramirez. Sistema integrado para recomendar dosis de fertilizantes en caña de azucar (SIRDF): Ingenio Pujiltic. Colégio de Postgraduados. Tabasco, México, 2006, p. 99.

30. R.W. Rice, R. Gilber, R.S. Lentini, Nutritional requirements for Florida sugarcane, University of Florida. Institute of Food and Agricultural Sciences Extension, Gainesville, 2006, 8p.
31. L. S. Chapman, M. B. C. Haysom and P. G. Saffigna, *Proceedings of the Australian Society of Sugar Cane Technologists*, 1992, 84.
32. P. C. O. Trivelin, R. L. Victoria and J. C. S. Rodrigues, *Pesquisa Agropecuária Brasileira*, 1995, **30**, 1375.
33. P. C. O. Trivelin, A. C. Vitti, M. W. Oliveira, G. J. C. Gava and G. A. Sarriés, *Revista Brasileira de Ciência do Solo*, 2002, **26**, 636.
34. G. J. C. Gava, P. C. O. Trivelin, A. C. Vitti and M. W. Oliveira, *Revista Brasileira de Ciência do Solo*, 2003, **27**, 621.
35. A. C. Vitti, *Adubação nitrogenada da cana-de-açúcar (soqueira) colhida mecanicamente sem a queima prévia: manejo e efeito na produtividade*, Centro de Energia Nuclear na Agricultura, Universidade de São Paulo, Piracicaba, 2003, 114 p, (Tese de Doutorado).
36. E. A. Ambrosano, P. C. O. Trivelin, H. Cantarella, G. M. B. Ambrosano, E. A. Schammass, N. Guirado, F. Rossi, P. C. D. Mendes and T. Muraoka, *Scientia Agrícola*, 2005, **62**, 534.
37. J. R. Freney, O. T. Denmead, A. W. Wood, P. G. Saffigna, L. S. Chapman, G. J. Ham, A. P. Herney and R. L. Stewart, *Fertil. Res.*, 1992, **3**, 341.
38. A. Dobermann, *Nitrogen-use efficiency – state of the art*. In: International Workshop on Enhanced-Efficiency Fertilisers. Frankfurt, 2005. Proceedings. Paris, International Fertiliser Industry Association. 2005, 16 p. CD-ROM.
39. R. M. Boddey, *CRC Crit. Rev. Plant Sci.*, 1995, **14**, 263.
40. S. Urquiaga, K. H. S. Cruz and R. M. Boddey, *Soil Sci. Soc. Am. J.*, 1992, **56**, 105.
41. R. M. Boddey, S. Urquiaga, B. J. R. Alves and V. M. Reis, *Plant Soil*, 2003, **252**, 139.
42. J. L. Donzelli, in: *A energia da cana-de-açúcar*, ed. I. C. Macedo, Única, São Paulo, 2007, p. 246.
43. G. H. Korndörfer and L. A. Oliveira, in: *Potássio na agricultura brasileira*, ed. T. Yamada and T.L. Roberts, *Potafós*, Piracicaba, 2005, p. 469.
44. R. Rossetto, F. L. F. Dias, A. C. Vitti and S. Tavares, in: *Cana-de-açúcar*, ed. L. L Dinardo-Miranda, A. C. M. Vasconcelos and M. G. A. Landell, Instituto Agronômico, Campinas, 2008, p. 289.
45. B. van Raij, *Ciência e Cultura*, 1974, **26**, 575.
46. G. H. Korndörfer, A. C. Ribeiro and L. A. B. Andrade, in: *Recomendações para o uso de corretivos e fertilizantes em Minas Gerais*, ed. A. C. Ribeiro, P. T. G. Guimarães and V. H. Álvares, Comissão de Fertilidade de Minas Gerais, Viçosa, 1999. 359p.
47. B. L. Legendre, *Sugarcane Production Handbook 2001*. Baton Rouge: Louisiana State University Agricultura Center, Louisiana Cooperative Extension. 2001, 54 p. (Pub. 2859).
48. F. J. Lampkowski, and G. Viera, Estratégia de produção: inovação tecnológica no processo de corte da cana de açúcar como fator competitivo. In: 11 SIMPEP, Bauru, São Paulo, Brasil, 2004, p. 1.

49. M. W. Oliveira, P. C. O. Trivelin, G. J. C. Gava and C. P. Penatti, *Scientia Agrícola*, 1999, **54**, 803.
50. W. A. N. Amaral, J. P. Marinho, R. Tarasantchi, A. Beber and E. Giuliani, in: *Sugarcane Ethanol*, ed. P. A. Zuurbier and J. Van Vooren, Wageningen Academic Publishers, Wageningen, the Netherlands, 2009, p. 113.
51. S. C. Rabelo, A. C. Costa and C. E. Rossel, in: *Cana-de-açúcar: Bionergia, açúcar e álcool*, ed. S. F. Borém and A. Caldas, Viçosa, 2010, p. 465.
52. I. C. Macedo, *Biomass Bioenergy*, 1998, **14**, 77.
53. I. C. Macedo, *Geração de energia elétrica a partir de biomassa no Brasil: situação atual, oportunidades e desenvolvimento*, Centro de Gestão e Estudos Estratégicos, 2001.
54. J. L. Olivério and A. G. P. Hilst, DHR – Dedini Hidrólise Rápida – Revolutionary process for producing alcolhol from sugar cana bagasse. XXV International Society of Sugar Cane Technogists Congress. Guatemala, 2005.
55. J. Finguerut, A. J. A. Meirelles, R. Guirardello and A. C. Costa, in: *Biomassa para Energia*, ed. L. A. B. Cortez, E. E. S. Lora and E. O. Gómez, Unicamp, Campinas, 2008, p. 435..
56. IEL. O Novo Ciclo da Cana: Estudo sobre a Competitividade do Sistema Agroindustrial da Cana-de-açúcar e Prospeção de Novos Empreendimentos- ed. IEL-NC. Brasília: IEL-NC;SEBRAE, 2005, 344 p.
57. I. C. Macedo, J. E. A. Seabra and E. A. R. Silva, *Biomass Bioenergy*, 2008, **32**, 582.
58. S. Urquiaga, B. J. R. Alves and R. M. Boddey, *Revista de Política Agrícola*, 2005, **14**, 42.
59. R. M. Boddey, L. H. B. Soares, B. J. R. Alves and S. Urquiaga, in: *Biofuels, Solar and Wind as Renew. Energy Systems*, ed. D. Pimentel, Springer, New York, 2008, p. 321.
60. L. H. B. Soares, B. J. R. Alves, S. Urquiaga and R. M. Boddey, *Mitigação das emissões de gases efeito estufa pelo uso de etanol de cana-de-açúcar produzido no Brasil*, Circular técnica, 26. Embrapa Agrobiologia, 2009, 14 p.
61. D. Pimentel, A. F. Warneke, W. S. Teel, K. A. Schwab, N. J. Simcox, D. M. Ebert, K. D. Baenisch and M. R. Aaron, *Adv. Food Res.*, 1988, **32**, 185.
62. D. Pimentel and T. Patzek, in: *Biofuels, Solar and Wind as Renew. Energy Systems*, ed. D. Pimentel, Springer, New York, 2008, p. 357.
63. M. D. de Oliveira, B. E. Vaughan and E. J. Rykiel Jr., *BioScience*, 2005, **55**, 593.
64. M. D. de Oliveira, in: *Biofuels, Solar and Wind as Renew. Energy Systems*, ed. D. Pimentel, Springer, New York, 2008, p. 215.
65. D. Pimentel, *CRC Handbook of Energy Utilization in Agriculture*, Boca Raton, CRC Press, 1980.
66. E. Lima, R. M. Boddey and J. Döbereiner, *Soil Biol. Biochem.*, 1987, **19**, 165.
67. T. Yoneyama, T. Muraoka, T. H. Kim, E. V. Decanay and Y. Nakanishi, *Plant Soil*, 1997, **189**, 239.

68. R. M. Boddey, J. C. Polidoro, A. S. Resende, B. J. R. Alves and S. Urquiaga, *Austral. J. Plant Physiol.*, 2001, **28**, 889.
69. www.mapa.gov.br/estatística, acessado em 05/01/2010.
70. A. M. Nassar, B. E. T. Rudorf, L. B. Antoniazzi, D. Alves de Aguiar, M. R. P. Bacchi and A. Adami, in: *Sugarcane Ethanol*, ed. P. Zuurbier and J. Van Vooren, Wageningen Academic Publishers, Wageningen, the Netherlands, 2009, p. 63.
71. WWF, 2009. O impacto do mercado mundial de biocombustíveis na expansão da agricultura brasileira e suas conseqüências para as mudanças climáticas. Programa de Agricultura e Meio Ambiente. Brasília, DF: WWF Brasil, 2009. 66 p.
72. C. V. Manzatto, E. D. Assad, J. F. M. Bacca, M. J. Zaroni and S. E. M. Pereira, *Zoneamento Agroecológico da cana-de-açúcar*, Embrapa Solos, Rio de Janeiro, RJ, 2009, 55 p (Documentos/Embrapa Solos, 110).
73. I. C. Macedo and J. E. A. Seabra, in: *Sugarcane Ethanol*, ed. P. Zuurbier and J. Van Vooren, Wageningen Academic Publishers, Wageningen, the Netherlands, 2009, p. 95.
74. S. Porder, A. Bento, A. Leip, L. A. Martinelli, J. Samseth and T. Simpson, in: *Biofuels: Environmental Consequences and Interactions with Changing Land Use*, ed. R. W. Howarth and Bringezy, Gummersbach, Germany, 2009, p. 233.

CHAPTER 6
Sugar Beet

MIKE MAY

Broom's Barn Research Centre, Higham, Bury St. Edmunds Suffolk, IP28 6NP, UK

6.1 Development of Sugar Beet Industries

Beet is a member of the *Amaranthaceae* family (subfamily *Chenopodiaceae*) and has been grown as a vegetable for over 2000 years. It was first grown in Europe as an agricultural crop, in the form of fodder beet (*Beta vulgaris*), in the 17th century.[1] Both the tops and roots of the beet were used mainly as fodder for cattle in France and Germany. Sugar beet was selected from fodder beet during the latter part of the 18th century. In central Europe, white-fleshed fodder beet varieties were cultivated for storage and use during the winter because they tended to have a slightly higher sugar content than other varieties. It was from these white-fleshed fodder beet that sugar beet varieties were eventually bred. An important milestone in the development of sugar beet occurred in the middle of the 18th century when Marggraf demonstrated that sugar crystals, identical to those from cane sugar, could be produced from beet.[2]

Achard, a student of Marggraf, took the work further, publishing his results in 1799.[3] Achard then successfully appealed to the King of Prussia for funding to further his work. The money was used to purchase an estate at Cunern in Silesia and to construct the world's first sugar beet factory in 1801. Around 250 t of beet were processed in the first campaign that year. Whilst only a small quantity of sugar was extracted (4% of the roots' fresh weight), it demonstrated that sugar production on a large scale was possible. This caused concern in England and its colonies, where the cane producers and trading firms realised

RSC Energy and Environment Series No. 3
Energy Crops
Edited by Nigel G Halford and Angela Karp
© Royal Society of Chemistry 2011
Published by the Royal Society of Chemistry, www.rsc.org

that competition was just over the channel. Unlike cane, sugar could also be produced directly from the beet roots without separate refining of raw sugar.

In the decades around 1800, cane sugar had a pivotal position in the world economy. During the Napoleonic wars Napoleon banned all British exports into Europe and Britain tried to cut France off from her colonies. The result was a severe shortage in Europe and the price of sugar increased enormously. In 1811, Napoleon published his decree for the introduction of sugar beet production in France and the countries under its administration. Within a short period more than 40 sugar beet factories were established. Most of these were in Northern France but there were also some in Germany, Austria, Russia and Denmark. However, this was short lived, as in 1813 Napoleon's power was reduced and the blockade removed. Beet sugar could no longer compete with cane and all sugar factories in Germany and Austria were shut down. France managed to carry on and improve its production techniques, helped by a duty imposed on imported cane sugar. The production in other countries, such as North and South America, Japan and Canada, started during the nineteenth century, with sugar production finally established in the UK in 1912. There had been earlier attempts to start processing factories in the UK, but these were uneconomic and soon closed.

Ever since those early days, sugar beet production has been subject to the vagaries of political influence in virtually all countries in which it is grown. This has had a major bearing on the use of beet for biofuel production.

6.2 Products from Sugar Beet

The main use of sugar beet was and is to produce sucrose and, by the mid-19th century, breeding had resulted in sugar contents of 18 to 20%. This is generally considered good for modern beet as most recent breeding has been for agronomic traits and disease resistance.

'Byproducts' of sugar beet processing have always been economically important. Ethanol and biogas production from beet are relatively recent introductions but rum and other alcoholic spirit have been produced from sugar beet for many years, especially in the Czech Republic and Germany. The earliest, and most obvious byproduct, is the beet "pulp" left after the extraction processes. It is usually used for animal feed, either as the wet pulp or as dried pellets. The economics of producing pulp for feed is related to the value of alternative feeds such as cereals and corn. Pulp becomes uneconomic if the value of these latter is too low. However, pulp has to be disposed of and in an increasing number of countries there is interest in using it in biodigesters for biogas production (see later).

Lime used in the sucrose-extraction process is recycled to help maintain soil pH balance in agricultural fields but it also contains a small but important amount of soil nutrients such as phosphorus, sulfur and magnesium. In some cases the lime is left as wet lime but in others, such as in the UK, it is dried and processed to achieve good storage and spreading properties.

Stones and soil delivered to the factory with the sugar beet are either extracted as the beet is unloaded and returned to the source farm on the delivering lorry (*e.g.* Northern USA) or sorted, processed and sold for use on landscaping of nonagricultural land (*e.g.* UK).

Small amounts of potassium, sodium and nitrogen are left after processing and these are sometimes sold as liquid fertiliser solutions to farmers.

In some cases, excess heat from the factory process is used to heat homes or other local buildings. In the UK, British Sugar's Wissington Factory uses excess heat and the carbon dioxide produced in the factory to enhance tomato production in specially-built glasshouses.

There is also a range of sugars and inert materials in sugar beet. One such compound is betaine (trimethylglycine or glycine betaine), which is used widely in the medical field, and in many countries, notably the USA and the UK, sugar beet factories have been set up to extract this from the low-sugar solutions created during the sugar production. Betaine concentration is usually highest in the crown of the roots but levels can be manipulated to a small extent by changes in nutrient management. Inositol (an antidepressant) is another medical compound that can be extracted from beet.

Pectins produced from beet have relative poor setting qualities, but this can be used to advantage in foods such as yogurts. Other food products have included breakfast-cereal-type products using part of the pulp remaining after sugar extraction.

The production of alternative sugars to sucrose has also been investigated. Doney and Theurer[4] suggested that breeding programmes could be altered to provide high levels of fermentable sugars coupled with different levels of impurities to the low levels that were traditionally required for sucrose production. In 1998, Sevenier[5] reported a line, created by genetic modification, in which carbohydrate metabolism had been altered so that the roots produced fructan rather than sucrose.

6.3 Growth of Beet

Typical sugar yields from beet are between 8 and 18 t/ha resulting from beet yields of 50 to 100 t clean beet per hectare and sugar concentrations of 17 to 18%. However, yields of greater than 36 t sugar/ha (200 t clean beet) have been recorded in countries with long growing seasons and adequate rainfall or irrigation. Dry matter accumulation is in the shoot during the early stages of growth and in the storage root (mainly as sugar) towards harvest.[6] However, there is a range of factors that govern the exact proportions, including seasonal weather, soil type, fertiliser (especially nitrogen), plant populations, the sugar-beet genotype that is sown and harvest date. Drought can have a major, negative effect on growth.[7] Therefore, the actual total dry matter in the root can vary between 47 and 77% and the sugar in the storage root from 72 to 78% on dry matter.[8] The main driver of high yields is radiation[9] and the longer the beet are allowed to grow, the higher the potential yield (providing other limiting factors are excluded or controlled).

One of the main factors limiting growth is cold periods (less than 12 °C) that cause the beet to change from the vegetative to reproductive phase. This vernalisation process limits yield and it is in warm climates (with adequate water), where it is absent, that the very high yields can be obtained, such as in California, southern Africa and Australia.

Whilst beet sugar yields vary according to the season, they continue to increase on a yearly basis in most producing countries. This is in contrast to yields of wheat and rape, which have plateaued for the last 10 years or more. Average EU beet sugar yields increased from around 3.5 t/ha in 1930 to 10.6 t/ha in 2007/8.[10] Jaggard *et al.*[11] suggest that much (56%) of the yield increases in the UK during the last 20 years can be attributed to changes in climate. However, Pidgeon *et al.*[12] suggest that there is likely to be a slowing of the increase as the most efficient countries now produce around 80% of estimated potential production.

6.4 Cane or Beet?

Sucrose is normally used in the food sector but it can be used on its own right or as a substrate or part substrate to produce other products. Examples are high-grade biodegradable plastics (especially for medical use), explosives, in the textile industry for sizing and finishing fabrics, cosmetics, toiletries and detergents, agrochemical formulations, inks and printing and the construction industry. Molasses are also used in industrial processes and can, because of their low freezing point and noncorrosive nature, be used instead of salt on sensitive roads such as bridges. These are in addition to fermentation into ethanol and other similar products. When products are produced from sucrose there is little or no difference in the processes involved whether the source of the sucrose is beet or cane. Therefore, there is need to consider the relationship between these crops on a global scale, but the specific use of cane for biofuels is discussed in the previous chapter.

Sugar beet is mainly grown in the northern hemisphere, between latitudes 30° and 60°N, whilst cane tends to be grown in the tropics. The main difference in cane and sugar production is that sugar beet is grown for one season, whereas cane is planted and harvested for *circa* five years. Cane grown in the tropics will typically use twice the amount of water compared to beet grown in its traditional northern hemisphere areas.[13] Cane is the main source of sucrose in the world.[10] In 2006/7, sugar beet produced approximately 26% of the world's sucrose, but this compares with 20 years earlier (1986/7) when approximately 37% was produced by sugar beet.[10] The main reasons for the change have been a reduction in beet production by the EU countries and an increase in cane production, especially in Brazil. World sugar production increased by 65 Mt between 1986/7 and 2007/8.[10] (Table 6.1)

Table 6.1 World sugar production from beet and cane 2007/8 (000 t white sugar).

Country	Total sugar production from cane	Number of countries	Total sugar production from beet	Number of countries
EU countries	0	0	17 405	20
Europe (non-EU)	0	0	9058	10
Africa	9173	32	1053	2
North & Central America	15 359	15	4451	2
South America	40 340	9	289	2
Asia	64 390	15	2883	8
Oceania	5220	3	0	0
World total	**134 482**	**76**	**35 139**	**44**

Source: F O Licht Sugar beet year book, 2009[10]

6.5 Factory Efficiency

Sugar beet is processed into sugar in factories especially designed and set up for this process. The efficiency of these factories depends partly on the quality of the sugar beet they receive and on the length of time they operate each year. The timing and length of the harvest campaign will depend on the weather, particularly the timing and duration of frosts that freeze and damage the beet in the ground. In areas such as the north mid-west of the USA or northern China, where the winters are severe, the harvest campaign has to be completed within a matter of 2 to 4 weeks in the autumn before the winter frosts. In continental Europe and Scandinavia, the campaign is usually around 12 to 15 weeks, whilst the maritime climate in the UK allows a harvest campaign of over 26 weeks (between mid-September and the end of February or early March). Where harvest campaigns are short, factory efficiency is usually low, but in the USA storage systems have been put in place that allow beet to be stored frozen or at low temperatures. The transport of beet from these storage areas into the factory is carefully managed to maximise factory processing from late September to the following May or even June. In Mediterranean climates, with mild wet winters, beet is usually drilled in the autumn and harvested before the hot summers. In most other countries, beet is drilled in spring and harvested in the autumn.

The physical state of the beet to be processed is important, such as the amount of soil attached and damage by rots or pests. In addition, the impurities within the beet root, such as amino nitrogen, potassium and sodium, also affect factory processes. As a consequence, breeders have focused on producing good-quality traits to facilitate sucrose extraction and changing the shape of the roots to reduce soil tares.[14]

The efficiency of the factory (daily intake and length of processing campaign) governs the amount of sugar beet that can be processed and, therefore, almost all sugar beet is grown under contract to a processing plant. In addition, most

Sugar Beet

countries operate a quota system for production and little more than 15% of sugar (from cane or beet) is actually traded on the world market.[10] Country quotas are not determined by factory capacity but rather sugar consumption or demand or by political considerations, which are often historically based. Each European country that grows beet has a quota system for sugar production and this is controlled and prices are set by Brussels. Excess sugar from Europe used to be exported onto the world market but other countries, particularly the USA and Brazil, challenged this under WTO rules as they considered these as subsidised exports. The result was a reform of the European sugar regime in 2007, with quotas and prices cut with the aim of reducing sugar production by 6 Mt. This led to many inefficient beet companies and countries ceasing production in Europe and for the remaining ones to consolidate and improve sugar production with fewer factories. Under the reform, the export of excess sugar is limited to 1.374 Mt of sugar per annum, a decrease of around 80% compared to previous. Excess sugar production therefore has to be managed carefully to minimise the expensive storage of "carry-over" stocks. This led many companies to become interested in ethanol production as an outlet for excess beet production. However, the vagaries of the growing season and the value of alternative crops means that the quota is not consistently attained and, if ethanol plants are in place, it is inefficient if they are not used. The management of quotas is something that has to occur in other countries and in the USA, in years of excess production, this is achieved by paying growers to destroy some of their crop, equivalent to the forecast excess.

Traditionally, the EU sugar price of €404.4/t was above the world market price but, as a result of production shortfalls in countries such as Brazil and India, the world market price rose in 2009 from around $300 to over $550/t (€408/t). In some countries (such as France) a separate contract is offered to growers for beet for ethanol production and in some (*e.g.* UK in 2009), beet produced in excess of contract is paid for at a different price for use in ethanol production: beet destined for ethanol rather than sugar production is typically paid for at a 30% lower price. This makes it very difficult for farmers to produce ethanol beet economically, especially if grown as a "standalone" crop.

6.6 Biofuels from Sugar Beet

As intimated above, there is a range of options for biofuel production that could use sugar beet as a source. If biofuels are to be produced from sugar or sugar solutions, then cane and other sugar-producing plants are alternatives to sugar beet but, in most cases, the biofuel production is likely to be closely linked to sucrose-processing facilities.

In sugar beet processing factories, the roots are washed and sliced. The slices are then boiled to extract the sugar, which is then crystallised and centrifuged out. Not all the sugar is extracted first time and it is usual for the liquid molasses to be reboiled and the remaining sugars crystallised at least twice more – often mixed in with fresh solutions. In many factories, these low-sugar solutions are diverted to fermenters that produce the ethanol. Almost all of the ethanol produced from beet is produced in sugar factories because of the ability to use low-concentration sugar solutions, which are expensive to transport. The techniques used are based on fermentation of sugars by yeasts.

With current sugar prices and the volatility of the world sugar market, it is generally uneconomic to set up new plant in developed countries to process sugar beet for sucrose production. As ethanol is less profitable than sugar production at the moment, there is no incentive in these countries to set up dedicated factories for ethanol production from beet. However, in some developing countries the situation is different. The cost of setting up factories, whilst still high, is less than in developed nations and, combined with an increasing demand for sugar, some new factories have been built in the last five years. In some countries (*e.g.* India), dedicated factories for ethanol production have been built.

The use of beet pulp as a substrate for biofuels is an option[15] but is also linked to sucrose processing. Production of dry pulp uses a lot of energy for the drying process, whilst wet pulp is costly to transport over long distances and difficult to store for any length of time. However, pulp contains sugars and therefore the production of biogas or ethanol from pulp could be a long-term economic option, especially if animal feed prices are depressed.

Biogas (biomethane) is an alternative option that was first investigated in Germany in the early 1990s. It has the advantage that it can be a standalone production process and not associated with sucrose production. For biogas production, sugar beet can act as the sole or part substrate. Biogas yield varies with substrate and estimates are for approximately $50\,\text{m}^3$ per tonne of sugar beet pulp, $430\,\text{m}^3$ per tonne of molasses and $400\,\text{m}^3$ per tonne of sugar beet tops. The nutrients remaining in the organic residues from the process can be used as fertilisers. Spreading pathogenic organisms is a consideration whenever factory products are used to treat agricultural soils. However, Haraldsson[16] suggests that where mesophilic anaerobic digestion is used, this will destroy bacteria and fungi, minimising the risk of these causing problems on treated agricultural land. The greatest interest is the use of sugar beet to provide a consistent source of sugars for biogas production where other, more variable, products (*e.g.* waste) are also to be included in the biodigestion.

Fodder beet silage has been used as a substrate for continuous biogas production[17] (Scherer *et al.*, 2003) but there has been debate as to whether fodder or sugar beet would be the preferable substrate for biofuel production. Much is likely to depend upon the enzymes and process systems used. Early work by Theurer *et al.*[18] reported that at a range of six locations in the USA sugar beet cultivars gave equivalent or slightly higher ethanol yields than fodder beet, and that sugar beet hybrids showed better promise than fodder beet because they had more extractable sugar per unit mass, were cheaper to transport and generally had better resistance to diseases. However, focused breeding for ethanol beet could possibly use either fodder or sugar beet, or even crosses with other forms of beet (*e.g.* red or table beet). However, the characteristics of a "bioethanol" beet would very likely be similar to those of sugar beet: high root yields with high sugar and dry matter contents.

Further processing to produce biobutanol is a long-term goal as an alternative to ethanol because of its better miscibility with petrol. If biobutanal is to be produced from ethanol, then the source of the ethanol is unlikely to be important but it would probably be produced in a single process through from the raw material rather than from ethanol produced separately.

Not all the alcohol produced from sugar beet is used for biofuels and, as well as the rum and other spirits mentioned earlier, alcohol for "alcopops" has been an important use of alcohol from sugar beet, especially in France in the early 21st century.

One of the main constraints of using sugar beet to produce biofuels is the ability to store it and hence produce fuels all year round. Wheat or maize are therefore seen as preferred options in many developed countries. The total input energy required to produce a hectare of sugar beet is very similar to that for wheat (19.4 GJ/ha for beet compared to 20.8 GJ/ha for winter wheat); delivery of sugar beet to the factory takes another 0.6 GJ/ha.[19] In 2006/7 the average yield of sugar from beet in the UK was 11.0 t/ha. This is a greater amount of energy and dry matter than the average winter wheat crop, even if the wheat were to yield 10 t/ha. However, wheat is easier to store and transport compared with beet and is therefore available for ethanol production all year round, whereas, at best, sugar beet would only be available from September to March under current optimum-length growing seasons (as in the UK). Therefore, the new ethanol plants being built in the UK are designed to operate on wheat. It will be interesting to see if they can reach the carbon intensity figure reported by British Sugar for its Wissington factory under the UK's Renewable Transport Fuel Obligation of 24.5 g CO_2/MJ of bioethanol produced. For very similar reasons, in the USA corn is preferred to beet for biofuel production and it is for corn that bioethanol processing plants have been built in that country, mostly in the beet- and corn-growing areas of the north.

6.7 Ways Forward and Research

There are three main areas of research that should lead to increased yields of biofuel from sugar beet: (i) changes to the beet plant; (ii) development of the processing systems; (iii) improved agronomy.

i) Changes to the beet plant for biofuel production

The area with greatest research input appears to be control of bolting as this is being studied by breeding companies and in industry-funded research. The ability to stop or alter the vernalisation process so that beet can maintain its vegetative stage for longer in the season, or to allow earlier sowing of the crop, is a holy grail for beet whether it is being used for sucrose, biofuel or even animal feed. Various different approaches are being used, from targeting giberellin pathways to investigating the B (or bolting) gene.

Another area of breeding that would help both sucrose and biofuel production is the reduction of root rots in harvested beet. One option is to use fungicides to control fungal rots, but those currently available do not give adequate control. The longer-term approach of stimulating the beet's host defence mechanism to combat fungal, viral and bacterial diseases using a GM approach offers a potentially more robust answer.

Beet continues to respire when harvested and loss of sugars is generally in the region of 0.12% per day during winter storage in the UK.[20] An option in countries with a maritime climate, such as the UK, is to leave the beet in the ground until harvest, even though this may risk damage by freezing during cold winters. This "just-in-time" harvesting approach is used to avoid the cost of lifting and storage and, if the winter is relatively mild and no freezing occurs, less loss of sugar from respiration than from those stored in clamps (heaps of beet, often protected against the weather by straw or plastic coverings). A major problem with clamps is that temperatures increase as a result of respiration and, unless outside temperatures are cool, clamp temperatures can increase such that the beet break down and rot. In some countries, such as the USA, sugar beet companies in the north use the cold winter air to freeze (or sometimes just cool) clamps or sheds of stored beet. Such systems are generally uneconomic in most other beet-growing areas owing to the vagaries of the weather. Breeders are considering producing beet that respires at a lower rate after harvest. Such beet could potentially be stored for longer than at present and, in the UK, encourage growers to harvest beet early and store for delivery rather than risk frost damage.

Changes in plant structure would be focused according to the endproduct required. One line of interest is the cell wall structure of sugar beet. This, as beet is a member of the *Amaranthaceae,* has an almost unique structure compared to other broad-leaved plants as a result of ferulic acid linkages. At present, the cell walls in beet hold many sugars that cannot be released easily during normal fermentation. If the cell walls could be altered or weakened sufficiently so that these sugars can be utilised, the productivity of beet for both sucrose and biofuel production would increase. Whilst care would need to be taken to ensure that the cell walls do not become too weak, such that the beet were too easily damaged or deteriorated in storage, this approach could offer great potential for reducing the energy required to extract biofuels (and probably sugars) from the beet. It is likely to provide the greatest benefit in biogas production, reducing energy inputs required to break down the root.

ii) Developments of processes

There is little information about research and developments underway to improve the processing of beet into biofuels. This lack of information (or activity) means that seed companies are unwilling to invest and breed beet specifically for biofuel until the technical requirements of such beet are identified.

iii) Agronomy

Without changes to the beet plant itself, modifications of agronomic practices are likely to lead to only minor changes in yield[12] and it is thought that the requirements are likely to be little different from those for sucrose production. However, changes in agronomy, particularly crop nutrition, can alter the composition of sugar beet roots (for instance, extra nitrogen will increase amino-nitrogen in the root and reduce sugar concentration). If alternative compositions of sugar beet for biofuel production could be identified, then changes to agronomy might provide an answer. At present this is not an area of active research.

Most developments should be focused on the roots and crown rather than the leaves because the leaves contain a high proportion of potassium and phosphorus, as well as some nitrogen and other nutrients. If these are removed from the field, they need to be replaced. This may require purchase of fertiliser and, even where a farm has its own organic manure available, this still entails use of energy to transport and spread it. Also, at present the yield of biofuels from tops is much lower (12%) compared to roots (see above).

6.8 Why Use Sugar Beet – Pros and Cons

Positive attributes include the good agronomy research and transport infrastructure that is in place for beet production in most countries. This could speed up adoption, particularly in areas where beet is already grown.

Beet is seen in many countries as a good break crop in rotations, especially those that are dominated by winter (autumn)-sown combinable crops (*e.g.* the UK). However, there are some misgivings as to the reduced yields of crops following beet being harvested under wet soil conditions. The fact that beet is a spring-sown crop in northern Europe and has a relatively open canopy early in its growth allows birds better access to the soil and this is seen as a particular benefit in the UK. It also has a range of other environmental benefits.[21,22]

Although yields are reduced by drought, the tolerance to drought that beet does have makes it more suitable than many other crops for production on light soils. It is also relatively salt tolerant and this attribute is important in some areas (*e.g.* India) where beet is produced for biofuel production. The relatively low requirement for nitrogen and other inputs compared to rape, wheat and corn is also a benefit.

The wide range of other products that can be obtained from sugar beet, plus the efficient use of all materials produced in the process, means that sugar beet could be one of the higher yielding and higher value (per hectare) crops of the future, which will be important as agricultural land becomes even more scarce. The options for development of sugar beet, like other crops, have not been fully investigated. There are projects focusing on unraveling the beet genome, such as the GABI beet collaborative project coordinated by the Max Planck Institute at Potsdam in Germany, which should provide fruitful information for future developments.

The main negative attribute of beet for biofuel production is the difficulty of producing the crop all the year round. The limited season and difficulties and expense of long-term storage and transport are the main reasons that wheat and corn are seen as more viable alternatives.

6.9 What Could Be Used in the Future?

Whilst sugar beet leaves could be used for biofuel production in the future, they have a different structure and composition from the roots and crown, so that fermenting processes may not be as efficient. However, the main reason for not using leaves (as mentioned above) is that the leaves are a good source of nutrients when left on the field and, given the potential shortages of these in the future, leaves are probably best not included as a target for biofuel production.

The crown is generally less than 10% of the total root but does contain a slightly different composition of chemicals compared to the main root. As removing the crown with the root is likely to be the most efficient means of harvesting, investigation of the potential exploitation of the crown would be useful. However, the main focus of research should be the roots. Unfortunately, there is relatively little information in the public domain on research being carried out to improve beet for bioethanol production. Part of this is because of the tendency of processing companies to keep their own research to themselves in order to gain an economic advantage over their competitors. In addition, the basic fermentation processes and the relatively low profit margin on beet for biofuel means that there is less incentive to invest in research compared with beet for sucrose.

References

1. J. Sneep, B. R. Murty and H. F. Utz, In: J. Sneep, A. J. T. Hendriksen, O. Holbek (ed.), *Plant Breed. Pespectives*. PUDOC, Wageningen, 1979, 203–215.
2. A. S. Marggraf, In: *Histoire de l'Académie Royale des Sciences Belles Lettres*, Haude, Berlin, 1749, 79–90.
3. F. C. Achard, *Ausführliche Beschreibung der Methode, nach welcher bei der Kultur der Runkelrübe verfahren werden muß*, 1799. C.S. Spener, Berlin (reprinted Akademie-Verlag, Berlin, 1984).
4. D. L. Doney and J. C. Theurer, *Crop Sci.*, 1984, **24**, 255–257.
5. R. Sevenier, R. D. Hall, I. M. Van der Meer, H. J. C. Hakkert, A. J. Van Tunen and A. J. Koops, *Nature Biotech.*, 1998, **16**(9), 843–846.
6. R. K. Scott and K. W. Jaggard, In: D. A. Cooke, R. K. Scott (ed.), *The Sugar Beet Crop*, Chapman & Hall, London, 1993, 179–237.

7. C. J. Bell, G. F. J. Milford and R. A. Leigh, In: E. Zamski, A. A. Schaffer (ed.), *Photoassimilate Distribution in Plants and Crops: Source-Sink Relationships*. Marcel Dekker Inc., New York, 1996, 691–707.
8. G. F. J. Milford, In: A. P. Draycott, (ed.), *Sugar Beet*, Blackwell Publishing, Oxford, 2006, 30–49.
9. A. R. Werker and K. W. Jaggard, *Agric. For. Met.*, 1998, **89**, 229–240.
10. Anon, *World Sugar Beet Year Book*, Agra Informa Ltd, Tunbridge Wells, 2009, pp. 134.
11. K. W. Jaggard, A. Qi and M. A. Semenov, *J. Agric. Sci.*, 2007, **145**, 367–375.
12. J. D. Pidgeon, A. R. Werker, K. W. Jaggard, G. M. Richter, D. H. Lister and P. D. Jones, *Agric. For. Met.*, 2001, **109**, 27–37.
13. G. James (ed.), *Sugar cane*, CAB International, Oxford, 2004.
14. J. M. McGrath and R. T. Lewellen, *Crop Sci.*, 2004, **44**, 1032–1033.
15. M. Hutnan, M. Drtil, J. Derco, L. Mrafkova, M. Hornak and S. Mico, *Polish J. Env. Studies*, 2001, **10**(4), 237–243.
16. L. Haraldsson, *Master thesis Swedish University of Agricultural Sciences, Dept. of Microbiology*, 2008, **4**.
17. P. A. Scherer, S. Dobler, S. Rohardt, R. Loock, B. Buttner, P. Noldeke and A. Brettschuh, *Water Sci. Technol.*, 2003, **48**, 229–233.
18. J. C. Theurer, D. L. Doney, G. A. Smith, R. T. Lewellen, G. J. Hogaboam, W. M. Bugbee and J. J. Gallian, *Crop Sci.*, 1987, **27**, 1034–1040.
19. J. Tzilivakis, D. J. Warner, M. J. May, K. A. Lewis and K. W. Jaggard, *Agr. Syst.*, 2005a, **85**(2), 101–119.
20. K. W. Jaggard, C. J. A. Clark, M. J. May, S. McCullagh and A. P. Draycott, *A. P. J. Agr. Sci.*, 1997, **129**, 287–301.
21. J. Tzilivakis, K. W. Jaggard, K. A. Lewis, M. J. May and D. J. Warner, *Agr. Ecosystems Environ.*, 2005b, **107**(4), 341–358.
22. CIBE, CEFE, *Environmental report – beet growing and sugar production in Europe*. Paris, 2003, pp. 40.

CHAPTER 7
Oilseed Rape

RICHARD WEIGHTMAN,[a] PETER GLADDERS[a] AND PETE BERRY[b]

[a] ADAS Boxworth, Boxworth, Cambridgeshire CB23 4NN, UK; [b] ADAS High Mowthorpe, Duggleby, Malton, North Yorkshire YO17 8BP, UK

7.1 Introduction

Oilseed rape (*Brassica napus* and others) is the principal oilseed crop of temperate zones and in 2007 accounted for almost 10% of the world's principal oilseed crops by weight and about 13% of vegetable oil production (FAOSTAT 2009; all area and production data in this chapter are from FAO unless otherwise stated). Production within warmer regions tends to be limited to higher altitudes and species other than *B. napus* are often cultivated. Globally, China, Canada and India produce nearly 60% of the total production, although with yields per hectare of 1.6, 1.5 and 1 t/ha respectively, this high production requires a little over 18 million hectares (Mha) of land. Such is the predominance of these major producers the average world yield per hectare was only 1.62 t/ha over the period 1998–2007. Higher yields are achieved in many European countries, *e.g.* in recent years production in the UK has been around 1 500 000 tonnes from 500 000 ha giving average yields of around 3 t/ha. Although seed yields are low compared with other crop species the amount of energy within oilseed rape crops is of a similar magnitude with many other species due to its high energy, oil-rich seed and the relatively large amount of nonseed biomass (Table 7.1). This indicates that oilseed rape species could become an even more important energy crop given that this species has generally received less breeding and research effort than other crop species.

RSC Energy and Environment Series No. 3
Energy Crops
Edited by Nigel G Halford and Angela Karp
© Royal Society of Chemistry 2011
Published by the Royal Society of Chemistry, www.rsc.org

Table 7.1 UK average yields and crop energy contents.

	Oilseed rape	Winter wheat
[a]Seed yield (t/ha)	2.8	6.7
[a]Straw yield (t/ha)	[b]6.5	[b]6.7
[c]Seed energy content (GJ/ha)	69	114
[c]Straw energy content (GJ/ha)	111	114
Energy content of whole crop (GJ/ha)	180	228

[a]All biomass yields expressed at 100% dry matter.
[b]Based on harvest indices (ratio of seed weight : total crop weight) of 0.3 for oilseed rape and 0.5 for wheat.
[c]Based on energy content of 24.7 MJ/kg for seed of oilseed rape and 17.0 MJ/kg for all other tissues.[2]

The group of Brassicae that have given rise to the main oilseed rape crops is quite diverse. The four widely cultivated oilseed species are *B. napus* (Swede rape), *B. rapa* (Turnip rape), *B. juncea* (Indian mustard) and *B. carinata* (Ethiopian mustard). The genetic basis of the primary species and the amphidiploids that arose from crosses were described by Downey and Rimmer.[3] *Brassica rapa* ($n = 10$: A), *B. nigra* ($n = 8$; B), and *B. oleracea* ($n = 9$; C) are the primary species, whilst *B. juncea* ($n = 18$; AB), *B. napus* ($n = 19$; AC) and *B carinata* ($n = 17$; BC) are the amphidiploid crosses. *Brassica napus* dominates global production and is predominant in Europe and China where the winter types are most common. In Canada and cooler northern latitudes winter or summer types of *B. rapa* are grown, or summer types of *B. napus*. Indian mustard (*B. juncea*) is grown in India and parts of China, whilst summer types of *B. napus* are grown in Australia where water is frequently limiting. Genetically modified (GM) crops are grown in Canada and China, but their products are not accepted by some markets, including Europe.

Following the introduction of the oilseed rape crop the market has demanded low erucic acid and low glucosinolate variants that meet the requirements of both the food and animal-feed markets. These so-called double-low varieties have been segregated from the high erucic types grown for the lubricant market. The widespread production of the double lows throughout the EU provided the basic feedstock for the production of rape methyl ester (RME) when the oil was treated with methanol. The product worked well in diesel engines and this biodiesel helped set the basic standards of the current fatty acid methyl ester (FAME) European standard EN 14214. However, such a standard can perhaps be viewed as a starting point in the development of biodiesel; the EU-funded Bioscope[4] project investigated six compositional parameters of biodiesel as the starting point for improving the standard. The oilseed rape crop is now one of many feedstocks being used to produce biodiesel, and can command a higher price than methyl esters from other oilseeds. Furthermore, material sourced from several EU countries can currently offer processors and end users good assurance documentation demanded by climate-change-mitigation programmes (*e.g.* UK Renewable Fuels Agency [RFA] 2009[5]). Recent data from the RFA show 42% of biodiesel feedstock comes from cropland, 25% from byproducts and 33% from unknown resources.[5] Oilseed rape productivity therefore has a very significant role to play in increasing the efficiency of biodiesel production and limiting land-use change that may be associated with biofuel production.

High yields, obtained within an overall carbon-efficient production process, are key to oilseed rape making a useful contribution to greenhouse-gas (GHG) reduction. Viewed globally, the highest yields per hectare are achieved in Western Europe, with Benelux countries and Germany averaging between 3.3 and 3.6 t/ha. The UK average yield between 1998 and 2007 was 3.1 t/ha, varying between 2.6 and 3.4 t/ha. Reviewing the yield of oilseed rape and constraints on its improvement, Berry and Spink[6] highlighted the relative lack of yield improvement in the UK and major EU producing countries in recent years. The scope for improving yields across the globe in the context of carbon-efficient biodiesel production is the theme of this chapter, which is essential if oilseed rape is to make a significant contribution as a low-carbon biofuel feedstock.

7.2 Current Crop and Use as Biofuel Feedstock

7.2.1 Area Yield and Production

Oilseed rape, in many world markets referred to as Canola or Colza, is the ninth most common crop in the world, occupying 2.2% of the cropped area. Although oilseed brassicas have been grown for many years, interest in oilseed rape started to develop globally and in the UK in the 1960s. Figure 7.1 shows

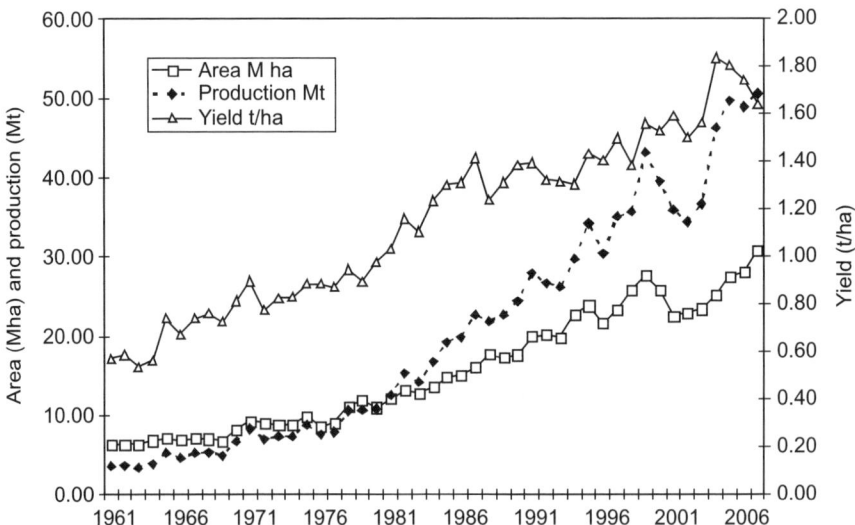

Figure 7.1 World area (Mha), production (Mt) and yield (t/ha) of oilseed rape (1961–2007).

that for most of the 1960s the world area of oilseed rape was relatively steady and was followed by increases at the end of the 1960s and again at the end of the 1970s.

From the start of the 1980s, the global production of oilseed rape has increased rapidly, as a result of an expanding crop area and greater yields per hectare (yields having increased linearly by 26 kg/ha since 1961; Figure 7.1). However, looking at the nine largest producers (Figure 7.2), this yield trend can be seen to have two different groups. The European countries of Germany, France, and the UK have yields increasing from about 1.5 to 3.5 t/ha, whereas the major global producers, China, India, and Canada, show increases from 0.25 to about 1.75 t/ha.

Although there is a lot of year-to-year variation in both groups, the rate of yield increase has been slower in the major producing countries. It has also been shown that the rate of yield increase has slowed in many European countries.[6] Climate and soil type will play a large part in determining final yields, whilst the application of technology to the production process gives future scope for yield improvements. The approaches to this are discussed in Section 7.5.

The principal countries producing oilseed rape are listed in Table 7.2, and the areas under the crop are given along with the yield per hectare and total production for 2007. The 2007 crop would have been used to produce 2008 biodiesel, which is also listed. However, the biodiesel production includes output from all oil sources, including crops, tallow and used cooking oil. Between 2002 and 2007, when the volumes of biodiesel production were

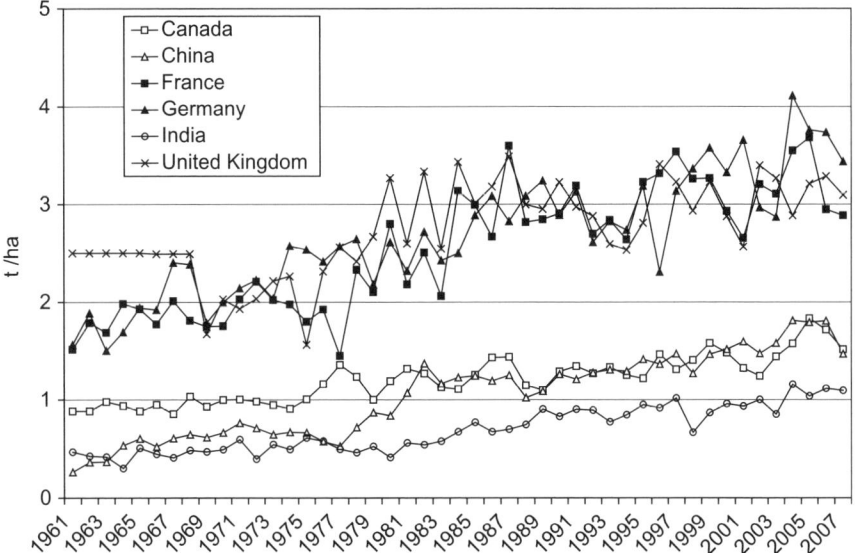

Figure 7.2 Oilseed rape seed yield in six of the main producing countries 1961–2007 (t/ha).

Table 7.2 Crop area yield and production data for oilseed rape and the biodiesel production (2008) and potential production capacity as of 2009.

	Area Harvested (Mha)	Yield (t/ha)	Production (Mt)	Biodiesel production (kt)	2009 biodiesel production capacity (Ml) (EU states only)
China	7.05	1.47	10.38	*93*[a]	
India	6.79	1.10	7.44	*81*	
Canada	6.28	1.52	9.53	*35*	
France	1.58	2.89	4.55	1815	2505
Germany	1.55	3.44	5.32	2819	5200
Australia	1.06	1.00	1.07	80	
Ukraine	0.80	1.31	1.05	150	
Poland	0.80	2.67	2.13	275	580
United Kingdom	0.68	3.10	2.11	192	609
Russian Federation	0.53	1.18	0.63	61	
United States of America	0.47	1.40	0.66	*1453*[c]	
Czech Republic	0.34	3.06	1.03	104	325
Denmark	0.18	3.33	0.60	231[b]	352[b]
Total	28.11		46.50	7197	8637
% of world production			92		

[a]Used oil;
[b]DK and S;
[c]Soy feedstock
Crop production data from FAO for harvest 2007, biodiesel data from EBB (2009)[8] for production in 2008, and Fischer et al. (2009).[7]

increasing, the world oilseed rape area climbed by 7.8 Mha, and the European rapeseed price increased by $204 per tonne.[1] Fischer et al. (2009)[7] suggest 95% of the current growth in demand for vegetable oils within the EU is due to biofuels. Estimates of the global biodiesel production vary between sources. The European Biodiesel Board[8] estimate that the global production of biodiesel in 2008 was around 12 million tonnes (Mt), with the EU accounting for about 65% of the total. If all the 2008 biodiesel production was from oilseed rape it would have needed about 9.3 Mha at an average yield of 3.1 t/ha or 17.0 Mha at the world average yield of 1.62 t/ha. To produce the EU target of about 33 Mt of biodiesel by 2020 entirely from oilseed rape will require 25 Mha at 3.1 t/ha, or 49.2 Mha at 1.62 t /ha. The latter area is just under 4% of the global arable area. The production of the 33 Mt of biodiesel would also produce 45.5 Mt of high protein oilseed rape meal.

7.3 Utilisation and Economics

7.3.1 Current Use for Food, Feed and Fuel

Oilseed rape is used for the production of oils for use in cooking and the food sector generally, whilst specialist oils are used as lubricants and slip agents. To produce biodiesel, rapeseed oil and other vegetable oils used for biodiesel production are transesterified with monohydric alcohols to give the corresponding monoalkyl esters. Rapeseed oil treated with methanol produces a mix of methyl esters that depend on the fatty acids that make up the oil. The effects of the oil composition on this process are discussed in Section 7.4. The cake from food oils left after extraction is used in livestock feed as a protein meal. In UK oilseed rape variety trials, oil contents of around 44 g/100 g are recorded,[9] whilst the total oil extracted, averaged across all world production in 2007, was equivalent to a net extraction rate of 34 g/100 g (total rapeseed oil/total seed production FAOSTAT).

The global world rapeseed oil production was 17.2 Mt in 2007 (FAO PRODSTAT), the year for which the IEA[10] estimated biodiesel production from the world's 21 leading producers at 11.45 billion litres (10.1 Mt). This was estimated by Bacovsky[10] to have dropped back a little to 10.9 billion litres (9.5 Mt) in 2008, which is less than the estimate by EBB (2009)[8] and reflects the uncertainty of biodiesel production rates. If a global production of 10 Mt is assumed and the rape oil inclusion is 18% of the total oil mix used for biodiesel production as reported for the UK (RFA, 2009),[5] then 1.8 Mt of biodiesel would be produced from about 1.9 Mt of rapeseed oil, *i.e.* about 11% of world rapeseed oil would have been used for biodiesel production. More comprehensive data on the provenance of treated oils will be needed before an accurate assessment can be made of the amount of oilseed rape used and the carbon inputs used in its production (see Section 7.3.2). Soya, palm oil, oilseed rape, tallow and used cooking oils and others all contribute to biodiesel feedstock, preference arising though price, local production and import/export strategies.

Crop-yield levels and oil-extraction rates will underpin the demands made on global resources for biodiesel production. If plant output of biodiesel per tonne of rapeseed oil is 0.95 (RFA, 2009)[5] Table 7.3 shows how crop yield per hectare and the percentage oil extraction affect how much of the world crop area would be needed to provide 100% of the 11.45 billion litres of biodiesel produced in 2007. Clearly this overstates the area as 100% of production is unlikely to come from one crop type.

At oil contents below 30 g/100 g of seed and yields of only 1 t/ha, the crop area would need to expand by more than 15%. At the current global yield and net extraction rate, between a half and two thirds of the world oilseed rape crop area would be needed to produce 100% of the present biodiesel production.

Forecasting the demand to produce 10% of all transport fuels by 2020 is more difficult; the proportion of rapeseed oil used may vary; vehicles are becoming more fuel efficient and although their replacement cycle for the UK is just under 15 years, many of the cars and lorries in use by 2020 will be more fuel

Table 7.3 Area (M ha) required for 100% of current biodiesel production as a proportion of current world oilseed rape area.

Oilseed rape yield (t/ha)	Oil content (g/100 g of seed)				
	25	30	35	40	45
1	**1.38**	**1.15**	0.98	0.86	0.77
1.5	0.92	0.77	0.66	0.57	0.51
2	0.69	0.57	0.49	0.43	0.38
2.5	0.55	0.46	0.39	0.34	0.31
3	0.46	0.38	0.33	0.29	0.26
3.5	0.39	0.33	0.28	0.25	0.22
4	0.34	0.29	0.25	0.22	0.19

Table 7.4 Area (Mha) needed for biodiesel as proportion of current world oilseed rape area assuming 18% of biodiesel feedstock is from oilseed rape and 10% of diesel fuel from biodiesel.

Oilseed rape yield (t/ha)	Oil content (g/100 g)				
	25	30	35	40	45
1	0.84	0.70	0.60	0.53	0.47
1.5	0.56	0.47	0.40	0.35	0.31
2	0.42	0.35	0.30	0.26	0.23
2.5	0.34	0.28	0.24	0.21	0.19
3	0.28	0.23	0.20	0.18	0.16
3.5	0.24	0.20	0.17	0.15	0.13
4	0.21	0.18	0.15	0.13	0.12

efficient; the gasoline: diesel mix of the car stock may change, and not all countries are agreed on a 10% biofuel inclusion rate like the EU.[11] China has set itself a 15% target. It is conceivable that diesel volume demand in 2020 could be similar to the present if vehicle efficiency improvements offset some of the increase in vehicle numbers. If the world biodiesel inclusion rates average 10% overall, and oilseed rape continues to provide 18% of biodiesel feedstock, and the volume of diesel used remains the same as in 2006,[12] then the figures in Table 7.4 show the proportion of the current oilseed rape area needed to meet these targets at the different crop yield and oil extraction percentages.

As mentioned above, the current biodiesel production is using about 11% of world rapeseed oil output. By 2020, if the world crop yields and oil contents are increased to slightly above the current European levels (lower shaded box), then the 10% target could be met with little need to increase the crop area. However, if the average world yield and oil contents remain the same (upper shaded box), the 10% biodiesel inclusion rate would require between 30% and 47% of the oilseed rape area. Given that soy and palm oil, on the same basis as these assumptions, would be contributing 34% and 10% of production, respectively, this high level of rapeseed oil consumption would lead to severe

Oilseed Rape

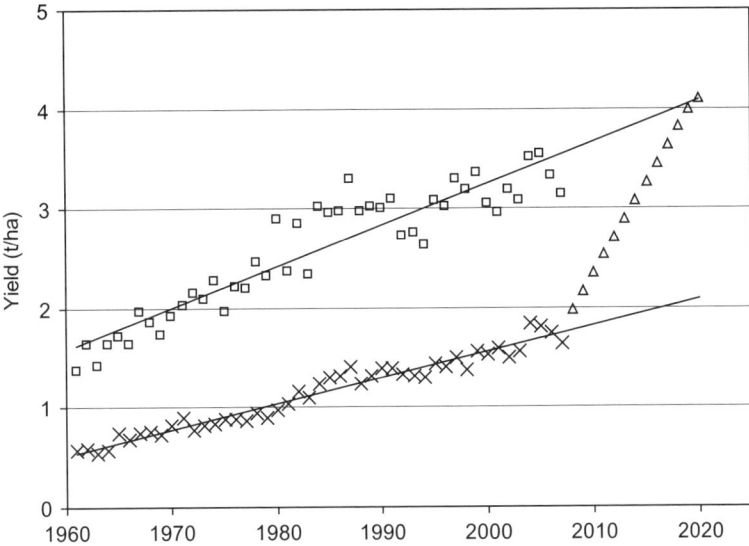

Figure 7.3 Linear best-fit regressions lines of the yield per hectare from 1961–2007 for the World (×), and Germany, France and UK (□) and the slope of interpolated figures for the required global yield increase to meet the 2020 10% biofuel inclusion target based on Table 7.3 (Δ).

market disruption. Improvements in production are clearly needed if targets are to be met. The highest-yielding variety on the current UK HGCA Recommended List of winter oilseed rape varieties yields over 5 t/ha and produces 2.4 t/ha of rapeseed oil.[9] This suggests that there is some scope for meeting these targets from genetic resources that are already available. But the yields of cultivars in trials are well above the net national outputs that affect the markets and supply and demand-based decisions.

Figure 7.3 shows the linear yield trends and linear best-fit regression lines from 1961 to 2007 for two groups, the three main EU producers and all world producers. The trends are projected to 2020, when the 10% target has to be met. If the yield levels in Table 7.4 that are associated with the percentage crop off-take of (0.12 to 0.16) shown in the lower right-hand cells are to be achieved, the necessary yield increase in world production is shown by the interpolated line (Δ). This rate of increase is about 4 times greater than that seen in the EU since 1961 and nearly 7 times greater than that seen across the global supply chain.

7.3.2 Global Trade and Assurance

Although the principal oilseed rape biodiesel producers shown in Table 7.2 produce significant tonnages of crop, the demands of fuel production are only evident in the import figures for Germany, which imports 1.8 Mt of oilseed rape

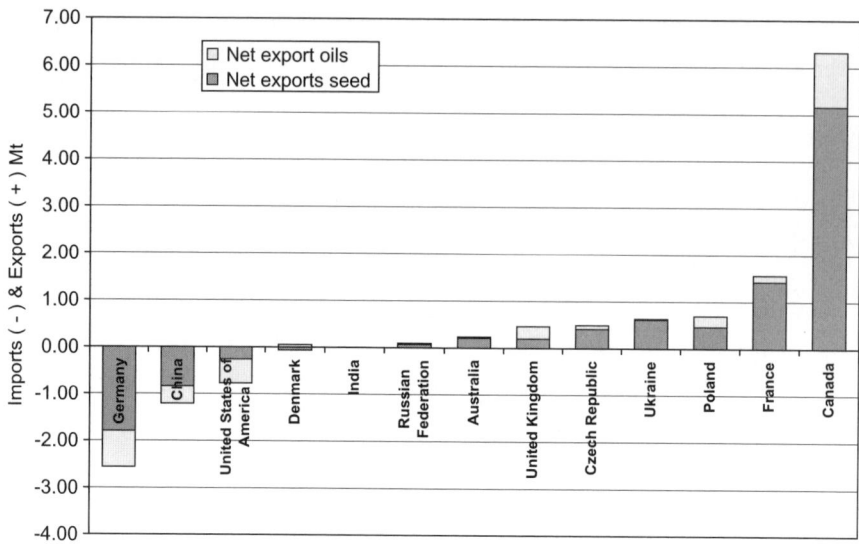

Figure 7.4 The net exports/imports of rapeseed and rapeseed oil for the principal crop and biodiesel-producing countries.[13]

and 0.8 Mt of oil, sufficient for just under half its total biodiesel production (Figure 7.4).

Most of the exports are as seed. The United Kingdom is the only country in those shown where the principal component of trade is a rapeseed oil rather than seed. The Netherlands, with its large crush capacity, also exports more oil than seed. China has a net import of rapeseed and oil of over 1 Mt. With a very small biofuel output, most is used for food. India, with the world's second largest area of crop, has production and consumption almost in balance, with a few hundred tonnes exported. Given the annual variation in production and the latent demand from a large and rapidly growing population, the Indian import/export balance is a key feature of world supply and demand. Canada, the world's second largest producer by crop output is by far the main supplier to world markets. With a population of just over 33 million, domestic demand is an order of magnitude lower than that of India and China.

Soy and palm oil are oilseed rape's main competitors on world markets. Both are grown in warmer, tropical and subtropical climates and their producers are under pressure to show environmentally sustainable production and compliance with the Roundtable on Sustainable Palm Oil[14] and the Basel Criteria of Responsible Soy Production,[15] which set out guidance on how crops, the environment and social issues should be managed. Oilseed rape crops, in contrast, are grown in many countries where established assurance schemes are in operation to meet existing market demands. The Assured Combinable Crops Scheme (ACCS) in the UK now complies with the RTFO meta-standard for biofuels[16] and in due course more organisations like GLOBALGAP and others will support assurance standards that add value to the crop at point of sale.

7.3.3 Prices and Crop Gross Margins

Over recent years, despite declines in profitability, autumn-sown oilseed rape has been the most profitable break crop grown in cereal rotations in the UK. The predictive budget figures in Figure 7.5 show how margins for OSR production declined over two decades from 1987, so much so that the net margins showed a loss by 2007. This decline mirrors the decline in rapeseed prices during the 1990s shown in Figure 7.6, alongside a less steep decline in wheat prices. The short-term peak in energy prices and the economic changes at the start of the global financial crisis of 2008–2009 caused a sharp rise in commodity prices alongside petroleum and many other commodities. Whatever underlying cause was behind this very widespread economic blip, it was sufficient to return net margins to profit, but did not return gross margins to 1987 levels. In the UK, oilseed rape gross output per hectare was, nevertheless, larger than all combinable crops except wheat.[17] Fertiliser is the main financial and energy cost, and increased from 0.51 to 0.66 of the total variable costs from 1987 to 2009, partially as a result of the 2008/9 increase in fossil energy prices, the principal feedstock for nitrogen (N) fertiliser production.

The future profitability of oilseed rape for the grower clearly relies on how commodities in general, and specifically oilseed rape, respond in price to rising energy costs. The effective management of fertiliser inputs, particularly N, will play a major role and this is discussed in Section 7.6. The future profitability of

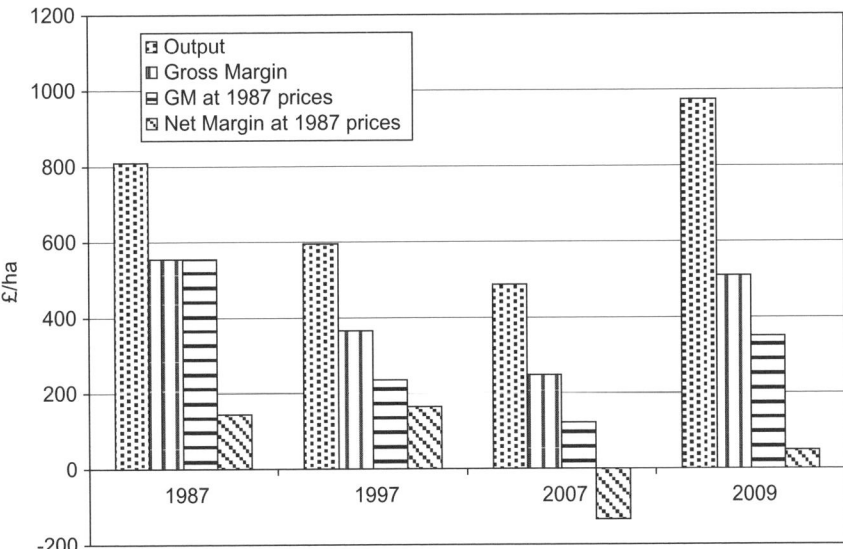

Figure 7.5 The output and gross margin (GM) per hectare for 1987, 1997, 2007 and 2009, and the GMs and net margins per hectare at 1987 prices for oilseed rape crops in the UK (data from The Farm Management Pocketbook, Dept of Agricultural Economics, Wye, and The Andersons Centre, Melton Mowbray).

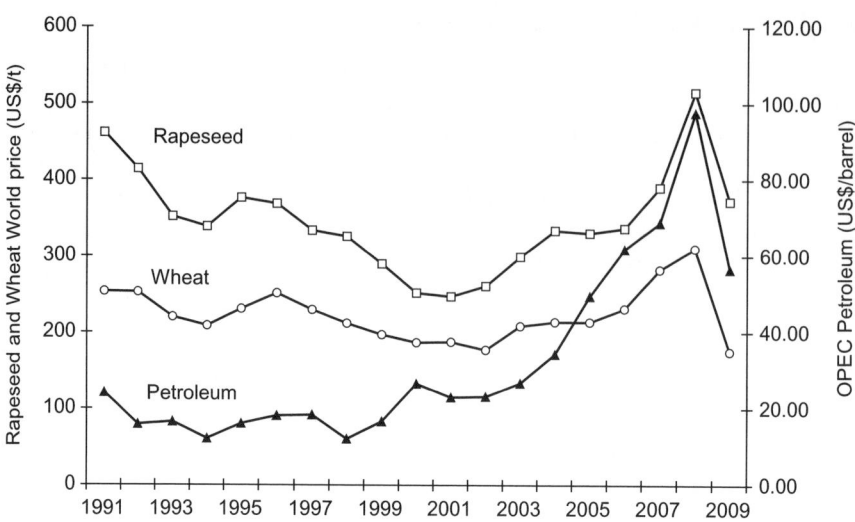

Figure 7.6 World oilseed rape, wheat and petroleum prices 1991–2009 (US$/t or barrel).

biodiesel production relies on the cost of feedstocks, but only a small percentage of the world's oilseed rape production is used for biodiesel, so the majority of market movements are a reflection of demand in the food market. A sensitivity analysis by Booth et al.[18] showed production costs of biodiesel to change by 1.8 p/litre for every £10 change in oilseed rape costs up to £170 per tonne. At the time of writing this would have put production costs at about 55 p/l, but at the peak prices of 2008 costs for biodiesel made exclusively from oilseed rape would have risen to 75 p/l or more. If sold as a 100% biodiesel fuel with the prevailing duty, margin and value added tax (VAT), this would have needed a pump price in excess of £1.35/l. However, being sold as a minority blend with petroleum-based-diesel cushions the impact of high food commodity prices on the final pump price.

7.3.4 Revenue from Support Measures

The decoupling of farm-support measures across the EU in theory allowed market forces to determine margins and consequent levels of production. The supply of potential foodstuffs for biofuel produces an interesting dilemma. If food supplies are to be adequate it is essential that the food market prices provide effective market signals to retain adequate supplies, otherwise edible rapeseed oil moves to the fuel market, resulting in possible food shortages. If production is in "surplus" as judged by the food market, prices will be low and the excess will provide cheap feedstocks for biofuel production. Yet, as Figure 7.3 shows, if biofuel supplies are to be met the rate of yield increase between 2010 and 2020 has to be about 4 times that previously achieved. The above market scenario of a relatively well-supplied food market will not incentivise the increases in crop yield required to meet the 10% transport fuel targets set out in the renewable energy directive (RED). Specific support measures from the biofuel market that may incentivise this increase in productivity could be the passing of payments from renewable fuel certification schemes back down the supply chain to producers of assured crops with proven GHG savings in their production. Low fertiliser nitrogen use in the production process (see Section 7.6) is likely to play a major role in securing the highest level of assurance for rapeseed.

7.3.5 Coproducts: Rapeseed Meal, Glycerol and Straw

In addition to the production of liquid biofuels the process of making biodiesel from oilseed rape also produces a significant supply of coproducts, rapeseed meal and glycerol. The nutritional quality of rapeseed meal (RSM), the principal coproduct of biodiesel production, has been considered extensively, since the oilseed rape crop was first developed in the 1970s (*e.g.* Wareham[19]). Recently, Cottrill *et al.*[20] reviewed the relevant literature in relation to the potential expansion in biodiesel production in the UK, and the impact of additional biodiesel coproducts on the market. Given that the type of oilseed rape that will be used for biodiesel production is likely to be the same as that used for the food market (*i.e.* canola or "double-zero" types), the RSM coproduct from crushing the seed is generally of high quality, with low erucic acid content and low glucosinolates. The rapemeal contains 320–400 g protein/kg, is considered to be well balanced in terms of amino acid composition, and is used widely as a protein feed in animal diets. Ewing[21] has suggested maximum inclusion levels of 250–300 g/kg for RSM in concentrates for a range of ruminant livestock types. As well as glucosinolates, other antinutritional factors are present, such as phenolic compounds and phytic acid. The fibre content can also give problems in monogastric animals, meaning that inclusion levels in the diet are generally limited. Typical inclusion limits for nonruminant diets are shown in Table 7.5.

It can be argued that a large proportion of RSM on the market will be incorporated at higher rates in livestock rations if the price is low enough, and one consequence of this would be that nitrogen and phosphorus excretion from animals may increase. Other options are available to mitigate some of these effects. For instance, phytase enzymes can be added to the feed[30,31] or, alternatively, engineered into the plant through a GM approach. Phytase serves to break down phytic acid and make the phosphorus more available to the animal.[32] Carbohydrase enzymes added to the feed can to a certain extent be used to improve the fibre digestibility.

Cottrill *et al.*[20] suggested that for each extra 150 kt of rapemeal produced associated with biodiesel production, there will also be 10 kt of glycerol. There are existing markets for glycerol (also called glycerine) in cosmetics, for example, but the expected production from biodiesel production will saturate this market and the most likely end use for the majority is either direct

Table 7.5 Recommended rates of inclusion of RSM in pig and poultry diets from published scientific literature (from Cottrill et al., 2007).[20]

Species	Glucosinolate content (μmol/g)	Rate of inclusion (g/kg)	Reference
Pig			
Weaner	4.0	250	King et al.,[22]
	10.5	150	Mullan et al.,[23]
Grower	2.2	260	Roth-Maier et al.,[24]
Finisher	0.9	108	Roth-Maier et al.,[24]
Grower/Finisher	4.0	300	King et al.,[22]
Gilts	6.6	120	Opalka et al.,[25]
Lactating Sows	4.0	202	King et al.,[22]
Poultry	NR	100	McNeill et al.,[26]
Broilers	14.3–16.4	100	Fasina and Campbell,[27]
	15.4	150	Roth-Maier,[28]
	NR	<100	Mawson et al.,[29]
Layers	15.4	150[a]	Roth-Maier,[28]

NR = not reported in the study.
[a] For layers producing white-shelled eggs.

combustion for energy or feeding to livestock. Glycerol is a high-energy feed (gross energy content 18.1 MJ/kg) that can be fed to both ruminant and non-ruminant animals. There is currently little experience of its use in animal feed, although it is known that appreciable quantities of glycerol can be found already in other biofuel coproducts such as distillers grains (where the glycerol arises as a byproduct of fermentation by brewers yeast[33]). While pure glycerol has a well-defined chemical composition, the byproducts of biodiesel production will be contaminated with varying amounts of methanol, sodium chloride and potassium chloride as well as variable quantities of water, and three grades of glycerol have been reported. In addition to its use as a feed ingredient, glycerol may also be used in compound feed manufacture where it improves pellet quality and dust control. Recommended rates of inclusion of 100 g/kg have been reported for glycerol in diets for pigs[34] and poultry[35–37] but further research is required to determine the maximum inclusion limits for different livestock types. Proposals to process glycerol to methanol or hydrogen have been put forward, but as yet are far from widely incorporated into process plants.

The ratio of seed dry matter to total above-ground dry matter at harvest has been measured at 0.25 to 0.33.[38] This means that a crop with a seed yield of 4 t/ha has between 11 and 14 t/ha of stem and pod wall dry matter. It is impractical to expect all of this dry matter to be recovered because crop stems are cut several centimetres above ground level and some of the plant material will be small and therefore difficult to collect. If it is assumed that 50% of the nonseed dry matter is recoverable, then this indicates that 5 to 7 t/ha of plant material could be available for use as a source of energy or for other uses (e.g. livestock bedding or building material). Currently, the nonseed biomass is collected on only a small proportion of oilseed rape crops and this therefore represents an area where oilseed rape could be more fully utilised as an energy crop.

7.4 Quality of Oilseed Rape Oil and Other Methyl Esters

Oilseed rape oil is rich in unsaturated fatty acids. There are two main forms of oilseed rape oil in widespread commercial use, with a third high-oleic type just entering commercial production. Early varieties contained a high proportion (0.46) of erucic acid (C22:1) in the total fatty acids, and a low proportion (0.14) of oleic acid (C18:1). These varieties, sometimes called "European varieties", have been superseded for use in the food market by varieties with low proportions of erucic acid in the oil (<0.05) and a high proportion of oleic acid (0.64). Some varieties also have reduced levels of the polyunsaturated fatty acid, linoleic acid (C18:2).[39] These improved varieties were originally known as Canadian low-acid varieties, and so were responsible for the name "Canola". This form is now the dominant type of rapeseed oil used worldwide, although some high erucic acid rape oil is used in industrial (nonfood) applications, where erucamide finds a use as a slip agent in plastic manufacture. New non-GM oilseed rape varieties with altered oil composition are now being developed that are still higher in oleic acid and low in the polyunsaturated fatty acid, linolenic acid (C18:3), hence generating so-called "HO" or "HOLL" (high-oleic–low-linolenic) types, which have up to 0.85 oleic acid in the total fatty acids.[40,41] The reputed advantage for HOLL types is high-temperature stability for food applications such as deep-fat frying. Such types are not yet widely grown but may have a particular benefit for the biodiesel market as, being lower in polyunsaturated fatty acids, they will be less prone to auto-oxidation. This is considered further below. The main disadvantage of HOLL types for biodiesel is currently the lower seed yield, and hence more expensive oil than standard oilseed rape oil.

7.4.1 Composition

During the biodiesel manufacturing process, the triglycerides are saponified to produce glycerol (a coproduct), and a series of individual fatty acids, each esterified to a molecule of methanol. Therefore, the methyl ester composition of a particular biodiesel sample will simply reflect the original spectrum of fatty acids in the native oil. The fatty acids in oilseed rape oil contain a relatively high proportion of oleic acid (0.58–0.62) compared to most other vegetable oils, *e.g.* palm, maize, soya, bean, sunflower (0.40, 0.27, 0.22, 0.23, respectively), close to that of groundnut (0.56), with only olive oil containing a higher proportion (0.73).[39] Given that the cetane number (CN; a measure of the combustion quality of diesel fuel during compression ignition) in the EN 14214 standard is set at a minimum of 51, a combination of methyl oleate, methyl linolate and methyl linolenate found in rapeseed biodiesel with CN values of 57, 38 and 23, respectively, give an acceptable overall cetane value of 51–55.[42]

With vegetable oil for biodiesel production, there is a trade-off with respect to the content of unsaturated fatty acids in that high content gives a low melting ('cold-filter plugging" or "cloud') point, but there is a tendency for unsaturated acids to polymerise under oxidative conditions. Native oilseed rape oil has a low cloud point ($-10\,°C$), which is lower than both palm and groundnut oils ($+35$ and $+3\,°C$, respectively), but higher than sunflower and soya oil (-17 and $-16\,°C$, respectively).[39] Similarly, since biodiesel manufactured from canola-type rapeseed oil will contain principally methyl oleate, methyl linolate and methyl linolenate esters with melting points of -19.5, -35 and $-52\,°C$, respectively,[42] clouding is very unlikely to be a problem in most road-transport applications. The biodiesel standard EN14241 also defines a range for kinematic viscosity (the ratio of the dynamic viscosity to the fluid density) of 3.5–$5.0\,mm^2/s$, which the three esters above comply with easily but methyl erucate from industrial oilseed rape oil would fail ($7.33\,mm^2/s$).[42]

Polymerisation can occur at ambient temperatures when an oil is stored for any length of time, but is of course accelerated under the high-temperature conditions experienced inside the vehicle engine, and so over time certain methyl esters with high levels of unsaturated fatty acids can crosslink and form gums inside the engine. Therefore, limits are set on the levels of unsaturated fatty acids, described in terms of iodine values (mass of iodine in g consumed by 100 g of oil). The EN14214 standard has a maximum iodine value (IV) of 120. Palm, olive, peanut and oilseed rape all fall below this limit (IVs of 54, 81, 93 and 98, respectively), whereas sunflower, and soybean fail to pass the standard with IVs of 125 and 130, respectively. Knothe[42] argues that the IV is a poor indicator of quality because it is only a crude measure of the degree of unsaturation of a sample and is affected by molecular weight of the esters (*i.e.* if ethyl rather than methyl esters had been manufactured), and prefers to use oxidative stability. Nevertheless, set against EN14214, rapeseed oil is an excellent feedstock for biodiesel production in terms of balancing a low cloud point and a medium-low iodine value, being the only one of the major vegetable oils to pass both criteria. When blending oils, quality issues associated with both their original make up and their blended values need to be considered, especially where the changes are nonarithmetic.

It is often suggested that to avoid competing with food uses and possibly from a GHG savings point of view as well, that waste cooking oils should be used for biodiesel production. However, it should be recognised that oil that has been heated to high temperature and become partially hydrogenated and polymerised will tend to have a higher cloud point, and then may fail to meet the EN14214 standard.

7.4.2 Genetic and Environmental Effects on Fatty Acid Composition

While being an excellent nutritional source of oils, and meeting the various biodiesel standards comfortably, oilseed rape oil still has relatively low oxidative stability in high-temperature applications due to its content of polyunsaturated fatty acids, linoleic acid (C18:2) and linolenic acid (C18:3). Replacing these with oleic acid would improve oxidative stability, while also increasing the cetane value of the fuel produced. Various (non-GM) options are available to raise the levels of certain fatty acids in oilseed rape and have met with some success. Mutation experiments started around 20 years ago, and in Germany the first mutants with increased oleic-acid content were identified by Rücker and Robbelen.[43] Genetic studies showed that the mutation was inherited as a monogenic trait, increasing the proportion of oleic acid in the total fatty acids by 0.11.[40] Further studies with doubled haploid populations and mapping of mutated genes showed that the oleic acid desaturase gene was affected, there being two forms in the plant in different tissues.[41] Further selection in mutant lines identified three other minor genes that had to be incorporated into the plant to achieve a proportion of 0.85 oleic acid (higher than that of olive oil), which complicated the breeding of such HO types.

In further experiments, it was found that while oil content in the seed and proportion of oleic acid were positively correlated,[44,45] there was unfortunately a negative correlation between oleic-acid content and seed yield. Möllers[41] reported that the increase in oil content of 0.6–1 g/100 g could not compensate for the reduction in yield (5%). In the meantime, such has been the progress in breeding OSR that the conventional variety Castille has both a higher oil content in the seed than the HOLL variety Splendor (45 vs. 43 g/100 g) and a higher gross output (Monsanto[46]). A second HOLL variety marketed in the UK, V141OL (Monsanto[47]), has an oil content ca. 1 g/100 g higher than Castille, but still underperforms in terms of yield.

As well as genetics, environmental factors change oil content and composition in the seed. Increasing N fertiliser applications tend to decrease oil content slightly by between 0.5 and 1.0 g/100 g of seed per 100 kg N/ha,[48,49] although the higher seed yield means that the gross output of oil per hectare is greater when using N (at the economic optimum). Geographical location has been shown to have the greatest effect on fatty acid profiles, with the proportion of oleic acid varying between locations from 0.760 to 0.806, the lowest oleic-acid content being found at the higher-latitude Scottish site,[49] or from 0.576 to 0.598.[50] Harvesting a crop before it was completely ripe has also been shown to decrease the oleic-acid content slightly.[49] However, such differences are likely to have only minor effects on biodiesel quality.

7.5 Yield Improvement

7.5.1 Estimating the Yield Potential of Oilseed Rape

From the figures in Table 7.2 it is clear that if oilseed rape is to be a significant feedstock for biodiesel then yields must be improved. The scope for improvements in yield were reviewed by Berry and Spink,[6] who estimated that yields of up to 9 t/ha could be a achieved in a UK environment if several crop characteristics could be improved to increase the two main components of yield: seeds/m^2 and seed size, as summarised in Table 7.6.

Table 7.6 Characteristics of crops with the average UK farm yield, potential yield for existing germplasm and with the ultimate potential yield following breeding advances.[6]

	Typical farm crop	Potential for existing germplasm	Ultimate yield potential
Date of midflowering	1 May	23 Apr	23 Apr
Proportion of flower cover	0.60	0.38	0.30
Number of seeds (no/m^2)	80 000	130 000	150 000
Seed filling traits			
Start date	23 May	18 May	18 May
Duration (days)	40	46	50
Light intercepted by green tissue	0.86	0.86	0.86
Assimilate used for pod growth	0.15	0.05	0.05
Contribution of stem reserves to yield	0.05	0.10	0.20
RUE during seed filling (g/MJ)	0.47	0.75	0.88
^{a}Individual seed weight (mg)	3.8	5.0	6.1
^{a}Yield (same oil content at each yield; t/ha)	3.0	6.5	9.2
Water use from start of stem extension (mm)	203	316	393

^{a}Expressed at 910 g/kg dry matter

Seeds/m^2

The size of the sink (defined as seeds/m^2) accounts for 0.85 of yield variation and is determined during a period of about 19–25 days after midflowering.[51,52] In thermal time this period is about 300 °Cd and covers the period in which pod and seed numbers are set. During this period the plants'' ability to intercept optimum radiation levels is affected by the layer of flowers, which, although essential for pod and seed formation, absorb and reflect solar radiation away from the primary photosynthetic capacity of the leaf canopy. Berry and Spink[6] concluded that the optimum fertile pod number was between 6000 and 8000 pods/m^2. Below that level the associated crop canopy size was too small to trap all the radiation and above that level the thickness of the flowering layer reduces the amount of radiation reaching the photosynthetic tissues that in turn reduces the number of seeds/pod and seeds/m^2.

If plant breeders can hasten the flowering date of crops into cooler but frost-free spring days, the number of days required to achieve the 300 °Cd thermal time period over which seeds are set will be increased, as will the radiation receipts. It is estimated that flowering 7 days earlier will increase radiation receipts during the period when seed numbers are being set by about 2%. Further improvements in radiation receipts and potential yield increases may be obtained by reducing the light interception by growing apetalous lines,[53] or by variation in petal size and flower number.[54] Agronomic measures can also be used to reduce the size of the flower canopy by using lower seed rates, applying plant growth regulators and avoiding early sowing.[55] Reducing the flower canopy size by 25% could increase the radiation transmitted to the photosynthetic tissues by a similar amount. Further gains are possible by increasing the leaf to stem ratio, since Major[54] showed leaf material had total solar radiation-use efficiency (RUE) three times greater than stem. Berry and Spink[6] suggest that increases in RUE of about 12% would be possible by this route, and that during the entire seed determination period a 39% increase in photoassimilate available for setting seeds should be feasible.

Seed Growth

Further increases in the production efficiency can be obtained by improving seed-growth processes. The physiological maturation of seeds has been shown to take 715 °Cd, above a base temperature of 4.2 °C, extending the whole yield formation period from late April to early July in normal UK conditions.[51] This produces a seed-filling period of around 43 days. Efforts to extend this period will help to increase seed filling.

Even after accounting for filling the oil-rich seed, the RUE of oilseed rape during seed filling is lower than that of many other C3 crops. A key reason for this is that the pods, which are now contributing to seed fill, have a photosynthetic capacity of approximately 0.60 that of leaves.[56,57] Therefore, prolonging leaf life and maximising the amount of radiation that reaches the leaves are important ways by which RUE may be improved. This may be achieved by producing less-dense crops that do not lodge and are disease free.[58,59] In particular, yield losses due to lodging can be between 16 and 50%;[60,61] therefore lodging resistance is a high priority. Finally, the proportion of yield that results from remobilisation of water-soluble stem carbohydrate is generally regarded to be low in oilseed rape. Therefore, efforts to increase this parameter towards levels of up to 0.3, as occurs in wheat, must be regarded as a priority.

Assessing the effects of RUE, dry matter partitioning and water availability and use led Berry and Spink[6] to produce a conservative estimate for potential seed production in an average UK environment on normal oilseed rape-growing soils of 6.5 t/ha (at 0.91 g DM/kg), by combining the best traits available within current germplasm, and about 6.0 t/ha on low moisture-retentive soils like the sandy loams. To explore the ultimate yield potential of oilseed rape, Berry and Spink[6] considered the effect of breeding new crop traits that would increase seed number to 150 000 seeds/m^2 and create a yield potential of 9.2 t/ha (Table 7.6). However, it should be recognised that water availability would be expected to limit the ability to achieve this yield on less water-retentive soil types in the drier east of the UK, and it is clear that further work is required to improve the rooting and water-use efficiency of oilseed rape to alleviate this constraint.

Although oilseed rape produces less dry matter than cereals per unit of water, its higher energy content means that, as yields increase, energy yields of processed biofuel are proportionately higher than for wheat producing ethanol.

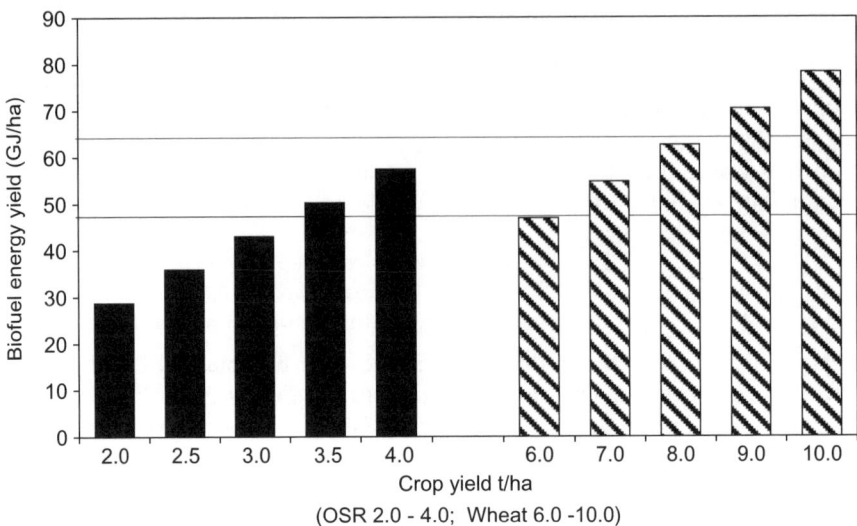

Figure 7.7 Comparison of energy yields from oilseed rape and wheat for low-, medium- and high-yielding crops.

Using yields of biofuel per tonne of input used, the RFA Technical Guidance[5] and an oil content of 0.4 g/100 g shows biodiesel produces 14.4 GJ/t of oilseed rape, whilst bioethanol produces 7.8 GJ/t. However, as Figure 7.7 shows, the total yield per hectare is higher from the cereals due to the higher grain yields. At the average 2008 UK harvest yields of 3.3 t/ha and 8.3 t/ha for oilseed rape and wheat, respectively, the latter produces an additional 17.6 GJ/ha. On this basis, the 6.5 t/ha crop of oilseed rape postulated as potential current production is equivalent to an 11.9 t /ha wheat crop. As Germany and the Netherlands are the only countries that have managed to approach 4 t/ha yields of oilseed rape, the challenges for the higher yields are significant.

7.5.2 Achieving the Yield Potential of Oilseed Rape

A comparison of yield improvements achieved in the UK variety testing system with the yield improvements on UK farms revealed that the gap between the yields achieved in the variety testing system and those achieved on farm had widened since the mid-1980s because farm yields had remained static.[6] Similarly static, or reduced, rates of increase have been observed for oilseed rape yields in other European countries.[6] This is an important observation because it indicates that improved germplasm alone will not necessarily result in greater yields onfarm, and improvements in crop management are also required. Several crop-management challenges that affect yield improvement have been identified and include rotation length, disease, pest and weed control, establishment methods, timing and rate of sowing, and fertiliser use. One of the conclusions from the review of Berry and Spink was that, in the UK, increasing costs of crop inputs relative to oilseed rape prices has reduced the level of crop management used to below that required to achieve the genetic potential of varieties. This may also be true in other countries. The remainder of this section describes some of the crop-management problems that may be limiting yield improvements on farms.

7.5.2.1 Rotation Length

The frequency with which oilseed rape is grown in arable rotations has increased because it is commonly the second most profitable combinable crop after first winter wheat and is an effective break for wheat. The frequency with which oilseed rape is grown averaged one year in three in the UK in 2007, when the crop area peaked at 681 000 ha. The frequency has intensified during the last decade from only 8% crops being grown one year in three in 1990.[62] Shorter rotations will increase the likelihood of diseases such as club root (*Plasmodiophora brassicae*). Sieling and Christen[63] showed that oilseed rape grown every 2 years yielded about 0.5 t/ha less than when it was grown once every 3 years.

7.5.2.2 Disease, Pest and Weed Control

Pest and disease problems have been reviewed previously by Weiss.[64] Whilst there are some important pests and diseases worldwide, the occurrence of problems varies greatly between regions or countries. Yield level, choice of cultivars and availability of crop-protection products also varies between the main production areas. In the higher-yielding areas of Europe, pesticides are widely used, *e.g.* an average of 2.1 fungicide spray rounds are applied in the UK,[65] but such inputs are difficult to justify when yields are lower. Phoma stem canker (blackleg) causes serious crop losses in the major production areas of Europe, Australia and North America.[66] Losses from stem canker have reached €147M in France, €56M in the UK and €30M in Australia.[66] Control relies mainly on the integration of hygienic measures to bury crop residues and use of resistant cultivars, with use of fungicides in some regions. Stem rot (*Sclerotinia sclerotiorum*) causes crop damage in Europe, Canada, India, China and Australia[67] and can reduce yields by up to 50%. Foliar fungicides are effective for stem-rot control if well timed, but resistant cultivars are not yet available. In the UK, following severe outbreaks,[68] many crops now receive two fungicide applications at flowering to control sclerotinia, giving a total of four fungicide sprays. In northern Europe, light leaf spot (*Pyrenopeziza brassicae*) is more prevalent and disease-resistant cultivars and fungicides are required to minimise its effects.

Clubroot is an important disease worldwide, particularly on acid soils. It has caused concern in Canada where it has spread since it was first reported in the Edmonton area of Alberta in 2003. Clubroot has been declared a pest under Alberta's *Agricultural Pests Act* in 2007 and a Management Plan has been produced,[69] which involves longer rotations, and sanitation measures to limit spread of infested soil are being used. Cultivars with resistance to clubroot are available in Canada (www.dekalb.ca) and Europe, but crop rotation and use of lime to raise soil pH above 7.0 remain important control measures.

Control of insect pests is targeted against cabbage stem flea beetle (*Psylliodes chrysocephala*) and aphids (*Mysus persicae*) in autumn and against pollen beetle (*Meligethes aeneus*) and cabbage seed weevil (*Ceutorhynchus assimilis*) in spring. *M. persicae* is the main vector of *Turnip yellows virus* that can cause yield losses of more than 20%.[70] Early pest attacks, seed-borne and damping-off diseases, are controlled by seed treatments. The recent introduction of seed treatments with a neonicotinoid insecticide component offers improved pest control and opportunity to reduce foliar sprays in autumn. In the UK, insecticides were used on 85% of the crop area in 2006 with an average of 1.9 applications per crop.[65] Molluscicides are also commonly used at establishment to minimise loss of plants due to slug predation.

Oilseed rape is vulnerable to weed competition when small but is a very effective weed suppressor when a full canopy has developed.[71] Some of the low-cost establishment techniques (see next section) can give patchy establishment, so, even when a full canopy has developed, open areas allow unhindered weed development as well as aiding crop predation by pigeons. In UK arable rotations, cereal volunteers, black grass (*Alopecurus myosuroides*) and cleavers (*Galium aparine*) can be problems, causing potential yield losses of 61%, 38% and 13% respectively.[72] Control of other broad-leaf weeds, such as poppy (*Papaver rhoeas*), can also be difficult, as can weeds related to the *Brassica* family, such as Charlock (*Sinapis arvensis*) and Runch (*Raphanus raphanistrum*). For example, four Runch plants/m^2 have been shown to reduce canola yields by 9 to 11%.[73] Some of the earliest applications of GM techniques were seen in oilseed rape crops and now just under 90% of the total oilseed rape production in Canada shows GM herbicide tolerance to glyphosate or glufosinate, and around a further 9% are conventionally bred broad-range herbicide-tolerant crops.

The disease, pest and weed problems described above can usually be minimised through cultural and chemical methods. However, increased pressure to make a profit often means that the optimum crop management methods cannot be used, *e.g.* lengthening rotations or the most effective dose of a pesticide. Furthermore, breakdown of varietal disease resistance, insensitivity of pests to pesticides and increased regulation of pesticides mean that the chemical options available are not always effective.

7.5.2.3 Establishment Methods

The use of techniques like autocasting seed directly into the preceding cereal crop prior to harvest, direct drilling or establishing after minimum tillage have become more commonly used alternatives to ploughing to a depth of approximately 20 to 25 cm because they are quicker and cost less. Ploughing has traditionally been the main method of cultivation on farms and is used in the UK variety testing systems which out-yield farm crops. The precultivation condition of the seedbed determines the overall success of the other, less-expensive establishment techniques. For example, direct drilling and autocasting result in lower yields when there are large amounts of poorly chopped straw from the previous crop, wet conditions or wet and poor soil structure.[74–76] Additionally, nonburial of the residues of the previous crop has been shown to increase lodging and phoma infection[74] and increase slug damage. In future, the costs of establishment and hence selection of techniques may become more important as energy costs increase.

7.5.2.4 Timing and Rate of Sowing

Early sowing and high seed rates increase the likelihood of exceeding the optimum canopy size and, probably as a result of this, have been shown to reduce yield.[51,77–80] They can also increase the risk of pests such as cabbage root fly (*Delia radicum* L.), light leaf spot, aphids and viruses. Oilseed rape is often sown early and at high seed rates in order to spread the farm workload at the time of sowing and to reduce the risk of failed establishment. If the risk of producing an overlarge canopy can be predicted, then azole fungicides with growth regulatory activity can be used to reduce canopy size and the risk of lodging.[81]

7.5.2.5 Fertiliser Use

In areas where soils or atmospheric depositions fail to supply plant sulfur (S) requirements, applications of fertiliser S may need to be routine. In the UK between 1970 and 2002, atmospheric SO_2 depositions decreased by 82%.[82] Many oilseed rape crops are responsive to S fertiliser, with yield increases of 0.7 to 1.6 t/ha recorded in response to 40 kg S/ha on a sandy soil.[83] Thus, oilseed rape crops yielding 3 t/ha now require about 30 kg S/ha.[83] The proportion of UK farm crops that had sulfur applied increased from 0.10 to 0.50 between 1993 and 2006.[6] In the many regions where atmospheric sulfur depositions have declined most, farmers probably still do not apply enough sulfur to prevent deficiency from limiting yield potential.

Recent research has shown that crops with a high yield potential have a greater requirement for N, but that it is possible to reduce yield potential by applying too much N to the crop too early because this increases the likelihood of producing an overlarge canopy that is prone to lodging.[81] It is therefore clear that both the amount and timing of N fertiliser must be chosen carefully in order to maximize yield potential. Nitrogen-fertiliser applications to commercial oilseed rape crops increased from 250 kg N/ha to 280 kg N/ha between 1980 and 1984, then decreased steadily to 180 kg N/ha by 1993 and were 189 kg N/ha in 2007.[84] The reduction in N fertiliser use during the late 1980s and early 1990s was largely a result of falling oilseed prices and is likely to have contributed to the static farm yields in the UK.

7.6 Carbon Footprint

The GHG emissions associated with growing and harvesting one hectare of oilseed rape have been calculated using the same methods as described for wheat by Berry et al.[85] and subsequently for oilseed rape by Mahmuti et al.[86] These include using the following standard crop inputs for a UK oilseed rape crop (per hectare): 202 kg N, 42 kg K_2O, 39 kg P_2O_5, 409 kg lime, 5 kg of seed, 0.42, 1.60, 0.03 kg active ingredient for fungicides, herbicides and insecticides, respectively, and 3971 MJ of energy used during field operations. GHG emission factors described in Berry et al.[85] have been used to calculate the GHG emissions associated with each input. The GHG emissions account for the different global-warming potentials of CO_2, NO_2 and CH_4 and are expressed as kg of CO_2 equivalent (e). The heat and electrical energy required to extract the oil from the seed and for the esterification process are assumed to come from a natural gas power plant and electricity from the national grid. GHG emissions associated with these processes has been estimated at 305 kg CO_2 e/t of oilseed. The proportion of the total GHGs that may be allocated to biodiesel and rape-meal production are then allocated based on the economic values of the two products. Based on prices at the time of writing, approximately 80% of the total GHGs associated with the production and processing of oilseed rape are allocated to biodiesel.

It is clear that GHGs associated with nitrogen contribute the greatest proportion of GHG emissions (Figure 7.8), with more than 80% of GHGs

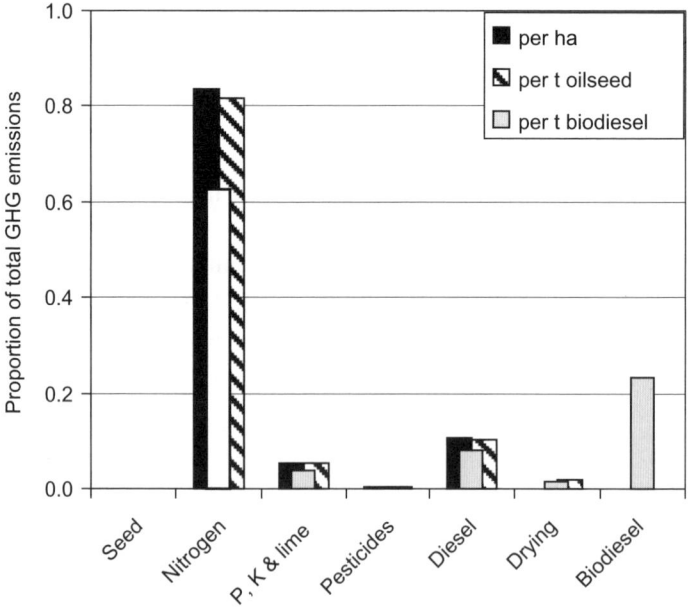

Figure 7.8 GHG emissions associated with the production of a typical oilseed rape crop yielding 3.2 t/ha.

associated with the production of oilseed rape and more than 60% associated with the production of biodiesel. GHG emissions associated with N come from two sources: (1) NO_2 and CO_2 emissions associated with the manufacture of N fertiliser; (2) NO_2 emissions from the soil and N fertiliser during crop growth. The extraction of oil and the esterification process account for almost 25% of the GHG emissions associated with producing biodiesel.

Crop yield also has a large impact on GHG emissions associated with producing each tonne of oilseed rape, if it is assumed that crop inputs remain the same (Figure 7.9). For example, increasing yield from 3 t/ha to 5 t/ha is calculated to reduce GHG emissions per tonne from 1059 to 647 kg CO_2e. Changes in yield of this magnitude are not always associated with changes in the level of crop inputs. Increasing yields using crop inputs with zero or negligible GHG costs such as selecting varieties with greater disease resistance or using pesticides reduces GHG emissions per tonne of oilseed. This was demonstrated by Mahmuti et al.[86] who calculated that improving varietal resistance to disease was associated with a reduction in GHG emissions of 56 kg CO_2e/t, and controlling disease with fungicides was associated with a reduction in GHG emissions of 100 kg CO_2 e/t. For crop inputs with large GHG costs, such as N fertiliser, the calculation of the net effect of reducing inputs is more complex.

The effect of reducing the rate of N fertiliser on yield and GHG emissions may be estimated using a typical yield response to N fertiliser taken from Berry and Spink[48] (Figure 7.10). This shows that the direct GHG emissions associated with each tonne of oilseed may be minimised by applying zero N fertiliser. However, it should be recognised that applying zero fertiliser will give a

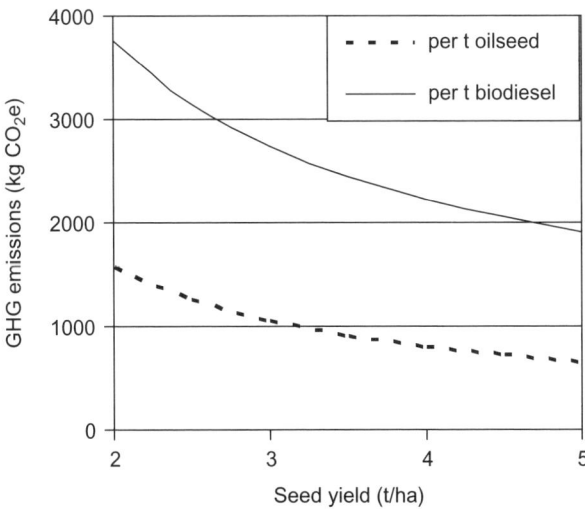

Figure 7.9 Effect of seed yield on GHG emissions if it is assumed that crop inputs remain the same.

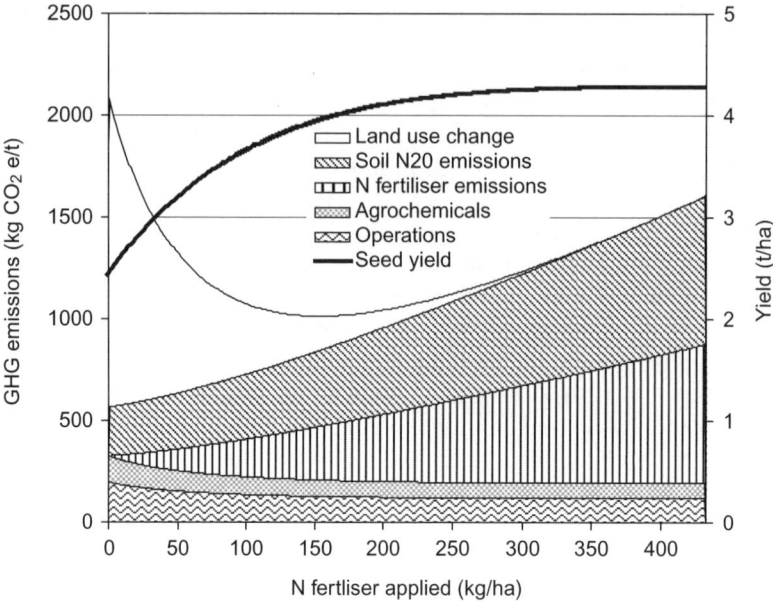

Figure 7.10 Effect of changing N fertiliser rate on seed yield and GHG emissions before and after accounting for possible indirect land-use change effects.

yield of 2.4 t/ha compared with a yield of 4.1 t/ha at the economic optimum N rate of 180 kg N/ha. If an inelastic demand for crop produce is assumed, then the lower yields from reduced N fertiliser rates would require greater cropped area to maintain oilseed production. If previously uncultivated land were brought into production, the expansion of cropped area could result in very large GHG emissions, as carbon would be released from vegetation and soil organic matter. For illustrative purposes, it was assumed that all of the foregone yield would be produced on land converted from temperate grassland. The area of indirect land-use change (ILUC) to meet this production may be estimated by dividing the foregone yield by the expected yield on the newly cropped land, taken as 3 t/ha. Over a 30-year period, the GHG emissions resulting from converting temperate grassland to arable crop land have been estimated at between 111 000 and 242 000 kg CO_2 e/ha.[87,88] In this example, we have assumed a midrange value of 180 000 kg CO_2 e/ha over a 30 year period for converting temperate grassland. See Kindred *et al.*[89] for further details on the methodology for estimating the effects of ILUC. Figure 7.10 shows that that GHG emissions resulting from ILUC are potentially considerable and when these are taken into account the optimum rate of N fertiliser for minimising GHG emissions per tonne is 150 kg N/ha, compared with the economic N rate of 180 kg N/ha. Other scenarios that result in greater GHG emissions from ILUC (*e.g.* conversion of rain forest to crop land) are estimated to give a greater optimum N rate for minimising GHGs than the economic N rate.

7.7 Future Prospects

In order to meet the increasing demands for food, feed and fuel without expanding the area of cropped land it is clear that oilseed rape yields must increase. Furthermore, yield increases must be achieved with the same or fewer inputs, particularly of N fertiliser, to reduce the carbon footprint of the crop. In several countries, farm yields have been shown to be either increasing at a slower rate or to have stopped increasing altogether, despite plant breeders introducing new varieties with a greater yield potential and estimates for the potential yield of oilseed rape being significantly greater than what is achieved on farm. The most likely reason for reduced farm yield increases since the mid-1980s is the employment of crop-management practices that do not allow the genetic potential of the crop to be realised. Since the 1970s, the real value of commodity crops, after accounting for inflation, has declined steadily.[90] Over the same period the cost of many of the inputs required for crop production have increased. This situation has reduced the economic level of inputs as illustrated by the uptake of autocasting seed, minimal tillage techniques and the decline in the rate of nitrogen fertiliser use since the mid-1980s. It is possible that the increase in genetic yield potential created by breeders has countered the effect of reducing inputs and has stopped farm yields from declining. If the price of oilseeds continue to decline in real terms then it is likely that farm yields will continue to increase at the current slow rate. However, if crop prices increase due to increasing demand for oilseeds for both food and fuel, this is likely to stimulate more attention to crop-management inputs, which could trigger a rapid increase in yields as farmers close the large gap between farm yields and the genetic potential.

If crop prices do not increase relative to the price of inputs, then, in order to increase yields, plant breeders must focus more on producing varieties that are able to yield well with fewer inputs, *e.g.* by increasing N-use efficiency, durable disease resistance and lodging resistance. However, these traits are probably under complex genetic control and will require significant effort to improve them. Nonetheless, there does appear to be significant potential for improving these and other traits. For example, it has been estimated that the fertiliser N-use efficiency of oilseed rape could be improved by 60% by increasing N uptake efficiency and reducing the amount of N stored in the stems and pod walls.[91] A current UK Defra LINK project (LK0979) "Breeding oilseed rape with a low requirement for N fertiliser" has found significant genetic variation for the key traits that determine N-use efficiency that could be exploited by plant breeders. Further work is required to better understand the physiological mechanisms that determine N-use efficiency and other complex traits, *e.g.* resistance to lodging, water-use efficiency and rooting, and to identify the most important traits for breeders to focus on. Once the key traits have been identified, methods for rapidly selecting them must be developed. These could be based on visual assessments, instrument-aided assessments or genetic markers. Genetic markers have been identified in oilseed rape for traits such as oil content,[92] while quantitative trait loci (QTL) have been found for traits associated with N requirement in maize, including late N uptake, senescence and stem remobilisation, Gallais and Hirel.[93] An alternative approach for improving N-use efficiency has been initiated by Good et al.,[94] who introduced barley alanine aminotransferase into wild-type canola and produced a 40% decrease in the amount of applied nitrogen required to produce equivalent yields to the wild-type plants.

Looking further ahead, climate change may have a significant effect on yield trends. Climate change predictions for the UK in 2050 (UKCP09) include a temperature increase of 2 °C, and a decrease in spring and summer rainfall. Assuming that the seed-filling period lasts 715 °Cd above 4.2 °C, then an increase in temperature of 2 °C would shorten the seed-filling period by 9 days and reduce the amount of radiation intercepted by the crop during seed filling by 20%. Biomass accumulation would be further reduced because of greater respiratory losses in warmer temperatures. It has been estimated that the combined effect of less summer rainfall and greater evapotranspiration may reduce the amount of water available to crops by 20 to 40 mm. These negative effects would only be partially compensated for by greater CO_2 levels and it is likely that varieties will need to be bred with different traits to minimise potential yield losses due to the changing environment. The IPPC forecast for temperature and rainfall indicates that some of the areas where spring rape now predominates may be able to switch to higher-yielding winter rape, where most breeding effort has been directed. Globally, the Hadcm3 model shows parts of China and Canada are forecast to have an increase in both spring and summer soil moisture contents following increased spring rainfall, possibly bringing prospects of higher yields from these major producers.

The temperature rises along with some changes in rainfall to give drier summer and wetter winters and more variable rainfall events will affect disease incidence and timing. In the case of phoma stem canker, severe epidemics are predicted to become more frequent and spread northwards in the UK by 2050.[95] For soil-borne pathogens, a rise in temperature is expected to increase the window for crop infection in autumn and spring for clubroot and verticillium wilt (*Verticillium longisporum*).[96] The stem-rot pathogen, *Sclerotinia sclerotiorum*, requires warm, moist soils for its sclerotia to germinate and produce apothecia. Warmer soils in the autumn and winter would therefore allow airborne inoculum (ascospores) to be produced by apothecia over longer periods.[96] Therefore, the importance of breeding for disease resistance as well as developing effective fungicides is likely to become more important.

In conclusion, oilseed rape has significant potential to provide an increasing share of the global energy requirements in a carbon-efficient way. Achieving this will require a sustained effort by plant breeders, crop physiologists, agronomists and farmers to increase crop yields towards its high potential without increasing the requirement for crop inputs.

Acknowledgements

The authors thank John Garstang of ADAS for his valuable guidance with writing this chapter.

References

1. *FAO Statistics* (2009). http://faostat.org/site/567/default.aspx.
2. T. R. Sinclair and C. T. De Wit, *Science*, 1975, **189**, 565–567.
3. R. K. Downey and S. R. Rimmer, *Adv. Agron.*, 1993, **50**, 1–65.
4. Bioscope 2008, Final report for Lot 1 of the Bioscopes project. EC project TREN/D2/44-LOT 1/S07.54676. Details at http://ec.europa.eu/energy/renewables/biofuels/doc/standard/lot1.pdf accessed September 2009.
5. Renewable Fuels Agency, Quarterly Report 5, 2009, 5 April – 14 July 2009.
6. P. M. Berry and J. H. Spink, *J. Agr. Sci., Cambridge*, 2006, **144**, 381–392.
7. G. Fischer, E. Hizsnyik, S. Prieler, M. Shah, H. van Velthuizen, *Biofuels and Food Security, Land Use Change and Agriculture Program*, International Institute for Applied Systems Analysis. Laxenburg, Austria. 2009 [www.iiasa.ac.at accessed August 2nd 2009].
8. European Biodiesel Board, 2009. http://www.ebb-eu.org/EBBpressreleases/ EBB%20press%20release% 202008%20prod%202009%20cap%20FINAL.pdf.
9. www.HGCA.com.
10. D. Bacovsky, J. Barclay, D. Bockey, R. Saez, L. Edye, T. Foust, P. Grabowski, L. Kujanpää, D. de Lang, P. Larsen, W. E. Mabee, T. Mäkinen, J. McMillan, A. Munack, J. Murphy, K. Øyaas, L. Pelkmans, J.-C. Pouet, B. A. Prior, S. Saka, M. Samejima, J. Sandquist, T. Sidwell, K. Werling, A. Wrobel, W. H. van Zyl, *Update on implementation agendas 2009*, ed. W. E. Mabee, J. Neeft and B. van Keulen IEA Task 39 Report T39-P5, 2009, 74 pp. +.
11. European Commission, *Directive 2009/28/EC on the promotion of the use of energy from renewable resources and amending and subsequently repealing Directives 2001/77/EC and 2003/30/EC*, 2009.
12. Eurostat, *Energy Pocket Book*, 2009.
13. *FAO TradeSTAT 2007*.
14. *Roundtable on Sustainable Palm Oil, RSPO Principles and Criteria for Sustainable Palm Oil Production*. Guidance Document. March 2006.
15. *Proforest, The Basel Criteria for Responsible Soy Production: Local interpretation for use in Brazil*. November 04, 2005, Version 16.2.2005.
16. Renewable Fuels Agency, *Carbon and Sustainability Reporting within the Renewable Transport Fuel Obligation, Technical Guidance Part Two, Carbon Reporting – Default Values and Fuel Chains*. Renewables Fuel Agency, 2008, Version 2.0. March 2009.
17. Defra. *Agriculture in the UK 2008*. Available on line at http://statistics.defra.gov.uk/esg/publications/auk/default.asp, 2009.
18. E. J. Booth, J, Booth, P, Cook, B, Ferguson, K. Walker, Economic Evaluation of Biodiesel *Production from Oilseed rape grown in North and East Scotland*, SAC October 2005.
19. C. Wareham, *Europroteins: Development of protein rich products of plant origin by plant breeding and (bio) technology for application in human and*

animal nutrition, Report for the Europroteins group produced by the University of Nottingham, UK, 1992.
20. B. R. Cottrill, T. C. Smith, P. M. Berry, R. M. Weightman, J. Wiseman, G. White, M. Temple, *Opportunities and implications of using the co-products from biofuel production as feeds for livestock*. Report prepared for: The Home-Grown Cereals Authority, the English Beef and Lamb Executive and the British Pig Executive. Research Review No. 66. July 2007. Home Grown Cereals Authority, Caledonia House, 223 Pentonville Rd, London, N1 9HY.
21. W. N. Ewing, *The Feeds Directory*, CONTEXT Products Ltd, Ashby-de-la-Zouch, Leics, UK, 1998.
22. R. H. King, P. E. Eason, D. K. Kerton and F. R. Dunshea, *Austral. J. Agricul. Res.*, 2001, **52**, 1033–1041.
23. B. P. Mullan, J. R. Pluske, J. Allen and D. J. Harris, *Austral. J. Agricul. Res.*, 2000, **51**, 547–553.
24. D. A. Roth-Maier, B. M. Böhmer and F. X. Roth, *Animal Res.*, 2004, **53**, 21–34.
25. M. Opalka, L. Dusza, M. Koziorowski, J. Staszkiewicz, K. Lipiñski and J. Tywoñczuk, *Livestock Prod. Sci.*, 2001, **69**, 233–243.
26. L. McNeill, K. Bernard and M. G. Macleod, *Br. Poultry Sci.*, 2004, **45**, 519–523.
27. Y. O. Fasina and G. L. Campbell, *Can. J. Animal Sci.*, 1997, **77**, 191–195.
28. D. A. Roth-Maier, *Investigations on feeding full-fat canola seed and canola meal to poultry*, 10th International Rapeseed Conference, Canberra, Australia, 1999.
29. R. Mawson, R. K. Heaney, Z. Zdunczyk and H. Kozlowska, *Nahrung*, 1995, **39**, 21–31.
30. V. Šašytė, A. Racevičiūtė-Stupelienė, R. Gružauskas and R. Mosenthin, *Biologija*, 2006, **1**, 69–72.
31. P. Guggenbuhl and C. Simões Nunes, *Livestock Sci.*, 2007, **109**, 261–263.
32. F. Georges, A. A. Hussain, W. A. Keller, *Method for reducing phytate in canola meal using genetic manipulation involving myo-inositol 1-phospathe synthase gene*. United States Patent 7148064, 2006.
33. M. Hazzledine, Chapter 14. In *Recent Advances in Animal Nutrition – 2008* ed. P. C. Garnsworthy and J. Wiseman, Nottingham University Press, Nottingham, UK, ISBN 978-1-904761-044, 2008.
34. P. Lammers, M. Honeyman, B. Kerr, T. Weber, (A.S. Leaflet R2224)- www.ans.iastate. edu/report/air/2007pdf/R2224.pdf, 2007.
35. J. Barteczko, J. Kaminski, [Polish], *Ann. Warsaw Ag. Univ.*, 1999, **36**, 197–209.
36. A. Simon, H. Bergner and M. Schwabe, *Arch. Animal Nut.*, 1996, **49**, 103–112.
37. A. Simon, M. Schwabe and M. Bergner, *Arch. Animal Nut.*, 1997, **50**, 271–282.
38. M. F. Dreccer, A. H. C. M. Schapendonk, G. A. Slafer and R. Rabbinge, *Plant Soil*, 2000, **220**, 189–205.

39. D. S. Robinson, Lipids. In: *Food Biochemistry and Nutritional Value*. Longman Scientific and Technical, Essex, England, 1987, pp. 254–255.
40. A. Schierholt, B. Rücker and H. C. Becker, *Crop Sci.*, 2001, **41**, 1444–1449.
41. C. Möllers, *Development of high oleic acid oilseed rape*. Vortrag auf dem Internationalen Fachkongress fur Nachwachsende Rohstoffe und Pflanzenbioetahnologie NAROSSA in Magdeburg, 10–11 June 2002; Proceedings.
42. G. Knothe, *Energy Environ. Sci.*, 2009, **2**, 759–766.
43. B. Rücker, G. Robbelen, *Development of high oleic rapeseed*. Proceedings of the 9th International Rapeseed Congress, Cambridge, UK, 1995, pp. 389–391.
44. A. Schierholt and H. C. Becker, *Plant Breed.*, 2001, **120**, 63–66.
45. C. Möllers and A. Schierholt, *Crop Sci.*, 2002, **42**, 379–384.
46. Monsanto. (2009). *Splendor speciality oilseed rape*. http://www.vistive.co.uk/splendor.html. Accessed 15th November 2009.
47. Monsanto. (2009). VI4101 speciality oilseed rape. http://www.vistive.co.uk/vi41ol.html. Accessed 15th November 2009.
48. P. M. Berry, J. H. Spink, Home-Grown Cereals Authority Project Report No 447. HGCA, London, 2009a, 212 pp.
49. J. Orson, E. Booth, C. Merritt, C. Lea, *Growing "high oleic low linolenic" (HOLL) oilseed rape for specialised markets*. HGCA final project report No. 442. HGCA, Stoneleigh, Kenilworth, Warwicks, CV8 2TL, 2008.
50. Ali H. Fayyaz-ul-Hassan, A. M. Cheema and A. Manaf, *J. Res (Sci)*, 2005, **16**, 65–72.
51. N. J. Mendam, P. A. Shipway and R. K. Scott, *J. Agr. Sci.*, 1981, **96**, 389–416.
52. P. Leterme, Modelisation du fonctionnement du peuplement de colza d'hiver en fin de cycle: elaboration des composantes finales du rendement. In: *Colza: Physiologie et Elaboration du Rendement CETIOM*, INRA/CETIOM, Paris, 1988, pp. 124–129.
53. G. D. Lunn, J. H. Spink, D. T. Stokes, A. Wade, R. W. Clare and R. K. Scott, *Canopy Management in Winter Oilseed Rape. HGCA Project Report No. OS49*, Home-Grown Cereals Authority, London, 2001.
54. D. J. Yates and M. D. Steven, *J. Agr. Sci.,* Cambridge, 1987, **109**, 495–502.
55. G. D. Lunn, J. H. Spink, A. Wade, R. W. Clare, HGCA Project Report No.OS64, London: Home-Grown Cereals Authority, 2003.
56. D. J. Major, *Agron. J.*, 1977, **69**, 541–543.
57. L. H. Gammelvind, J. K. Schjoerring, V. O. Mogensen, C. R. Jensen and J. G. H. Bock, *Plant Soil*, 1996, **186**, 227–236.
58. S. C. McWilliam, J. A. Stafford, R. K. Scott, G. Norton, D. T. Stokes and R. Sylvester-Bradley, *Proceedings of the 9th International Rapeseed Congress, Cambridge, UK*, ed. D. Murphy, Volume 2, Dorset Press, Dorchester, UK, 1995, pp. 491–493.
59. J. A. Stafford, University of Nottingham, Ph.D. thesis, 1996.
60. A. D. Baylis and I. T. J. Wright, *Ann. Appl. Biol.*, 1990, **116**, 287–295.
61. E. L. Armstrong and H. I. Nicol, *Austral. J. Exp. Agr.*, 1991, **31**, 245–250.

62. J. A. Turner and N. V. Hardwick, *Proceedings of the 9th International Rapeseed Congress, Cambridge, UK*, In D. Murphy, ed. Volume 2, Dorset Press, Dorchester, UK, 1995, pp. 640–642.
63. K. Sieling and O. Christen, *Eur. J. Agron.*, 1997, **7**, 301–306.
64. E. A. Weiss, *Oilseed Crops*, Longman, London, 1983, 660pp.
65. D. G. Garthwaite, M. R. Thomas, E. Heywood and A. Battersby, *Pesticide Usage Report 213. Arable Crops in Great Britain 2006*, Defra and SEERAD, 2007, 116pp.
66. B. D. L. Fitt, H. Brun, M. J. Barbetti and S. R. Rimmer, *Eur. J. Plant Pathol.*, 2006, **114**, 3–15.
67. L. Buchwaldt, In: *Compendium of Brassica Diseases*, S. R. Rimmer, V. I. Shattuck, L. Buchwaldt, ed. APS Press, St Paul, MN, 2007, pp. 43–47.
68. P. Gladders, D. Ginsburg, J. A. Smith, *HGCA Project Report No. 433*, 2008, 44pp.
69. Anon., http://www1.agric.gov.ab.ca/$department/deptdocs.nsf/all/agdex 11519#Management, Government of Alberta, Canada, 2007.
70. M. Stevens and W. Clark, *Aspects Appl. Biol.*, 2009, **91**, *Crop Protection in Southern Britain*, 109–114.
71. J. T. O'Donovan, *Weed Sci.*, 1994, **42**, 385–389.
72. J. Clarke, S. Wynne, P. Berry, S. Twining, S. Cook, S. Ellis and P. Gladders, *Home-Grown Cereals Authority Research Review No 70*. HGCA, London, 2009, 131 pp.
73. R. E. Blackshaw, R. Lemerle, R. Mailer and K. R. Young, *Weed Sci.*, 2002, **50**, 344–349.
74. G. Sauzet, R. Reau and J. Palleau, in *Proceedings of the 11th International Rapeseed Congress*, H. Sprensen, ed. Volume 3, The Royal Veterinary and Agricultural University, Copenhagen, Denmark, 2003, pp. 863–864.
75. P. Bowerman, B. J. Chambers and A. E. Jones, in *Rapeseed Today & Tomorrow: Proceedings of the 9th International Rapeseed Congress*, ed. D. Murphy, Volume 2, Cambridge, UK, 1995, pp. 220–222.
76. O. Christen, B. Hofmann & J. Bischoff, Oilseed rape in minimum tillage systems. In *Proceedings of the 11th International Rapeseed Congress, Volume 3* (ed. H. Sprensen), The Royal Veterinary and Agricultural University, 2003, Copenhagen, Denmark, pp. 762–764.
77. P. D. Jenkins and M. H. Leitch, *J. Agr. Sci., Cambridge*, 1986, **107**, 405–420.
78. J. E. Leach, R. J. Darby, I. H. Williams, B. D. L. Fitt and C. J. Rawlinson, *J. Agr. Sci., Cambridge*, 1994, **122**, 405–413.
79. J. E. Leach, H. J. Stevenson, A. J. Rainbow and L. A. Mullen, *J. Agr. Sci., Cambridge*, 1999, **132**, 173–180.
80. A. Baer and M. Frauen, In: *Proceedings of the 11th International Rapeseed Congress*, H. Sprensen, ed. Volume 3, The Royal Veterinary and Agricultural University, Copenhagen, Denmark, 2003, pp. 887–889.
81. P. M. Berry and J. H. Spink, *J. Agr. Sci., Cambridge*, 2009b, **147**, 273–285.
82. The National Environmental Technology Centre (NETCEN), 2009, at http://www.airquality.co.uk/.

83. S. P. McGrath and F. J. Zhao, *J. Agr. Sci., Cambridge*, 1996, **126**, 53–62.
84. *British Survey of Fertiliser Practice*, DEFRA, York, UK, 2008.
85. P. M. Berry, D. R. Kindred and N. D. Paveley, *Plant Pathol.*, 2008, **57**, 1000–1009.
86. M. Mahmuti, J. S. West, J. Watts, P. Gladders and B. D. L. Fitt, *Int. J. Agr. Sustain.*, 2009, **7**(3), 189–202.
87. IPPC 2006. C. A. M. De Klein, R. S. A. Novoa, S. M. Ogle, K. A. Smith, P. Rochette, T. C. Wirth, B. G. McConkey, A. Mosier, K. Rypdal, Chapter 11: N_2O Emissions from managed soils, and CO_2 emissions from lime and urea application. In *2006 IPCC Guidelines for National Greenhouse Gas Inventories; Volume 4: Agriculture, Forestry and Other Land Use*: International Panel on Climate Change.
88. T. Searchinger, R. Heimlich, R. A. Houghton, F. Dong, A. Elobeid, J. Fabiosa, S. Tokgoz, D. Hayes and T. Yu, *Sci. Exp.*, 2008, **319**, 1238–1240.
89. D. Kindred, P. Berry, O. Burch, R. Sylvester-Bradley, In: Effects of climate change on plants: implications for agriculture. *Aspects Appl. Biol.*, 2008, **88**, 2008, UK.
90. J. Morris, E. Audsley, I. A. Wright, J. McLeod, K. Pearn, A. Angus, S. Rickard, Agricultural Futures and Implications for the Environment. Main Report. Defra Research Project IS0209. Bedford: Cranfield University, 2005.
91. P. Berry, J. Foulkes, J. Spink, G. Teakle and P. White, Breeding for improved nitrogen-use efficiency in oilseed rape, In: *Resource capture by crops: integrated approaches*, AAB 10–12 September 2008, University of Nottingham, UK, Sutton Bonington, 2008b.
92. J. Zhao, C. Heiko, C. Becker, D. Zhang, Y. Zhang and W. Ecke, *Crop Sci.*, 2005, **45**, 51–59.
93. A. Gallais and B. Hirel, *J. Exp. Bot.*, 2004, **55**, 295–306.
94. A. G. Good, S. J. Johnson, M. De Pauw, R. T. Carroll, N. Savidov, J. Vidmar, Z. Lu, G. Taylor and V. Stroeher, *Canad. J. Bot.*, 2007, **85**, 252–262.
95. N. Evans, A. Baierl, M. A. Semenov, P. Gladders and B. D. L. Fitt, *J. of the Royal Soc. Interfac.*, 2008, **5**, 525–531.
96. J. A. Smith and P. Gladders, *Aspects Appl. Biol.*, 2009, **91,** Crop Protection in Southern Britain, 159–162.

CHAPTER 8
Soybeans

ANTHONY J. KINNEY[1] AND TOM E. CLEMENTE[2]

[1] DuPont Experimental Station, Wilmington, Delaware 19807, USA; [2] Center for Plant Science Innovation, University of Nebraska, Lincoln, Nebraska 68588, USA

8.1 Introduction: A Brief History of Soybean Production

Soybeans have been grown for more than 3000 years and for most of that time they have been used almost exclusively for food use. It is only in the last decade or so that soybeans have begun to gain momentum as an energy crop. Since soybean is a legume, it fixes its own nitrogen and does not require nitrogen fertiliser. This improves the net energy balance when the oil is used for fuel, a strong attribute for a biodiesel feedstock.

Domesticated soybean (*Glycine max* L. Merrill) has its origins in China during the Shang dynasty (1700–1100 BC).[1,2] It is a member of the Fabacae family (Phaseoleae tribe) and was domesticated from the wild annual soybean *Glycine soja*. Both *soja* and *max* are annuals but the founding members of the *Glycine* genus are all perennials. The wild, perennial ancestor of *G. soja* is not known but was probably an ancient tetraploid (palaeopolyploid) *Glycine* species.[3]

By the late 16th century, when European trade with China and Japan was beginning to be established, soybean protein had become a key dietary component of most of East Asia. Despite an active trade in soy sauce between East and West, soybeans were not grown in Europe until the late 18th century, and then mostly by botanical gardens for scientific purposes.[1] Soybeans were first grown in the New World as early as the late 1700s but it was almost another 100 years before agricultural societies began to send seeds to farmers around the US.[2] Around the same time soybeans first began to be cultivated in South America, especially Brazil and Argentina.[4]

Throughout the early part of the 20th century many genotypes of soybean were introduced into the US from Asia by the United States Department of Agricultire (USDA) and commercial soybean processing (separating oil from the protein-rich meal) was established. By 1925 US soybean production was the equivalent of about 150 000 tons.[1] Initially, soybeans were grown in the US mainly for forage and hay, but by 1940 the agricultural focus had shifted to the seed itself as processed soybean meal began to be incorporated in animal-feed formulations and soybean oil began to replace other fats and oils for food uses.[5]

Land use for soybean production in the US increased from about 0.6 million ha in 1925 to over 20 million ha in 1970.[1] In the mid-1970s, soybean prices soared as demand increased. The US failure to meet export demands to Europe and Japan triggered a ten-fold increase in soybean production in Brazil to over 10 million tons.[6]

In the past decade total South American soy production has reached 50% of the total world production. Currently, combined soybean production in the New World (US, Canada, Brazil, Argentina, and other South American countries) is close to 190 million tons, which represents 88% of the total world production.[7]

In 1992 US soybean producers with funds generated through the national check-off program founded the National Biodiesel Board to promote the use of soybean as a feedstock for biodiesel and began to fund soy biodiesel research. Federal tax incentives passed in 2005 accelerated the use of soybean oil for biodiesel, from about 200 million litres in 2005 to over 1200 million litres in 2007.[7] In Brazil, blending biodiesel with all petroleum diesel sold became mandatory in 2008, which resulted in a minimum of 2% biodiesel blends in all retail diesel sold on the market. In the US it is estimated that over half the farms use soy biodiesel in their operations.[8,9] Soy biodiesel is also used extensively in marine engines and in mass transit throughout the Americas. Total use of soybean oil in the US for biodiesel is estimated to represent 12% of total soybean oil usage and has been projected to be as much as 1700 million litres by the end of 2010.[7] New legislation in Brazil that will mandate the increased blending level to 5%, with plant-derived biodiesel, in petroleum diesel, will translate to soy biodiesel production in excess of 2500 million litres in the next few years.[9]

8.2 Soybean-Seed Composition and Uses

Soybean seed contains 38% protein and an average of about 19% oil, although protein and oil levels can vary, being impacted by both genotype and environment, with oil contents ranging from less than 17% to as much as 21.5%.

During processing the oil is separated from the meal. A bushel of soybeans (27.2 kg) will yield about 21 kg of high protein meal that is toasted and used for animal feed. About 25% of all soy meal is processed into poultry feed, most of the rest is used for pork, cattle and pet-food formulations. Soy protein is also used in many food and industrial products, although these uses represent only about 2% of the total soybean crush. During 2008 and 2009 international soybean meal prices averaged US $350 per tonne and have been as high as US $450 in 2009.[10] Demand for high-quality soy protein for animal production plays an important role in the cost of soy biodiesel production in North and South America. In Brazil, for example, both soybean and castor oil are used for biodiesel production. The toxic meal of castor greatly restricts its use and thus the price of producing castor oil for biodiesel cannot be offset by selling the meal for animal feed. Once issues of scale and additional processing costs are factored in the cost of producing castor oil ($1000 per tonne) is twice that of soybean oil ($500 per tonne) even though castor has twice the total oil content of soy.[9]

Once the oil is extracted from the meal it is refined, bleached and deodorised to produce RBD oils. During the deodorisation process high-value tocopherols are recovered from the oil. Tocopherols are powerful antioxidants and are used in foods as natural preservatives. Soybean α-tocopherol is also the most important commercial source of vitamin E. Another important byproduct from soy-oil refining is lecithin, used as an emulsifier by the food industry. RBD soybean oil is used in many food applications and may be further modified by hydrogenation or other processes for use in frying, baking and margarine manufacture. RBD soybean oil is also the substrate for biodiesel manufacture.

The most common method of biodiesel manufacture is through trans-methylation using methanol and sodium hydroxide, which produces fatty acid methyl esters (FAME) and glycerol. Ethanol may also be used in the reaction to produce fatty acid ethyl esters.[11] One litre of FAME biodiesel will be produced for every litre of oil. The properties of the resultant biodiesel will be impacted by the fatty acid composition of the oil used for transmethylation.[11] Soybean oil is primarily composed of five fatty acids; palmitic acid ($\sim 13\%$), stearic acid ($\sim 4\%$), oleic acid ($\sim 18\%$), linoleic acid ($\sim 55\%$) and linolenic acid ($\sim 10\%$) and the resultant FAME composition will reflect these proportions. The cost of producing soy FAME biodiesel is in the region of $1.20 per litre, depending on the price of soybean oil. This is about 25% more than the cost of producing a litre of petrochemical diesel.[12]

Almost all biodiesel currently in commercial use consists of monoalkyl esters produced from methanol or ethanol transesterification. However, recent work has demonstrated that vegetable oil biodiesel may also be produced by a process of hydrodeoxygenation (HDO) using sulfur catalysts.[13] Under the right conditions RDB soybean oil can be converted in this way to long-chain alkanes and propane. The propane may be recycled to fuel further oil hydrodeoxygenation, thus potentially reducing the cost of the process. In the HDO process only the chain length of the original fatty acids (C16 or C18) will be reflected in the resultant alkane biodiesel, since all the double bonds will be eliminated during the reaction. Since soybean oil contains 87% C18 fatty acids, the resulting HDO biodiesel will contain a relatively homogenous population of alkanes. The removal of oxygen and double bonds will presumably increase the quality of soy biodiesel made by HDO although no actual engine-testing reports have yet been published. An attractive aspect of alkane biodiesel is that it would be able to enter directly into the current petrochemical alkane production chain.

8.3 Properties of Soybean Biodiesel

RDB soybean oil consists predominantly of triacylglycerols (TAG). The TAG molecule has three acyl chains esterified to a glycerol backbone. Vegetable oils can be used directly in engines, without prior conversion to FAME or HDO alkanes. However, since RBD soybean oil is highly viscous it is impractical in cool environments and can also lead to poor engine performance.[11] Vegetable oils with shorter-chain fatty acids esterified to glycerol, such as those found in some exotic oilseeds such as *Cuphea* species, would be considerably less viscous than soybean oil.[14] In theory, these oils would function more effectively than soybean oil[15] The genes that control the fatty acid chain length in plants oils are known and it would therefore be possible to create soybean oil that consisted mostly of medium-chain TAG.[16] Such oils have not been commercialised because the regulatory compliance and development costs of doing so outweigh the current net projected value of such a product.[17]

FAME biodiesel from soybean oil, while more costly to produce than petroleum diesel, offers a number of advantages over all petrochemical fuels, such as increase in fuel performance and significant environmental benefits. With respect to the former, biodiesel blends offer improved flashpoint and increased lubricity, while environmental attributes with biodiesel use include reduced toxicity, lower emissions, and biodegradability.[11,18,19]

Flashpoint and lubricity are important parameters of a liquid transportation fuel. Flashpoint is the fuel-ignition temperature, which for petroleum diesel is 71 °C. Soybean biodiesel has a flashpoint greater than 100 °C, greatly reducing the possibility of unintended ignition.[20] The lubricity of a fuel has a direct effect on engine wear and the limited lubricity of sulfur-free petroleum diesel fuels accelerates engine deterioration. Importantly, vegetable FAME biodiesel blends, including soybean biodiesel, all significantly increase lubricity and reduce engine wear when compared with petroleum diesel.[21]

The most significant advantage of soybean biodiesel, both neat and in blends, over petroleum diesel is its environmental impact. Soybean biodiesel, unlike petroleum diesels, has zero hydrocarbon particulate emissions and greatly reduced sulfur emissions.[22] The reduced content of mutagenic molecules in biodiesel exhaust, even in the presence of scrubbers and catalytic converters, is another environmental attribute over petroleum fuels.[22] This is especially important for engines that run in confined spaces, common in the mining industry, but should be promoted across the board for all applications since there is a direct correlation between reduction in chemical mutagens in engine exhausts and reduced mutagenic rates in biological assays.[23,24] For example, the mutagenic capacity of a diesel bus engine exhaust, monitored by implementing a bioassay, demonstrated significant reduction in mutagen emission rates of soybean biodiesel fuels.[24]

For marine applications the most important advantage of soy biodiesel is biodegradability. Marine engines release fuel into the water surrounding them, which is potentially damaging to aquatic environments. Soy biodiesel degrades very rapidly when compared with petroleum diesel and in biodiesel blends the FAME biodiesel may actually promote degradation of the petroleum diesel in the mix.[19]

Despite the above advantages, soybean biodiesel made from commodity RDB soybean oil does have some significant drawbacks. One downside is the oxidative instability of soybean oil biodiesel. Soybean oil is rich in polyunsaturated fatty acids; more than 60% of the fatty acid species in soybean TAG contain two or more double bonds. The oxidative instability of soybean will impact long-term storage. Moreover, oxidation of fuel occurring in storage or in engine run promotes fuel gumming and an increased viscosity. Double bonds are also believed to be the root cause of the relatively higher emissions of oxides of nitrogen (NO_x) from biodiesel when compared with petroleum diesel.[18] A soy biodiesel blend of 20% (B20) can lead to around 5% higher NO_x emissions than petroleum diesel. The use of 100% biodiesel results in even higher NO_x emissions.[27] NO_x gases are thought to be detrimental to the environment since they are indirect precursors to greenhouse gases.[25]

The use of antioxidants and synthetic additives can alleviate the oxidative stability issues, thereby providing a route to improved storage life of soy biodiesel and reduced gumming, although they have limited application in the reduction of NO_x emissions.[26,27] Perhaps the most effective means of reducing NO_x emissions has been the use of advanced exhaust control technology that incorporates more effective NO_x filters.[28] However, spiking soy biodiesel with methyl oleate was also shown to be effective at mitigating the NO_x effect, bringing emissions of this gas in B20 blend down to those of petroleum diesel.[27] This latter observation leads to the suggestion that increasing the oleic content of soybean oil would also be an effective route for NO_x control. In addition, data has shown that oils high in oleic acid esters also result in biodiesel fuels with an improved cetane number (an ignition quality parameter of a fuel: with lower cetane value, the longer ignition delay time). While the presence of double bonds in fatty acids will lower the cetane value and reduce ignition time delay.[29]

As noted above, soybean oil contains about 17% saturated acids of 16 and 18 carbons. While they make a positive contribution to the oxidative stability of the oil and biodiesel made from it, saturated fatty acids reduce the cold-flow properties of biodiesel, resulting in poor engine performance at lower temperatures.

Thus, the ideal soybean oil for use in biodiesel would appear to be one rich in oleic acid. Fortunately, the ideal oil for engine fuel may also be the best oil for human food. Since the oxidative instability of RBD soybean limits its utility in food applications, oil processors have historically used partial hydrogenation as a means to improve shelf life and cooking properties. Currently, about half of the RBD soybean produced in the US is hydrogenated to some degree.[31] Hydrogenation chemically shifts the fatty acid profile towards increased saturated and monounsaturated, and concomitantly reduces the polyunsaturated fatty acids. The *trans* configuration of a large proportion of the monounsaturated fatty acids produced by hydrogenation results in oils that are solid at room temperature. These hydrogenated oils are less suitable for biodiesel than liquid oils because of their inferior low-temperature properties, high viscosity, and poor lubricity.[32] Hydrogenated oils are also finding great disfavour for food use because of the well-established correlation with their consumption and coronary heart disease in humans.[30,31] The trend by the food industry towards alternative vegetable oils to replace hydrogenated soybean oil has led to various strategies to improve the fatty acid composition of soy. Hence, the rationale for the development of a high-oleic soybean oil, which has been a trait target for soybean breeders and plant biotechnologists for a number of years.[3,30]

8.4 Breeding Approaches to Improving Soybean-Oil Quality

As soybeans were introduced and spread throughout North and South America, plant breeders found they were well adapted for growth under a diverse range of environments and latitudes. For over 70 years soybean breeding has focused on improving and protecting yield, *i.e.* tolerance to biotic and abiotic stresses, with much success.[33] In the US in the early 1970s soybean genotypes with altered seed fatty acid composition were described, which ultimately led to the commercial development of low linolenic acid varieties.[3,34] Moreover, phenotypic and genotypic analyses of these soybean genotypes led to the identification of alleles that govern the control of oleic acid abundance in seed.[35]

The enzymes that control the synthesis and desaturation of fatty acids in the developing soybean seed have been elucidated (Figure 8.1).[16] Oleic acid is synthesised in the plastid of the developing seed as an oleoyl-acyl carrier protein and released into the cytoplasm as oleoyl-coenzyme A. The oleic acyl group is transferred to the sn-2 position of the membrane lipid phosphatidylcholine where it is metabolised to linoleic acid by a single desaturation step carried out by a Δ12-desaturase enzyme. This enzyme is encoded by a *FAD2* gene.[36] In soybean there are at least six *FAD 2* genes that fall into two classes, *FAD2-1* and *FAD2-2*. The *FAD2-1* class is primarily embryo specific, while the *FAD2-2* class is generally constitutive, expressed during both vegetative and seed-developmental stages.[37] The seed palmitate content is controlled by a palmitoyl-ACP thioesterase that releases palmitate from the plastid before it can be converted to stearoyl-ACP and then oleoyl-ACP. This thioesterase is encoded

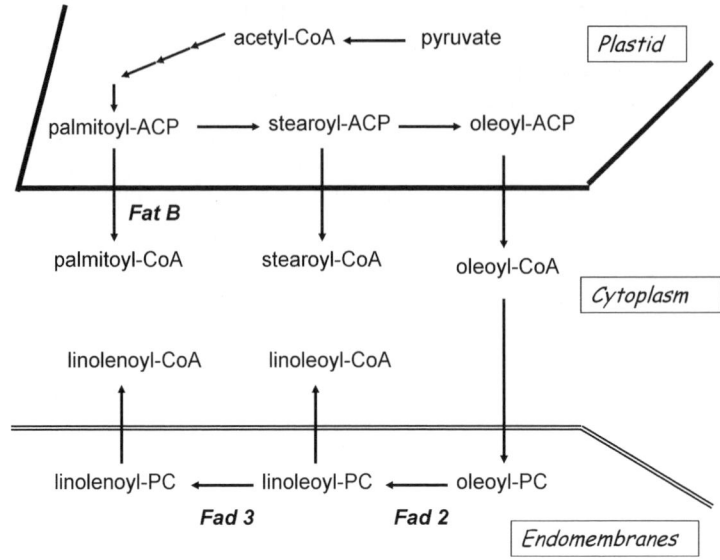

Figure 8.1 Simplified diagram of fatty acid biosynthesis in developing soybean seeds. The acyl-CoA pool in the cytoplasm is the result of reactions in the plastid and endomembranes and activities that move acyl chains among these cell compartments. The final acyl composition of triacylglycerol (the main component of soybean oil) is a reflection of the relative abundance of the different acyl-CoA species in the cytoplasm. The abundance of palmitoyl-CoA in the cytoplasm and hence palmitate in the TAG, is controlled mostly by the product of *Fat B* genes, which are acyl-ACP thioesterases. Desaturation of monounsaturated acyl chains occurs in the endomembranes, catalysed by the desaturase gene products of the *Fad 2* and *Fad 3* gene families.

by a *FatB* gene (Figure 8.1) and there are four members of this gene family in the soybean genome.[38]

Many of the increased oleic acid and decreased palmitate genotypes reported in soybean have been associated with allelic variations in these structural genes encoding fatty acid biosynthesis enzymes.[39,40] Soybean breeders have generally been successful in moving these allelic variants into elite soybean germplasm to create genotypes with increase seed oleic acid and decreased linolenic, linoleic and palmitic acids.[41-43] The resulting oleic-acid content of these genotypes ranges from 30% to 70% of the total seed fatty acid content. Low palmitic acid soybean lines have also been made in this way.[44]

These high oleic, low palmitic genotypes have been a challenge to commercialize for a number of reasons, not least the large genotype by environment impact on the fatty acid phenotype, which means the actual oleic and palmitic content of the seed can vary considerably depending upon the location, season and genetic background of the line.[45] Genotypes that are 50% oleic acid in one environment can be 25% in another. Such variation will complicate product specifications leading to unacceptable performance standards required by food processors and biodiesel producers.

In addition to these phenotype stability issues, a number of other factors have blocked high oleic acid and/or low palmitate soybean genotypes from becoming commercialised. In all cases multiple genetic loci need to be stacked to create a high oleic acid or low palmitate phenotype that, when yield and stress resistance loci are taken into account, can make for a highly complex and resource-intensive breeding strategy.[42,43] In addition, and not surprisingly since these variant fatty acid alleles are often derived from mutangenised populations, these fatty acid lines often have a considerable yield drag.[45] The underlying biology triggering this yield drag is undoubtedly related to the whole plant growth effects resulting from some of the variant genes being constitutively expressed in soybean, thereby altering the fatty acid composition of vegetative tissue and inevitably leading to developmental problems in the plant.[46]

Fortunately, the availability of genes encoding fatty acid biosynthetic enzymes and the development of effective genetic transformation of soybean has allowed the creation of single locus, environmentally stable, high yielding, transgenic elite genotypes with very high seed oleic acid and low palmitate contents.[31] The genetic transformation of soybean is a key enabling technology for the production of these lines and will be briefly reviewed next.

8.5 Genetic Transformation of Soybean

The successful introduction of cloned genes directly into the soybean genome was first accomplished towards the end of the 1980s by the use of *Agrobacterium tumefaciens* infection of cotyledon explants[47] or microprojectile-mediated DNA delivery into apical meristems.[38] A few years later a protocol using microprojectile-mediated DNA delivery to soybean somatic embryos was developed.[48] The majority of transgenic soybeans in existence today have been made by methods targeting either cotyledonary explants or somatic embryos, although numerous other protocols have been described and used on a laboratory or production scale.[49] The advantage of the *Agorobacterium*-mediated method is that it results in a high percentage of low-copy transgenic events with a higher frequency of intact, nonrearranged transgenic insertions when compared with direct DNA delivery methods. The power of the soybean somatic embryo system is that somatic embryogenesis mirrors zygotic embryogenesis. Most importantly, protein and oil deposition during somatic embryogenesis is similar to that of zygotic embryos.[50] This permits rapid screening of events for the intended embryo-specific change in metabolite composition imparted by transgene expression, without waiting for whole plant regeneration, development and seed set.[50]

The time from transformation through plant regeneration to seed set is about eight months for *Agrobacterium* infection of cotyledon explants and twelve months for particle bombardment of somatic embryos.[47,50] Selection systems for both transformation methods include antibiotics, such as hygromycin, for research-related transformations and herbicide tolerance to glyphosate, glufosinate or sulfonureas for production transformations.[51–53]

8.6 Biotechnology Approaches to Improving Soybean-Oil Quality

One of the problems associated with breeding for seed-oil composition, as we have discussed above, is that changes in fatty acid composition of mutant genotypes can often be found in nonseed tissues. This compromises the health of the plant and leads to impaired agronomic yield. In contrast, using biotechnology approaches to produce novel oil traits in soybean results in seed-specific changes to fatty acid content. This specificity is achieved by the use of tissue-specific promoters. These promoters are derived from genes that are normally only expressed in the seed, such as storage protein genes. Examples of these promoters include beta-conglycinin, Kunitz trypsin inhibitor 3 and glycinin, although a large toolbox of soybean-seed-specific promoters is now available.

The gene families controlling the relative abundance of soybean seed fatty acids are well understood (Figure 8.1). Key to controlling the palmitate and oleic-acid content of the seed oil are the *Fat B* and *Fad 2* gene families described above. Downregulation of the *Fat B* gene family results in reduced palmitate and increased oleic acid. This is because the *Fat B* gene controls a palmitoyl-ACP thioesterase that, in effect, makes palmitate available for incorporation into TAG. When the thioesterase activity is removed, palmitoyl-ACP stays in the cell plastid and is converted, first to stearoyl-ACP and then to oleoyl-ACP. The oleoyl-ACP is released into the cytoplasm as oleoyl-CoA that is then available for incorporation into TAG or further desaturation to linoleic and linolenic acids. Downregulation of the *Fad 2* gene family will result in reduced flux from oleic acid into polyunsaturated fatty acids, thus enhancing the monounsaturated fatty acid content at the expense of linoleic and linolenic acids. Blocking *Fad 2* also causes a reduction in palmitate in the oil, probably as a result of the changed oleoyl-CoA and palmitoyl-CoA pools in the cell cytoplasm.[50,55]

Both *Fad 2* and *Fat B* genes can be stacked in a single transgene construct, resulting in genetically modified plants with a fatty acid modification controlled by a single, dominant locus.[54] Having a single locus greatly simplifies the breeding process and permits the fatty acid trait to be stacked with numerous agronomic traits.

Downregulating endogenous gene expression in plant cells may be achieved by inducing posttranscriptional gene silencing through the use of RNA interference (RNAi) or the use of targeted artificial microRNA (amiRNA).[56–58]

In RNAi, small pieces of the endogenous target gene are reintroduced into the plant genome in a sense, antisense or hairpin configuration. This induces a silencing of the endogenous target gene in the plant *via* pathways that have probably evolved as viral defense mechanisms. Indeed, silencing of endogenous genes has been shown to be induced by plant viruses (VIGS) by a similar mechanism.[56] For amiRNA, endogenous microRNA, which are endogenous regulators of gene expression, are redesigned to target a gene of choice rather than their usual target. These amiRNA ultimately mediate silencing through a mechanism common to RNAi and VIGS. Silencing of *Fad 2* genes may also be induced through artificial *trans*-acting siRNA (TAS).[57] This involves the engineering of a gene (TAS1c) that controls the silencing of an endogenous miRNA in *trans*. Replacing the small interfering RNA (siRNA) target of TAS1c with a Fad 2 target in *Arabidopsis* led to an increase in oleic acid.[57]

A number of increased oleic, reduced palmitic acid transgenic soybean lines have been described. The first of these, Event 260-05, dates back to the early 1990s.[55] This event was made by using a *Fad 2* cosuppression construct under the control of a seed-specific promoter, beta-conglycinin. The gene was introduced into the soybean genome by particle bombardment of meristem.[59] The resulting seed oil had an increase in oleic acid from 20% to 85% and a reduction in palmitic acid from 10% to about 6%.[55]

This change in seed fatty acid composition was stable across environments and growing seasons and the seed fatty acid modification did not result in any impaired agronomics of the soybean plant.[50]

The creation of a second event was described in 2002. In this case, a silencing genetic element was used to target both the *Fad 2* and the *Fat B* genes in the soybean seed. The *Fat B* and *Fad 2* silencer was under the control of a seed-specific ß-conglycinin promoter. The gene construct was introduced into the soy genome using *Agrobacterium* infection of cotyledon tissue. The resulting seed oil had an oleic-acid content of greater than 90% and a palmitic acid content of less than 3%.[54] Extensive field testing of this event in different genetic backgrounds was conducted over several years under both irrigated and dryland conditions. In all cases the phenotype remained stable and agronomic yield was not in any way compromised.[18]

The development and characterisation of these two events demonstrated that it was possible to create commercially viable soybean genotypes with drastically altered seed fatty acid content. These alterations in seed content result in oils that can replace many hydrogenated oils for food use and, at the same time, result in oils that are theoretically a better feed stock for biodiesel production. Recently, new high oleic, low palmitic acid soybean events have been introduced by two major seed companies, under the names Plenish™ and Vistive Gold™, and oil from these events will be commercially available over the next few years.[60]

8.7 Properties of High-Oleic Soybean Oil in Biodiesel Applications

Biodiesel from high-oleic, low-palmitic acid soybean oils meets the theoretical criteria for improving biodiesel fuel. High-oleic-acid soybean has been shown to have a much greater oxidative stability than commodity soybean oil that should lead to reduced NO_x emissions, higher cetane values and better ignition quality.[61] The reduction in saturated fatty acids should result in improved cold flow.[61]

Early studies with high oleic acid soybean oil from the 260-05 event measured NO_x emissions from a 1992 Detroit Diesel Corporation Series 60 engine. Results from neat (B100) high-oleic diesel were compared with commodity B100 soybean diesel and with commercially available on-road petroleum diesel. In these studies the commodity soy biodiesel showed an increase of around 20% in brake-specific NO_x ($BSNO_x$) emissions when compared with petroleum diesel.[63] In contrast, the high-oleic diesel showed increases in $BSNO_x$ emissions of less than 10% when compared with petroleum diesel.[63]

Biodiesel derived from the oil of soybean event 335-13[54] has been more extensively tested. Initial studies with a John Deere 4045T engine showed a significant improvement in $BSNO_x$ emissions of the high oleic B100 soy biodiesel when compared with commodity B100 soy biodiesel. As in the 260-05 study the increase in $BSNO_x$ emissions of the high-oleic diesel over petroleum diesel was about half that of commodity soy diesel.[22] Other than that, commodity and high oleic fuels were very similar in the tests performed in this study. Both commodity and high-oleic-acid biodiesel showed a very large reduction in unburned hydrocarbon and smoke emissions when compared with petroleum diesel, reflecting the environmental attributes associated with the use of biodiesel.

A similar study of methyl esters from 335-15 oil with a 5.9-l Cummins diesel engine compared No. 2 petroleum diesel, soy biodiesel, 335-13 soy biodiesel and canola biodiesel. At low and high loads. At low loads all the biodiesel fuels had less $BSNO_x$ emissions than the petroleum diesel. At high loads the commodity biodiesels had higher $BSNO_x$ than petroleum diesel but the 335-13 soy biodiesel had lower $BSNO_x$ emissions than all the other fuels. In parallel experiments with a John Deere 3150 engine $BSNO_x$ emissions were higher for all fuels tested and there was very little difference between petroleum diesel, commodity soy biodiesel, canola biodiesel and 335-13 biodiesel fuels.[18] Thus, engine design appears to play an important role in the control of NO_x emissions and makes comparisons between different emission studies conducted with different engines difficult. Biodiesel manufacture can also make a difference. For example, isopropyl esters of high oleic acid soy biodiesel had significantly improved properties when compared with the more common methyl esters.[18] It should also be noted that the $BSNO_x$ improvements with high-oleic-acid soy biodiesel are usually more dramatic in neat (B100) biodiesel formulations. Only small differences have been reported for the more commonly used B20 blends.[18,22,62]

Reducing polyunsaturates should also improve the ignition quality parameter, cetane number, of a biofuel that is the case with methyl and isopropyl esters derived from 335-13 oil. But reducing double bonds should negatively impact cold-flow properties since oleic acid has a lower melting point than linoleic and linolenic acids. However, high oleic acid soybean oils are all lower in saturated fats than commodity soy oil and this has the larger effect. In methyl

Table 8.1 Cold-flow properties of biodiesel and petroleum diesel test fuels. The results demonstrate that the key properties determining engine performance in cold conditions (the cloud point and pour point values) of high oleic acid soybean diesel (event 335-13) are comparable to, or better than, petroleum diesel. The biodiesel fuels rich in oleic acid (made from HO soy and canola oils) had improved cetane values compared with commodity soy biodiesel. Data from *Graef et al.*[18] Abbreviations: HO, high oleic; FAME, fatty acid methyl esters; FAIE, fatty acid isopropyl esters.

Fuel Type	Cetane Number	Cloud Point (°C)	Pour Point (°C)
No. 2 Diesel	0.846	−6	−9
Soybean FAME	0.883	−1	0
HO Soybean FAME	0.881	−5	−9
HO Soybean FAIE	0.870	−10	−18
Canola FAME	0.879	−2	−9

esters derived from 335-13 oil the major parameters of cold flow, cloud point and pour point were significantly improved over commodity soy and very close to the values for petroleum diesel (Table 8.1).[18] This observation is quite significant for soybean biodiesel usage in colder climates. For example, the northern US state of Minnesota, which currently requires all diesel to contain 5% biodiesel, has recently passed legislation mandating the use of B10 biodiesel blends by 2012 and B20 blends by 2015.[63] However, this requirement will only be enforced for seven months of the year, with a fall-back to B5 in the winter months due to cold flow issues. Indeed, in a recent very cold winter (2009), even the requirement for B5 was suspended.[64] Use of high-oleic soy biodiesel with properties similar to that of the 335-13 soybean oil would allow these blends to be used year-round, potentially increasing soy biodiesel usage in that one state alone by over 100 million litres per year.[63]

8.8 Future Possibilities: Increasing Soybean Oil Content

While demand, legislation and environmental concerns are driving a portion of soybean production from food and feed uses to energy use, there are two major limitations to the rate and extent of this expansion of soybean as a bioenergy feedstock. Clearly, the current soybean oil supply cannot meet the market demand for total displacement of diesel fuel worldwide. It has been estimated that the entire annual US soybean oil production would replace only 6% of the current US consumption of petroleum diesel fuel, which is about 150 million metric tons.[11] Indeed, this rate of annual diesel consumption in the US alone could not be met by the entire world production of soybean oil.[65] For this reason arguments have been made for targeting the use of soybean biodiesel.[18] Limiting its use to public transport, for example, where 100% conversion could be mandated, might have a broader and more immediate effect on reducing particulate carbon in an urban environment.[66] It may also have a more direct impact on human health since the interior air quality of buses running on biodiesel is measurably superior to those running on petroleum diesel.[67] Targeting soybean biodiesel to marine applications would directly benefit marine environments because of the improved biodegradability of fatty acid methyl esters when compared with petroleum products.[68]

The other related factor that limits the expanded use of soybean biodiesel is cost. Even with the economy of scale in countries like US and Brazil (where soybean oil is a commodity), the low nitrogen input requirement of soybean and the meal credit obtained from the use of soy protein in animal feed, the cost of the soybean oil feedstock still accounts for nearly 90% of the cost of soy methyl ester biodiesel.[65] Nevertheless, in terms of carbon balance, soybean biodiesel compares very favourably to corn ethanol. It has been calculated that corn ethanol yields about 25% more energy than the total energy invested to produce it.[12] Soybean, on the other hand, yields about 95% more biodiesel energy than the carbon energy used to produce it, which is a strong environmental argument for increasing soy biodiesel supply to help meet demand.[12] In addition, there is the ever-growing demand for soybean oil for food uses, which is still by far the dominant market. Thus, an important goal for the soybean industry is to increase the supply of soybean oil to meet the increasing demand for biodiesel.

One means of increasing the supply of soybean oil would be to increase the oil yield per hectare. Soybean currently produces a little less than 500 l of oil per hectare, which is about half the oil yield of canola and about one fifth the oil yield of palm.[65]

There have been a number of attempts to increase the oil content of soybean seed through metabolic engineering.[69] The source of carbon for oil biosynthesis in the seed is pyruvate from glycolysis, which is converted to fatty acids *via* acetyl-CoA. The fatty acids are assembled onto a glycerol backbone by the action of acyltransferases, one for each carbon position on the glycerol molecule, to form TAG, which is the major component of seed oil (Figure 8.2). These reactions are common to both membrane lipid and TAG synthesis with the exception of the final acylation to TAG, a reaction catalysed by a diacylglycerol acyltransferase (DGAT).[69] Expression of a fungal DGAT gene in soybean seeds led to a small but significant increase in the total oil content of seed, from an average of 20.5% to an average of 22%, caused by a 20-fold increase in total DGAT activity in the seed.[70] Larger increases, of 3 to 5% points, have been reported using other DGAT genes and related acyltransferase genes.[69,71] The other successful approach to increasing oil in some oilseeds has been the transgenic expression of global regulators of carbon supply such as wrinkled 1 (*wr1*) and leafy cotyledon 2 (*lec 2*).[72] While the transgenic expression of this type of regulator has not been reported for soybean, in *Arabidopsis* increases in oil content of several percentage points have been observed with the *wr1* gene.[72] Transgenic expression of combinations of DGAT and *lec 2* has also been reported in tobacco leaves, leading to high accumulation of TAG in

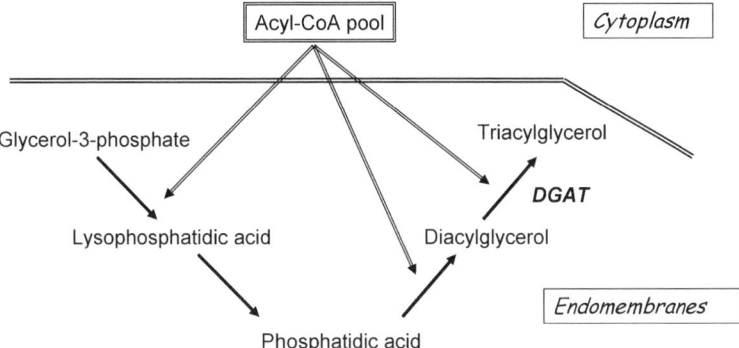

Figure 8.2 Simplified diagram of triacylglycerol biosynthesis in developing soybean seeds. The acyl-CoA pool, which consists of palmitoyl, stearoyl, oleoyl, linoleoyl and linolenyl-CoAs, provide the acyl chains for assembly onto the glycerol backbone. All of these reactions are common to phospholipid as well as triacylglycerol biosynthesis except for the final acyl addition, catalysed by a diacylglycerol acyltransferase (DGAT).

tissues that do not normally make oil. This combination of pathway genes and pathway flux regulator genes suggests a strategy for making large changes in the oil content of soybeans. In theory, it should be possible to double the oil content of the seed, at the expense of protein and carbohydrates, essentially converting soybean to an oilseed crop.[73,74] While the genes for this have not yet been identified the recent availability of the complete soybean genome, when combined with informatics, genetics and biochemistry, may provide a means to do so.[75]

Considerable research effort is currently being expended on increasing ethanol yields from corn by using leftover corn-stover material as a feedstock.[76] By analogy, one other possible means of extracting more energy from a soybean hectare would be to convert a soybean agricultural residue, such as the hulls, directly to a biofuel. Recent work has demonstrated that it is possible to engineer microbes to convert cellulose material from plants to simple sugars and then on to long chain fatty alcohols and wax esters that can be used directly as biodiesel.[77] Using residual soybean biomass for biodiesel production would not only increase the energy yield of a soybean crop but also begin to meet the increasing need for plant biodiesel without competing with the food supply.

References

1. T. Hymowitz, in *Advances in New Crops*, ed. J. Janick and J. Simon, Timber Press, Oregon, 1990, p. 159.
2. T. Hymowitz, *Crop Sci.*, 2005, **45**, 473.
3. E. R. Cober, S. R. Cianzo, V. R. Pantalone and I. Rajcan, in *Oil Crops*, ed. J. Vollmann, I. Rajcan, Springer, Dordrecht, 2009, p.57.

4. K. J. Smith and W. Huyser, in *Soybeans: Improvement, Production and Uses*, ed. J. R. Wilcox, American Society of Agronomy, Madison, WI, 1987, p.1.
5. FAOSTAT http://faostat.fao.org, accessed January 4th, 2010.
6. W. Shurtleff and A. Aoyagi, in *History of Soybean and Soyfoods in South America*, Soyinfo Center, California, 2009.
7. USDA Oils Crops Yearbook 2009 http://usda.mannlib.cornell.edu/MannUsda/viewDocumentInfo.do?documentID = 1290, accessed January 4th, 2010.
8. USB, http://www.unitedsoybean.org/programs/soy_biodiesel.aspx, accessed January 4th, 2010.
9. E. Johnson, *Biodiesel Magazine*, August/September Issue, 2005.
10. Index Mundi, http://www.indexmundi.com/commodities/?commodity = soybean-meal&months = 60, accessed January 4th, 2010.
11. A. J. Kinney and T. E. Clemente, *Fuel Proc. Technol.*, 2005, **86**, 1137.
12. J. Hill, E. Nelson, D. Tilman, S. Polasky and D. Tiffany, *Proc. Natl. Acad. Sci. (USA)*, 2006, **103**, 11206.
13. O. I Senol. Ph.D Thesis, Helsinki University of Technology, Finland, 2007.
14. S. A. Graham, *Crit. Rev. Food Sci. Nutr.*, 1989, **28**, 139.
15. D. P. Geller, J. W. Goodrum and C. C. Campbell, *Trans. ASAE*, 1999, **42**, 859.
16. T. Voelker and A. J. Kinney, *Ann. Rev. Plant Physiol. Plant Mol. Biol.*, 2001, **52**, 335.
17. N. Kalaitzandonakes, J. M. Alston and K. J. Bradford, *Nature Biotechnol.*, 2007, **25**, 509.
18. G. Graef, B. J. LaVallee, P. Tenopir, M. Tat, B. Schweiger, A. J. Kinney, J. H. Van Gerpen and T. E. Clemente, *Plant Biotech. J.*, 2009, **7**, 411.
19. X. Zhang, C. Peterson, D. Reece, D. Haws and G. Möller, *Trans. ASAE*, 1998, **41**, 1423.
20. M. S. Graboski and R. L. McCormick, *Prog. Energy Combust. Science*, 1998, **24**, 125.
21. D. C. Drown, K. Harper and E. Frame, *J. Am. Oil Chem. Soc.*, 2001, **78**, 579.
22. M. E. Tat, P. S. Wang, J. H. Van Gerpen and T. E. Clemente, *J. Am. Oil Chem. Soc.*, 2007, **84**, 805.
23. S. T. Bagley, L. D. Gratz, J. H. Johnson and J. F. McDonald, *Environ Sci. Technol.*, 1998, **32**, 1183.
24. N. Y. Kado, P. A. Kuzmicky, *Bioassay analyses of particulate matter from a diesel bus engine using various biodiesel feedstock fuels Report 3*. National Renewable Energy Laboratory, Golden, CO (2003).
25. D. Fowler, C. Flechard, U. Skiba, M. Coyle and J. N. Cape, *New Phytol.*, 1998, **139**, 11.
26. R. O. Dunn, *Fuel Proc. Tech.*, 2005, **86**, 1071.
27. J. P. Szybist, A. L. Boehman, J. D. Taylor and R. L. McCormick, *Fuel Proc. Tech.*, 2005, **86**, 1109.
28. P. Saiyasitpanic, T. C. Keener, M. Lu, F. Liang and S. J. Khang, *J. Air Waste Manag Assoc.*, 2008, **58**, 1311.

29. G. Knothe, A. C. Matheaus and T. W. I. Ryan, *Fuel*, 2003, **82**, 971.
30. E. B. Cahoon, T. E. Clemente, H. G. Damude and A. J. Kinney, in *Oil Crops*, ed. J. Vollmann, I. Rajcan, Springer, Dordrecht, 2009, p. 31.
31. A. J. Kinney, in *Functional Foods and Biotechnology*, ed. K. Shetty, G. Paliyath, A.L. Pometto, R.E. Levin, CRC Press, Florida, 2007, p. 85.
32. B. R. Moser, M. J. Haas, J. K. Winkler, M. A. Jackson, S. Z. Erhan and G. R. List, *Eur. J. Lipid Sci. Tech.*, 2007, **109**, 17.
33. J. R. Wilcox, *Crop Sci.*, 2001, **41**, 1711.
34. S. E. Hawkins, W. R. Fehr and E. G. Hammond, *Crop Sci.*, 1983, **23**, 900.
35. J. W. Burton, R. F. Wilson, C. A. Brim and R. W. Rinne, *Crop Sci.*, 1989, **29**, 1583.
36. E. P. Heppard, A. J. Kinney, K. L. Stecca and G. H. Miao, *Plant Physiol.*, 1996, **110**, 311.
37. G. Q. Tang, W. P. Novitzky, H. C. Griffin, S. C. Huber and R. E. Dewey, *Plant J.*, 2005, **44**, 433.
38. A. J. Kinney, in *Genetic Engineering*, ed. J. K. Setlow, Plenum Press, New York, 1997, p.149.
39. J. R. Byrum, A. J. Kinney, K. L. Stecca, D. J. Grace and B. W. Diers, *Theor. Appl. Genet.*, 1997, **94**, 356.
40. A. J. Cardinal and J. W. Burton, *Crop Sci.*, 1997, **47**, 1804.
41. Y. Takagi and S. M. Rahman, *Theor. Appl. Genet.*, 1996, **92**, 179.
42. J. L. Alt, W. R. Fehr, G. A. Welke and D. Sandu, *Crop Sci.*, 2005, **45**, 1997.
43. J. L. Alt, W. R. Fehr, G. A. Welke and J. G. Shannon, *Crop Sci.*, 2005, **45**, 2005.
44. D. M. Bubeck, W. R. Fehr and E. G. Hammond, *Crop Sci.*, 1989, **29**, 652.
45. V. S. Primomo, D. E. Falk, G. R. Ablett, J. W. Tanner and I. Rajcan, *Crop Sci.*, 2002, **42**, 37.
46. M. Miquel, D. James, H. Dooner and J. Browse, *Proc. Natl. Acad. Sci. USA.*, 1993, **90**, 6208.
47. M. A. Hinchee, D. V. Connor-Ward, C. A. Newell, R. E. McDonnell, S. J. Sato, C. S. Gasser, D. A. Fischhoff, R. T. Fraley and R. B. Horsch, *Biotechnology*, 1988, **6**, 915.
48. D. E. McCabe, W. F. Swain, B. J. Martinell and P. Christou, *Biotechnology*, 1988, **6**, 923.
49. W. A. Parrott and T. E. Clemente, in *Soybeans: Improvement, Production, and Uses*, ed. H. R. Boerma and J. E. Specht, American Society of Agronomy, Madison, WI, 2004, p. 265.
50. A. J. Kinney, *J. Food Lipids*, 1996, **3**, 273.
51. T. Clemente, B. J. LaVallee, A. R. Howe, D. Conner-Ward, R. Rozman, P. Hunter, D. L. Broyles, D. Kasten and M. A. Hinchee, *Crop Sci.*, 2000, **40**, 797.
52. P. M. Olhoft, L. E. Flagel, C. M. Donovan and D. A. Somers, *Planta*, 2003, **216**, 723.
53. Z. Zhang, A. Xing, P. E. Staswick and T. E. Clemente, *Plant Cell Tiss. Org. Cult.*, 1999, **56**, 37.

54. T. Buhr, S. Sato, F. Ebrahim, A. Xing, Y. Zhou, M. Mathiesen, B. Schweiger, A. Kinney, P. Staswick and T. Clemente, *Plant J.*, 2002, **30**, 155.
55. A. J. Kinney and S. Knowlton, in *Genetic Modification in the Food Industry*, ed. S. Roller and S. Harlander, Blackie, London 1998, p.193.
56. J. M Watson, A. F. Fusaro, M. Wang and P. M. Waterhouse, *FEBS Lett.*, 2005, **579**, 5982.
57. M. de la Luz Gutiérrez-Nava, M. J. Aukerman, H. Sakai, S. V. Tingey and R. W. Williams, *Plant Physiol.*, 2008, **147**, 543.
58. J. P. Alvarez, I. Pekker, A. Goldshmidt, E. Blum, Z. Amsellem and Y. Eshed, *Plant Cell*, 2006, **18**, 1134.
59. P. Christou, W. F. Swai, N. S. Yang and D. E. McCabe, *Proc. Natl. Acad. Sci. USA*, 1989, **86**, 7500.
60. www.dupont.com; www.monsanto.com (last accessed January 4th, 2010).
61. G. Knothe, *Fuel Proc. Tech*, 2005, **86**, 1059.
62. D. Lyons and G. Thompson, personal communication.
63. S. R. Schill, *Biodiesel Magazine*, July 2008 issue.
64. J. Lane, *Biofuels Digest*, January 14th, 2010 issue.
65. T. P. Durrett, C. Benning and J. Ohlrogge, *Plant J.*, 2008, **54**, 593.
66. A. Vijayan, A. Kumar and M. A. Abraham, *J. Transport. Res. Board.*, 2008, **2058**, 68.
67. A. Kumar, V. Kumar and V. Nerella, *Open Environ. Eng. J.*, 2009, **2**, 81.
68. X. Zhang, C. Peterson, D. Reece, D. Haws and G. Möller, *Trans. ASAE*, 1998, **41**, 1423.
69. K. Meyer and A. J. Kinney, in *Lipids in Photosynthesis*, ed. H. Wada and N. Murata, Springer, Dordecht, 2009, p. 407.
70. K. Lardizabal, R. Effertz, C. Levering, J. Mai, M. C. Pedroso, T. Jury, E. Aasen, K. Gruys and K. Bennett, *Plant Physiol.*, 2008, **148**, 89.
71. S.S. Rao and D. Hildebrand, *Lipids*, 2009, Sept. 19th, Epub ahead of print.
72. A. Cernac and C. Benning, *Plant J.*, 2004, **40**, 575.
73. V. Andrianov, N. Borisjuk, N. Pogrebnyak, A. Brinker, J. Dixon, S. Spitsin, J. Flynn, P. Matyszczuk, K. Andryszak, M. Laurelli, M. Golovkin and H. Koprowski, *Plant Biotechnol. J.*, 2009, **8**, 1.
74. A. S. Carlsson, *Biochimie*, 2009, **91**, 665.
75. J. Schmutz, S. B. Cannon, J. Schlueter, J. Ma, T. Mitros, W. Nelson, D. L. Hyten, Q. Song, J. J. Thelen, J. Cheng, D. Xu, U. Hellsten, G. D. May, Y. Yu, T. Sakurai, T. Umezawa, M. K. Bhattacharyya, D. Sandhu, B. Valliyodan, E. Lindquist, M. Peto, D. Grant, S. Shu, D. Goodstein, K. Barry, M. Futrell-Griggs, B. Abernathy, J. Du, Z. Tian, L. Zhu, N. Gill, T. Joshi, M. Libault, A. Sethuraman, X. C. Zhang, K. Shinozaki, H. T. Nguyen, R. A. Wing, R. P. Cregan, J. Specht, J. Grimwood, D. Rokhsar, G. Stacey, R. C. Shoemaker and S. A. Jackson, *Nature*, 2010, **463**, 178.
76. F. Torney, L. Moeller, A. Scarpa and K. Wang, *Curr. Opin. Biotechnol.*, 2007, **18**, 193.
77. E. J. Steen, Y. Kang, G. Bokinsky, Z. Hu, A. Schirmer, A. McClure, S. B. del Cardayre and J. D. Keasling, *Nature*, 2010, **463**, 559.

CHAPTER 9
Perspectives on Sunflower as an Energy Crop

ZINA FLAGELLA AND MASSIMO MONTELEONE

Department of Agro-Environmental Science, Chemistry and Crop Protection, Foggia University, via Napoli, 25-71122, Foggia, Italy.

9.1 Worldwide Scenario of Sunflower Cultivation and Production

Sunflower is one of the leading oilseed crops cultivated for the production of oil, mainly for human consumption. It is also considered as an important crop for biodiesel production. During the period 2003–2007, the average annual world production of sunflower seed was about 28.5 million t from a cultivated area of around 22.6 million ha (Table 9.1). Europe is the most important sunflower producer in the world, with 17.0 Mt of total seed production representing approximately 60% of the sunflower seed harvested globally. The Americas are the second most important, producing 5.3 Mt of sunflower seed, whilst Asia approaches this, with 5.2 Mt of seeds produced. Africa is the smallest producer, achieving 0.9 Mt. Sunflower cultivation is mainly located in Eastern Europe, in South America, in Eastern and Southern Asia and in Eastern and Southern Africa (Table 9.1). Considering single countries, the Russian Federation, Ukraine, Argentina and India have the largest areas under production, representing over 60% of world cultivation (Table 9.2). In the European Union (EU) sunflower is mainly cultivated in Southern Countries: Romania (969 000 ha and 1295 kt), Spain (658 000 ha and 666 kt), Bulgaria

Table 9.1 Cultivated surface, seed production and yield of sunflower in different geographical regions, over the 2003–2007 period.[1]

	Surface (ha × 10^3)	Production (kt)	Yield (t/ha)
WORLD	22 643	28 475	1.3
EUROPE	13 116	16 968	1.3
Eastern Europe	11 384	13 946	1.2
Northern Europe	–	–	–
Southern Europe	1041	1409	1.4
Western Europe	690	1611	2.3
EUROPEAN UNION	3752	6166	1.7
AMERICA	3350	5340	1.6
Central America	82×10^{-3}	120×10^{-3}	1.5
Northern America	910	1362	1.5
South America	2439	3977	1.6
Caribbean	–	–	–
ASIA	5189	5173	1
Eastern Asia	1035	1774	1.7
Central Asia	490	337	0.7
Southern Asia	2440	1639	0.7
South-Eastern Asia	569	380	0.6
Western Asia	653	1040	1.6
AFRICA	935	945	1
Eastern Africa	326	249	0.7
Middle Africa	15	11	0.7
Northern Africa	110	124	1
Southern Africa	483	561	1.2
Western Africa	–	–	–

(648 000 ha and 913 kt), France (626 000 ha and 1457 kt), Hungary (508 000 ha and 1099 kt) and Italy (135 000 ha and 275 kt).

With respect to other oil crops that compete with sunflower as oil raw material and feedstock, the Americas are mostly characterised by the production of soybean (189.6 Mt in 2007), Asia is largely dominated by the production of oil palm fruit (164.5 Mt in 2007), while in Europe rapeseed is the primary oilseed crop with a production of 20.5 Mt.

The most important sunflower seed exporting countries, in 2007, were Hungary, France, Romania, Ukraine and Bulgaria, with commercial trades greater than 300 kt each, while the most relevant importing countries (in the same year) were Turkey (596 kt), Spain (359 kt), the Netherlands (356 kt), Pakistan (318 kt), Italy (261 kt) and Germany (243 kt). It can be estimated that approximately 11–12% of sunflower-seed production is traded on the international market, the remainder being processed in the country of cultivation.

The average world oil production obtained from sunflower seeds was about 12 Mt, over the period 2003–2007 (Table 9.3). Europe is the leading producer and sunflower-oil production in 2007 amounted to 7.3 Mt, corresponding to almost 63% of the entire world production. The Americas and Asia, each share

Table 9.2 Cultivated surface, seed production and yield of sunflower in the main producing countries over the period 2003–2007.[1]

Country	Area ($ha \times 10^3$)	% of world surface	Production (kt)	% of world production	Seed yield (t/hd)
Russian Federation	5190	22.9	5704	21.2	1.1
Ukraine	3649	16.1	4301	16	1.2
Argentina	2258	10	3558	13.2	1.6
India	2194	9.7	1249	4.6	0.6
China	1035	4.6	1774	6.6	1.7
Romania	969	4.3	1295	4.8	1.3
United States of America	833	3.7	1250	4.6	1.5
Spain	658	2.9	665	2.5	1.0
Bulgaria	648	2.9	912	3.4	1.4
France	626	2.8	1457	5.4	2.3
Turkey	549	2.4	929	3.5	1.7
Myanmar	534	2.4	352	1.3	0.6
Hungary	508	2.2	1099	4.1	2.2
South Africa	477	2.1	554	2.1	1.1
Kazakhstan	420	1.8	259	1	0.6
Pakistan	280	1.2	314	1.2	1.1
Moldova	254	1.1	318	1.2	1.2
Serbia and Montenegro	185	0.8	364	1.4	1.9
Uganda	157	0.7	172	0.6	1.1
Italy	135	0.6	275	1	2.0
Bolivia	105	0.5	108	0.4	1.0
Uruguay	104	0.5	137	0.5	1.3
Slovakia	97	0.4	201	0.7	2.1
United Republic of Tanzania	81	0.4	27	0.1	0.3
Canada	76	0.3	112	0.4	1.5
Islamic Republic of Iran	73	0.3	59	0.2	0.8
Brazil	65	0.3	83	0.3	1.3
Morocco	55	0.2	43	0.2	0.8
Kyrgyzstan	51	0.2	64	0.2	1.2
Others	362	1.6	825	2.9	2.3
Total	22 643		28 475		

17–18% of the world oil production, accounting for approximately 2.0 Mt of oil each. The greatest oil amounts were achieved by the countries producing the highest seed quantity: the Russian Federation, Ukraine and Argentina (Table 9.3) accounting together for 45% of global production.

On the world scene, the amount of oil derived from sunflower corresponds to only 8.9% of the overall oil production; palm oil and palm kernel oil together reach an impressive 33.4%, while soybean accounts for 28.1% and rape 13.2%.[1] Therefore, sunflower oil ranks as the fourth available vegetable oil in

Table 9.3 Sunflower oil production in the main countries over the 2003–2007 period.[1]

Country	Production (t)	% of total production
Russian Federation	2 116 388	17.5
Ukraine	1 834 190	15.2
Argentina	1 452 260	12.1
Turkey	542 843	4.5
France	468 756	3.3
China	451 375	3.8
India	433 680	3.6
Romania	338 020	2.9
Spain	324 563	2.7
United States of America	239 040	1.9
Hungary	235 700	1.9
South Africa	218 160	1.8
Netherlands	192 860	1.6
Italy	148 200	1.2
Bulgaria	141 247	1.2
Kazakhstan	128 279	1.0
Myanmar	121 803	1.0
Serbia	115 831	0.9
Moldova	104 860	0.9
Pakistan	104 261	0.9
Uganda	60 040	0.5
Slovakia	47 053	0.4
Others	2 218 772	18.4
Total	12 038 181	

the world. In 2007, Europe produced 6.5 Mt of rapeseed oil, a lower value but not too far from that of sunflower oil. The higher European sunflower-oil production, with respect to rapeseed oil, can be accounted for by considering the amounts of imported sunflower seeds. The most important sunflower-oil-exporting countries, in 2007, were Ukraine (1292 kt), Argentina (853 kt), Russian Federation (614 kt), the Netherlands (419 kt) and France (332 kt); with respect to imports, the most relevant countries (in the same year) were all European, and particularly Germany, United Kingdom and France (with amounts greater than 300 kt each), followed by the Netherlands, Belgium, Spain and Italy (with amounts larger than 200 kt each). The European Union (EU27) thus absorbs about two thirds of the whole import trade. Considering both import and export trades, the amount of sunflower oil that reaches the international markets is approximately 35% of the total oil produced.[1]

Strictly coupled with sunflower-oil production is the attainment of the press-cake or press-meal, a very useful byproduct. Press-cake is still rich in oil but also has a remarkable protein content and, for this reason, is mostly used in animal feeding. Assuming, very roughly, an average oil yield of 40% (highly variable according to the geographical area of cultivation) and an annual sunflower-oil production of 10 Mt, a world value of approximately 15 Mt of press-cake can be estimated. The import/export trade on the international market concerns 3.9 Mt of press-cake (about 26% of the estimated overall production). The most important oil-exporting countries (Ukraine, Russian Fed., Argentina, the Netherlands) are clearly also involved in press-cake exportation; on the other hand, the European Union (EU27) imports almost 60% of the traded product, in order to satisfy the internal demand for animal feed.[1]

The mean world yield of sunflower seed is 1.3 t ha^{-1}, the highest yield being registered in America, the lowest in Africa. Considering single countries, the mean annual seed yield ranged from 0.57 t ha^{-1} in India to 2.3 t ha^{-1} in France during the 2003–2007 period (Table 9.2). These seed yields were much lower than the potential yield of the sunflower crop. However, it might be possible to increase seed yield through a better control of limiting factors. Sunflower seed yield and oil content both depend on their potential values and on the availability of water and nutrients. They are also affected by diseases, pests and weeds as well as by environmental constraints such as high lightning, flooding, frost, and hail.[2] In particular, the water available to the plant is the most limiting factor of sunflower seed yield in the dryland agriculture of the semiarid regions[3] where this crop is mainly grown.

Sunflower may be irrigated, but irrigation water is often used for the production of crops with greater economic return. For instance, in France only 4% of the area cultivated under sunflower in 2006 was irrigated.[4] However, sunflower is a drought-tolerant crop, being able to withstand short periods of water deficit through deep rooting.[2,5] Supplementary irrigation has been shown to have a key role at critical growth stages, particularly sensitive to water stress[6] resulting in similar yields as for periodic replenishment of evapotranspiration calculated on the basis of a meteorological approach.[7] So, the potential exists for sunflower to become an important crop also in subarid environments and whenever irrigation water is limited.

9.2 Survey of Biodiesel Production and Trade

From the beginning of the new century, the production of biodiesel, together with bioethanol, has been marked by an impressive and extraordinary expansion, which some observers have defined as a "biofuel frenzy".[8] Actually, the growth rates of biodiesel production have been even more striking than bioethanol, although its absolute production is still much less than the latter. At world scale, biodiesel production increased sixfold from 2 billion liters in 2004 to at least 12 billion liters in 2008.[9]

In the four-year period 2003–2006 the biodiesel world production was equal to 3.89 Ml, but in 2007 it reached 10.58 Ml.[10] Europe is still the world-leading aggregate country in biodiesel production (with 3.16 Ml in 2003–06 and 6.36 Ml in 2007), although the extra-European countries have recently improved their production rate considerably: North America reached 1.96 Ml in 2007; Asia and Pacific 1.32 Ml; Latin America 0.99 Ml. Today, on a world scenario, these greater expansion rates compared to Europe, result in a lower relative influence of Europe with respect to only a few years ago. The European share of biodiesel world production was 60.1% in 2007, compared to 81.4% in 2003–06; in a complementary way, North America reached 18.5% (9.7% in 2003–06), Asia and Pacific 12.5% (7.1 in 2003–06), Latin America 8.5% (1.5% in 2003–06). The EU is thus responsible for about two-thirds of world biodiesel production, with Germany, France and Italy being the top three EU producers. According to the European Biodiesel Board (EBB), 2008 production was increased by 35.7% compared to 2007; that of 2007 was increased by 16.8% compared to 2006; whilst in 2006 production was increased by 54% compared to 2005.

By the end of 2008, EU biodiesel production capacity reached 16 billion liters per year. Outside of Europe, the top biodiesel producers include the United States, Argentina, Brazil, and Thailand.[9]

Biofuel industries expanded dramatically during 2007/2008, especially in North America and Latin America, and to a lesser extent in Europe. Argentina became a major biodiesel producer in 2008, with 18 commercial plants in operation, all producing for export; another 16 plants were expected during 2009 to bring capacity to 1.8 billion liters per year. In Europe, more than 200 biodiesel production facilities are currently operating.[9]

There are several oil-producing plants and crops that can be used as feedstock for biodiesel production; it should be noted that not only virgin vegetable oils but also used fryer oil, animal fats and pond algae (as well as micro- and macroalgae exclusively cultivated) can be used to produce the same biofuels, with minor differences in fuel characteristics. Limiting our interest to oil-bearing crops only, the main feedstocks used worldwide for biodiesel production, based on a rough estimate, changed significantly their relative contribution in recent years. Comparing the first (2005) and the second edition (2008) of Pahl's book "Biodiesel: growing a new energy economy",[11] it is possible to notice that the contribution of sunflower oil to biodiesel decreased, in relative terms, from 13% in 2005 to 5% in 2007. A similar trend was also observed for rapeseed oil, which was the global leader in 2005 (84% of the biodiesel feedstock), but lowered its value to 59% in 2007. These remarkable reductions are exactly counterbalanced by the relative increase of both palm oil and soybean (from 1 to 10% and from 1 to 25%, respectively). These latter trends are closely related to the increased capacity of North America and Asia and Pacific to produce biodiesel from their domestic and most representative crops.

Assuming that the fraction of sunflower oil used for biodiesel manufacturing is approximately equal to 5% and also considering a 1:1 ratio between oil and biodiesel weight, an overall biodiesel production of 8.8–9.7 Mt (10–11 000 Ml) corresponds to approximately 0.45–0.50 Mt of sunflower oil used globally to produce biodiesel. This quantity, in relative terms, represents about 4–5% of the overall production of sunflower oil.

More recently, the demand for biodiesel has increased even more due to recent developments such as the fluctuation of petroleum prices and the development of government measures. These include directive 2009/28/EC and the amending and repealing directives 2001/77/EC and 2003/30/EC, aimed at promoting renewable fuels for transport and the US Energy policy act (EOAct) of 1992, aimed at accelerating the use of alternative fuels in the transportation sector. The ambitious targets defined in Europe for the development of biofuels are all aimed at improving European domestic energy security, overall CO_2 balance and European competitiveness.

9.3 Biodiesel and Oil Quality in Relation to Genetic and Environmental Influence

Biodiesel, defined as the monoalkyl esters of vegetable oils or animal fats,[12] has clear benefits in comparison with diesel fuel. It is a renewable fuel with a favourable energy balance that constitutes an alternative to petroleum-based diesel fuel. It is nontoxic and biodegradable and has a low emission profile all of which is better for environmental sensitive areas. Furthermore, biodiesel is compatible with commercial diesel engines and practically no engine modifications are required. Biodiesel is produced from vegetable oils, which are derived from the seeds or the pulp of a range of oil-bearing crops. These oil crops can be annual (rapeseed, sunflower, groundnut, soybean) or perennials (oil palms, coconut palms, physica nut, Chinese tallow tree). Oil from rapeseed was the first type used for biodiesel production. Biodiesel from rapeseed oil shows good stability and winter performance. Over time, many other oils have been used successfully as biodiesel feedstocks. They include sunflower oil in southern France and Italy; soybean oil in the USA and palm oil in Malaysia. Today, in Europe, rapeseed is still the main feedstock for biodiesel production. Rapeseed is grown throughout Europe, while sunflower seed crops are grown in the warmer areas only. Due to their physical and chemical properties, biodiesel fuels are appropriate for combustion in compression-ignition engines (diesel engines). Moreover, due to the increasing interest and use of biodiesel the assurance of fuel properties and quality has a key role for successful commercialisation and market acceptance requires. So, biodiesel standards have been established or are being developed in different countries and regions at a global level, including the United States (ASTM D 6751), Europe (EN 14214), Brazil, South Africa and Australia.[13]

Some of the main specifications in the ASTM biodiesel standard and for the European standards EN14214 are compared in Table 9.4. These standards are based on the physical and chemical properties needed for satisfactory diesel-engine operation.

Several quality parameters of biodiesel such as density, kinematic viscosity, cetane number and iodine value are highly dependent on fatty acid composition. Viscosity, is higher for biodiesel than for standard diesel. The reduction in viscosity is the major reason why alkyl esters of vegetable oils—biodiesel—are used as fuel and not the neat oil. The cetane number depends on the ignition delay time of the fuel upon injection into the combustion chamber. Generally, the shorter the ignition delay time, the higher the cetane number and the higher

Table 9.4 Comparison between some European and American Biodiesel Standards (modified from Knothe[13]).

Property	Europe				America		
	Test Method	Limits		Units	Test Method	Limits	Units
		EN 14214	EN 14213				
Density; 15 °C	EN ISO 3675, EN ISO 12185	860–900	860–900	kg/m³			
Viscosity; 40 °C	EN ISO 3104, ISO 3105	3.5–5.0	3.5–5.0	mm²/s	D445	1.9–6.0	mm²/s
Cetane number	EN ISO 5165	51 min	—		D 613	47 min	
Heating value	DIN 51900-1 DIN 51900-2	—	32 min	MJ/kg			
Flash point	EN ISO 3679	120 min	120 min	°C	D 93	130.0 min	°C
Sulfur content	EN ISO 20846; EN ISO 20884	10.0 max	10.0 max	mg/kg	D 5453	0.0015 max (S 15) 0.05 max (S500)	% mass
Acid value	EN 14104	0.50 max	0.50 max	mg KOH/g	D 664	0.50 max	mg KOH/g
Oxidative stability, 110 °C	EN 14112	6.0 min	4.0 h	h			
Iodine value	EN 14111	120 max	130 max	g I₂/100 g			
Linolenic acid content	EN 14103	12.0 max	—	% (mol/mol)			

the fuel quality. The minimum cetane numbers prescribed in ASTM D 6751 and EN 14214 exceed those in petrodiesel standards.

The iodine value is a measure of total unsaturation of a lipidic material. Furthermore, the restrictions on the fatty acid profile, contained mainly in EN 14214, are present to exclude those components of biodiesel with less desirable properties, for instance, with respect to oxidative stability. For example, the content of methyl linolenate is restricted in EN 14214 because of its propensity to oxidize. However, the limit (12%) is set so as not to exclude high-oleic rapeseed oil, the major biodiesel source in Europe.

Biodiesel can also be used as heating oil. Accordingly, a separate standard (EN 14213) exists in Europe for biodiesel that is to be used as heating oil. While most specifications are the same or very similar, the requirements on oxidative stability are lighter in the heating-oil standards. Moreover, there is no restriction on the amount of linoleic acid although the iodine value is set at 130. Finally, the cetane number, for example of the specifications of group II and group I metals, such as methanol and phosphorous, are not included.

9.3.1 Genetic, Environmental and Agronomic Influences on Sunflower-Oil Quality

With respect to sunflower oil used for biofuel production, the importance of maintaining a high quality of raw materials is similar to any fuel used in modern diesel engines. Knowledge of the genetic and environmental influences on oil quality is helpful to evaluate if biodiesel produced from a given raw material could meet the specification of a given quality standard.

Oil quality and yield are both dependent on the genotype and its interaction with the environment. In the case of high-oleic sunflower, the genotype can be the main determinant.[6,14] In contrast, for traditional sunflower genotypes, oil quality and yield largely depend on environment, which is highly variable among years, locations and sowing dates within a single year. It should also be stressed that intraspecific variation in oil quality can be larger, in some cases, than interspecific differences. In particular, mid- and high-oleic-acid oils combined with a high content of *in vitro* antioxidants are usable for biodiesel and other applications requiring high-temperature processing (*e.g.* deep frying and biodegradable lubricants).

9.3.1.1 Effect of Genotype

Sunflower oil is among the highest quality vegetable oils on the market. Oil quality is determined by the fatty acid composition and the levels of tocopherols, sterols, carotenoids and other compounds. Standard sunflower oil is predominantly composed of linoleic acid (C 18:2) and oleic acid (C 18:1). These two acids account for about 90% of the total fatty acid content of sunflower oil. The remaining 8–10% consists of palmitic and stearic acid (C 16:0 and C 18:0, respectively) and of several other fatty acids usually found only in traces.[15,16] During the last 30 years these components have been extensively modified in sunflower through conventional selection from naturally occurring variation and through mutagenesis.[17] A wide range of sunflower lines with contrasting fatty acid profiles have been developed, for example with high and low levels of saturated fatty acids, mid and high levels of oleic acid, high concentrations of linoleic acid, as well as different ranges of intermediate levels of the different fatty acids and combinations among them.[17] Besides the genetic manipulation of the higher fatty acid composition, lines with high levels of β-, γ- and δ- tocopherol have also been developed. In standard sunflower oil α-tocopherol accounts for more than 95% of the total tocopherols.[18] Large variation for total tocopherol content has been reported in sunflower, in particular, from 480 to 1128 mg/kg oil,[19] and from 534 to 1858 mg/kg oil.[18] For α-, β- and total tocopherol content the effect of the genotype was larger than that of the environment, whereas the latter had a greater effect on γ-tocopherol content.[20] Genetic improvement has resulted in more variability for tocopherol profiles in sunflower oil than in any other oilseed crop. The combination of several quality traits in a single phenotype will enable the tailoring of speciality oils for specific uses in the food and nonfood industries. The novel fatty acids and tocopherol traits are governed by a reduced number of genes and can be easily managed in breeding programs aimed to develop cultivars incorporating these traits. High-oleic hybrids are suitable for biofuel production. According to Fernandez Martinez et al.[17] different modes of inheritance of a high oleic-acid content have been reported. In one study, a strong positive correlation between a δ12 RFLP marker and the high-oleic mutant was revealed.[21]

9.3.1.2 Effect of Temperature, Sowing Date and Water Regime

Temperature is the main environmental factor affecting sunflower oil quality, especially during grain filling.[22] *In vivo* experiments with developing sunflower seeds demonstrated that oleate desaturase activity was stimulated by low temperature, repressed by high temperature and rapidly restored when seeds were once again exposed to low temperature.[23] Further studies[24] confirmed the influence of temperature on fatty acid composition in both standard and high oleic sunflower hybrids. Izquierdo and Aguirrezàbal[25] found that increasing temperature increased the sum of oleic and linoleic acid and reduced the concentration of saturated fatty acid. Genetic variability in the response of these fatty acid to temperature was also observed. A larger variation in oleic-acid content was found in standard hybrids compared with high-oleic ones.[14] Roche et al.[26] noted an increase both in oleic acid and phytosterols under high temperature.

Due to the ability of sunflower to tolerate short periods of water deficit,[5,27] the potential also exists for it to become an important crop in subarid environments and wherever irrigation water is limited. In such environments, early sowing allows the crop to benefit from late winter rainfall and water can be given only as supplementary irrigation to sustain yield.

The sowing date may also influence fatty acid oil composition, probably by modifying ontogenesis. In particular, Unger[28] and Jones[29] found that linoleic and oleic acid concentrations of the oil consistently increased and decreased, respectively from the first (spring) to the last (summer) sowing date, when cultivated in Texas. Cilardi et al.[30] found the same results when sunflower was grown in Southern Italy as a catch crop in summer. Flagella et al.[6] evaluated the effect of two spring sowing dates: end of March and mid-April; they found a decrease in oleic acid and a considerable increase in linoleic acid in the earlier sowing date, both in standard and high-oleic hybrids. In contrast, in Florida, Robertson and Green[31] found that oleic-acid content was intermediate for the February planting, highest for the April planting and lowest for the late July planting; the linoleic-acid content varied inversely with the oleic-acid content. The contrasting results reported in the literature are due to the different experimental conditions adopted. In fact, the diverse planting dates caused flowering and seed development to occur in periods with different air temperatures. Consequently, the highest oleic-acid content was obtained when the mean temperature was higher during grain filling. This effect is probably due to the inhibition of oleate desaturase at high temperatures.

In relation to irrigation, despite the positive effects observed on yield parameters, contrasting results were obtained for oil fatty acid composition. In particular, Talha and Osman[32] reported an increase in the oleic/linoleic acid ratio under water stress. Conversely, Unger[33] found a positive correlation between oleic-acid content and water use at the vegetative stage, while Salera and Baldini[34] observed no effect of water management on oleic-acid content. When water stress occurred during the grain-filling period on standard and high-oleic genotypes, an increase in the oleic/ linoleic acid ratio was observed with respect to more favourable water regimes in Southern[6,35] and Northern Italy.[36]

Water quality can also influence sunflower yield and quality. Salt stress was shown to significantly affect fatty acid oil composition of high-oleic and standard sunflower hybrids.[37,38] In particular, an increase in oleic acid and a consistent decrease in linoleic acid were observed on increasing salinity level. A possible inhibition of oleate desaturase under water deficit and salt stress might be hypothesised.

9.3.1.3 Simulation Models Predicting Sunflower-Oil Quality in Relation to Environment and Genotype

Models predicting both yield and quality under different environmental conditions can be useful to aid crop management in the choice of the best location or sowing date to obtain a specific oil quality with highest yield. Pereyra and Aguirrezàbal[39] developed and validated a simple model based on published relationships, which is able to estimate not only yield and its components, but also grain and oil-quality aspects that are of relevance for industrial processes or human health. Differences in potential yield and grain and oil quality between locations, were explained in relation to differences in incident radiation, mean or minimum temperature. The results obtained showed that the relative low yields obtained at low-latitude locations, were compensated for by the higher nutritious value and oxidative stability of the sunflower oil, while at higher latitudes, high-linoleic acid oil production should be compatible with high yield potentials.

Fatty acid composition affects several quality parameters of biodiesel (*e.g.*, density, kinematic viscosity, heating value, cetane number and iodine value). Moreover, there are several well-validated equations in the literature that enable the prediction of the above parameters from the oil fatty acid composition. Relationships between temperature and fatty acid composition, and between fatty acid composition and fuel-quality parameters were integrated into a sunflower crop simulation model,[40] which allowed the effect of cultivar selection and other crop-management practices (sowing date, sowing location) on the final quality of biodiesel to be simulated. The authors found that the choice of cultivars is crucial to ensure the production of high-quality biodiesel from sunflower oil. The results of model simulations underline the facts that sunflower oil does not have one fixed quality, but different qualities depending on weather conditions and agricultural practices and that intraspecific variation in biodiesel quality can be larger than interspecific differences.

9.4 Energetic Uses of Sunflower Other than Biodiesel

Besides biodiesel production, there are several other different possibilities to effectively utilise the energy contained within sunflower biomass. Whatever the energy-conversion processes involved, the general purpose should be to significantly enhance the energy yield and the overall efficiency of the process itself. In order to achieve this kind of aim, two broad but divergent alternatives may be pursued: the first one pertains to the exploitation of the whole crop biomass, without separating its diverse components of different energetic value, but instead considering the global energy potential of the entire crop in relation to a single energy-conversion process; the second possibility involves the setting-up of a composite technological platform, able to integrate different energy-conversion processes according to the specific composition of the diverse biomass components. The former solution may be considered as the conventional energy approach, based on the concept of a single energy supply chain, while the latter is in line with the more advanced technological idea of the "biorefinery" concept.

9.4.1 Anaerobic Digestion and Biogas Production

The process of anaerobic digestion of biomass for biogas production is today considered as a vital and strategic mature technology for the sustainable use of agricultural biomass as a renewable energy source. Since economic biogas production requires high biogas yield,[41] specifically of methane, it is of the uppermost importance to gain an accurate optimisation of the whole process, fine tuning each single factor affecting the outcome of the process. Crop species and cultivar, cultivation practices, time of harvesting with respect to crop development stage, pretreatment operations and storage conditions, different raw-material mixing properties, digestion technology and many other factors, are all very influential.

An essential facet of anaerobic digestion relates to the opportunity to use digestion residues as a valuable fertiliser for agricultural crops: thus, not only can the CO_2 cycle be considered on balance (theoretically no CO_2 net emissions are generated) but also there is efficient nutrient recycling, in which the nutrients taken up by the crop are restored to the soil. As far is possible, these kind of "closed" solutions should be encouraged.

When it is considered that sunflower can be also cultivated as a forage crop, its destination as silage for anaerobic digestion (likewise maize or sorghum) is reasonable but, as yet, only applied to a limited extent. In comparison with maize, sunflower requires a shorter growing season, is more drought tolerant and generally produces silage of a lower yield but often slightly superior quality. Modern biomass hybrids are specifically tailored to maximize whole-crop biomass rather than seed production: the vegetative vigour is reinforced and the crop growth rate increased, while grain replenishment is depressed. The crop cycle is significantly shortened with respect to the traditional oilseed crop since harvesting time is anticipated, approximately 20–25 days after flowering and there is no reason to wait till seed maturity. This particular feature enables a second crop cycle to be grown in the same year, for instance after the harvesting of a main traditional cereal crops (wheat, barley, oats, triticale).

An evaluation carried out in Austria[41] reported sunflower methane productivity to be in the range of 2600–4550 $Nm^3\,ha^{-1}$, which is lower compared to maize (7500–10 200 $Nm^3\,ha^{-1}$, with maize harvested at the "wax ripeness" stage). In sunflower, the specific methane yield ($Nm^3\,kg^{-1}$ volatile solids) progressively decrease from the stages of "floral button" to "full flowering" and onwards.

With respect to the Mediterranean climate, under limited water availability due to the semiarid summer conditions, the biomass weight of sunflower has been shown to be very similar to sorghum or maize, provided that a supplementary irrigation of about 2000 $m^3\,ha^{-1}$ is performed, preferably split into two different periods: soon after sowing and at the stage of "floral button". The actual yield level of sunflower suitable to biogas production is in the range of 12–14 $t\,ha^{-1}$ of dry matter, but a dry yield of 16–18 $t\,ha^{-1}$ is attainable with early sowing and optimal growth conditions.

The optimal cutting time is approximately 3–4 weeks after flowering, when the inflorescence (anthodium) starts yellowing and the cut-green biomass reaches 25–30% dry matter. Although oilseed yield is not the relevant part of production, the higher sunflower biomass oil content promotes methane generation and enhances the crop energy yield. It can be estimated that the oil fraction of biomass enhances methane productivity by about 5%. Each fresh biomass ton (20–30% of dry matter) can produce 230–270 Nm^3 of biogas, corresponding to 500–650 Nm^3 per ton of dry matter with an average methane concentration of about 58%.

9.4.2 Thermochemical Conversion Processes

Biomass sources, and particularly agricultural residues, are considered suitable raw materials for the production of different kinds of biofuels. The recovery of energy from these solid wastes, mostly lignocellulosic residues, commonly utilises thermochemical processes such as direct combustion, gasification and pyrolysis. Pyrolysis has been receiving increasing attention in recent years as an interesting approach to biomass energy valorisation because the operational conditions of the process can be optimised to better exploit the production of either gases (biosyngas), liquid (bio-oil) and solid (biochar) biofuels, depending on the product required.

Biochar is a carbon-rich byproduct and is also considered a soil amendment: it improves soil texture and increases the ability of soil to retain fertilisers and release them slowly; in this way it reduces the risk of nitrate leaching and groundwater contamination; it naturally contains many of the micronutrients needed by plants to grow. As a result of its long persistence in soil, due to very slow microbial degradation and chemical oxidation, biochar is also of great interest for the sequestration of atmospheric carbon into soil, with the aim of mitigating global warming. All these interesting properties can be used effectively to address some urgent environmental problems: soil degradation and food insecurity, water pollution from agrochemicals and climate change. A possible key for securing environmental benefits is thus the production of a biochar byproduct during pyrolysis that can then be applied to soil.[42–44]

An experimental comparison of rape and sunflower residues as pyrolysis feedstocks,[45] showed that H_2 was the main gas produced by rape residues (48.7%), CO (28.9%) and CH_4 (10.5%) being the other most important components of the gas; 78% of syngas is reached. Sunflower residues showed a higher methane concentration (16.9%) but a still interesting concentration of H_2 (15.6%) and CO (30.3%). The lower heat value (LHV) of the syngas thus generated was equal to 13.64 MJ m^{-3} for rape and 13.80 MJ m^{-3} for sunflower. With respect to bio-oil, the heating value measured using rape residues was equal to 37.2 MJ kg^{-1}, while for sunflower residues it was equal to 36.6 MJ kg^{-1}. The biochar from rape contained: 0.76% N, 0.36% P_2O_5 and 4.40% K_2O, whilst for sunflower the content was the following: 1.19% N, 0.44% P_2O_5, 7.26% K_2O.

Sunflower press-cake may also be a potentially valuable source of fuel or chemical feedstocks. The effects of heating rate, final pyrolysis temperature, particle size and pyrolysis atmosphere on the pyrolysis product yields have been investigated in several experimental trials. A maximum bio-oil yield of 52.1% in weight was experimentally obtained at a pyrolysis temperature of 550 °C and in a nitrogen (N_2) atmosphere with a flow rate of 50 ml min^{-1} and a heating rate of 5 °C s^{-1}.[46] In another trial, a maximum oil yield of 23% in weight was obtained in a N_2 atmosphere at a pyrolysis temperature of 550 °C and a heating rate of $7\,°C\,min^{-1}$.[47]

Sunflower husk, obtained from the achene decortication, can be considered an interesting fuel for industrial boilers; it has the advantages of a negligible ash content (less than 4% in weight) and the absence of sulfur compounds, a low total water content (about 10% in weight) and a considerable low heating value (LHV, superior to $15\,MJ\,kg^{-1}$). On the opposite side, the main drawbacks are: low energy density and energy concentration (demanding large storage space and fuel flow rates) besides a low ash fusion temperature.[48]

9.5 Energy Balance of Sunflower Cultivation and Processing

To determine whether alternative fuels provide benefits over the fossil fuels they should displace requires detailed accounting of both the direct and indirect energy inputs consumed in the process and of the energy outputs that are generated. To be a viable substitute of a fossil fuel, an alternative fuel should not only have superior environmental benefits, be economically competitive and producible in sufficient quantities to make a meaningful impact on energy demands, but it also should provide a net energy gain over the energy sources used to produce it.[49] The latter conditions (positive energy balance) raised great concern about biofuels just a few years ago and still the debate is going on.

Biofuel production requires energy to grow crops and convert them into biofuel. The energy balance may be related to a single unit of cropping area (hectare of cultivated sunflower, for instance) or, alternatively, to a single unit of product obtained (kilogram of biofuel produced). This balance is strictly associated with the farming operations required in the field in order to promote crop growth and yield and to overcome potential environmental constraints. The industrial transformation process has to be considered also, together with transportation and warehousing. Calculation of the energy input is not straightforward; direct and indirect energy input must be accounted for. All the operations of the farmer have to be recognised and their energy expenditure precisely calculated. Most often, appropriate energy equivalents from the literature are taken into account. On the other hand, with respect to energy outputs, not only the energy value of the final product (biofuel) but also that of all possible simultaneously generated coproducts have to be considered.

Concerning the energy inputs strictly associated with sunflower cultivation, several estimates are available in the literature, manly related to Mediterranean conditions (Greece and Italy) or with respect to the USA. An evaluation performed in Greece,[50] calculated the total energy input to be $10.49\,GJ\,ha^{-1}$, without irrigation and with fertilisers being the major input. A fertiliser application rate of $60\,kg\,ha^{-1}$ of N accounted for 42.4% of the energy consumption; fuel (33.9%) and indirect machinery (13.3%) were the other principal energy costs; phosphorus (4.1%) and herbicides (4.1%) applications corresponded to lower energy expenditures.

From a wide survey[51,52] conducted in Northern Italy (400 farms, during a four-year period), the cultivation energy costs were, on average, equal to 21.6 GJ ha^{-1}; this value is mainly accounted for by a high nitrogen fertilisation rate (80 kg ha^{-1} of N; 26.8% of the total energy expenditure) and a relevant fuel consumption (49.0% of energy costs), indicating an intensive mechanisation of the farms. Still higher is the energy cost attributed to sunflower cultivation in the USA.[53] The total energy inputs related to farm operations was equal to 25.62 GJ ha^{-1}; fuel explains 29.4% of the total energy, while nitrogen accounts for a slightly lower percentage (28.6%); another important cost being that of seeds (9.1%), significantly higher than in the other evaluations.

With reference to an evaluation carried out in central Italy, on seven different farms employing different agrotechnical inputs,[54,55] the total energy cost of sunflower cultivation was in the range of 12.89–17.30 GJ ha^{-1}, with an average value of 15.43 GJ ha^{-1} and a standard deviation of 1.40 GJ ha^{-1}. As usually observed, nitrogen fertilisation accounted for the highest fraction of the total energy costs, in this case in the range of 52.7–61.1%; as far as soil mechanical cultivation practices are concerned, the incidence on total energy costs ranged between 6.8% and 16.5%. Sunflower is generally a rainfed crop but in a drought climate a supplementary irrigation is required in order to gain a reasonable yield. When irrigation is supplied, the energy cost related to cultivation increases dramatically. With respect to sunflower cultivated in Southern Italy,[56] the total energy cost amounted to 12.65 GJ ha^{-1} without irrigation but 31.64 GJ ha^{-1} with an irrigation supply of 2000 m^3 ha^{-1}. Without irrigation, fertilisation energy costs amounted to 56% and mechanical 43%, whereas with irrigation, the main energy fraction pertained to water supply (59.7%); fertilisation (22.4%) and mechanisation (17.5%) having remarkably lower values.

With respect to energy outputs, grain energy content per unit of cultivated area is extremely variable, and it depends on yield as affected by location (agroenvironmental factors), and technical inputs (rate of fertilisation and irrigation supply, first), as well as their reciprocal interaction. It is generally observed[57] that, in sunflower, inputs have a lower range of variation in comparison with outputs, with a factor of 2 between minimum and maximum values. The very large seed-yield variability and, as a consequence, the great differences in energy output data are related to the effects of greatly diversified environmental and farming conditions, even considering a relative narrow geographical area. Referring to sunflower cultivated in the northeast of Italy,[51,52] energy outputs in the range of 14–124 GJ ha^{-1} were observed, with an average value of 79 GJ ha^{-1}. A similar range, 25–110 GJ ha^{-1}, has been reported with respect to another trial,[57] still in the north of Italy; both the surveys considered a wide array of farming systems. Since the specific energy content of sunflower seed is about 26–27 GJ t^{-1}, an output of 14–124 GJ ha^{-1} corresponds to a very wide seed-yield range, 0.5–4.5 t ha^{-1}, with an average value of approximately 3 t ha^{-1}. This average yield value is remarkably higher than the commonly average yield registered in Central and Southern Italy, as well as in the majority of the European cultivation regions.

An analysis of sunflower energy balance, with respect to the cultivation phase, has showed quite clearly[52] that the energy outputs do not vary greatly with the corresponding input and that the energy gains (outputs minus inputs) tend to decrease with increasing agrotechnical cultivation inputs. Higher energy gains are thus observed in correspondence to lower inputs, generally between 10 and 20 GJ ha^{-1}. This observation confirms the well-known ability of sunflower to adapt to poor environmental conditions of cultivation, for instance characterised by low water availability. As a general guideline, it can be argued that an energy input lower than approximately 10 GJ ha^{-1} exposes the grower to a high risk of crop failure, while inputs higher than approximately 30 GJ ha^{-1} significantly reduce the marginal energy gains.

Postharvest inputs relate to the industrial process of seed crushing/pressing and oil extraction; the energy cost of this operation is related to the amount of seed processed and it approximately corresponds to about 0.45 GJ t^{-1} of seed.[54] The seed oil content is another greatly variable feature in oilseed crops; as regards sunflower, it is generally in the range 40–48%.[57] An overall mechanical and chemical oil extraction yield of 41%[54] should be considered as a proper value. The specific energy content of sunflower oil is 39.6 GJ t^{-1}, approximately equal to that of sunflower biodiesel (LHV, lower heat value), while LHV of press-cake can be estimated approximately 18.24 GJ t^{-1}. The whole process of biodiesel production (seed transportation, seed drying and warehousing, industrial oil processing, refinery, distribution) result in a total energy cost of about 8.51 GJ t^{-1}.[58]

At this point, we have all the necessary data to outline an energy balance with seed yield as the only unknown quantity. Assuming biodiesel as the only final energy product and a cultivation energy input, for instance, equal to 15 GJ ha^{-1}, sunflower yield should be superior to 1.18 t ha^{-1} in order to have a positive energy balance; in the case of a higher cultivation input, 30 GJ ha^{-1} for instance, the minimum threshold seed yield increases to 2.35 t ha^{-1}, a considerably higher target to reach.

Otherwise, considering sunflower oil (and not biodiesel) the final energy product, the minimum acceptable seed yield should be 0.95 t ha^{-1}, with respect to a farming energy input of 15 GJ ha^{-1}, or 1.89 t ha^{-1}, if the farming energy input is equal to 30 GJ ha^{-1}. This last result explains the increasing interest in the straight use of oil in power generation or transportation biofuel, as an alternative to biodiesel, especially with respect to a small farming industrial plant, mostly designed to satisfy internal energy needs.

Finally, considering the energy values of both biodiesel and press-cake the seed threshold yield to even out energy balance is equal to 0.64 t ha^{-1} with respect to a farming energy input of 15 GJ ha^{-1} and 1.28 t ha^{-1} when the farming input is 30 GJ ha^{-1}. In this way, the byproducts are placed into value and the importance of including the energetic (and economic) contribution of all the possible byproducts is highlighted.

9.6 Criticism and Perspectives on the Use of Sunflower as an Energy Crop

Sunflower cultivation is mainly devoted to the production of oil which is one of the major edible oils in the world, particularly appreciated for human nutrition.

Approximately 9% of the world total oil production derives from sunflower, the greater part originating from palm, soybean and rape. A limited amount of sunflower global oil production (4–5%) is actually used for fuel refining, the greater part is still being used traditionally for food. Moreover, sunflower oil accounts for only 5% of global biodiesel production despite its potential to meet the biodiesel standards that have been established for successful commercialisation and market acceptance. In particular, it is known that several biodiesel quality parameters depend on fatty acid oil composition that might be strongly affected both by genotype and environmental conditions. Sunflower high-oleic hybrids are suitable for biodiesel production due to the higher oxidative stability of the oil regardless of the cultivation conditions. On the other hand, sunflower oil composition from traditional genotypes may be highly affected by the environment giving the best performance in warmer regions where a higher oleic/linoleic ratio is obtained.[40]

While the chemical properties of the oil lend themselves well to biodiesel manufacture, the high cost of sunflower oil casts doubts as to whether it can ever be a significant feedstock for biodiesel production; in simple words: it is suitable as a fuel, but not necessarily cost effective. For this reason, we can expect that, in the near future, sunflower will continue to have a greater relevance as a food crop than as a fuel crop. A rising demand for vegetable oils, is indeed projected to feature heavily over the next years, although[59] the demand for vegetable oil in biodiesel production could also increase notably. According to the same projection, an increasing world livestock production will continue to be the driving force behind the consumption of oilseed-derived protein meal.

Besides economic profitability (arising from the financial balance) another decisive factor to be met is that the overall energy balance has to be positive. The crop-energy balance is a useful approach to check the actual efficiency of each agrotechnical input with the aim to maximize the cultivation energy gain. In order to improve the energy performance of the crop, the operations where energy savings could be realised without markedly decreasing seed yield should be identified, so that effective modifications to the cultivation practices can be made.

In comparison with other crops, sunflower is able to sustain lower energy inputs without considerably reducing energy outputs, as a result of its broad adaptability to even poor environments. Major energy savings should be addressed toward the most energy expensive operations. Thus, a better management of fertilisation (for instance, improving the crop recovery of the supplied nitrogen and enhancing soil fertility by efficient agronomical crop rotations) and an optimisation of soil mechanical cultivation practices (for instance by reducing the depth of tillage or avoiding ploughing) are clearly needed.

In rural areas where cultivation lands are scarce and competitiveness among crops is very high, intensive cultivation practices are adopted; in these conditions, there is the need to finely tune all the technical operations in order to enhance the resource use efficiency. When conditions of intensive farming prevail, sunflower cultivation is frequently pushed on to more marginal lands, where specific constraints notably reduce the crop productivity. Below a minimum yield threshold, sunflower for energy production can no longer be grown, since its energy balance could be negative. As a general reference, an acceptable sunflower minimum yield is placed around $2\,t\,ha^{-1}$.

On marginal farming systems, where sunflower is frequently cropped, the bioenergy supply chain could mainly have a local connotation, the prevalent feature being that of a "short" supply chain, where energy production, distribution and use are spatially coupled and self-consumption should be the first requirement to be met. In this respect, for instance, the use of "straight vegetable oil" (SVO) as a fuel represents a positive way to shorten the energy supply chain by avoiding the final step of biofuel production and, thus, by reducing the energy inputs and maximising the internal farm energy gains.

Conversely, in agricultural areas where the land availability is ample and the cultivation pressure on land is low, extensive farming prevails and low-input agriculture is the reference model. When vast land surfaces are taken into consideration, albeit with low yield, it is the total amount of production that is relevant, providing sufficient justification for a "large" energy supply chain, based on high-capacity refinery plants, thus better exploiting the emerging scale economies and directly competing on the global market.

The two contrasting models of energy production can exist together, side by side, in so far as they interpret different specific social and economic requirements and are considered appropriate with respect to different conditions of rural development. The first one is oriented in a more "agricultural" way, while the second is more oriented towards an "industrial" endpoint.

In the large-scale biofuel economy, agribusinesses are most assured to profit, since mechanised harvesting and production chains are the easiest option for rapidly scaling up biofuel production.[60] Large-scale agricultural processors and distributors will be responsible for supplying most of the refined fuel as well. In contrast, a biofuel industry that is locally oriented, in which farmer-owners produce fuel for their own use, is more likely to guarantee benefit to a rural community. In these marginal conditions, farmers may risk bad seasons and a poor harvest but, by adding value to their own products and using these goods locally, they become less vulnerable to external exploitation and disruptive market fluctuation.[60]

Another relevant way to greatly increase the energy gains of the whole transformation process is to emphasize the energy content of all the crop biomass components, thus considering every byproduct and residue generated along the energy supply chain. Considering the large diversity of biomass entering the process, the high flexibility and reliability in biomass supply is one of the most interesting features of agroenergy; another one being the ability to adjust the process to the peculiar agricultural traits of different rural areas and farm managements. Of particular relevance, in this regard, is the opportunity to integrate different technological processes in a unique biorefinery "platform", thus creating a diversified multiended energy production supply.

The sunflower crop will, probably, never succeed in gaining the relevant product quantities necessary to influence significantly the biodiesel world scenario. Far from representing a threat with respect to the "food vs. fuel" diatribe, the crop, in the agroenergy context, will preserve its mainly food relevance. However, it could provide an interesting complementary strategy of "product diversification", the maintenance of a good level of demand and a good price, both being conditions that would help to secure a good and stable level of production over time.

Acknowledgements

Thanks are due to Dr. Luigia Giuzio and Dr. Marianna Pompa for the skilful cooperation in drawing up the tables.

References

1. FAO-STAT http://faostat.fao.org/site/291/default.aspx (last update: 23 June 2009).
2. F. P. C. Blamey, R. K. Zollinger and A. A. Schneiter, in *Sunflower Technology and Production*, A. A. Schneiter, ed. Agronomy Monograph 35. ASA, CSSA and SSSA, Madison, WI, USA, 1997, 595.
3. G. Petcu and E. Petcu, *Helia*, 2006, **29**, 135.
4. F. Flénet, P. Debaeke, P. Casadebaig in *Proc. 17th Int. Sunfl. Conf.*, Cordoba, Spain. Int. Sunfl. Assoc., Paris, France. 2008, 13.
5. A. Merrien and L. Grandin, in *La tournesol et l'eau. Adaptation a la secheresse. Response a l'irrigation*, ed. CETIOM, Paris, 1990, 75.
6. Z. Flagella, T. Rotunno, E. Tarantino, R. Di Caterina and A. De Caro, *Eur. J. Agr.*, 2002, **17**, 221.
7. R. D'Andria, F. Quaglietta Chiarandà, V. Magliuolo and M. Mori, *Agron. J.*, 1995, **87**, 1122.
8. K. K. Klein and D. G. Le Roy, *The Biofuels Frenzy: What's in it for Canadian Agriculture?* Green Paper Prepared for the Alberta Institute of Agrologists Presented at the Annual Conference of Alberta Institute of Agrologists Banff, Alberta March 28, 2007, Department of Economics. The University of Lethbridge, Lethbridge, Alberta.
9. Renewable Energy Policy Network for the 21st Century (Ren21). Renewables: Global Status Report, 2009.
10. Energy Information Administration (EIA), Official energy statistics from The USA Government. http://www.eia.doe.gov/emeu/international.
11. G. Pahl and B. Mckibben, *Biodiesel: Growing a New Energy Economy*, Chelsea Green Public Company, VT, USA, 2008.
12. G. Knothe, J. Van Gerpen and J. Krahl, *The Biodiesel Handbook*, AOCS Press, Champaign, Illinois, 2005, 302.

13. G. Knothe, *AOCS*, 2006, **83**, 823.
14. N. Izquierdo, L. Aguirrezàbal, F. Andrade and V. Pereyra, *Field Crops Res.*, 2002, **77**, 115.
15. W. Friedt, M. Ganssmann and M. Korell, in *Proc of EUCARPIA – Symposium on Breeding of Oil and Protein Crops*, Albena, Bulgaria, 1994, 1–30.
16. D. Škorić, S. Jocić, Lečié, Nada, in *Genetics and Breeding for Crop Quality and Resistance*, Kluwer Academic Publishers, Dordrecht/Boston/London, 1998, 339.
17. J. M. Fernandez-Martinez, L. Velasco and B. Perez-Vich, in *Proc. of the 16th International Sunflower Conference*, 2004, **Vol. I**, 1–15.
18. D. Dolde, C. Vlahakis and J. Hazebroek, *JAOCS*, 1999, **76**, 349.
19. V. Marquard, *Fett Wiss. Technol.*, 1990, **92**, 452.
20. L. Velasco, J. M. Fèrnandez-Martìnez, R. Garcìa-Ruiz and J. Domìnguez, *J. Agr. Sci.*, 2002, **139**, 425.
21. S. Lacombe, A. Berville, in *Proc. of the 15th International Sunflower Conference. Toulouse.* 2000, Tome I: PI.D-16-PI.D-26.
22. Trèmolières, J. P. Dubacq and J. P. Drapier, *Phytoc.*, 1982, **21**, 1.
23. C. Sarmiento, F. Garcè and M. Mancha, *Planta*, 1998, **205**, 595.
24. D. Rondanini, R. Savin and A. Hall, *Field Crops Res.*, 2003, **83**, 79.
25. N. G. Izquierdo, L. A. N. Aguirrezàbal, F. Andrade and V. Pereyra, *Field Crops Res.*, 2002, **77**, 115.
26. J. Roche, A. Bouniols, Z. Mouloungui and M. Cerny, *Eur. J. Lipid Sci. Technol.*, 2006, **108**, 287.
27. M. J. Hattendorf, M. S. Redelfs, B. Amos, L. R. Stone and R. E. Gwin, *Agron. J.*, 1988, **80**, 80.
28. P. Unger, *Agron. J.*, 1986, **78**, 507.
29. O. Jones, *Agron. J.*, 1984, **76**, 229.
30. M. Ciliardi, D. Ferri, F. Lanza, N. Losavio and P. Santamaria, *Riv. Agron.*, 1990, **24**, 250.
31. J. A. Robertson and V. E. Green, *JAOCS*, 1981, **58**, 698.
32. M. Talha and F. Osman, *J. Agric. Sci.* Cambridge, 1974, **84**, 49.
33. P. W. Unger, *Soil Sci. Soc. Am. J.*, 1982, **46**, 1072.
34. E. Salera and M. Baldini, *Helia*, 1998, **21**, 55.
35. Z. Flagella, T. Rotunno, R. Di Caterina, G. de Simone, A. De Caro, in *Proc of XV International Conference, Toulouse*, 2000, Vol I, C 139.
36. M. Baldini, R. Giovanardi, G. Vannozzi, in *Proc of XV International Conference, Toulouse*, 2000, Vol I, A 79.
37. Z. Flagella, M. M. Giuliani, T. Rotunno, R. Di Caterina and A. De Caro, *Eur. J. Agr.*, 2004, **21**(2), 267–272.
38. R. Di Caterina, M. M. Giuliani, T. Rotunno, A. De Caro and Z. Flagella, *Ann. Appl. Biol.*, 2007, 145.
39. G. Pereyra-Irujo and L. Aguirrezàbal, *Agric. For. Meteorol.*, 2007, **143**, 252.
40. G. Pereyra, N. G. Izquierdo, M. Covi, S. M. Nolasco, F. Quiroz and L. A. N. Aguirrezàbal, *Biom. Bioen.*, 2009, **33**, 459.

41. T. Amon, B. Amon, V. Kryvoruchko, A. Machmuller, K. Hopfner-Sixt, V. Bodiroza, R. Hrbek, J. Friedel, E. Potsch, H. Wagentristi, M. Schreiner and W. Zollitsch, *Biores. Technol.*, 2007, **98**, 3204.
42. J. Lehmann, J. Gaunt and M. Rondon, *Mitig. Adapt. Strat. Glob. Change.*, 2006, **11**, 403.
43. J. Lehmann, *Nature*, 2007, **447**, 143.
44. J. Gaunt and J. Lehmann, *Environ. Sci.*, 2008, **42**, 4152.
45. M. E. Sanchez, E. Lindao, D. Margaleff, O. Martinez and A. Moran, *J. Anal. Appl. Pyrol.*, 2009, **85**, 142.
46. S. Yorgun, S. Sensöz and Ö. M. Kockar, *J. Anal. Appl. Pyrolysis.*, 2001, **60**, 1.
47. S. Yorgun, S. Sensöz and Ö. M. Koçkar, *Biom. Bioen.*, 2001, **20**, 141.
48. M. Radovanovic, T. Zivanovic, D. Stojiljkovic, V. Jovanovic, N. Jerinic, S. Knezevic, M. Guzijan, B. Umicevic, European Conference N°5, Espinho-Porto, Portugal (11/04/2000). INFUB, Rio Tinto, ETATS-UNIS. 2000. ISBN 972-8034-04-0.
49. J. Hill, E. Nelson, D. Tilman, S. Polasky and D. Tiffany, *PNAS*, 2006, **103**, 11206.
50. L. Kallivroussis, A. Natsis and G. Papadakis, *Bios. Eng.*, 2002, **81**, 347.
51. S. Bona, *Riv. Agron.*, 2001, **35**, 219.
52. S. Bona, G. Mosca and T. Vamerali, *Renewable Energy*, 1999, **16**, 1053.
53. D. Pimentel and T.W. Patzek, *Natural Resourse Res.*, 2005, **14**, 65.
54. F. Pedretti, E. Toscano, G. Scrosta, V. AIIA 2005: Catania, 27–30 giugno 2005. *L'ingegneria agraria per lo sviluppo sostenibile dell'area mediterranea.* Codice lavoro: T7A1 7029.
55. G. Riva, E. Pedretti, G. Toscano, R. Cerioni, D. Duca, 2006. *Agroenergie: filiere per la produzione di energia elettrica da girasole.* Regione Marche, Italy. Comitato Termotecnico Italiano (CTI).
56. M. Monteleone, *Secondo Rapporto dello Studio per la Valorizzazione Energetica di Biomasse e la Diffusione di Colture Energetiche nelle Aree Agricole e Boscate della Provincia di Foggia.* Chapter 3. Aforis, AmbienteItalia, Università di Foggia.
57. P. Venturi and G. Venturi, *Biomass Bioenergy*, 2003, **25**, 235.
58. Nomisma 2008. I biocarburanti in Italia. Opportunità e costi. Rapporto all'Unione Petrolifera.
59. OECD FAO *Agricultural Outlook* 2008–2017.
60. *Biofuel for transportation.* 2006. Worldwatch Institute in cooperation with the Agency for Technica Cooperation (GTZ) and the Agency of Renewable Resources (FNR) to the German Federal Ministry of Food, Agriculture and Consumer Protection (BMELV).

CHAPTER 10
Palm Oil as an Energy Crop

KEAT TEONG LEE AND KOK TAT TAN

School of Chemical Engineering, Universiti Sains Malaysia, Engineering Campus, Seri Ampangan, 14300 Nibong Tebal, Pulau Pinang, Malaysia

10.1 Introduction

Palm oil (PO) is derived from the oil-palm tree (*Elaeis guineensis*), which is one of the most important species in Palmae family. It is believed to be indigenous to West Africa but now is planted in all tropical countries throughout the world, particularly in South East Asia countries including Malaysia, Indonesia and Thailand.[1] A typical oil-palm plantation in Malaysia is shown in Figure 10.1 that consists of young and mature oil-palm trees. Oil palm is a perennial crop that requires hot and humid climate conditions for optimum growth. Mature oil-palm trees can grow up to 20 m tall with pinnate leaves measuring between 3 and 5 m long. Figure 10.2 shows the inflorescences of male and female for oil palm that is a monoecious crop. Palm fruit, which is reddish, is about the size of a plum and grows in huge bunches of 1500 individual fruits, weighing up to 40 kg per bunch as shown in Figure 10.3. Each fruit consists of a hard kernel, which is surrounded by fleshy mesocarp and encapsulated by a shell (endocarp). Oils are extracted from the pulp and kernel to obtain PO and palm kernel oil (PKO), respectively. Comparatively, PO is superior to PKO in terms of nutritional content and consequently has higher commercial value. For instance, it is a rich source of carotenoids and tocopherols, which are valuable micronutrients in human diet. Hence, PO is commonly processed and refined to produce edible oil, while PKO is generally used to manufacture soap-related products.[1]

Figure 10.1 Oil-palm plantation in Malaysia.

Figure 10.2 Male (left) and female (right) inflorescences of oil palm.

Crude PO is orange-red in colour due to the presence of high carotene content. Typically, PO consists of 45% palmitic acid, 40% oleic acid, 10% linoleic acid and 5% stearic acid. The high percentage of palmitic acid that is a saturated fatty acid leads to its natural semisolid state at room temperature.

Figure 10.3 Palm fruit that grow in bunches on oil-palm tree.

This characteristic allows PO to have high oxidation and thermal stability even at elevated temperature. Hence, it is widely utilised in most industrial processes for multiple purposes with products ranging from edible cooking oil to inedible soap. Recently, with fluctuating petroleum price in the world's market and inevitable efforts to curb global warming, PO application has been extended to renewable energy with the production of palm-based biodiesel. Biodiesel is defined as fatty acid alkyl esters, and is derived *via* a transesterification reaction between triglycerides and alcohol. It has similar physicochemical properties to petroleum-derived diesel but is superior in terms of biodegradability, toxicity, renewability and sustainability.

10.2 Potential of Palm Oil as Energy Crop

At present, the main vegetable oil used in the industries to produce biodiesel is rapeseed oil which comprised nearly 84% of global biodiesel production as shown in Table 10.1.[2] From the table, it can be seen that palm-based biodiesel is relatively insignificant with only 1% of total world production, but it remains the most attractive choice of feedstock due to its high yield per hectare as shown in Figure 10.4.[2] From the figure, it is obvious that the average yield of PO is approximately 4.2 tons per ha per year, far exceeding other types of vegetable oils. For instance, two other major oils like rapeseed and soybean produce only 1.2 and 0.4 tons per ha per year, respectively. Furthermore, the production cost of PO is the least expensive oil among all vegetable oils. For instance, as shown in Table 10.2, the production cost for PO is approximately US$228/ ton, while for soybean and rapeseed, the cost is relatively higher at

Table 10.1 Percentage of biodiesel produced worldwide from various vegetable oils.[2]

Source of biodiesel	Percentage
Rapeseed oil	84
Sunflower oil	13
Palm oil	1
Soybean oil and others	2

*(Republished with permission from Elsevier)

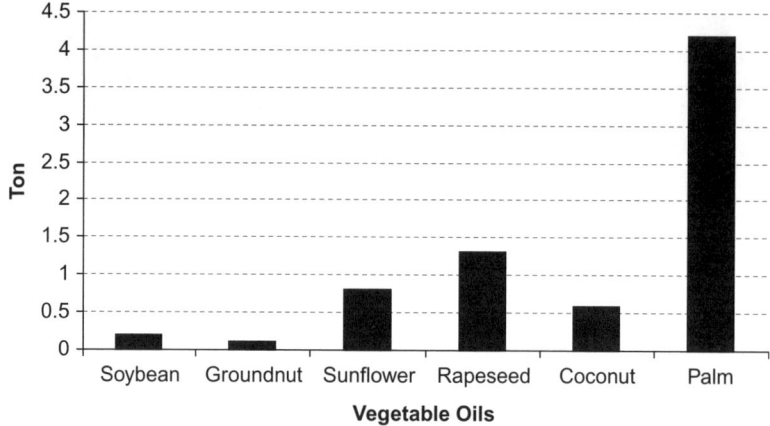

Figure 10.4 Oil yield for selected vegetable oils.[2] (Republished with permission from Elsevier)

Table 10.2 Comparative production cost of selected oils (US$/ ton).[3]

Source	Cost
Rapeseed oil	648
Palm oil	228
Soybean oil	400

*(Republished with permission from Elsevier)

US$400/ ton and US$648/ ton, respectively.[3] As mentioned previously, oil palm is a perennial crop that can produce a consistent yield for up to 25 years, while annual crops such as soybean and rapeseed require high energy input and expenditure yearly for ploughing, sowing and fertiliser.

Although oil palm is grown on less than 5% of the world's agriculture land for major edible oils production, it accounts for a massive 25% of the global market share. Consequently, it is not surprising to note that annual global production of PO is the highest among all major vegetable oils in the world, as shown in Table 10.3.[3] From the table, it can be seen that the total PO production is 38.5 million tons, surpassing soybean oil, which only has 36.9 million tons productivity. Furthermore, in terms of land usage, oil palm utilises only a

Palm Oil as an Energy Crop

Table 10.3 Production capacity and plantation area for selected oils in 2007.[3]

Source	Capacity (mt)	Area (mha)	Percentage of area
Soybean	36.9	94.2	43
Sunflower	10.8	23.9	11
Rapeseed	18.5	27.2	12
Palm oil	38.5	10.6	5

*(Republished with permission from Elsevier)

mere 10.5 million ha to produce this staggering amount of yield, while soybean requires an enormous plantation area of nearly 95 million ha. Hence, this shows that oil palm is approximately 10 times more efficient than soybean in terms of land utilisation, thus ensuring optimised productivity to meet the growing demand in the world market. Consequently, PO has proven its capacity to be the largest source of oil that was dominated previously by soybean oil. This is important to cater for the growing demand of biodiesel in the market, particularly in the European Union (EU), which is the current largest consumer of biodiesel. Furthermore, with the increased target of 5.75% biofuels in 2010 set by EU commission, PO has been seen as the ideal feedstock to meet this mounting biodiesel demand.

In addition, the oil-palm tree has the potential to mitigate carbon-dioxide (CO_2) emission due to its high efficiency in absorbing this harmful greenhouse gas during photosynthesis compared with tropical rainforest. In a recent study, it was reported that net assimilation of CO_2 for oil-palm plantation is higher than that of rainforest, with values of 64.5 and 42.4 tons CO_2 per hectare annually, respectively.[2] This proves that oil palm is superior in terms of carbon sequestration and thus has the ability to alleviate the greenhouse effect and global warming. Besides, the oil-palm tree is competent in biomass accumulation as well, with an annual increment of 8.3 tons compared with 5.8 tons for rainforest.[2] Biomass generated by oil palm has several applications and uses, ranging from natural fertiliser derived from empty fruit bunch to renewable energy sources such as bioethanol and hydrogen. Hence, the importance of biomass is significant, since it minimises the usage of inorganic chemical fertiliser and generates valuable commodities of renewable energy. Consequently, it is not surprising to note that palm-based biodiesel has the lowest CO_2 emission compared with other crops such as rapeseed and soybean.[2] For instance, PO companies adopt best management practice (BMP) in their operations; these are environmentally benign approaches including zero burning, integrated pest management (IPM) and waste minimisation.[2] The zero-burning method is applied to ensure that clearing activities will not degrade the environment, especially the air and water system. This is achieved by returning unwanted biomass such as oil palm fronds and empty fruit bunch (EFB) to the land as fertilisers. In IPM, instead of chemical pesticides, biological pest-control methods are adopted to eliminate unwanted crop pests. For example, barn owls are used to control rat populations in oil-palm plantations rather than inorganic pesticides, which are harmful to the environment and the plant itself. Apart from that, fibres from EFB can be utilised in the pulp and paper industries to produce organic papers. This "waste-to-wealth" approach in waste management is important to minimise wastes generated, while contributing side income to PO companies as well. All these BMP practices ensure that PO companies are operating sustainably in terms of production and environment conservation.

10.3 Economic Consideration

Economic analysis for the production of palm-based biodiesel is crucial to investigate its feasibility in terms of investment, technical and environmental. As mentioned previously, the cost of PO per ton is the least expensive among all vegetable oils in the market. Consequently, biodiesel derived from PO is also the most economical with the least production cost of US$684/ ton compared with rapeseed and soybean with the costs (per ton) of US$996 and US$751, respectively.[3] Oil palm, being a perennial crop with substantially high yield per hectare is the most ideal and economical candidate for biodiesel production. Affordable biodiesel cost is crucial to encourage its marketability and competitiveness with petroleum-derived diesel in terms of application and pricing. Hence, palm-based biodiesel has the potential to be a price-competitive source of energy in the global market and help to fulfil the escalating demand of energy throughout the world.

Apart from financial considerations, technoeconomic analysis is also one of the most important factors in determining the feasibility of a technology. In this context, the technoeconomic consideration concerns the total amount of energy consumed to produce one unit of palm-based biodiesel based on life-cycle assessment (LCA) analysis. This analysis is vital to investigate the sustainability and efficiency of a renewable energy source in terms of energy requirement. The boundary of this analysis covers the lifespan of biodiesel that begins from crop cultivation to the combustion of biodiesel in engine. It has been reported that the amount of energy required annually to produce palm-derived biodiesel is substantially lower compared with rapeseed-based biodiesel, with total energies consumed of 17.2 GJ/ ton and 74.7 GJ/ ton, respectively. This demonstrates that PO is more efficient and effective in energy utilisation to produce biodiesel. Furthermore, the output to input energy ratio for the production of 1 ton of palm biodiesel is significantly higher at 3.53, far superior to rapeseed oil with a value of only 1.44.[4] Hence, palm-derived biodiesel is proven to be an economical source of energy that only requires a small amount of input energy.

10.4 Palm-Oil Issues and Challenges

Although PO has been shown to have remarkable potential to be an affordable and sustainable energy crop, its application as biodiesel feedstock has been surrounded by various environmental and ethical issues. For instance, oil-palm cultivation is highly associated with environmental issues such as deforestation and animal extinction. In addition, it is claimed that utilisation of PO as an energy feedstock has caused the price to be inflated on the global market, leading to increases in the price of PO-based products. Furthermore, converting a food source such as PO to fuel has received stern criticism as being unethical since there are millions of people around the world who suffer from malnutrition and hunger, leading to a food versus fuel debate. Consequently, these numerous allegations have smeared the image of PO as an environmentally friendly and sustainable oil crop.

However, accusations of environmental issues related to PO have been well responded to by the Malaysian Palm Oil Council (MPOB), which is one of the leading research centres of PO in the world. It was reported that although PO productivity increased substantially from 1990 to 2007, particularly in Malaysia and Indonesia, most of the plantation area was converted from other agricultural crops. For instance, although the plantation area in Malaysia within the period of 1990–2007 increased substantially from nearly 2 million to more than 4 million hectares, the increase in cultivation area was achieved by replacing crops such as rubber and cocoa, instead of clearing virgin tropical forests as claimed. Furthermore, Malaysia, as one of the leading producers of PO in the world, only utilised a mere 4 million hectares plantation area to produce an overwhelming amount of 16 million tons of PO in 2007, while having 76% of its land area or 25 million hectares available as tropical forests.[5] Hence, as far as oil-palm plantation is concerned, deforestation of virgin forests is not an issue at all.

In addition, the ecological sustainability of PO plantations was also under scrutiny due to accusations of it causing animal extinction. There were claims that as a monoculture crop, oil-palm plantation could affect the ecosystem of flora and fauna and subsequently lead to extinction. On the contrary, it has been reported that biodiversity in oil-palm plantations is not affected and that some rare species of animals can be observed in the plantation area. In addition, livestock crop integration (LCI) has been integrated with oil-palm plantation to generate side income by animal rearing and short-term cropping.[6] For instance, livestock such as cattle not only provide a source of protein but help to reduce weeding cost by grazing unwanted weeds and improve soil fertility through addition of organic matter. In addition, planting of short-term crops such as sugarcane, banana and pineapple in immature palm plantations is also one of the ways of diversifying the ecosystem while generating some side income for farmers. Besides, these crops are planted in between rows of young oil palm in space that is usually filled with weeds and grasses, thus ensuring optimised land utilisation. Consequently, the LCI system has proven the sustainability and diversity of oil-palm plantations that not only produce PO but livestock as well.

On the other hand, in the quest to produce a sustainable energy source such as biodiesel, converting a food source such as vegetable oil into biofuel leads to an inevitable debate of food versus fuel. It has been claimed that extensive conversion of food into fuels could result in food shortages and cause an escalation in the number of malnourished people around the world. Besides, it could lead to substantial increases in food prices and higher inflation rates in countries that are heavily dependent on imported food. However, based on the PO inventory in Malaysia in 2006 and 2007, concerns about the food versus fuel issue were found to be exaggerated. For instance, in line with collective efforts to increase biodiesel production in Malaysia, total export of palm-based biodiesel increased from nearly 50 000 tons in 2006 to more than 89 000 tons in 2007, an increment of 86%. However, the year-end stock of PO increased from 1.5 million tons to nearly 1.7 million tons within the same period, testifying that there is sufficient PO stock to fulfil both food and fuel demands.[7] Hence, it can be concluded that conversion of PO to biodiesel does not necessarily lead to a shortage in the supply of food as PO has the capacity to satisfy both market demands simultaneously.

Furthermore, converting PO into biodiesel is not the main reason for the price to increase steadily in the global market over the past few years. For instance, although the price of PO increased from USD400/ ton in 2006 to nearly USD800/ ton the following year, the same trend was also observed for other major edible oils such as soybean and rapeseed oil. This scenario was caused by escalating demand of edible oils due to the booming population in the world and greater affluence, particularly in China and India. In addition, snowstorms and droughts struck major countries that produce soybean and rapeseed oil, leading to lower crop yields and a tightening in the supply to the world market. Consequently, with limited available supply, the price of edible oils including PO increased substantially during this period. Hence, the price of PO would have increased inevitably, and was not influenced by utilising it as a biodiesel feedstock. Furthermore, only a mere 1% from the total PO production in the world was utilised to produce palm-based biodiesel, which is insignificant and not sufficient to affect the price of this commodity. Consequently, it is concluded that the price of PO which augmented recently was not related to biodiesel production.

10.5 Current Research and Development

Continuous efforts to improve the quality and quantity of PO are crucial to allow PO to be the leading feedstock for biodiesel production and leading edible oil in the world. Realising this, numerous companies have been actively involved in extensive research and development in all aspects of oil palm and PO. One of the major breakthroughs was the successful DNA decoding of the oil-palm tree, which allowed the invention of improved oil-palm clones that have higher yield per hectare. It has been claimed that these genetic clones could produce more than 10 tons per hectare, a significant increment of more than 100% compared with conventional oil-palm species. Besides, these new clones have a shorter maturity period, allowing them to produce fruits earlier. Apart from that, it has been reported that the new clones are shorter in height, simplifying the harvesting process significantly. Furthermore, the successful identification of oil-palm DNA could help in recognising genetic traits to improve oil-palm resistance towards diseases and pests.

On the other hand, PO has a high percentage of saturated compounds, causing palm-based biodiesel to suffer from a high cloud point, leading to difficulty in igniting the engine at low temperature. Consequently, palm biodiesel is not suitable to be utilised during the winter season without any fuel additives to enhance its performance. In view of this constraint, MPOB has successfully developed a new type of winter-grade palm biodiesel that has lower cloud point and pour point to accommodate winter season market demand. This breakthrough research could open up vast market opportunities for palm biodiesel, especially to European countries, and subsequently establish itself as the major feedstock for biodiesel throughout the world.

Apart from biodiesel, oil-palm biomass is another valuable renewable energy source that could be utilised for optimised application. For instance, oil-palm biomass including EFB can be utilised to produce steam and subsequently generate electricity. In fact, most PO mills are self-sustaining in terms of electricity generation, which reduces the operating cost of PO mills. In addition, oil-palm biomass contains valuable lignocellulose that can be hydrolysed and fermented to produce bioethanol, the main transportation fuel in Brazil.[8] Hence, the versatility of oil palm as an energy crop is not limited to biodiesel only as its biomass can be fully utilised to produce energy-related products such as steam and bioethanol.

Acknowledgements

The authors would like to acknowledge Universiti Sains Malaysia and Elsevier for their contribution towards this chapter.

References

1. S. Sumathi, S. P. Chai and A. R. Mohamed, *Renew. Sustain. Energ. Rev.*, 2008, **12**, 2404–2421.
2. K. T. Tan, K. T. Lee, A. R. Mohamed and S. Bhatia, *Renew. Sustain. Energ. Rev.*, 2009, **13**, 420–427.
3. M. K. Lam, K. T. Tan, K. T. Lee and A. R. Mohamed, *Renew. Sustain. Energ. Rev.*, 2009, **13**, 1456–1464.
4. K. F. Yee, K. T. Tan, A. Z. Abdullah and K. T. Lee, *App. Energ.*, 2009, **86**, S189–S196. Y. Basiron, *Symp. Sust. Develop.*, 2006. Available at: (http://www.mpoc.org.my/upload/POS_-_Sustainable_Palm_Oil_Production_in_Malaysia.pdf).
5. F. Ahmad, *Regional Workshop on Integrated Plant Nutrition System (IPNS)*, 2001. Available at: (http://banktani.tripod.com/faridah.pdf).
6. Malaysian Palm Oil Board, *MPOB report*, 2007. Available at: (http://econ.mpob.gov.my/economy/overview07.htm).
7. S. H. Shuit, K. T. Tan, K. T. Lee and A. H. Kamaruddin, *Energy*, 2009, **34**, 1225–1235.

CHAPTER 11

Jatropha curcas: A Source of Energy and Other Applications

SATYAWATI SHARMA AND ASHWANI KUMAR

Centre for Rural Development & Technology, Indian Institute of Technology, New Delhi – 110016, India

11.1 Introduction

The major sources of energy are fossil fuels, nuclear fuels and others including solar energy, tidal energy, wind energy, bioenergy, *etc.* Fossil fuels are non-renewable and getting depleted. Further, the transport and distribution bottlenecks add to the steep hike in their prices. Nuclear fuels generate radioactive wastes, severely limiting their scope. In this context, biomass appears to be a trouble-free and attractive alternative. Each and every component of biomass could prove an inexhaustible source of energy, provided we generate the capability to tap it in a sustainable manner.

India is sixth in the world in energy demand accounting for 3.5% of the world's total commercial energy consumption. Diesel is India's main liquid fuel and the trains consume a large quantity of it.[1] Domestic supply can presently satisfy only 22% of the demand and hence, dependence on crude oil import ($18 billion/annum) is on the rise. Nonetheless, the demand–supply gap is further increasing. It is argued that locally produced biofuel can reduce this gap. Significant quantities of energy from fossil may be replaced if biofuel development is carefully planned and implemented in a sustainable manner. Furthermore, from a variety of oil-bearing crops the most suitable ones, matching local soil, water and climatic conditions along with the economic

---

RSC Energy and Environment Series No. 3
Energy Crops
Edited by Nigel G Halford and Angela Karp
© Royal Society of Chemistry 2011
Published by the Royal Society of Chemistry, www.rsc.org

returns are to be chosen. The improved agronomical practices of well-managed biomass plantations may also provide a basis for environmental improvement by stabilising certain soils and avoiding desertification already occurring rapidly in tropical countries.

Jatropha a multipurpose, drought-resistant plant, widely distributed in the wild or semicultivated areas in Central and South America, Africa, India and South East Asia,[2,3] has considerable potential for biodiesel production. As *Jatropha* has the ability to grow on poor soils, using marginal land for *Jatropha* cultivation is therefore attractive since it would not displace food-producing crops. In many tropical regions land degradation and soil erosion have been identified as major threats to existing land-use patterns. *Jatropha* is known to fix the soil with its root system, thus reducing wind and water erosion of the soil. The *Jatropha* roots allow entry of surface water into the ground, thus increasing water absorption and reducing runoff, which also improves the soil and thereby enhances soil quality, potentially recovering degraded land for other uses. Also, *J. curcas* L. ranks second in land efficiency in a list of oil crops (both edible and nonedible) (Table 11.1). It is not grazed by animals and grows in poor and stony soil. Furthermore, it is disease resistant, multipurpose and yields high-quality biodiesel.[4] With respect to these traits, *Jatropha* proved to be superior over other nonedible oil plants. It grows relatively quickly and is easy to establish.

Nearly 40% of the land area in India is wasteland. Importance is given on plantation of *Jatropha* species on wastelands for the protection of the environment and fulfilling future energy requirements.

The Planning Commission of India suggested large-scale *Jatropha* cultivation, starting with 400 000 hectares of land in the first phase, 2.5-million hectares in the second phase, and then extending to 13.4-million hectares. The National Mission on Biodiesel in India had started a demonstration project in 2006–2007, by planting *Jatropha* in 4000 km^2 of wastelands, namely unoccupied dry lands near understocked forests, dry lands held by landlords, cultivable fallow lands, public lands along railway tracks, roads and canals and marginal lands with unsuitable conditions for crop production due to soil and climatic constraints.[5] A number of projects on various aspects of *Jatropha* are being implemented in India.

The latest estimates suggest that there are now 2.5 million hectares of *J. curcas* in India and China alone, with plans for an additional 23 million acres by 2010. The Philippines and some other countries have also initiated large-scale plantation of *Jatropha*.[1]

Table 11.1 Land efficiency of different energy crops.[1]

Crop	Litres oil/ha*
Oil palm	2400
Jatropha	1300
Rapeseed	1100
Sunflower	690

At present, there are three main biofuel producers in the world – the US, Brazil and EU. About 96 per cent of the biofuel is bioethanol and the rest is biodiesel. According to FAO estimates, world ethanol production is dominated by the US followed by Brazil and these two account for more than 73 per cent of world ethanol production. The share of the US in the world biofuel market is 40 per cent of the total production, basically from corn (maize), followed by Brazil's 34 per cent (mainly from sugarcane). Given the mandatory fuel-blending requirements legislated in many countries, annual global production of bioethanol is set to increase to 120 billion litres. The world annual biodiesel production is estimated to increase to 12 billions litres by 2020. Biodiesel is mainly produced in the EU and the US with the EU being the largest biodiesel producer and consumer in the world (66%). The EU produces about 95 per cent of the global biodiesel. Other countries have also started producing ethanol and biodiesel but their individual shares in overall production are not very significant. Major increases in production volumes are expected in countries such as Brazil, the US, the EU, China, India, Indonesia and Malaysia.[6]

It has been observed that the global biodiesel production has a steady growth rate of 30% per year, and it will easily reach 12 MT per year by the end of 2010, with a 3-fold larger capacity[7] (Figure 11.1).

In this chapter, the information about the potential and current development in the field of *Jatropha* research is discussed. It is hoped that the state-of-the-art information provided here will stimulate research and development leading to more intensive, efficient, and sustainable cultivation and utilisation of *Jatropha*.

11.2 Taxonomy and Botanical Description

The genus *Jatropha* belongs to tribe Joannesieae in the Euphorbiaceae family and contains approximately 170 known species. Among them, 12 species are recorded in India. Linnaeus was the first to name the physic nut *Jatropha* L. in "Species Plantarum" and this is still valid today.[8] The genus name *Jatropha*

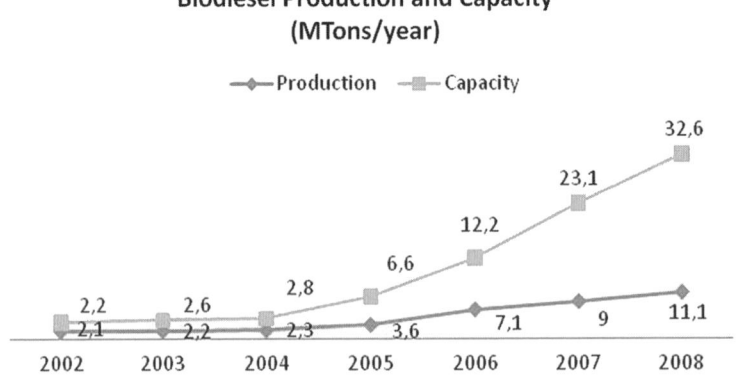

Figure 11.1 World biodiesel production and capacity (million tons/year).

Table 11.2 Various names of *Jatropha* in different countries.[9]

Common Name	Country
American purging nut, Barbados purging nut, Big purge nut, Black vomit nut, Physic nut, Purging nut	South Africa
Ba dau me, Lohong, Pe-fo-tze, Tong-chou	Vietnam
Bagbherenda (India), Physic nut bush, Wiriwiri	Fiji
Jarak pagar, Ma feng shu	Indonesia
Kananeranda, Nepalamu, Pindi, Punne-tang, Ratanjyot, Seemanepaalam	India
Kasla, Tubang-bakod	Philippines
Medisiyen blen, Perchnut, Purge nut bush, White physic nut	West Indies
Mupfure-donga	Venda
Physic nut	Ghana, Guam, Guyana, Nepal, Nigeria, South Africa, Thailand, Virgin Islands,
Physic nut bush	Fiji, West Indies
Tubaag-bakod	Cape Verde Islands
Tubatuba, Physic nut	Guam
Udukaju, Purging nut, Sabbu dam	Thailand
Piao brancho, Pinnao de purga	Brazil
Pignon d'inde	Rodrigues Islands
Pinon botija	Cape Verde Islands
Purging physic	Nicaragua

derives from the Greek word *jatrós* (doctor) and *trophé* (food), which implies medicinal uses. Table 11.2 shows various vernacular names of *Jatropha* in different countries. The physic nut, by definition, is a small tree or large shrub, which can reach a height of three to five metres but under favourable conditions it can attain a height of eight or ten metres. The branches contain latex. *Jatropha* cultivated with seeds normally and five roots are formed from each seedling, one central and four peripheral. Leaves are green to pale green and alternate. The phyllotaxy is spiral and venation is palmate. Petiolate 2–20 cm long stipulate with ciliate glands usually represents the minute stipules.

Pollination of the physic nut flowers is by insects, especially honey bees. When insects are excluded from the greenhouse, seed set does not occur without hand pollination.[10] The rare hermaphrodite flowers can be self-pollinating. Each inflorescence yields a bunch of approximately 10 or more ovoid fruits. With good rainfall conditions nursery plants may bear fruits after the first rainy season, while directly sown plants do so after the second rainy season. On the plant, *J. curcas* fruits change colour from green to yellow and then to brown. Three, bivalved cocci are formed after the seeds mature and the fleshy exocarp dries. The seeds mature about three to four months after flowering. Fruit is usually a three-chambered and schizocarpic capsule splitting into three one-seeded cocci that look like black beans with similar dimensions of about 18 mm long (range 11–30 mm) and 10 mm wide (range 7–11 mm). The seeds are black

Figure 11.2 Different parts of *Jatropha curcas*.

and the seed weight per 1000 is about 727 g. On average, there are 1375 seeds per kg.[11] Singh described the microscopical anatomy of fruits.[12] Gupta investigated the anatomy of other plant parts.[13] Figure 11.2 shows *Jatropha* tree, inflorescence, longitudinal and transverse section of fruit.

11.3 Distribution and Ecological Requirement

It is highly probable that the centre of origin of the physic nut is in Mexico (and Central America), although the "true" centre of origin is still to be established. To elucidate this, the original collecting sites in Mexico and Central America would have to be revisited and the existing diversity assessed, preferably by molecular techniques. At present, its native distributional range is considered to be Mexico, Central America, Brazil, Bolivia, Peru, Argentina and Paraguay. Nowadays, it has a pantropical distribution[14] with distinct seed provenances. From the Caribbean, this species was probably distributed by Portuguese seafarers *via* the Cape Verde Islands and former Portuguese Guinea (now Guinea Bissau) to other countries in Africa and Asia.[4] In India, Portuguese navigators introduced it in the 16th century. It occurs in almost all parts of India including Andaman Island and is generally grown as live fence. The genus is cosmopolitan in distribution except for the arctic region, being found in the tropical and subtropical Himalayas, the mountains of Western and Eastern Ghats and plains of South India.

The current distribution of physic nut shows that introduction has been most successful in drier regions of the tropics with an average annual rainfall of between 250 and 1200 mm.[15] As physic nut occurs mainly at lower altitudes (0–500 m), it can be concluded that it is adapted to higher temperatures, however, it withstood slight frost in the Chã das Caldeiras, Fogo (approximately 1700 m altitude).[16] It is not sensitive to day length.[4] It grows on well-drained soils with good aeration and is well adapted to marginal soils with low nutrient content.[17] It is susceptible to collar rot disease in high-rainfall, humid areas, or fields receiving excessive irrigation or prone to water logging.[18] Soil depth should be at least 45 cm[19] and surface slope should not exceed 30°.[20] It has low nutritional requirements, but the soil pH should not exceed 9.[21] On very acidic soils *Jatropha* might require some Ca and Mg fertilisation. Though it is adapted to marginal soils with low nutrient content, to support a high biomass production the crop shows a high demand for nitrogen and phosphorus fertilisation.[17] It has been demonstrated that soil structure improved significantly after growing *J. curcas* for 18 months under semiarid conditions in India. Macroaggregate stability was increased by 6–30% and soil bulk density was reduced by 20%.[22,23]

The leaves shed during the winter months form mulch around the base of the plant and the organic matter from shed leaves enhances earthworm activity in the soil around the root-zone of the plants, which improves the fertility of the soil. In heavy soils, root formation is hampered.[4] Clay soils are unsuitable for *Jatropha* if water logging or saturation occurs due to climatic conditions. In general, "working soils" are not good because root-system development is impaired. Acid (pH < 6) or alkaline (> 9) soils are not fit for *Jatropha*.[17] Also shallow soils (less than 50–80 cm soil depth) are not suitable for *Jatropha* cultivation from seeds. Cuttings can be grown on shallow soils but only with sufficient rainfall over the year or irrigation. Sandy to loamy soils seem to be a best fit for *Jatropha* cultivation. Mycorrhiza assisting the uptake of phosphorus and microelements was found on the root system.[19,24–26] Mycorrhiza-inoculated *Jatropha* showed a 30% increase in both biomass and seed production 7 months after plantation of 1-year-old saplings.[20]

Jatropha is well adapted to arid and semiarid conditions. In low-rainfall areas and in prolonged rainless periods, the plant sheds its leaves as a counter to drought. Its water requirement is comparatively low and it can stand long periods of drought by shedding most of its leaves to reduce transpiration loss.

11.4 *Jatropha* Cultivation

There are various methods to cultivate *Jatropha*, which vary from region to region and climate to climate. These are; direct seeding, precultivation of seedlings (nursery raising), transplanting of spontaneous wild plants and direct planting of cuttings. Although it grows readily from seeds or cuttings, wider spacing (3 m 3 m) is reported to give larger yields of fruit, also in the early years. The *Jatropha* plants propagated by cuttings show a lower longevity and possess a lower drought and disease resistance than those propagated by seeds.[4] This might be due to the fact that trees produced from cuttings do not produce true taproots (hence less drought tolerant), rather they produce pseudotaproots that may penetrate only 1/2 to 2/3 the depth of the soil. According to estimates the use of direct seeding is a reliable method due to genetic divergence among individuals to perpetuate those plants with high yield in seeds and oil.

Since flowers of *Jatropha* are developed terminally, the plants with good ramification after pruning can produce many fruits. Also, the pruning of the plants keeps them low enough to facilitate the harvest of the fruits, because the time needed to harvest a certain number of kg of seeds, is a very important factor of the economic feasibility of *Jatropha* oil production. At the same time, pruning provides cuttings to produce new plants with identical genetic material.[19]

In better rainfall or good moisture condition the plantation could also be established by direct seeding. The survival rate depends not only on sowing time and depth of sowing, but also on the trial year (year with different rainfall). The application of biofertilisers containing beneficial microbes showed a promoting effect on the growth of *Jatropha*. The effect of vermi-compost on *Jatropha* was better than farmyard manure.[27] *Jatropha* fruit should be harvested when it changes colour from green to yellow to get high lipid content. Despite the interest in *Jatropha* the available yield performance data for this plant species are limited and somewhat uncertain. However, in different countries and regions the seed yield of *Jatropha* ranges from 0.1 to 15 tons/ha/yr.[28]

11.5 Composition of *Jatropha*

The fruit of *Jatropha* contains approximately 35% of the dry fruit husk and 1 to 3 seeds. An individual seed weighs from 0.4 to over 1 g. Seed lipid contents vary from 22–48%. The general chemical compositions of *Jatropha* cake is shown in Table 11.3. On average, 35–38% of the seeds are shells. Kernel weight takes a proportion of $\sim$65% of seed weight and has lipid content close to 60%. Completely deoiled kernel yields a meal of high protein ($\sim$60%) with an excellent amino acid profile.[30] Figure 11.3 shows the average proportions on a dry weight basis starting with one tonne of *J. curcas* fruits/capsules. *Jatropha*

Table 11.3 Composition of *Jatropha* cake.[29]

Constituent	Jatropha
Nitrogen/protein (wt%)	4–6/25–40
Carbohydrates (%)	15–20
Fibre (wt%)	1–20
Ash (wt%)	3–5
Phosphorus (wt%)	1.5–3
Potassium (wt%)	1–2
Calcium (wt%)	<1
Magnesium (wt%)	<1
Zinc, copper, magnesium and boron (ppm)	<100
Sulfur (ppm)	<3000

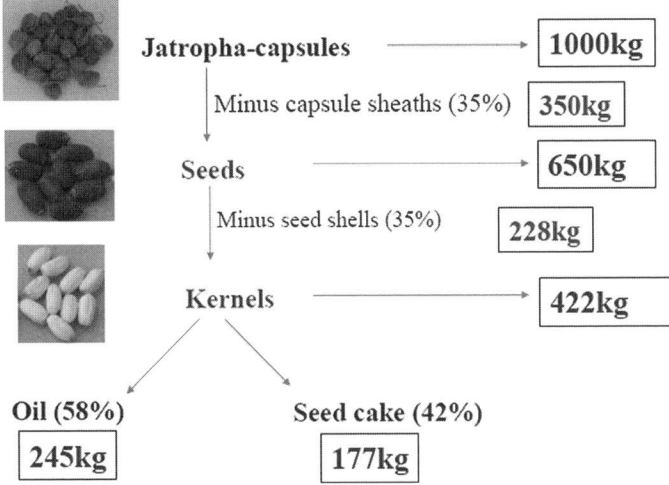

Figure 11.3 Different fractions of *J. curcas*.[30]

oil contains approximately 24.60% of crude protein, 47.25% of crude fat, and 5.54% of moisture content.[31]

The chemical composition pertaining to dry matter, crude protein, lipid, ash, neutral detergent, acid detergent fibre, acid detergent lignin, gross energy, total phenols and tannins of kernel, shell and husk of *curcas* varieties are shown in Table 11.4. *J. curcas* kernel is composed mainly of lipid and protein, with very little moisture and ash. There were varietal differences in CP (crude protein) in the kernels (22.2–27.2%). The low moisture content of the shell (<10%) and kernel (< 6%) could be partly responsible for the nondeterioration of seeds over a long period.

The presence of antinutritional factors is also likely to increase the shelf life of the seeds. The shell of *J. curcas* seed is composed mainly of fibre (>83% NDF and >74% ADF) and lignin (>45%) with very little protein (<6%) indicating poor nutritional value, but can be a good source of fuel because of its high gross energy.

11.6 Toxicity of Jatropha Seeds

The chemicals isolated from different parts of the plant can be used in various industrial applications[11] (Table 11.5). The different active components present in different parts of *Jatropha* plants are summarised in Figure 11.4. Depending on the variety, the decorticated seeds contain 40–60% of oil,[25,33–35] which is used for many purposes such as lighting, as lubricant, for making soap[36] and more importantly for biodiesel production. The seeds of *J. curcas*, in general, are toxic to humans and animals. Curcin, a toxic protein isolated from the seeds, was found to inhibit protein synthesis in *in vitro* studies.

Table 11.4 Chemical composition of kernel, shell and husk of Jatropha curcas varieties.[32]

Item	Variety								
	Cape Verde		Nicaragua		Ife-Nigeria			Nontoxic Mexico	
	Kernel	Shell	Kernel	Shell	Kernel	Shell	Husk	Kernel	Shell
Dry matter (DM)	96.6	90.3	96.9	90.4	95.7	91.9	91.3	94.2	89.8
Analysis,% in DM:									
Crude protein	22.2	4.3	25.6	4.5	27.7	5.8	6.3	27.2	4.4
Lipid	57.8	0.7	56.8	1.4	53.9	0.8	1.1	58.5	0.5
Ash	3.6	6.0	3.6	6.1	5.0	4.6	15.4	4.3	2.8
Neutral detergent fibre	3.8[a]	83.9	3.5[a]	85.8	4.1	89.6	65.9	3.8	89.4
Acid detergent fibre	3.0[a]	74.6	3.0[a]	75.6	2.6	79.8	61.3	2.4	78.3
Acid detergent lignin	0.2[a]	45.1	0.1[a]	47.5	0.00	47.4	14.4	0.00	45.6
Gross energy (MJ kg^{-1})	30.7	19.5	30.5	19.5	29.7	19.5	15.6	31.1	19.5
Total phenols (% tannic acid equivalent)	0.36	3.0	0.29	2.8	0.31	3.1	0.18	0.22	4.4
Tannins (% tannic acid equivalent)	0.04	2.2	0.03	2.0	0.0	2.2	0.01	0.02	2.9

[a]Calculated from values obtained for fat-free samples since high lipid content interferred with fibre determination.

Table 11.5 Chemicals isolated from different parts of the Jatropha plant.

Various parts	Chemical components	References
Aerial parts	Organic acids (o and p-coumaric acid, p-OH-benzoic acid, protocatechuic acid, resorsilic acid, saponins and tannins	37
Stem bark	β-amyrin, β-sitosterol and taraxerol	38
Leaves	Cyclic triterpenes stigmasterol, stigmast-5-en-3β, 7 β-diol, stigmast-5-en-3β,7α-diol, cholest-5-en-3β,7β-diol, cholest-5-en-3fl,7α-diol, campesterol, β-sitosterol, 7-keto- β-sitosterol as well as the β-D-glucoside of β-sitosterol, flavonoids apigenin, vitexin, isovitexin.	38–40
	Leaves also contain the dimer of a triterpene alcohol ($C_{63}H_{117}O_9$) and two flavonoidal glycosides	39
Latex	Curcacycline A, a cyclic octapeptide	41
	Curcain (a protease)	42
Seeds	Curcin, lectin	43
	Phorbolesters	35,44
	Esterases (JEA) and Lipase (JEB)	45
Kernel and press-cake	Phytates, saponins and a trypsine inhibitor	34,46,47
Roots	β-sitosterol and its β-D-glucoside, marmesin, propacin, the curculathyranes A and B and the curcusones A–D. diterpenoids jatrophol and jatropholone A and B, the coumarin tomentin, the coumarino-lignan jatrophin as well as taraxerol	48

Despite the toxicity of the *J. curcas* seeds, edible varieties also exist in Mexico (Table 11.4). The high concentration of phorbolesters present in *Jatropha* seed has been identified as the main toxic agent responsible for *Jatropha* toxicity.[35,44] The content of phorbolesters appears to vary from provenance to provenance, and, therefore, there is a need to promote the collection of nontoxic varieties of *Jatropha*. The phorbolesters content of oil from the Cape Verde toxic and the Mexican nontoxic variety was found to be 2.49 and 0.27 mg ml^{-1}, respectively. Therefore, long-term toxicological studies by feeding diets containing oil from this nontoxic *Jatropha* variety need to be conducted on rats or other laboratory animals before it is recommended for human consumption. Phorbolesters were found to be responsible for purgative, skin-irritant effects and tumour promotion.[44,49]

The phorbolesters are found in plants belonging to the families Euphorbiaceae and Thymelaeaceae.[51] Several cases of *J. curcas* nut poisoning in humans after accidental consumption of the seeds have been reported with symptoms of giddiness, vomiting and diarrhoea and in the extreme condition even death has been recorded.[52] A comparative analysis of edible and nonedible seed varieties revealed that edible seeds lacked phorbolesters. The presence of phorbolesters

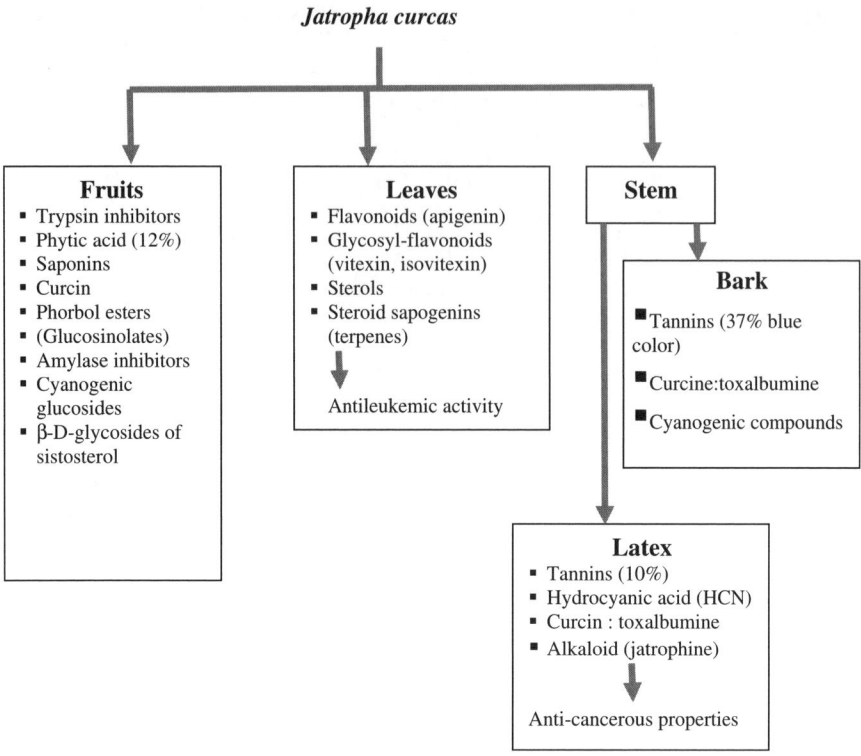

Figure 11.4 Active components of different parts of *Jatropha*.[50]

in *J. curcas* seeds has been known for some time.[44,49] Six phorbolesters are described in *J. curcas* and all are derivatives of a single-core molecule (Figure 11.5). The structure of six phorbolesters has now been determined using NMR.[53] These are usually found either in the seeds or the latex, which is exuded from these plants after wounding. The phorbolesters are analogues of diacylglycerol, an activator of many isoforms of protein kinase C (PKC).[54] The phorbolesters are acutely toxic, and oils containing phorbolesters are known purgatives.[55] Toxicity markers have been extensively studied using different molecular techniques.[56] Also, the genes responsible for the production of phorbolesters have been determined and very precise and abundant markers using single-nucleotide polymorphisms (SNP) differentiating toxic from nontoxic varieties have been obtained.[57] Ionising radiation treatment could serve as a possible additional processing method for inactivation or removal of certain antinutritional factors such as phorbolesters, phytates, saponins and lectins.[58] Martinez-Herrerra, studied the nutritional quality and the effect of various treatments (hydrothermal processing techniques, solvent extraction, solvent extraction plus treatment with $NaHCO_3$ and ionising radiation) to inactivate the antinutritional factors in defatted *Jatropha* kernel meal of both toxic and nontoxic varieties from different regions of Mexico.[59]

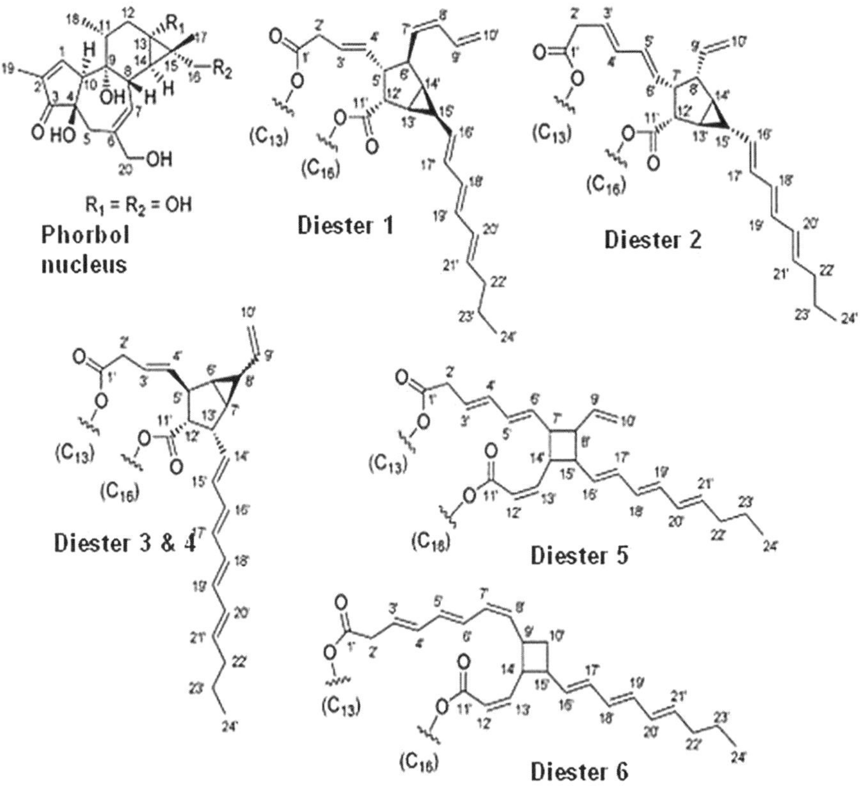

Figure 11.5 Phorbolesters present in *Jatropha curcas* L.[49]

It is not clear at present whether an economically feasible method of eliminating the phorbolesters from *J. curcas* meal is developed. Complete removal of the toxins is therefore necessary before *Jatropha* oil is used in industrial applications or in human medicine. The oil must be completely innocuous before it is used commercially for different applications.

11.7 Potential Applications of *Jatropha*

11.7.1 Fuel from *Jatropha*

Numerous sources are available on the fatty acid composition of physic nut oil originating from different countries.[11] The oil fraction of *Jatropha* contains 22.3% saturated fatty acids, comprising mainly (14.5%) palmitic acid (16:0) and 6.3% stearic acid (18:0) and 77.5%. unsaturated fatty acids The unsaturated fatty acid composition of *Jatropha* oil is dominated by oleic acid (C18:1) with 42% and linoleic acid (C18:2) with 35.3% (Figure 11.6). The maturity stage of the fruits at the moment of collection is reported to influence the fatty

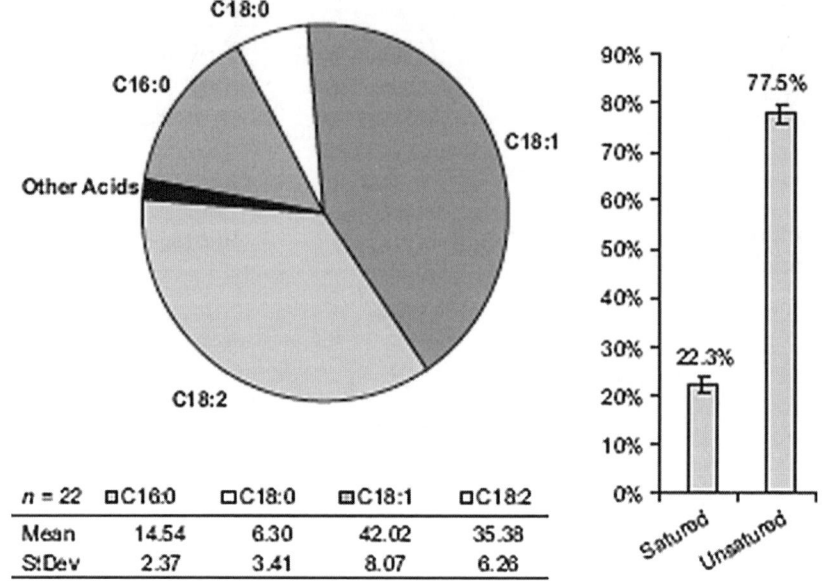

Figure 11.6 Fatty acid composition (%). C16:0 = palmitic acid; C18:0 = strearic acid; C18:1 = oleic acid; C18:2 = linoleic acid. Other acids containing capric acid, myristic acid (C14:0), palmitoleic acid (C16:1), linolenic acid (C18:3), arachidic acid (C20:0), behenic acid (C22:0), cis-11-eicosenoic acid (C20:1) and cis-11,14-eicosadienoic acid (C20:2). n = number of observations used.[61]

acid composition of the oil.[60] The composition and characteristics of the crude *Jatropha* oil are given in Table 11.6. The extraction of *Jatropha* oil does not require sophisticated technology. Simple equipment, similar to that being used for production of groundnut oil or palm-oil extraction, can be used. The ripe fruits are collected from the trees and decorticated manually or with the help of a decorticator. The seeds should be dry before they are pressed as this improves the efficiency of the extraction process.

The oil from *Jatropha* seeds can be extracted using different methods, including mechanical extraction with a screw press and solvent extraction. The extracted oil can then be purified by sedimentation or filtration. Extraction using a mechanical screw press is the simplest method and has found wide use in rural areas in developing countries. *Jatropha* oil can be used either directly as *Jatropha* fuel (pure plant oil) or as biodiesel through transformation of the oil.

There is widespread experience in developing countries with the utilisation of pure *Jatropha* oil to fuel in small one-cylinder Lister-type diesel engines that power cereal-grinding machines, electric generators, mechanical presses, *etc*. In addition, pure plant oil can also be blended in a certain percentage with diesel to run modern diesel engines (although manufacturers normally only guarantee engines with a use of 5–10% biodiesel blends). But the utilisation of *Jatropha*

Table 11.6 *Jatropha curcas* oil composition and characteristics.[61]

Characteristics	Range	Mean	S.D.	n
Specific gravity (g cm^{-3})	0.860–0.933	0.914	0.018	13
Calorific value (MJ kg^{-1})	37.83–42.05	39.63	1.52	9
Pour point (1 °C)	−3			2
Cloud point (1 °C)	2			1
Flash point (1 °C)	210–240	235	11	7
Cetane value	38.0–51.0	46.3	6.2	4
Saponification number (mg g^{-1})	102.9–209.0	182.8	34.3	8
Viscosity at 30 °C (cSt)	37.00–54.80	46.82	7.24	7
Free fatty acids % (kg kg^{-1}×100)	0.18–3.40	2.18	1.46	4
Unsaponifiable % (kg kg^{-1}×100)	0.79–3.80	2.03	1.57	5
Iodine number (mg iodine g^{-1})	92–112	101	7	8
Acid number (mg KOH g^{-1})	0.92–6.16	3.71	2.17	4
Monoglycerides % (kg kg^{-1}×100)	n.d.–1.7			1
Diglycerides % (kg kg^{-1}×100)	2.50–2.70			2
Triglycerides % (kg kg^{-1}×100)	88.20–97.30			2
Carbon residue % (kg kg^{-1}×100)	0.07–0.64			3
Sulfur content % (kg kg^{-1}×100)	0–0.13			2

S.D. = standard deviation; n = number of observations used; n.d. = no data.

oil in modern diesel engines requires some modifications that consist mainly of heating the *Jatropha* oil to reduce its viscosity.

Due to these technical changes required, some people promote the conversion of pure *Jatropha* oil into biodiesel through a transesterification process. Figure 11.7 shows a block diagram of the biodiesel-production process. First, pure oil that has been produced by solvent extraction is treated in a process called transesterification to produce biodiesel. The transesterification process involves mixing of alcohol (usually methanol that is toxic but cheap) with the oil in the presence of a catalyst and then separation of lighter methyl ester phase by gravity from the heavier glycerol. The free fatty acid content of nonedible-grade oils are cheap feedstock for economic production of biodiesel.

Jatropha oil contains about ∼14% free fatty acid (FFA), which is far beyond the limit of 1% FFA level that can be converted into biodiesel by transesterification using an alkaline catalyst. Hence, an integrated optimised procedure for converting *Jatropha* oil, which contains high FFA%, into biodiesel is very much required. A few researchers have worked with feedstocks having higher FFA% levels using alternative processes. These include a pretreatment step to reduce the FFAs of these feedstocks to less than 1% followed by a transesterification reaction with an alkaline catalyst.[62,63] This procedure yielded more than 95% biodiesel.

Although the transesterification process is quite straightforward, in order to reach optimal biodiesel production results, the optimal inputs for the transesterification of JCL (*J. curcas* L.) oil (3.1% free fatty acids and acid number 6.2 mg KOH g^{-1}) are identified to be 20% methanol (by mass on oil basis) (molar ratio methanol:oil E5.5:1) and 1.0% NaOH by mass on an oil basis. Maximum ester yield is achieved after 90 min reaction time at 60 °C.[64] Optimal

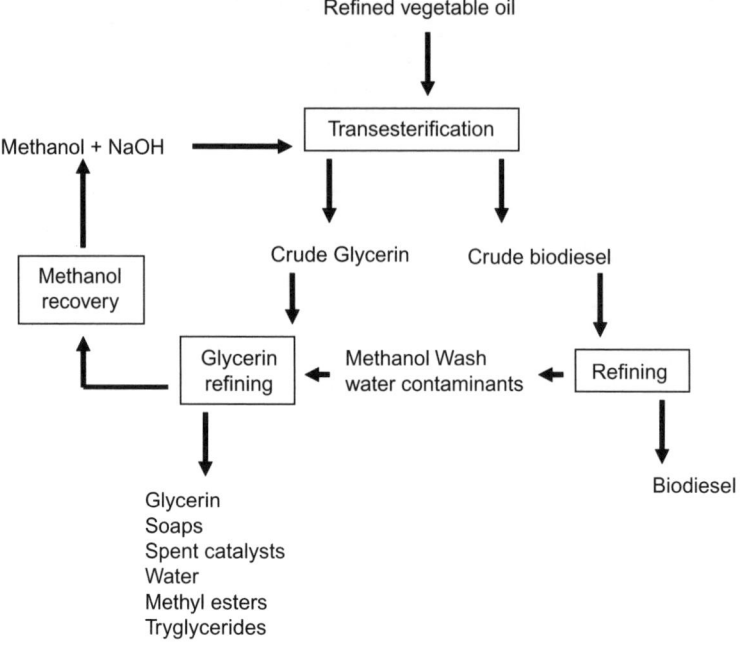

Figure 11.7 Block diagram of biodiesel-production process.[5]

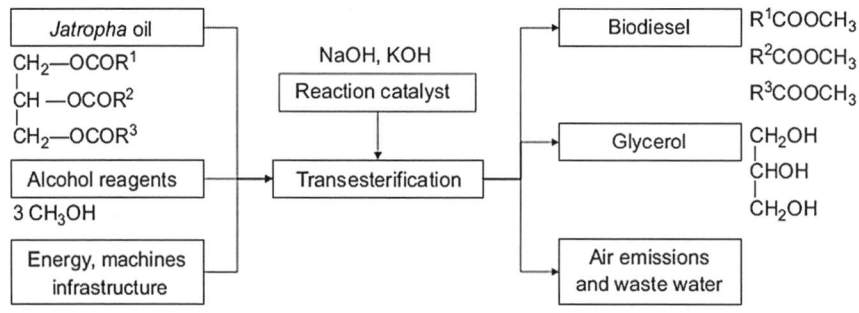

Figure 11.8 Input and output of transesterification process.

conversion of JCL oil with high free fatty acids (14%) and high acid number (28 mg KOH g^{-1}) needs a pretreatment reaction with methanol (molar ratio methanol:oil E6.5:1) using H_2SO_4 as the catalyst (1.43%) for 88 min at 60 °C. After pretreatment a maximal conversion rate of more than 99% is achieved by transesterification with methanol (molar ratio methanol:oil E4:1) and 0.6% KOH by weight during 24 min.[65]

Figure 11.8 showing the input and output of the transesterification process. The free fatty acid content is the key parameter for identifying the process of biodiesel preparation. The acid value of *Jatropha* oil ranges from 3 to 38 mg KOH/g.[66] In a typical biodiesel preparation, 2000 g of the crude *Jatropha* oil is transesterified with a solution of 30 g KOH in 331 g methanol. The reaction is carried out in a batch reactor in two steps at 30 °C. The oil is mixed with two parts of the methanolic KOH solution and the reaction mixture stirred for 30 min and the glycerol layer allowed to separate. The upper organic layer is mixed with one part methanolic KOH and stirred for a further 30 min. After 5 h settling time, the glycerol layer is separated and the ester layer is washed with warm water and passed over Na_2SO_4, which results in 92% theoretical yield of the methyl esters. Biodiesel prepared on a pilot scale has been reported with 99.5% purity of the methyl esters.[17]

A number of factors determine the suitability of a particular feedstock for the production of biodiesel and many countries have defined specification standards. Although it may be preferable to produce and use the biodiesel locally, the biodiesel specifications are not available for many of the countries in which *Jatropha* is being grown.[67] Unfortunately, no one single FAME is ideal when matched against all these parameters, but oils containing high levels of oleate and palmitoleate are desirable.[68]

Interestingly, temperature is known to have an effect on the fatty acid composition of other oilseed crops.[69] Berchmans and Hirata and Tiwari *et al.*, have developed a technique to produce biodiesel from *Jatropha* with high free fatty acids contents (15% FFA), in which a two-stage transesterification process was selected to improve methyl ester yield.[17,70] In order to reduce the cost of biodiesel fuel production from *Jatropha*, the lipase-producing cells of *Rhizopus oryzae* immobilised onto biomass support particles were used and found to be a promising biocatalyst for producing biodiesel.[71] The CN value is perhaps the most important factor for biodiesel and is a measure of delay in the combustion of the fuel from ignition. The actual cetane values of *Jatropha* biodiesel have been determined by a number of groups, and have so far been within the range of 50–57.[17,72,73] Sarin *et al.*, have examined the blends of *Jatropha* and palm biodiesel for their physicochemical properties and to get the optimum mix of them to achieve better low-temperature and improved oxidation stability needed for South Asian and South East-Asian countries.[73]

11.7.2 Other Uses

The potential uses of *J. curcas* are summarised in Figure 11.9. Initially, the commercial applications of this plant were reported from Lisbon, where the oil imported from Cape Verde was used for soap production and for lamps.

In addition to producing oil, the plant has other important roles, such as: (i) land reclamation and additional agroecological advantages; (ii) provision of seed meal after detoxification as animal feed;[52] (iii) provision of chemicals with potential in medicine, pharmaceutical and biopesticide applications (*e.g.*, phorbolesters present in the oil kill the vector snail of schistosomiasis – the second most important human disease in the tropics – at an extremely low

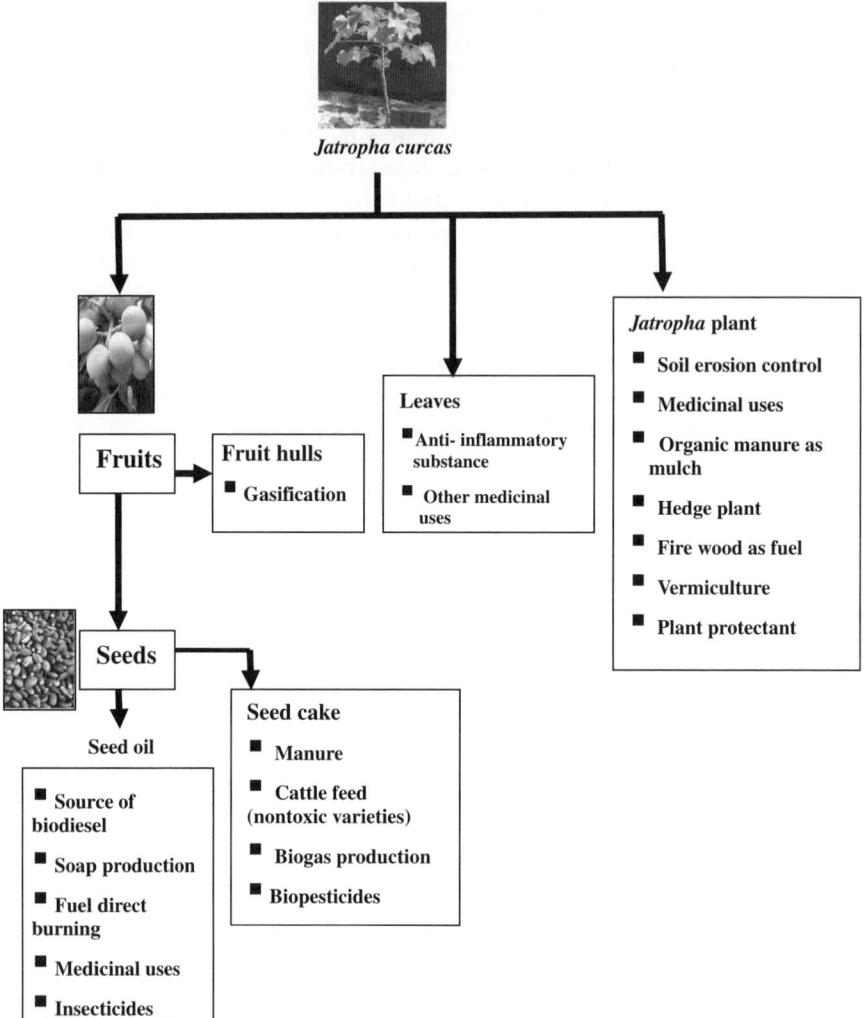

Figure 11.9 Utilisation of *J. curcas*.

concentration and without harming fish and other creatures in the same water body); (iv) carbon-dioxide emission impact (GHG reduction); (v) contribution to human welfare and to the whole world economy.

Jatropha is an excellent hedging plant and is generally grown in most parts of India and Africa as a live fence for the protection of agricultural fields against damage by livestock as it is unpalatable to cattle and goats. *Jatropha* seed-cake, a byproduct of oil extraction, contains curcin, a highly toxic protein similar to ricin in castor, making it unsuitable for animal feed. However, it does have potential as a fertiliser or in biogas production[74,75] The defatted meal has been found to contain a high amount of protein in the range of 50–62%, and the level of essential amino acids except lysine, which is higher than the FAO reference protein.[32] In a preliminary experiment, *Jatropha* seed-cake was utilised as feedstock for biogas production.[76,77] Besides substituting biofuels for fossil fuel, seed shells of *Jatropha* also have a high energy value (l18–19 MJ kg^{-1}). Both husks and shells are not suitable as substrates in biogas digesters because of their very low digestibility. Recently, experimentation on solid-state fermentation of *Jatropha* seed-cake showed that it could be a good source of low-cost production of industrial enzymes.[78]

The use of *Jatropha* meal (cake) as animal feed improves the economics of *J. curcas* production. The seeds of *J. curcas* contain a range of toxins and antinutrients. The toxicity of *Jatropha* seeds has been mainly attributed to the presence of a protein (curcin) and phorbolesters (diterpenoids). As a consequence, the protein-rich seed meal of *Jatropha* is not used as animal feed. However, a nontoxic variety of *Jatropha*, reported to exist in some provenances of Mexico and Central America, is said to lack toxic phorbolesters.[32] Sujatha *et al.*, have established the protocols for *in-vitro* propagation of the nontoxic variety through auxiliary bud proliferation and direct adventitious shoot bud regeneration from leaf segments.[79]

The byproduct of biodiesel, glycerin can be used to make soap. The soap can also be produced from *Jatropha* oil itself. In either case the process produces a soft, durable soap well adapted to household or small-scale industrial applications. *Jatropha* stem is a very light wood and burns too rapidly and is not popular as a fuel wood source. The work concluded that *Jatropha* wood would not be of much value for either charcoal or as firewood.[80] The oil and aqueous extract from oil has potential as an insecticide. For instance, it has been used in the control of insect pests of cotton including cotton bollworm, and on pests of pulses, potato and corn.[81]

Jatropha has various uses in different countries. In India, leaves are applied near horses' eyes to repel flies. Nuts can be strung on grass and burned like candlenuts. The oil has been used for illumination, making candles and soap, adulterating olive oil, and making Turkey red oil. *Jatropha* has been found to have strong molluscicidal activity and the latex to be strongly inhibitory to watermelon mosaic virus.[82] Substances such as phorbolesters, which are toxic to animals and humans, have been isolated and their molluscicidal, insecticidal and fungicidal properties have been demonstrated in lab-scale experiments and field trials.[83,84] These characteristics along with its versatility make it of vital importance to developing countries.[85]

All parts of *Jatropha* (seeds, leaves and bark) have been used in traditional medicine and for veterinary purposes for a long time.[86,87] Some compounds (Curcacycline A) with antitumour activities were reportedly found in this plant.[41] The seed oil can be applied to treat eczema and skin diseases and to soothe rheumatic pain.[4,87] The leaves and latex are used in healing of wounds, refractory ulcers, and septic gums and as an antiseptic in cuts and bruises. A proteolytic enzyme (curcain) has been reported to have wound-healing activity in mice.[88,89] Investigation of the coagulant activity of the latex of *Jatropha* showed that whole latex significantly reduced the clotting time of human blood.

Diluted latex, however, prolonged the clotting time, at high dilutions, the blood did not clot at all.[90] The methanol extract exhibited systemic and significant anti-inflammatory activity in acute carrageenan-induced rat paw edema.[91]

At SPRERI (Sardar Patel Renewable Energy Research Institute) a holistic approach has been taken to utilise all components of the *Jatropha*, *i.e.* fruit shell for combustion, hull/husk for gasification, oil and biodiesel for running CI engines, cake for production of biogas and spent slurry as manure. It has been found that all components of *Jatropha* fruit can be utilised efficiently for energy purposes.[92]

11.8 Potential of *Jatropha* in Wasteland Reclamation

Jatropha, a drought-resistant plant with many economical and ecological attributes, has the ability to grow on degraded and poor soils and can be used to reclaim eroded and other poor sites. It is highly efficient with respect to nutrient and water use. Most of the crops grown today, including oilseeds, are annuals. Perennial crops have deeper root systems that help store more carbon, maintain soil quality, and manage water and nutrients more conservatively. The cultivation of perennial plants have therefore been advocated as potentially more efficient ways of farming, especially on marginal soils.[93,94] Globally, *Jatropha* is an important candidate as a source of nonedible oil. Since it is already a part of traditional agricultural crops in Africa, Asia and Latin America, it does not require high technology for its cultivation and is a very important source of subproducts for rural communities.[1,75,95]

As a perennial crop, *Jatropha* invests a decreasing fraction of its carbohydrates into the wooden standing biomass over time, and if properly pruned, the seasonal requirements for the nutrients are only needed for the seasonal formation of branches, leaves, flowers, fruits and seeds. If senescent plant material, like leaves, flowers and pruned branches are left in the field or incorporated in the soil as mulch, they are slowly decomposed, resulting in the release of the nutrient back into the soil where they are available again for crop uptake.

Jatropha thrives on unproductive lands with limited water supply and poor soil and could yield oilseed during the first year of cultivation, albeit on a small scale. Table 11.7 shows claims and facts associated with *Jatropha curcas*.[96]

11.9 Pests and Diseases

There are different pests and diseases observed on physic nut plants by different authors, which reduces economic yield of the crop in many areas (Table 11.8).

11.10 Food *vs.* Fuel

Jatropha is seen as a very promising option for producing biofuel from degraded areas, generating rural employment, increasing environmental quality and providing primary energy carriers to energy-deficient areas. The suitability

Table 11.7 Beliefs/claims and facts associated with *Jatropha*.[96]

Property	Belief/claim	Fact
Soil and climate	Grows anywhere	Grows, but very low seed yield
Fertiliser application	Not required	Seed production will be low
Irrigation	Hardy, needs no water	Survives, but seed yield very low
Gestation period	Seed yield from first year	Small quantities for first 2–3 years
Seed-yield potential	8.0–10 tons per ha per year	Not under average conditions
Oil content	As high as 50% or more	Usually about 25%
Multipurpose species	Bark, leaf, root and latex have uses	If these parts are harvested, seed yield will be low or absent
Varieties and genotypes	High-yielding varieties and propagules	Not certified or tested under diverse conditions

Table 11.8 Pest and diseases observed on physic nut plants.[81]

Pest	Diseases/Damage
Fusarium moniliforme	root rot
Phytopthora spp.	damping off, root rot
Pythium spp.	damping off, root rot
Fusarium spp., *etc.*	damping off, root rot
Helminthosporium tetramera	leaf spots
Pestalotiopsis paraguarensis	leaf spots
Pestalotiopsis versicolour	leaf spots
Spodoptera litura	larval feeding on leaves
Julus sp. (millipede)	total loss on seedlings
Lepidoptera larvae	galleries in leaves
Oedaleus senegalensis (locust)	leaves, seedlings
*Cercospora Jatroph*a-*curcas*	leaf spots

of promoting *Jatropha* cultivation on a commercial basis on fertile land replacing other food and cash crops in the tropics has been questioned. Less controversial and more desirable would be the cultivation of native species of *Jatropha* on degraded lands in respective countries. The production of biofuels out of agricultural products is more profitable and, therefore, there is a risk of the price of a raw material used for biofuel production to increase beyond the price offered by the food industry, then, this raw material will be converted into fuel. According to a report presented by Wahenga "a rise in the price of food can be expected mainly because of two reasons: the high prices of agricultural productions (mainly because today agriculture is the main consumer of fossil fuels) and the influence the biofuels has on the grain world prices".[97]

In contemplating introduction of biofuels as a full-scale alternative to conventional fossil fuels, it is necessary to take the potential competition over land between biofuels and food crops into consideration. It is predicted that the world population will reach a peak of about 9.2 billion around 2050.[98] The question is whether the world's agricultural lands will be able to supply the demands for both biofuel and food production at that time. The total area of land in the world is 145 billion ha, of which present agricultural land accounts for about 10% of the total area.[99] Based on the United Nations' projections for the demand for cereals, it will be necessary to increase the food producing area by 240 million ha by 2050 when the world population is expected to reach its peak. Therefore, if the increase of 240 million ha in food producing area is subtracted from the total increase in agricultural land, the biofuels producing area is estimated at 260 million ha in 2050. Considering the expected population growth, which by the year 2050 must reach about 9.2 billion inhabitants,[100] the problems regarding hunger and nourishment insecurity must continue or even increase dramatically in some regions, unless urgent measures are taken. The main causes of food insecurity are poverty, in terms of income, access to education, agricultural resources, technology and credit lines for food production. In most of the countries that suffer from food insecurity the most vulnerable populations depend mainly on the local agriculture.[99] In this regard, rural development is an important path towards the reduction in poverty and food insecurity. Thus, countries with a better climatic and land potential for the development of biofuels have significant possibilities of developing their agricultural regions, which can improve the population's quality of life substantially, by increasing their income.

It is now confirmed that clearing of land in favour of biofuel crops and the consequent loss of forests, peat lands and grasslands would actually aggravate global warming and climate change.[101] The possibility of converting land to grow *Jatropha* has captured the imagination of researchers, NGOs and policy makers and has helped in bringing over 11 million ha under *Jatropha* cultivation. But, as the amount of land in the world available for agriculture is limited, it is necessary to define the fraction of farmland that could be used for the production of biofuels.

11.11 Crop Improvements

Jatropha curcas is an introduced plant for many countries and now there are some systematic efforts for improvement of this crop. Improved varieties with desirable traits for specific growing conditions at present are not available, which makes growing *Jatropha* a risky business.[25] The objectives for genetic upgrading of the crop should aim at more female flowers or pistillate plants, high seed yield with high oil content (%) having improved fuel property, early maturation, resistance to pests and diseases, drought endurance, reduced plant height and high natural ramification of branches. The fact that *Jatropha* has adapted itself to a wide range of edaphic and ecological conditions, suggests that there exists a considerable amount of genetic variability to be exploited for potential realisation. Conventional plant-breeding techniques can be used to improve *Jatropha* yield, but before initiating breeding programmes it is useful to have some understanding of the natural genetic variation that is present within a species. Many studies have come to the conclusion that the genetic variability within *Jatropha* in Asia is very low.[56,102,103] Genetic variation in seed morphology and oil content of *Jatropha* is of great potential in improvement programs, unfortunately not much work has been done on germplasm conservation. However, Ginwal *et al.*, reported seed-source variability in central India,[104] and Kaushik *et al.*, studied the variation in seed trait and oil content in 24 accessions collected from Haryana state, India.[105]

Plant transformation is another valuable method for the development of improved plant varieties. The transformation of *J. curcas* cotyledon discs using *Agrobacterium tumefaciens* has recently been reported.[106] This recent development means that the full range of plant-breeding techniques are now available for the development of *Jatropha* as a robust commercially viable crop. *J. curcas* is a diploid species with a 2n chromosome number of 22.[107] A recent study has estimated the genome size (1C) to be 416 Mbp.[108] This is relatively small for a plant genome,[109] and could make *J. curcas* an attractive candidate for genome sequencing. There are a number of traits that could be targeted for improvement in *Jatropha* including seed yield, oil content, and seed toxicity (phorbolester content). The seed yield per plant could be enhanced by employing biotechnological tools like marker-assisted selection of quality planting material.

Improvements in seed yield could be achieved in a number of ways. The first advancement came with the introduction of RFLP markers. This helped in assessing the molecular diversity of *Jatropha* germplasm and can be used in breeding programs.[110,111] Recently, a new full length cDNA of stearoyl-acyl carrier protein desaturase was obtained by RT-PCR and RACE techniques from developing seeds of *Jatropha* and the gene was functionally expressed in *E. coli*.[112] It is an important enzyme for fatty acid biosynthesis in higher plants and also plays an important role in determining the ratio of saturated fatty acid to unsaturated fatty acids in plants.[113] *Jatropha* is monoecious, and has a male:female flower ratio of around 29:1. Increasing the ratio of female flowers may lead to increases in the seed yield. Increasing the number of branches on *Jatropha* may lead to an increased number of inflorescences, and, ultimately, the number of seeds produced per plant. Increasing the oil content in seeds can be achieved by altering the expression levels of enzymes in the triacylglycerol biosynthetic (Kennedy) pathway. Overexpression of diacylglycerol acyltransferases has been shown to increase oil content in *Arabidopsis*[114] and soybean.[115] The regulation of seed development and triacylglycerol biosynthesis in seeds has been studied in some depth.[116]

At present, a large number of *Jatropha* and *Pongamia* accessions are being collected by various research organisations in India under biodiesel network programs funded by the Department of Biotechnology and National Oilseeds and Vegetable Oils Development Board. The collections are being characterised for their oil content and fatty acid composition by ICRISAT, The Energy Research Institute (TERI), and other institutions in India. Seed oil content ranges from 28% to 40% in *Jatropha* and *Pongamia* accessions that are being maintained and characterised at ICRISAT.[117] As a result of out-crossing, large variability in seed yield and oil content between individual plants is observed. The appropriate kind of planting material (vegetative propagation/tissue culture seedlings) needs, therefore, to be standardised, to ensure that the true breeding nature of the best clone can be identified or developed through concerted research efforts. The advances in biotechnology provide opportunities to significantly reduce the cost of biofuel production by genetic manipulation of feed stocks in a way that improves biofuel yields. Biotechnological tools hold promise for altering fatty acid composition, one of the key traits for improving productivity of biodiesel from *Jatropha* and *Pongamia* seed oils. Addressing more complex traits such as reducing toxins/antinutrients in *Jatropha* oilseed-cake to increase its value as an animal feed also requires the use of plant breeding or biotechnology.

The success stories on the use of molecular-marker-assisted selection to improve the equally complex characteristic of oil concentration in maize kernels or fatty acid composition of soybean oils provide optimism for the potential of biotechnological tools to improve crop traits important for biofuel production in *Jatropha*. Molecular diversity analyses were carried out with random amplified polymorphic DNA (RAPD), simple sequence repeat (SSR), inter simple sequence repeat (ISSR) and amplified fragment length polymorphism (AFLP) markers.[56] In more recent studies, several genes such as *JcERF* (showing enhanced resistance to salt and frost), curcin gene and stearoyl-acyl carrier protein desaturase genes have been isolated and characterised.[112,118] Li *et al.* described a complete system of *Agrobacterium*-mediated transformation using cotyledon discs.[106] The combined study using proteomic approaches and chlorophyll fluorescence measurements indicated that the early stage acclimation of photosystem II (PSII) and the late-stage H_2O_2 scavenging might be involved in the cold response mechanisms of *J. curcas* seedlings.[119]

11.12 Energy, Environment and *Jatropha*

The oil from *Jatropha* is regarded as a potential fuel substitute. Diesel is a hydrocarbon with 8–10 carbon atoms per molecule, while *Jatropha* oil has 16–18[25] and thus has a lower ignition quality (cetane number). The type of fuels that can be obtained directly from the *Jatropha* are wood, the whole fruit and parts of the fruit namely seed coat (nut shell) and kernel. Different useful products can be made from the plant, such as oil, seed-cake and charcoal (from wood or nutshell). Processing increases the energy value of the product[75,120] (Table 11.9).

The energy content calculated from the chemical composition is reported in Table 11.10. The total life-cycle energy balance of the biodiesel production from *Jatropha* is reported to be positive.[61,95,122,123]

Achten *et al.*, on the basis of two case studies on life-cycle assessment (LCA) of biodiesel production system using intensive cultivation and applying

Table 11.9 Energy value of various fuel from *J. curcas*.120,121

Fuel	Ash content (%)[a]	Moisture content (%)[b]	Energy value (MJ/kg)[c]	Composition of the fruit (%)				Recovery percentage[d]
				Coat	Shell	Kernel		
Wood[e]	1	15	15.5					95–100
Whole fruit[f]	6	8	21.2					95–100
Whole nut	4	5	25.5					67–70
Coat	13	15	11.1	100				28–30
Shell	5	10	17.2		100			23–24
Kernel	3	3	29.8	30	24	46		44–46
Wood	3	5	30.0[h]	0	34	66		15–25
Charcoal				100	0	0		
Shell charcoal	15	5	26.3[h]	0	100	0		15–25[k]
Plant oil[g]	<0.1	0	40.7[i]	0	0	100		11–18 [23–38][l]
Seed-cake[g]	4	3	25.1[j]					29–35 [62–77][l]

[a] Ash content given as a percentage of dry weight (0% moisture content). All the ash can be used as a fertiliser.
[b] Moisture content given as a percentage of the wet weight (moisture content wet basis – mcwb).
[c] This is the low heat value. It is the energy that is practically available. For oxygenated (biomass) fuels, the difference between the high heat value and the low heat value is about 1.3 MJ/kg at 0% moisture.
[d] The recovery percentage is in relation to the air-dry wood raw material or the whole fruit.
[e] Energy value of green wood (50% mcwb), 8.2 MJ/kg.
[f] Energy value of fresh whole fruit (43% mcwb), 12.8 MJ/kg.
[g] The plant oil and the oil cake are from the kernel only not the whole nut.
[h] Fully carbonised charcoal.
[i] Energy value per litre, 37.4 MJ (specific gravity 0.92).
[j] Assume 70% seed-cake and 30% oil from kernel.
[k] Recovery percentage in relation to the shell input, not the whole fruit.
[l] The figures in brackets refer to the recovery percentage from the kernel.

Table 11.10 Gross energy content of *J. curcas* components calculated from their chemical composition.[124]

Sample	Total protein (%TS)	Total carbohydrates (%TS)	Lipid (%TS) literature data[131]	Calculated energy[a] (MJ kg^{-1})	Energy (MJ kg^{-1}) literature data[a]
Mature leaf	14.0	57.7	NR	12.0	N.R
Fruit hulls	5.3	72.4	NR	13.0	11.1
Seed entire	18.2	36.8	35.0	22.4	20.8–25.5
Deoiled cake	9.4	64.2	1.5	12.9	NR
Oil	ND	ND	NR	ND	37.8–45.8

NR – Not reported, ND – Not determined
[a]Calculation based on 16.7 MJ kg^{-1} for carbohydrates and protein and 37.7 MJ kg^{-1} for fat.

fertilisers and irrigation[61,95,125] and another study using low-input cultivation,[122] reported the life-cycle energy balance of the biodiesel from JCL to be positive. The LCA of the system using intensive cultivation and applying fertilisers and irrigation resulted in a less-positive energy balance compared to the study using low-input cultivation.

The transesterification has been found to be the biggest contributor to the energy requirement of the final biodiesel product. The use of pure *Jatropha* oil would significantly improve the energy balance. Although the use of pure oil is less energy efficient[61] and causes some engine problems,[126] it shows some opportunities for local use.

11.13 Energy Flow

From Figure 11.10, based on utilisation of residual biomass obtained from *Jatropha* for methane production (through anaerobic digestion), it can be seen that 4 t total solids (TS) ha^{-1} yr^{-1} of *Jatropha* would produce 54 GJ energy.[124] In addition to deoiled cake, pruned leaves and fruit hulls would produce 36 GJ with a total of 90 GJ ha^{-1} yr^{-1}.

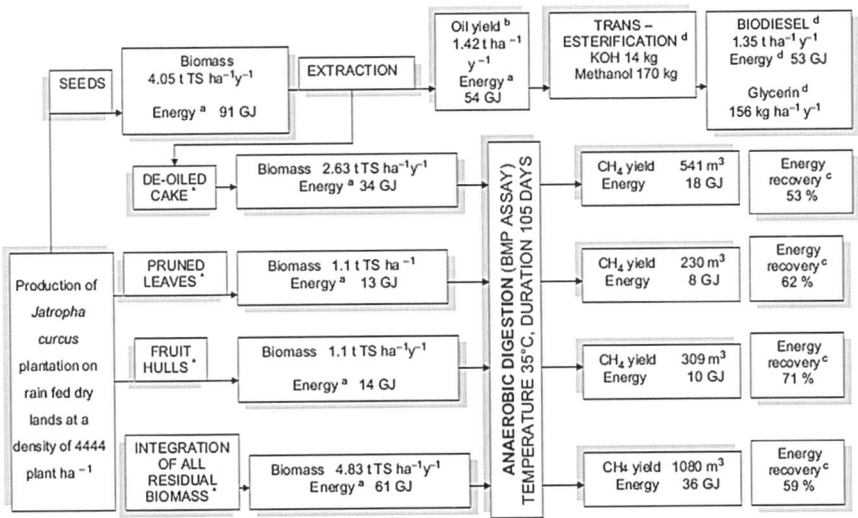

a Energy content in biomass calculated;

b Based on 35% oil in seed; 1m³ CH₄ 33.81 MJ;

c Energy recovery is defined as the energy value of CH₄ divided by the energy value of biomass added to digester;

d Heating value of biodiesel 38.93MJ kgL1; d Methanol @ 120 g kgL1, KOH @ 10g kgL1, Glycerin @ 110g kgL1 and biodiesel @ 950g kgL1 oil.

Figure 11.10 Schematics of energy flow during biodiesel and methane production from *J. curcas*.[124]

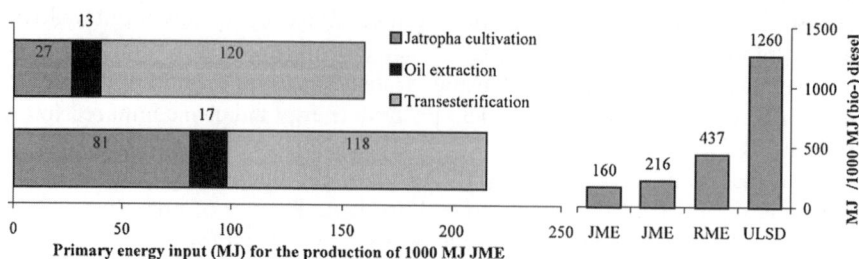

Figure 11.11 Primary energy input for the production of 1000 MJ JME after pro-rata allocation of the total energy requirement of the whole production process over the JME product and the byproducts based on the energy content of the JME product and the byproducts.[122,125] Comparison with reference systems rapeseed methyl ester (RME) and ultralow-sulfur diesel (ULSD) from crude oil.[128]

The energy recovery in methane (ratio of energy content in CH_4 to the energy conserved in the biomass from solar source) varied between 53 and 71%. The TOC content of the residual biomass would be about $1.2\,t\,ha^{-1}$. Anaerobic digestion of the biomass would yield $1080\,m^3$ CH_4 with the amount of carbon being 0.58 t. The carbon recovery in methane (ratio of mass of carbon in CH_4 to the mass of total organic carbon conserved in the biomass from solar source) accounts for 42–59%.[124]

In Figure 11.11, the energy input is distributed among the JME (*Jatropha* methyl esters, endproduct), glycerin (byproduct of transesterification) and seedcake (byproduct of oil extraction). The other byproducts (wood, fruit husks) are not included in this allocation because the use of these byproducts in an energy-efficient way is not common practice. When none of the byproducts are used, the energy balance will be only slightly positive (886 MJ input for 1000 MJ JME output)[125] or even negative.

On the other hand, if all byproducts (including wood and fruit husks) were used efficiently this total input of 886 MJ results in a total output of 17235 MJ, resulting in an allocated energy input of 160 MJ per 1000 MJ JME.[125] Figure 11.11, shows the energy balance of high and low JCL cultivation input. The difference between the two applied cultivation intensities is clear. In the low-intensity system the JCL cultivation step stands for 17% of the total primary energy input, while the cultivation step accounts for 38% of the total energy input in the high-intensity system. Irrigation and fertiliser are the most energy-intensive cultivation practices. Irrigation stands for 46% of the total energy input in the JCL cultivation, while fertiliser consumes 45%.[125] In both available studies the transesterification step is shown to be a big energy consumer. The oil-extraction step accounts for a similar share (78%) of the total life-cycle primary energy requirement in both studies (Figure 11.11). The transesterification process is the biggest contributor in the allocated energy requirement for the biodiesel endproduct. This would mean that the use of the crude oil as an endproduct would improve the energy balance significantly.

However, the engine combustion of pure JCL oil is less energy efficient and still causes some engine problems.[126,127] Based on the available results it can be expected that the life-cycle energy balance of biodiesel is generally positive. How positive the balance is in reality, will mainly depend on how efficiently the byproducts of the system are used.

11.14 *Jatropha curcas* and Clean Development Mechanism

Biofuels can have a better GHG balance than fossil fuels as the combustion of biofuel releases only the amount of CO_2 that the plant removed from the atmosphere when it was growing. However, the net balance of CO_2 savings depends on the amount of energy used for cultivating, harvesting, transporting and converting the plants. Thus, the choice of the crop and the technology pathways affects the CO_2 balance. For example, local production of *Jatropha* pressed into straight vegetable oil allows for maximum CO_2 savings as this pathways requires no additional chemical entrants or water and reduces CO_2 emissions linked to transport due to the decentralised nature of operations, and has very low energy and CO_2 reductions associated with its use. Through its suitability for wasteland recultivation, *J. curcas* provides two mechanisms for greenhouse-gas (GHG) abatement: substitution of fossil fuel and CO_2 sequestration through increasing carbon stocks above and below ground.

GHG Balance: Life-cycle assessment exercises showed positive results on the GHG requirement of the production of biodiesel from *Jatropha* in comparison to fossil diesel. The largest GHG contributors of the production process are irrigation (if applied: 26%), fertiliser (if applied: 30%) and transesterification (24%)[125] and 70%[122] depending on the applied cultivation intensity. Prueksakorn and Gheewala found that 90% of the total life-cycle GHG emissions are caused by the end use.[125] They calculated that the global-warming potential of the production and use of *Jatropha* biodiesel is 23% of the global-warming potential of fossil diesel. Detailed life-cycle assessments for the different pathways would help decision making. It should be noted that if virgin rainforest has been destroyed to make way for biofuel feedstock plantations, and the oil has been transported to distant markets, the biofuel produced starts out with a negative CO_2 balance.

11.15 Economics

The cost of production and profitability is shown in Table 11.11.[24] Seed production from plants propagated from seeds is expected within 3–4 years. However, branch cuttings can bear fruit within one year of planting.[129] Whilst *Jatropha* grows well in low-rainfall conditions requiring only about 200 mm of rain to survive, it can also respond to higher rainfall (up to 1200 mm) particularly in hot climatic conditions. The plant does not thrive in wetland conditions. In equatorial regions where moisture is not a limiting factor (*i.e.*

Table 11.11 Some economic benefits of biodiesel production from *Jatropha* seeds grown on wastelands in India.[24]

	Year 2010	2020	2050
Wasteland to be cultivated (million ha)[a]	0.4	2	10
Production of biodiesel (million tons/year)[b]	0.20	1.01	5.07
Foreign exchange saving by fuel substitution (million US$/year)[c]	67	334	1672
Employment generation (man-days)[d]	200 000	1 000 000	5 000 000
Saving of CO_2 consumption by the use of the produced biodiesel as automobile fuel (million tons/year)[e]	0.5	2.7	13.4
CO_2 sequestration in the biomass (million tons/year)[f]	0.9	4.6	22.9
Possible income from CO_2 reduction from emission trading (M US$)[g]	14.5	72.5	362.5

Notes:
[a] Assuming current production patterns don't change.
[b] Assuming production of 5831 per ha year.
[c] Assuming an average International price of US$45/barrel of crude oil.
[d] Assuming average employment of one person for two ha.
[e] Assuming that the end use of biodiesel reduces life-cycle CO_2 emissions by 85% compared to use of petrodiesel and a production of 2.7 kg of CO_2 per litre of diesel and a density of 0.87 for diesel.
[f] Except seeds taking an average of 2.5 metric ton of biomass increment per ha per year containing 25% C, thus sequestering 3.66 metric tons of CO_2 per metric ton of C.
[g] Calculating an average market value of US$10 per ton of CO_2 in international carbon markets.

continuously wet tropics or under irrigation) *Jatropha* can bloom and fruit all the year. However, a drier climate has been found to improve the oil yields of the seeds. In Mali, where *Jatropha* is planted in hedges, the reported productivity is from 0.8–1.0 kg of seeds per metre of live fence.[130] *J. curcas* seed yield varies and depends on a number of factors, namely site characteristics (precipitation, soil texture and soil fertility), age of the plant, variability in the germplasm and management. Annual seed production can range from 0.2 kg to 2 kg per plant. But in semiarid areas and cultural wasteland it can range from 2–3 t ha^{-1} yr^{-1}.[4,20,24] When good soil and average annual rainfall of 900–1200 mm are claimed and optimal management practice is used, 5 t dry seed ha^{-1} yr^{-1} can be achieved.[20,24] Jongschaap *et al.*, reported a potential yield range of 1.5–7.8 t dry seed ha^{-1} yr^{-1}.[131]

11.16 Costs and Returns

An estimate of costs and returns from cultivation of *Jatropha* plantations/hedgerows scenarios, however, is crucial to analysing its role in rural development. Costs, as well as returns are involved at different stages of the growing, harvesting and the manufacture/use of different products. It includes both tangible and intangible components of each. For instance, an estimate of the following cost heads would be required for *Jatropha* economic analysis;

Costs

Cultivation and oil extraction of Jatropha
a. Planting costs
b. Establishment costs
c. Tending costs
d. Other costs

Wood
a. Pruning
b. Thinning
c. Felling
d. Firewood production
e. Charcoal production
f. Pole production
g. Other products
h. Storage costs of products
i. Transport costs
j. Other costs.

Fruit
a. Collection
b. Removal of flesh
c. Removal of shell
d. Extraction of oil
e. Charcoal production from shells
f. Storage of products, (oil, cake, shells, flesh, *etc.*)
g. Transport costs
h. Other costs

Capital costs
   Building
   Machinery and equipment

Returns

Similarly several types of returns from the growing and use of the products from *J. curcas* need to be carefully estimated. The obvious returns pertain to sale or market prices of the different products should be recorded and then compared to the cost of the growing plus management of the plants and the manufacture of the products to arrive at profitability of various products. Some of the costs and returns have been indicated in the literature, while others may have to be extrapolated from estimates of similar crops. However, the cost of production and profitability of *J. curcas* in India calculated by Saxena in 2008 is given in Table 11.12.[132]

11.17 Conclusions

The current rate of Indian development of biofuels, particularly biodiesel, is just a drop in the bucket when compared to its potential. If 10 million hectares (100 000 square kilometers or 38 000 square miles) of India's vast wastelands are used for biodiesel production, with a modest estimate of 1.5 tons of seeds per hectare, 4 million tons of biodiesel would be produced. If one person is employed per hectare, that would mean 10 million new jobs. And, for use or

Table 11.12 Cost of production and profitability (in INR/Ha).[132]

S.No.	Cost	Year I	Year II	Year III	Year IV	Year V	Total
i	Land preparation and pit digging (15 man days @ Rs.80/day = 1200)	2400	0	0	0	0	2400
ii	FYM	1000	0	0	0	0	1000
iii	Fertilisers (urea 50 kg + SSP 50 kg + MOP 40 kg @ Rs. 5 kg for 3 years = 1200)	1200	1200	1200	0	0	3600
iv	Wages (fertiliser application)	240	240	240	0	0	720
v	Cost of saplings	2500	0	0	0	0	2500
vi	Pruning	800	800	0	0	0	1600
vii	Weeding	320	320	0	0	0	640
iv	Harvesting and cleaning of seeds	160	800	1200	1200	1200	4560
	Total	**8620**	**3360**	**2640**	**1200**	**1200**	**17 020**

Returns during the period

S.No.	Productivity and returns	Year I	Year II	Year III	Year IV	Year V	Total (Rs.)
	Jatropha curcas seeds						
i	Quantity (kg)/ha	0	1250	2500	5000	7500	16 250
ii	Unit price Rs/kg	0	6	6	6	6	6
iii	Total sale amount	0	7500	15 000	30 000	45 000	97 500

Unit cost/ha for 5-year period	Rs. 17 020
Returns for 5-year period	Rs. 97 500
Gross profit/ha for 5-year period	Rs. 80 480

sale, 11 million tons of organic seedcake fertiliser or livestock feed and 0.4 million tons of technical grade glycerol would be produced.

Research focused on different stress-tolerant ecotypes, improvement of seed quality, plantation techniques, botany, agronomy, ecology, flowering and fruiting characteristics, and harvest and postharvest handling of seeds is required to help *Jatropha* growers to realize its full potential. Commercial utilisation of the seed-cake as animal feed in addition to the oil may contribute to increasing the economic viability of *Jatropha* production system. To help mitigate GHG-emissions *Jatropha* needs to be supplied from degraded and abandoned lands. *J. curcas* would be cost effective if the cost of global warming is taken into consideration and fossil fuels have a carbon tax on them. Currently, growers are unable to achieve the optimum economic benefits from the plant, especially for its various uses. The markets of different products from this plant particularly from cake have not been properly explored or quantified. Therefore, more research is needed on medicinal and biopesticidal aspects of active principles present in different parts of *Jatropha*.

References

1. D. Fairless, *Nature*, 2007, **499**, 652–655.
2. L. M. Cano-Asseleih, Ph.D. Thesis, University of London, UK, 1986.
3. L. M. Cano-Asseleih, R. A. Plumbly and P. J. Hylands, *J. Food Biochem.*, 1989, **13**, 1–20.
4. J. Heller, 1996. <http://www.ipgri.cgiar.org/Publications/pdf/161.pdf>.
5. Planning Commission, *Planning Commission, Govt. of India, New Delhi*, 2003.
6. Centad, **IV** Issue 3, 2008.
7. Biodiesel 2020: 2008. www.emerging-markets.com.
8. C. Linnaeus, *Species plantarum*, 1753, 1006–1007.
9. A. Ross Ivan, *Medicinal Plants of the World*, Humana Press Inc., NJ, Totowa, 2003.
10. B. Dehgan, G. L. Webster, *University of California Publications in Botany*, 1979, **74**.
11. A. Kumar and S. Sharma, *Ind. Crops Prod.*, 2008, **28**, 1–10.
12. R. P. Singh, *Beitr. Biol. Pflanz.*, 1970, **47**, 79–90.
13. R. C. Gupta, *Indian Acad. Sci. (Plant Sci.)*, 1985, **94**, 65–82.
14. R. K. Henning, http://www.Jatropha.de/ S, February12, 2006, last modified on January 30, 2006.
15. R. P. S. Katwal and P. L. Soni, *Indian Forester*, 2003, **129**, 939–949.
16. J. Kiefer, Diploma thesis University Hohenheim, Stuttgart, 1986.
17. N. Foidl and P. Eder, *Biofuels and Industrial Products from Jatropha curcas*, DBV Graz. 1997, 88–91.
18. L. Singh, S.S. Bargali, S.L. Swamy, *Proceedings of the biodiesel conference toward energy independence—focus of Jatropha, Hyderabad, India, New Delhi: Rashtrapati Bhawan*, June 9–10, 2006, 252–67.

19. K. Gour, *Proceedings of the biodiesel conference toward energy independence—focus of Jatropha, Hyderabad, India, New Delhi: Rashtrapati Bhawan*, June 9–10, 2006, 223–51.
20. D. N. Tewari, *Jatropha and Biodiesel*, New Delhi, Ocean Books Ltd., India, 2007.
21. S. Biswas, N. Kaushik, G. Srikanth, *Proceedings of the biodiesel conference toward energy independence—focus of Jatropha, Hyderabad, India, New Delhi: Rashtrapati Bhawan*, June 9–10, 2006, 03–30.
22. D. R. Chaudhary, J. S. Patolia, A. Ghosh, J. Chikara, G. N. Boricha, A. Zala, *Expert seminar on J. curcas L. agronomy and genetics*, 26–28 March 2007, Wageningen, The Netherland, Published by Fact Foundation.
23. J. O. Ogunwole, J. S. Patolia, D. R. Chaudhary, A. Ghosh and J. Chikara, *Expert seminar on J. curcas L. agronomy and genetics*, 26–28 March 2007, Wageningen, The Netherland, Published by Fact Foundation.
24. G. Francis, R. Edinger and K. Becker, *Natural Resources Forum*, 2005, **29**, 12–24.
25. K. Openshaw, *Biomass Bioenergy*, 2000, **19**, 1–15.
26. TERI, http://www.teriin.org/Jatropha, December 19, 2006, last modified Dec. 19, 2006.
27. A. Kumar, S. Sharma, *ICPQR*, Dec. 2005, IIT, New Delhi, India.
28. K. Daey Ouwens, G. Francis, Y. J. Franken, W. Rijssenbeek, A. Riedacker, N. Foidl, R. Jongschaap, P. Bindraban, URL/http://www.factfuels.org/media_en/Position_Paper_on_Jatropha_curcas, 2007.
29. *International Conference on Biofuel 2008*, Winrock International, www.winrockindia.org/.
30. K. Becker and H. P. S. Makkar, *Lipid Technol.*, 2008, **20**, 104–107.
31. E. T. Akintayo, *Bioresour. Technol.*, 2004, **92**, 307–310.
32. H. P. S. Makkar, K. Becker and A. O. Aderibigbe, *Food Chem.*, 1998, **62**, 207–215.
33. G. D. Sharma, S. N. Gupta and M. Khabiruddin, In: *Biofuels and Industrial Products from Jatropha curcas*, G. M. Gubitz, M. Mittelbach, M. Trabi (ed.), DBV Graz., 1997, 19–21.
34. M. Wink, C. Koschmieder, M. Sauerwein and F. Sporer, In: *Biofuels and Industrial Products from Jatropha curcas*, G. M. Gubitz, M. Mittelbach, M. Trabi (ed.), DBV Graz., 1997, 160–166.
35. H. P. S. Makkar, K. Becker, F. Sporer and M. Wink, *J. Agr. Food Chem.*, 1997, **45**, 3152–3157.
36. J. A. Rivera-Lorca and J. C. Ku-Vera, In: *Biofuels and Industrial Products from Jatropha curcas*, G. M. Gubitz, M. Mittelbach, M. Trabi, (ed.), DBV Graz. 1997, 47–52.
37. A. Hemalatha and M. Radhakrishnaiah, *J. Econ. Taxonom. Bot.*, 1993, **17**, 75–77.
38. C. R. Mitra, S. C. Bhatnagar and M. K. Sinha, *Ind. J. Chem.*, 1970, **8**, 1047.
39. S. M. Khafagy, Y. A. Mohamed, N. A. Abdel and Z. F. Mahmoud, *Plant. Med.*, 1977, **31**, 274–277.

40. C. D. Hufford and B. O. Oguntimein, *Lloydia.*, 1987, **41**, 161–165.
41. A. J. Van den Berg, S. F. Horsten, J. J. Kettenes van den Bosch, B.H. Kroes, C. J. Beukelman, B. R. Loeflang and R. P. Labadie, *FEBS Lett.*, 1995, **358**, 215–218.
42. L. K. Nath and S. K. Dutta, *J. Pharm. Pharmacol.*, 1991, **43**, 111–114.
43. F. Stirpe, Pession, A. Brizzi, E. Lorenzoni, P. Strochi, L. Montanaro and S. Sperti, *Biochem. J.*, 1976, **156**, 1–6.
44. W. Adolf, H. J. Opferkuch and E. Hecker, *Phytochem.*, 1984, **23**, 129–132.
45. R. Staubmann, I. Ncube, G. M. Gubitz, W. Steiner and J.S. Read, *J. Biotechnol.*, 1999, **75**, 117–126.
46. E. M. Aregheore, H. P. S. Makkar and K. Becker, In: *Biofuels and Industrial Products from Jatropha curcas*, G. M. Gubitz, M. Mittelbach, M. Trabi, (ed.), DBV Graz., 1997, 65–69.
47. H. P. S. Makkar and K. Becker, In: *Biofuels and Industrial Products from Jatropha curcas*, G. M. Giibitz, M. Mittelbach, M. Trabi (ed.), DBV Graz., 1997, 190–205.
48. W. Naengchomnong, Y. Thebtaranonth, P. Wiriyachitra, K. T. Okamoto and J. Clardy, *Tetrahed. Lett.*, 1986, **27**, 2439–2442.
49. Hirota, *et al.*, *Cancer Res.*, 1988, **48**, 5800–5804.
50. http://bioveg.auf.org/IMG/pdf/Campa.pdf.
51. Y. Ito, S. Yanase, H. Tokuda, M. Krishishita, H. Ohigashi, M. Hirata and K. Koshimizu, *Cancer Lett.*, 1983, **18**, 87–95.
52. K. Becker and H. P. S. Makkar, *Vet. Human Toxicol.*, 1998, **40**, 82–86.
53. W. Haas and H. Sterk M. Mittelbach, *J. Natural Prod.*, 2002, **65**, 1434–1440.
54. G. Zhang, M. G. Kananietz, P. M. Blumberg and J. H. Hurley, *Cell*, 1995, **81**, 917–924.
55. V. M. Gandhi, K. M. Cherian and M. J. Mulky, *Food Chem. Toxicol.*, 1995, **33**, 39–42.
56. D. V. N. S. Pamidimarri, S. Singh, S. G. Mastan, J. Patel and M. P. Reddy, *Mol. Biol. Rep.*, 2008, doi:10.1007/s11033-008-9320-6.
57. I. Graham, *Presentation at the International consultation on pro-poor Jatropha development*, in Rome 10–11 April 2008.
58. P. Siddhuraju, H. P. S. Makkar and K. Becker, *Food Chem.*, 2002, **78**, 187–205.
59. J. Martínez-Herrera, P. Siddhuraju, G. Francis, G. Dávila-Ortíz and K. Becker, *Food Chem.*, 2006, **96**, 80–89.
60. A. K. Raina and B. R. Gaikwad, *J. Oil Technol. India*, 1987, **19**, 81–85.
61. W. M. J. Achten, L. Verchot, Y. J. Franken, E. Mathijs, V. P. Singh and R. Aerts, *et al.*, *Biomass Bioenergy*, 2008, **32**, 1063–1084.
62. M. P. Dorado, E. Ballesteros, J. A. Almeida, C. Schellert, H. P. Lohrlein and R. Krause, *Trans. ASAE*, 2002, **45**, 525–529.
63. S. V. Ghadge and H. Raheman, *Biomass Bioenergy*, 2005, **28**, 601–605.
64. P. Chitra, P. Venkatachalam, A. Sampathrajan, *Energy for Sustainable Development* 2005, **9,** 13–18.

65. Alok K. Tiwari, A. Kumar and H. Raheman, *Biomass Bioenergy*, 2007, **31**, 569–575.
66. E. Münch, J. Kiefer, *Gesellschaft für Technische Zusammenarbeit (GTZ)*, TZ Verlagsgesellschaft mbH, Roßdorf, Germany. 1986.
67. A. J. King, W. He, J. A. Cuevas, M. Freudenberger, D. Ramiaramanana and I. A. Graham, *J. Exp. Bot.*, 2009, doi:10.1093/jxb/erp025.
68. G. Knothe, *Energy Fuels*, 2008, **22**, 1358–1364.
69. H. C. Harris, J. R. McWilliam and W. K. Mason, *Austral. J. Agricul. Res.*, 1978, **29**, 1203–1212.
70. H. J. Berchmans and S. Hirata, *Bioresour. Technol.*, 2008, **99**, 1716–1721.
71. S. Tamalampudi, M. R. Talukder, S. Hama, T. Numata, A. Kondo and H. Fukuda, *Biochem. Eng. J.*, 2008, **39**, 185–189.
72. K. M. Senthil, A. Ramesh and B. Nagalingam, *Biomass Bioenergy*, 2003, **25**, 309–318.
73. R. Sarin, M. Sharma, S. Sinharay and R.K. Malhotra, *Fuel*, 2007, **86**, 1365–1371.
74. R. Staubmann, G. Foidl, N. Foidl, G. M. Gubitz, R. M. Lafferty, V. M. Arbizu and W. Steiner, *Appl. Biochem. Biotech.*, 1997, **63**, 457–467.
75. G. M. Gubitz, M. Mittelbech and M. Trabi, *Bioresour. Technol.*, 1997, **67**, 73–82.
76. A. D. Karve, URL /http://www.bioenergylists.org/en/compactbiogas, 2005.
77. J. Visser and T. Adriaans, *Report produced for FACT*, Ingenia Consultants & Engineers, Eindhoven. 2007.
78. N. Mahanta, A. Gupta and S. K. Khare, *Bioresour. Technol.*, 2007, **99**, 1729–1735.
79. M. Sujatha, H. P. S. Makkar and K. Becker, *Plant Growth Regul.*, 2005, **47**, 83–90.
80. M. Benge, www.echotech.org. 2006.
81. N. Kaushik, S. Kumar, *Jatropha curcas L. Silviculture and Uses*. Agrobios (India) Jodhpur. 2004.
82. J. P. Tewari and I. K. Shukla, *GEOBIOS*, 1982, **9**, 124–126.
83. M. O. Nwosu and J. L. Okafor, *Mycoses*, 1995, **38**, 191–195.
84. A. D. Solsoloy and T. S. Solsoloy, In: *Biofuels and Industrial Products from Jatropha curcas*, G. M. Gubitz, M. Mittelbach, M. Trabi, (ed.), DBV Graz., 1997, 216–226.
85. N. Foidl, A. Kashyap, *Exploring the potential of Jatropha curcas in Rural Development and Environmental Protection*. Rockefeller Foundation, New York. 1999.
86. J. M. Dalziel, *The Useful Plants of West-Tropical Africa*. Crown Agents for Oversea Governments and Administration, London. 1955, 147.
87. J. A. Duke, *CRC Handbook of Medicinal Herbs*, CRC Press, Boca Raton, FL, 1988, 253–54.
88. L. K. Nath and S. K. Dutta, In: *Biofuels and Industrial Products from Jatropha curcas*, G. M. Gubitz, M. Mittelbach, M. Trabi, DBV Graz,. 1997, 2–86.

89. L. F. Villegas, I. D. Fernandez, H. Maldonado, R. Torres, A. Zavaleta, A. J. Vaisberg and G. B. Hammond, *J. Ethnopharmacol.*, 1997, **55**, 193–200.
90. O. Osoniyi and F. Onajobi, *J. Ethnopharmacol.*, 2003, **9**, 101–105.
91. A. M. Mujumdar and A. V. Misar, *J. Ethnopharmacol.*, 2004, **90**, 11–15.
92. R. N. Singh, D. K. Vyas, N. S. L. Srivastava and Madhuri Narra, *Renew. Energy*, 2008, **33**, 1868–1873.
93. T. S. Cox, J. D. Glover, D. L. van Tassel, C. M. Cox and L. R. DeHaan, *Bioscience*, 2006, **56**, 649–659.
94. J. D. Glover, C. M. Cox and J. P. Reganold, *Sci. Am.*, 2007, **297**, 82–89.
95. W. M. J. Achten, E. Mathhijs, L. Verchot, V. P. Singh, R. Aerts and B. Muys, *Biofuels, Bioprod. Biorefin.*, 2007, **1**, 283–291.
96. J. N. Daniel, *Future Energy*, 2005.
97. Wahenga, 2007. http://www.wahenga.net/uploads/documents/briefs/Brief_11.pdf. Acessed: 14/07/2007.
98. United Nation: "*World Population Prospects: The Revision*" http://esa.un.org/unpp/, 2006.
99. Food and Agriculture Organization of the United Nations, "FAOSTAT": http://faostat.fao.org/site/377/default.aspx. 2007.
100. NREL, http://205.168.79.26/research_review/pdfs/2005/38668a.pdf., July 2005.
101. F. Christopher, *Time to move to a second generation of biofuels*, World Watch Institute, Washington D.C., 13 February 2008.
102. S. D. Basha and M. Sujatha, *Euphytica*, 2007, **156**, 375–86.
103. Q. B. Sun, L. F. Li, Y. Li, G. J. Wu and X. J. Ge, *Crop Sci.*, 2008, **48**, 1865–1871.
104. H. S. Ginwal, S. S. Phartyal, P. S. Rawat and R. L. Srivastava, *Silvae Genetica*, 2005, **54**, 76–80.
105. N. Kaushik, K. Kumar, S. Kumar, N. Kaushik and S. Roy, *Biomass Bioenergy*, 2007, **31**, 497–502.
106. M. Li, H. Li, H. Jiang, X. Pan and G. Wu, *Plant Cell Tiss. Org. Cult.*, 2008, **92**, 173–81.
107. B. Dehgan, *Syst. Bot.*, 1984, **9**, 467–478.
108. C. R. Carvalho, W. R. Clarindo, M. M. Praça, F. S. Araújo and N. Carels, *Plant Sci.*, 2008, **174**, 613–7.
109. B. J. M. Zonneveld, I. J. Leitch and M. D. Bennett, *Ann. Bot.*, 2005, **96**, 229–244.
110. M. Mohan, S. Nair, A. Bhagwat, T. G. Krishna, M. Yano, C. R. Bhatia and T. Sasaki, *Molec. Breed.*, 1997, **3**, 87–103.
111. S. L. Kumar, *Biotechnol. Adv.*, 1999, **17**, 143–182.
112. L. Tong, Shu-Ming. Peng, Wu-Yuan. Deng, Ma Dan-Wei, Ying. Xu, Meng. Xiao and Fang. Chen, *Biotechnol. Lett.*, 2006, **28**, 657–662.
113. Y. Lindqvist, W. Huang, G. Schneider and J. Shanklin, *EMBO J.*, 1996, **15**, 4081–4092.

114. C. Jako, A. Kumar, Y. Wei, J. Zou, D. L. Barton, E. M. Giblin, P. S. Covello and D. C. Taylor, *Plant Physiol.*, 2001, **126**, 861–874.
115. K. Lardizabal, R. Effertz, C. Levering, J. Mai, M. C. Pedroso, T. Jury, E. Aasen, K. Gruys and K. Bennett, *Plant Physiol.*, 2008, **148**, 89–96.
116. M. Santos-Mendoza, B. Dubreucq, S. Baud, F. Parcy, M. Caboche and L. Lepiniec, *The Plant J.*, 2008, **54**, 608–620.
117. S. P. Wani, M. Osman, E. D'silva and T. K. Sreedevi, *Asian Biotechnol. Develop. Rev.*, 2006, **8**, 11–29.
118. M. Tang, J. Sun, Y. Liu, F. Chen and S. Shen, *Plant Mol. Biol.*, 2007, **63**, 419–28.
119. Y. Liang, H. Chen, M. J. Tang, P. F. Yang and S. H. Shen, *Physiol. Plant*, 2007, **131**, 508–517.
120. K. Openshaw, *Concepts and methods for collecting and compiling statistics on biomass used as energy*, UN Statistical Office, New York, 1986.
121. G. M. Gubitz, *et al.*, *Proceedings from a Symposium held in Managua, Nicaragua*, Technical University of Graz, Graz, Austria, 1997.
122. J. Tobin, D. J. Fulford, *MSc dissertation*, The University of Reading, 2005.
123. R. K. Henning, In: *Biofuels and industrial products from Jatropha curcas-Proceedings from the symposium "'Jatropha 97,'"* Managua, Nicaragua, G. M. Gübitz, M. Mittelbach, M. Trabi (ed.), Dbv-Verlag, Graz, Austria, 1997, 92–7.
124. V. N. Gunaseelan, *Biomass Bioenergy*, 2009, **33**, 589–596.
125. K. Prueksakorn, S. H. Gheewala, In: *Proceedings of the second joint international conference on "'Sustainable energy and environments (SEE 2006),Bangkok, Thailand, November 21–23*.
126. L. C. Meher, D. Vidya Sagar and S. N. Naik, *Renew. Sustain. Energy Rev.*, 2006, **10**, 248–268.
127. C. M. V. Prasad, M. V. S. M. Krishna, C. P. Reddy and K. R. Mohan, *Proc. Inst. Mech. Eng. Part D – J. Auto. Eng.*, 2000, **14**, 181–187.
128. M. A. Elsayed, R. Matthews, N. D. Mortimer, http://www.dti.gov.uk/files/ file14925.pdf?pubpdfdload 1/4 03%2F836S. 2003.
129. N. Jones, J. H. Miller, *Jatropha curcas: A Multipurpose Species for Problematic Sites*, The World Bank, Washington DC USA. 1992.
130. R. Henning *Rothkreuz* 11, D-88138 Weissensberg, Germany, 1996.
131. R. E. E. Jongschaap, W. J. Corre, P. S. Bindraban, W.A. Brandenburg, *Wageningen: Plant Research International B.V.* 2007, 1–42.
132. A. P. Saxena, In: *Proceedings: 5th International Conference on Biofuels*, Feb. 7–8, 2008, New Delhi, India.

CHAPTER 12
Pongamia pinnata, a Sustainable Feedstock for Biodiesel Production

STEPHEN H. KAZAKOFF, PETER M. GRESSHOFF AND PAUL T. SCOTT

Australian Research Council Centre of Excellence for Integrative Legume Research (CILR), The University of Queensland, St Lucia, Brisbane, Australia 4072

12.1 Introduction

The current interest in biofuels has emerged in recent years following the recognition of a number of prominent environmental, economic and societal developments. First and foremost is the realisation that the demand for fossil fuels is increasing at such a rate that current mining, exploration and production technologies will soon not be able to meet demand. This strong demand for liquid fuels is from both the already established first-world economies and also the emerging new economies of countries such as India and China. The current demand for oil from fossil fuels is around 85 millions barrels per day (~13.5 billion litres). The forecast demand for oil in 2030 is expected to be around 106 millions barrels per day (~16.9 billion litres).[1] While the extent and status of "peak oil" is arguable, it is undeniable that at some time in the foreseeable future readily accessible sources of fossil fuels will be exhausted.

The potential exists for biofuels to meet at least some of any future unmet demand. The establishment of a sustainable biofuels industry will also provide to any country energy independence and security from the importing of oil, particularly from those countries that may be politically unstable. Complementing any future energy independence is the opportunity for rural development through the creation of employment and wealth in establishing the projected new industries. Finally, the broad consensus amongst much of the scientific community is that the industrialisation of human activities and the consequent consumption of fossil fuels have contributed to significant increases in the atmospheric concentration of greenhouse gases (*e.g.*, CO_2, CH_4 and NO_x) and observed global climate change.[2] A simple model put forward by proponents of biofuels is that any emissions of CO_2 through combustion of fossil fuels will be fixed by biofuel crops throughout their life cycle and hence lead to a carbon neutral production and utilisation scenario.[3-5]

While it may be possible to argue in general that biofuel crops provide a carbon-neutral alternative to fossil fuels, in the context of a more complete life-cycle analysis other inputs and outputs need to be considered.[6,7] In particular, the nitrogen cycle and its importance in assessing the suitability of biofuel crops has been neglected to date. Nitrogen is an essential element for all plants and in many agricultural soils is a limiting factor to sustained growth. To supplement the macro- and microelements provided by soil many agricultural crops are provided with nitrogen fertilisers, which are often in the form of nitrate and ammonium salts. Importantly, nitrogen fertilisers are produced from fossil fuels *via* the Haber–Bosch process and are an added cost to both the production and net energy balance of a biofuel crop. For example, life-cycle analysis indicates that about a third of the inputs for oil production from canola is accounted for by fertiliser, particularly nitrogen.[8] In addition, many bacteria residing in soil are capable of converting nitrogen fertilisers to volatile nitrous oxides, greenhouse gases approximately three hundred times more potent than CO_2.[9] With the exception of soybean, all first-generation biofuel crops are nonlegumes. Likewise, the vast majority of the proposed second-generation biofuel crops are also nonlegumes. Thus, they require nitrogen to produce the amino acids needed to build the photosynthesis machinery required to capture CO_2.[10] The importance of legumes in the sustainability of candidate biofuel crops is *via* the energy savings from the symbiosis with soil bacteria, collectively called rhizobia, which through their nitrogen-fixing capabilities negate the need for nitrogen-fertiliser applications to crops.

12.2 Legume Nodulation and Symbiotic Nitrogen Fixation

Most, but not all, legumes enter a mutually beneficial symbiosis with rhizobia. Distinct host specificity exists in many cases so that a soybean symbiont (*e.g.*, *Bradyrhizobium japonicum*) will not nodulate clover; in turn clover nodulates with *Rhizobium leguminosarum biovar. trifolii*, which does not interact with soybean. Occasionally, this specificity is absent; for example, strain NGR234 nodulates many legume species, and legumes such as siratro (*Macroptilium atropurpureum*) have a broad host range. Preliminary experiments in our laboratory indicate that *Pongamia pinnata* is able to form functional spherical nodules with a broad range of rhizobia (data not yet published).

It is believed that the specificity of host range is controlled by numerous molecular signals. For example, rhizobia are decorated with exo- and lipopolysaccharides. Specific legumes secrete into their root environment "cocktails" of flavones and isoflavones, needed for rhizobia chemotaxis (attraction towards the plant) and induction of so-called nodulation genes in the bacterium. Complete genome sequencing, insertional mutagenesis and careful inoculation studies have revealed nearly 50 genes from rhizobia needed for nodule induction. Another 50 genes may be needed for subsequent establishment and functioning of the nitrogen-fixing symbiosis.

Rhizobia, in response to flavonoid stimulation, synthesise a lipo-oligosaccharide molecule, called the Nod Factor. This has different decorations depending on the genetic capability of the bacterium. Nod Factors are perceived by the plant in the root hairs, epidermis and cortical regions by a LysM-type receptor kinase.[11-13] This receptor complex (two proteins seem to interact) communicates the infection signal so that root hairs curl, thereby entrapping the rhizobia colony. Subsequent digestion and resynthesis cycles, coupled with bacterial proliferation lead to the formation of an infection thread, which grows through the inside of the root hair, then traverses the cortex, where it further stimulates cell divisions. These cell divisions, together with concomitant pericycle divisions build the body of the future nodule. Bacteria from infection threads escape to enter, *via* endocytosis, the plant cytoplasm. In this form, the rhizobia differentiate to become nitrogen-fixing bacteroids.[14] Contained in the plant-derived symbiosome the bacteroids import photosynthate (as malate[15]), iron (for the nitrogenase enzyme complex), nitrogen gas to release ammonia (in water as ammonium), hydrogen gas and CO_2. The ammonia is exported to the plant cytoplasm where it is assimilated by a nodule-specific glutamine synthetase, making "fixed" nitrogen available for amino acid, nucleotide and other metabolite synthesis.

Nodule number is controlled by a number of environmental stresses. It now appears possible that many of these stresses target the Nod Factor perception cascade and possibly genetic modification may lead to stress tolerance and improvement of agronomic performance. Moreover, legumes, including *Pongamia pinnata*, regulate nodule organ numbers through an internal autoregulation system (called AON; autoregulation of nodulation).[14,16] Here developmental stages close to early cell division activation send a signal to the leaf vascular tissue.[17] It is now believed that this signal is a CLE-type peptide.[14,18-20] It is perceived by an LRR receptor kinase (GmNARK for soybean;[21] LjHAR1 for *Lotus japonicus*;[22,23] MtSUNN for *Medicago*[24]), which regulates the production of a nodulation inhibitor that migrates to the root, blocking further nodulation.[25] Mutations in the LRR receptor kinase lead to increased nodulation (super- or hypernodulation), altered nitrate sensing, and effects on lateral root growth.

The processes of nodulation and nitrogen fixation are progressively being understood at the genetic and biochemical levels; this leads to increased opportunities to optimise their performance in agricultural situations. The close developmental connection of nodule initiation and lateral root induction (same source tissue in the pericycle off the xylem poles) and their physiological controls (ethylene, nitrate, auxin, cytokinin) further adds to the value of analysing this critical plant process. It has been calculated that one third of all the nitrogen atoms in all humans on this planet have passed through legume-nodule associated nitrogenase!

Emphasis of nodulation research and nitrogen metabolism is paramount for bioenergy crops. Energy-negative fertiliser inputs can be negated by optimised nitrogen fixation through use of legume bioenergy species. For the legume tree *Pongamia pinnata* such studies and capability developments are only at an embryonic stage. However, genomic advances and technology transfer have now illustrated that common legume processes are functional (Figure 12.1).

In determining the suitability of a candidate feedstock species several environmental and sustainability issues need to be addressed. Perhaps most important of all are the production inputs and outputs that through life-cycle analysis will determine the net energy balance.[6-8,26,27] There has been much debate in recent times on the relative net energy balance of first-generation biofuel feedstocks and whether in fact there is a net positive energy balance and advantage over fossil-derived fuels. Life-cycle analyses should be interpreted with some caution as any data from such studies reflect outcomes under the defined production systems and geoclimatic conditions of the study site, which can make translation of such outcomes to alternative locations and agricultural systems difficult. The allocation of land to new biomass and bioenergy crops therefore requires analysis that integrates the biophysical and environmental characteristics (*e.g.*, soil, water, climate and landscape) with the societal and socioeconomic (*e.g.*, previous land usage and cultural heritage). For example, a candidate feedstock will be regarded highly as a crop if it is able to maintain above- and below-ground carbon, maintain and improve soil fertility and support plant growth through sustainable water-use efficiency.

12.3 *Pongamia pinnata*, an Emerging Biodiesel Feedstock

As the importance of biofuel production starts to impact the energy economies of the world it is clear that this new industry will rely on the reliable supply of feedstock from sustainable fuel crops.[28,29] From the long list of candidate species that have been proposed, a short list of strong candidates is beginning to emerge. In the context of the major tropical and subtropical agricultural production zones of the world *Pongamia pinnata* (a.k.a. *Millettia pinnata*) has attributes that place it in a good position for adoption as a future feedstock for biodiesel production. In the last five to ten years another oilseed plant, *Jatropha curcas*, (Chapter 11) has been considered by many in the biofuels industry to be

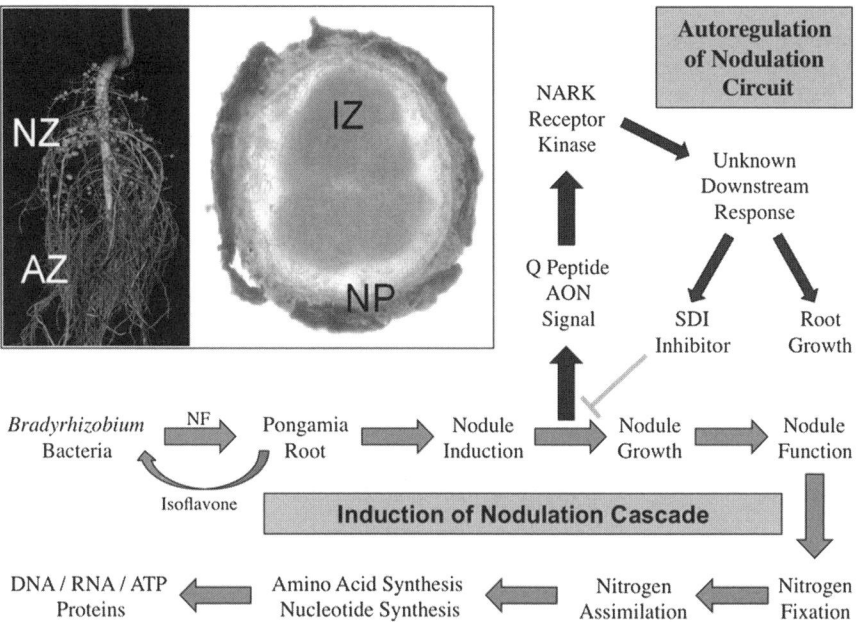

Figure 12.1 Nodulation and symbiotic nitrogen fixation in *Pongamia pinnata*: Two interacting cascades control the initiation and further development of a nodulated root system. Characteristically, *Pongamia* roots show strong "autoregulation of nodulation" (AON) facilitated by a signaling loop involving root–shoot interaction. As a result, the *Pongamia* root has a "nodulated" zone (NZ) and an "autoregulated" zone (AZ). *Pongamia* nodules are determinate and thus almost spherical. The interior is characterised by the "infected" zone (IZ), harbouring plant cells specialised to assimilate the ammonia produced by thousands of encased *Bradyrhizobium* bacteroids. Transport functions in and out of the nodule are contained in the nodule parenchyma and vascular vessel zone (NP). *Bradyrhizobium japonicum*, a common symbiont with soybean, is one of the bacteria that induces *Pongamia* nodules. In response to isoflavone exudate stimulation, Nod Factor (NF) is perceived by the *Pongamia* root epidermis and root hairs, leading to functional nodules that provide ammonia for protein, nucleic acid, ATP, vitamin, flavone and other metabolite synthesis. Some genetic components of this process for *Pongamia* have now been cloned and partially characterised. For example, P. Scott (unpublished data) has sequenced part of the genes encoding the PpNFR5 Nod Factor receptor protein kinase and part of the PpNARK AON receptor kinase.

the most promising feedstock for biodiesel production, particularly in the less-developed countries of the tropical and subtropical regions of the world.[30–33] This medium-sized perennial shrub is a member of the *Euphorbiaceae* and is perhaps equally noteworthy for the production of toxic compounds in both the fruit and the milky sap (*e.g.*, phorbolesters).[34] Its common name is vomit or purge oil. The initial excitement for *Jatropha*, based on the promise of high oil yields, even on what was considered low agriculturally productive land, is beginning to lessen amongst industrialists and investors.[35] The growth of *Jatropha* and its associated yields of oil, particularly on poor-quality soils, have not been at the levels initially promised. In addition, the seeds of *Jatropha* need to be harvested by hand sporadically throughout the year. No means of mechanical harvesting have yet been implemented on a commercial basis. As a result, *Jatropha* still retains some interest in countries with low labour costs. Furthermore, Jatropha as a nonlegume, will "mine" residual nitrogen from underused soils, then "drift" into lower productivity caused by nitrogen deficiency. For *Pongamia* to become a successful biodiesel feedstock there are clear lessons to be learnt from the experience with *Jatropha*.

To summarise the botanical characteristics of *Pongamia*, it is a medium-size tree capable of reaching 10 to 20 metres in height (Figure 12.2a). The flowers

Figure 12.2 Characteristics of *Pongamia pinnata*: (a) *Pongamia* as a mature tree growing on the streets of Brisbane, Australia; (b) During mid-November, the *Pongamia* trees of Brisbane begin to flower. Approximately 25–35% of these flowers set seed. Seed maturation takes about 10 months; (c) Characteristic legume flower morphology; (d) *Pongamia* seed cluster. Each seed weighs about 2 g and the dried pod wall also weighs about 2 g; (e) After harvesting, the oil from *Pongamia* seeds can be extracted by cold pressing and used to produce biodiesel; (f) Light micrograph section of a mature *Pongamia* seed. Letters indicate key cell components: O – oil body, P – protein body and S – Starch grain. This photograph is presented courtesy of Dr. Xin-Ding Wang and Prof. Ray Rose (CILR, The University of Newcastle, Australia).

are borne on a fluorescence, which resembles the ornamental plant Wisteria, in a major flush during late spring and early summer (Figure 12.2b). In less-common circumstances a minor flush may occur in late autumn. Each flower, of white to light pink colouration, has an arrangement typical of other legumes (*i.e.* a standard petal, two identical wings and two identical keel petals; Figure 12.2c). Each seed pod (Figure 12.2d) begins development with two embryos. However, in the vast majority of cases only one of the embryos reaches full maturity, with the other embryo aborted. The period for full development of the seeds can be as long as ten months, with seed pods often still attached to the tree by flowering of the following year.[36]

Whilst anecdotal evidence over many years suggested that *Pongamia* produces good-quality oil from its large oil-rich seeds (*e.g.*, use of oil as heating fuel in villages of India), it is only in the last five to ten years that there has been recognition in the scientific literature that the potential exists to exploit this oil for industrial purposes.[36–40] In describing the attributes that make *Pongamia* such a promising energy crop it is important to make two statements regarding the current biological status of this plant. First, *Pongamia* is yet to undergo any domestication that has accompanied the development of modern annual and perennial crops. This presents obvious challenges if *Pongamia* is to become an important oilseed crop in the near future. In order to initiate a domestication program, desirable traits need to be established in concert with functional genomics studies to identify and characterise the relevant traits (Table 12.1). The suggested domestication traits should include not only those associated with oil yield and composition, but also those associated with important agronomic traits that will support the development and broad-acre planting of well defined cultivars. Secondly, evolution has fortuitously provided us with a diverse range of germplasm with attributes that have already contributed in a positive way to traits, at least in part, that will lead to the development of an ideal energy crop. In other words, any plant improvement will begin from material that already exhibits traits highly desirable for sustainable oil production.

Table 12.1 Traits for *Pongamia* Domestication.

Repeated annual cropping
Crop uniformity
Seed mass per tree (yield)
Seed oil content (extractable)
Oil composition and stability
Growth vigour at seedling and adult stage
Erect growth and architecture
Seed abscission
Resistance to insect, nematode, fungi and bacteria infection
Flowering time
Nitrogen fixation efficiency
Water-use efficiency
Hardiness to cold, acid soils, drought and salinity

Pongamia pinnata is an arboreal legume that has a wide distribution in tropical and subtropical regions of the world.[36] The region for which *Pongamia* has the most well-documented history is the Indian subcontinent. However, its distribution also includes south-east Asia, southern China, Indonesia, southern Japan and northern Australia. Given the size and nature of its seeds and seed pods it would not be at all surprising to find wild stands of *Pongamia* in coastal regions of tropical and subtropical Africa and the Americas as well as other regions of Asia and the Pacific. To this extent and to reduce the threat of *Pongamia* becoming a weed, any future plantations should be located away from waterways that could act as a means of dispersal of seed pods.

The tree has low weed potential based on a number of properties. First, seeds germinate only under warm and wet conditions. Secondly, the frequency of germination is reduced within one year, preventing build-up of a soil-based seed bank. Thirdly, though the tree forms suckers from roots at times, this appears to be noninvasive and moderate. Finally, the seed appears to have low attraction as a forage for resident native animals, leading to restricted transmission; huge "carpets" of ungerminated seeds still retained in pods are often found under *Pongamia* trees.

Public debates surrounding biofuels have in recent times suggested that first-generation crops have adversely affected both the supply and price of some staple food crops (*e.g.*, maize in Mexico). This argument is losing traction, evidenced through the fall in commodity prices following the current global financial crisis and more importantly by the relatively small proportion of crop harvests that are currently destined for the biofuels market. Nonetheless, if biofuel feedstocks are to make a significant contribution to future liquid-fuel demands then they should be sourced from nonfood crops grown on land unlikely to be used for the cultivation of food crops. Land currently not under cultivation, and therefore candidate lands for biofuel feedstocks, may be so because it is often regarded as low agriculturally productive land or so-called "marginal land". This may be due to poor-quality soils (*e.g.*, salinity, acidity, nutrient deficiencies) and/or low and unreliable supplies of water.

Pongamia is reportedly both saline and drought tolerant, which places it in a good position when compared to other candidate feedstock species.[41,42] Research being undertaken in our laboratory supports the claims of the drought and saline tolerance of *Pongamia*. In pot trial experiments saplings grew well at saline levels up to 20 dS/m and nodulation of *Pongamia* was not adversely affected until irrigated with water of 10 dS/m, levels well beyond the saline levels for good-quality irrigation water in Australia. Preliminary experiments indicate that seedlings are capable of withstanding extensive periods of water deprivation (25 days to 55% relative water content) without significantly affecting growth and biomass production. In the context of extensive saline soils and drought periods, the vast Australian landscape should not provide an impediment to cultivation of *Pongamia*. It remains unclear as yet how such stresses influence tree seed and biomass productivity.

The value of a candidate feedstock species in the future will not only be measured by the yield of biofuel but also by any additional byproducts. In the case of *Pongamia* these additional byproducts might include the environmentally valuable improvement of soils through nitrogen fixation or the many potential economically valuable products. For example, once oil is pressed from the seeds the remaining seed-cake has been shown to have value as an animal-feed supplement.[43–45] At this point in time, due to problems associated with unpalatable and toxic components from contaminating seed oil, the seed-cake may only be provided as approximately 50% of any feed supplement. This problem is likely to be solved by selection of lines of *Pongamia* that have reduced levels of these undesirable compounds or repression of the gene(s) encoding these compounds through targeted genetic manipulation. Prior to seed-oil extraction the seed pods have potential value as a combustible fuel for the cogeneration of energy that may be used during biodiesel production. Unpublished data indicates that the seed pods have a calorific value of low- to medium-grade coal. Finally, in the production of biodiesel (*i.e.* fatty acid methyl esters, FAMEs) from *Pongamia* oil glycerol is an economically low-value byproduct.

12.4 *Pongamia* Seed Oil for Biodiesel Production

The seeds of *Pongamia* usually contain around 35 to 45% (w/v) fatty acids and triglycerides (commonly called "oil"). However, not all this oil can be extracted by mechanical cold pressing. This oil can be easily converted to biodiesel (FAMEs) by transesterification with methanol in the presence of a potassium or sodium hydroxide catalyst to meet current industry standards: European EN 14214 and US ASTM D 6751–02. Despite meeting these standards, the oxidative stability, cold-weather performance ('cloud point') and ignition properties of *Pongamia* FAMEs are problematic and as such will not ensure best diesel-engine performance.[28] Poor fuel-flow character, especially under low ambient temperatures, and subsequent solidification is a major problem associated with the use of many biodiesels.[46] Blockages of fuel lines and filters have been known to lead to problems with engine start-up and operation, and eventual market failure. Whilst *Pongamia* FAMEs are satisfactory for use in tropical and temperate climates, improvements in the fuel-flow properties will need to be addressed for this oil to find a market in colder regions.[36] One viable option is to introduce *Pongamia*-derived FAME to markets as a mixture with crude-oil-derived diesel fuel (*i.e.* B20), which would overcome limitations of cloud point.

When analysing *Pongamia* seed-oil content and composition, one needs to recognise the variant genetic nature of the seed. Its individual ancestry is undetermined, even if derived from the same mother tree, that is highly heterozygous, itself a product of genetic outcrossing. Moreover, pollen donors could be any *Pongamia* tree within a 1–3 km radius, being governed by the flight performance of local bees. It is possible that some *Pongamia* trees self-fertilise using genetically divergent sporophyte-gametophyte genetics. This aspect needs to be carefully tested in insect-proof cages and using DNA molecular markers.

Table 12.2 List of key enzyme and product by abbreviation, as described by Figure 12.1.

Plastidial enzymes and products		Endoplasmic and cytosolic enzymes and products	
PDH	Pyruvate Dehydrogenase (EC 1.2.4.1)	GPD	G3P Dehydrogenase (EC 1.1.1.8)
ACC	Acetyl-CoA Carboxylase (EC 6.4.1.2)	GPAT	G3P Acyltransferase (EC 2.3.1.15)
MCAT	Malonyl-CoA Acyl-transferase (EC 2.3.1.39)	LPAT	LPA Acyltransferase (EC 2.3.1.51)
KAS	Keto-acyl-ACP synthase (EC 2.3.1.41)	PAP	Phosphatidic Acid Phosphatase (EC 3.1.3.4)
SAD	Stearoyl-Δ9-Desaturase (EC 1.14.19.2)	PDCT	PC:DAG CPT (EC 2.7.8)
FatA	Thioesterase Fat A (EC 3.1.2.14)	CPT	Cholinephosphotransferase (EC 2.7.8.2)
FatB	Thioesterase Fat B (EC 3.1.2*)	PDAT	Phospholipid:DGAT (EC 2.3.1.158)
ACP	Acyl Carrier Protein	DGAT	DAG Acyltransferase (EC 2.3.1.20)
PEP	Phosphoenolpyruvate	G3P	Glycerol-3-Phosphate
		LPA	Lysophosphatidic Acid
		PA	Phosphatidic acid
		DAG	1,2-*sn*-diacylglycerol
		PC	Phosphatidylcholine
		Lyso-PC	Lyso-phosphatidylcholine
		TAG	Triacylglycerol

Table 12.3 *Pongamia pinnata* seed oil composition.

Tree	C16:0*	C18:0	C18:1	C18:2	C18:3	C20:0	C20:1	C22:0	C24:0	%
C8	46	67	340	119	13	12	8	16	10	36
H1	49	21	227	111	14	8	11	39	18	37
R22	105	115	691	236	33	37	19	91	25	34

*shown as milligram/seed chloroform extracted; GC-MS determination; C17:0 internal quantification control); % = percentage of total mg/seed per total dry weight per seed×100.

The genetic variability is also seen in trees of different mother genetics growing in different environments (Table 12.3). We chose to show the analytical results of three trees from the South-East Queensland region of Australia. All grow in areas free of winter frost, and extreme high-temperature summers (28–33 °C) are common for a 6-month period. Eight to 10 seeds per tree were solvent extracted and oil components were quantified by GC-MS. Extractable oil averaged around 35% (v/v) but individual seeds reached as high as 43%. The oleic acid (C18:1) content in general ranged from 46% (H1 tree) to 56% (R22 tree). Strong variation appeared for C18:0 (*e.g.*, 4.2% *vs.* 10.6%), whilst some fatty acids (C20:0; C18:3, C18:2) seemed to be relatively constant.

Engine-performance tests carried out with *Pongamia* FAMEs concluded that blends with mineral diesel could be successfully made up to 40% (v/v) to provide a reduction in exhaust emissions together with increases in torque, brake power, thermal efficiency and a reduction in brake-specific fuel consumption.[47] However, as the concentration of *Pongamia* FAMEs in the blend is raised, deterioration in the dynamic viscosity, cloud and pour points (as measures of cold-weather performance) of the fuel has been detected.[48] The pour point of a fuel refers to the lowest temperature at which the oil will flow, and the cloud point is the temperature at which dissolved solids will separate out from the liquid. There are many reports of the physicochemical properties of various plant-derived FAMEs[49–52] with the pour and cloud points of *Pongamia* FAMEs 2.1 °C and 8.3 °C, respectively, consistent with the presence of saturated oils such as palmitic acid (C16:0) and stearic acid (C18:0). Although the cloud point of *Pongamia* FAMEs compares favourably with biodiesel derived from sources such as palm oil (10 °C) and beef tallow (13 °C), biodiesels made from soybean (− 1 °C), rapeseed (− 7 °C) and sunflower (1 °C) have lower and more desirable cloud points.[50] It has been shown that biodiesel quality depends on the fatty acid composition of corresponding feedstock. As a starting point, to improve the low-temperature flow properties and other physicochemical properties of biodiesel requires improvements in the oil composition of the raw materials used in biodiesel manufacture.[53] Further, in order to manipulate the fatty acid composition of a plant-derived oil feedstock for optimal diesel-engine performance requires an understanding of how different fatty acids alter the combustion properties of a fuel.

During seed formation and maturation most plant seeds accumulate storage products to provide nutrients and energy (*i.e.* oil, protein and starch) for seedlings to grow competitively during the early stages of development. Seed protein provides the nitrogen (often a limiting factor in soil), and oil or starch is used as an energy store for seedling establishment. Oilseed crops such as soybean, canola and sunflower, accumulate triacylglycerol (TAG; a major class of glycerolipid) instead of starch. Similarly, *Pongamia* accumulates large stores of oil upon maturity of the seed (Figure 12.2f). Yet, the contribution of all enzymes to TAG accumulation in plant seeds (or any other plant tissue), even in model plant species, is currently incomplete. Biodiesel can be easily made from plant TAG and represents an important and growing renewable fuel resource. The ideal composition of TAG for biodiesel production has been described as having a high proportion of monounsaturated fatty acids, a reduced amount of polyunsaturated fatty acids and a controlled saturated fatty acid content.[28] The cloud point of a fuel has been observed to be mostly dependent on its saturated ester content, while the effects of unsaturated esters are thought to be negligible.[50] In this sense, saturated fatty acids, such as palmitic and stearic acids, increase the cloud point of a fuel as these molecules have less mobility. Polyunsaturated fatty acids such as linoleic (C18:2) and linolenic (C18:3) acids are also less desirable as they are prone to oxidation (*i.e.* rancidity occurs when the double bonds of these fatty acids react with atmospheric oxygen). Monounsaturated fatty acids have since been cited as highly favourable for biodiesel production.[54,55]

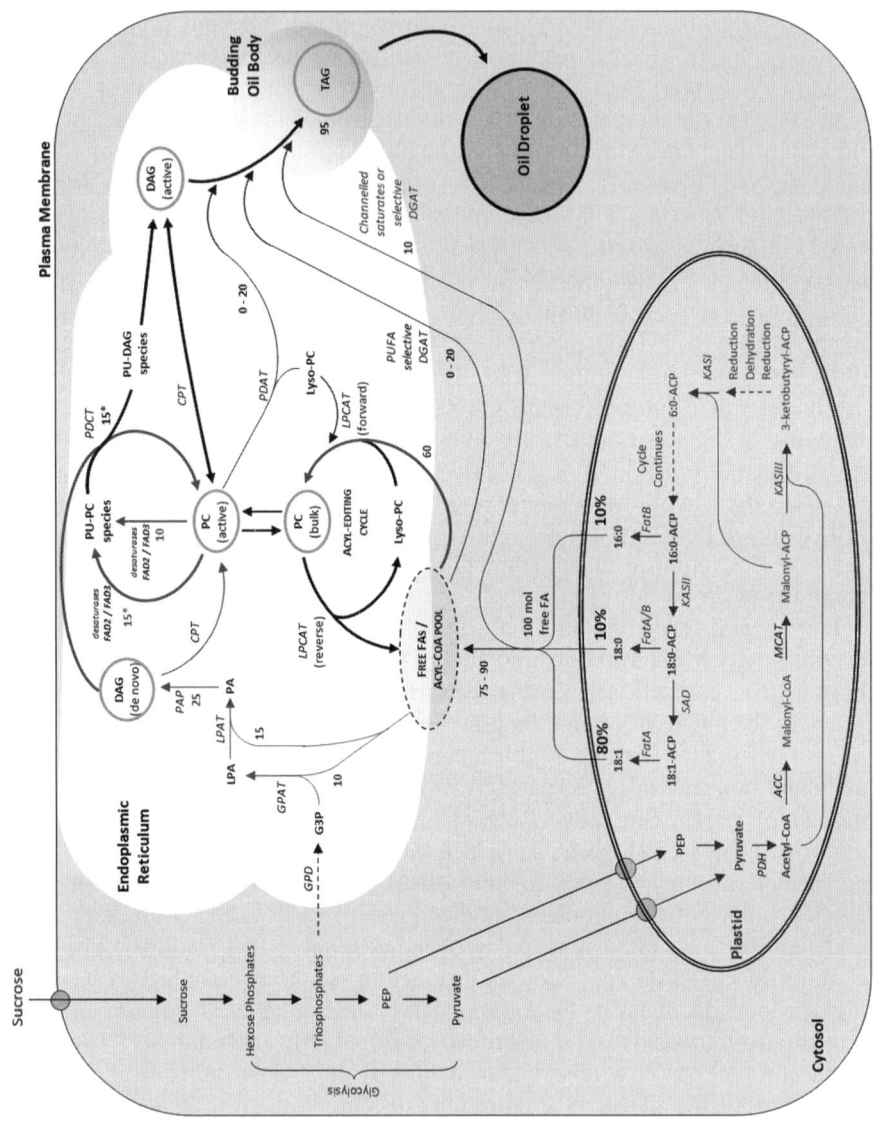

One such monounsaturated fatty acid, common to many plant species, is oleic acid. Oleic acid (C18:1Δ9) makes up roughly 55% of the total oil content of *Pongamia* seeds and is most advantageous because it is capable of producing a low cloud point biodiesel with high oxidative stability. Essentially all acyl chains for plant membrane and storage lipids are produced in the plastid by acyl carrier protein-dependent *de novo* FA synthesis.[56–58] The major fatty acids of plants (and most other eukaryotic organisms) have a chain length of sixteen or eighteen carbons and contain from zero to three *cis*-double bonds. Five fatty acids (C18:1, C18:2, C18:3, C16:0 and in some species C16:3) make up over 90% of acyl chains of structural glycerolipids of almost all plant membranes.[57] As mentioned previously, the main storage products of seeds are protein and oil. Protein mostly accumulates in the form of storage proteins in protein bodies and oil accumulates as TAG in oil bodies. The amino acids asparagine and glutamine provide the nitrogen and initial carbon skeletons for protein deposition and sucrose provides the energy for protein biosynthesis and the energy and hydrocarbon precursors for TAG accumulation.[59]

During seed development leaf-derived photosynthate is imported and converted into the precursors required for fatty acid biosynthesis (Figure 12.3). Cleavage of sucrose results in the formation of hexose phosphates that can be readily metabolised in a sequence of ten reactions, glycolysis. During this process, the hexose phosphates are further cleaved into triosphosphates. Dihydroxy acetone phosphate is a very common triosphosphate that can be reduced to a glycerol-3-phosphate (G3P) backbone for membrane lipid and TAG synthesis by an endoplasmic G3P dehydrogenase (GPD; EC 1.1.1.8).

Figure 12.3 Model of fatty acid and triglyceride biosynthesis in *Pongamia pinnata*: Integrated model of FA synthesis (Dyer and Mullen,[65]), modification, flux, assembly (Bates *et al.*,[85]), and desaturation and theoretical flux (based on *Pongamia* seed oil content) into PU-PC and -DAG species (Lu *et al.*,[94]) into oil bodies of soybean plant cells (Durrett *et al.*,[53]). The model follows 100 mol of newly synthesised fatty acids in the plastid (in *Pongamia* seed, this is believed to be approximately 80% C18:1, 10% C:18:0 and 10% C16:0) through the major fluxes of lipid metabolism such that the final accumulation of fatty acids in TAG is 95 mol. The arrows indicate the transfers and/or fluxes, with the major metabolite pools proposed from kinetic analysis circled in grey. The arrows highlighted in blue and red represent the major initial fluxes of newly synthesised fatty acid. Blue arrows represent fluxes where nascent and endogenous acyl groups mix, while red arrows indicate only nascent FA. Green arrows depict the production of PUFAs and green numbers represent the proposed flux in *Pongamia* seed tissue (to produce about 25% PUFA acyl groups). The red numbers represent the initial accumulation of nascent FA into the major labeled glycerolipids PC, DAG, and TAG, which together accumulate approximately 85% of initial fatty acids. Approximately 15% of nascent fatty acids initially accumulate in lipids other than PC, DAG, and TAG. For simplicity, not all possible reactions are shown; multiple reactions may be represented by single arrows. Key substrates are depicted in bold. Abbreviations of enzymes and products involved are listed in Table 12.2.

Triosphosphates can also be oxidised to 3-phosphoglycerate and this is isomerised to phosphoenolpyruvate (PEP) and then to pyruvate (the final glycolytic product). Both pyruvate and PEP enter the plastid where pyruvate is dehydrogenated to acetyl-CoA (a common precursor to many biochemical molecules) by a plastidial pyruvate dehydrogenase complex (PDH; EC 1.2.4.1). Although plastidic and cytosolic isoforms of the enzymes involved in glycolysis exist,[60] it is believed that at least the later glycolytic reactions relevant to fatty acid biosynthesis proceed in the plastid.[53] It is also believed that pyruvate synthesis and/or transport to plastids may be a limiting step in fatty acid biosynthesis and accumulation in oil[59] (Figure 12.3).

The first committed step of fatty acid biosynthesis is the formation of malonyl-CoA from acetyl-CoA and this process is highly regulated by a plastidic acetyl-CoA carboxylase (ACC; EC 6.4.1.2).[61] A malonyl-CoA acyltransferase (MCAT; EC 2.3.1.39) then translocates the malonyl moiety to acyl carrier protein (ACP) producing malonyl-ACP. De novo synthesis of fatty acids in the plastids then proceeds, where malonyl-ACP is elongated in cycles adding two carbon units at a time. Four different enzymes are involved with each cycle, starting with a condensation, followed by a series of reduction, dehydration and final reduction reactions.[62] In this manner, acyl chains containing up to eighteen carbon units can be synthesised. The condensation reactions are catalysed by enzymes known as 3-ketoacyl-ACP synthases or KAS enzymes (EC 2.3.1.41). The first condensation reaction that converts acetyl-CoA to 3-ketobutyrate is catalysed by KAS III, the reaction from butyryl (C4:0)-ACP to palmitoyl (16:0)-ACP by KAS I and from palmitoyl (C16:0)-ACP to stearoyl (C18:0)-ACP by KAS II. The first desaturation step also occurs in the plastid where the acyl chain remains conjugated to ACP and SAD (stearoyl-Δ9-desaturase) converts stearoyl-ACP to oleoyl-ACP. The reactions in most plant species stop at C16:0, C18:0 and C18:1 fatty acids not only by virtue of the specificity of the KAS and SAD enzymes, but also by the action of acyl-ACP thioesterases that hydrolyse the acyl-S-ACP thioester bonds.[63,64] This process of fatty acid biosynthesis, desaturation and export across the plastid membrane and into the cytosol has been well reviewed.[65] Palmitic, stearic and oleic free fatty acids are hence the common plastidial endproducts of most plant species.

Two main types of thioesterases are found in plants. The FatA (EC 3.1.2.14) class preferentially removes oleate from ACP, whereas FatB thioesterases (EC 3.1.2*) are more specific for saturated acyl-ACPs.[66] After release from ACP, the free fatty acids are exported from the plastid and converted to acyl-CoA esters through the activity of an acyl-CoA synthase located on the outer plastid membrane.[67,68] Plants have multiple acyl-CoA synthases with differential specificities that participate in lipid metabolism[69,70] and these are encoded by different genes as required for particular fatty acids.[71] Nascent fatty acids pool in the cell cytosol so that they can be incorporated into TAG during seed development. The nature of the acyl composition of the TAG is dependent on the availability of the fatty acids from the acyl-CoA substrate pool, as well as the selectivity of the acyltransferases and transacylases involved in this process.[72]

Fatty acids in cells are almost never in a "free" molecular form. Instead, their carboxyl group is usually esterified. In membranes, almost all of the fatty acids are found esterified to glycerol and these are called glycerolipids. Membrane glycerolipids have acyl groups attached to both the *sn*-1 and *sn*-2 positions of the glycerol backbone and a polar head group attached to the *sn*-3 position. If all three positions on the backbone are esterified with fatty acid chains, a TAG structure results. Of course, combinations with other fatty acid chains are possible; however, such structures are not suitable for membranes. These TAGs instead constitute the major form of lipid storage in seeds. In oilseeds, over 95% of the fatty acids synthesised in the plastid are exported for the assembly of eukaryotic glycerolipids.[57,73] Until recently, the biosynthesis of plant TAG was proposed to take place in the endoplasmic reticulum by the relatively straightforward Kennedy pathway.[74] In this process, fatty acids are sequentially transferred to the *sn*-1, *sn*-2 and *sn*-3 positions on the glycerol backbone.[75] The series begins with the acylation of the G3P backbone producing lysophosphatidic acid (LPA), catalysed by G3P acyltransferase (GPAT; EC 2.3.1.15). A second acylation of LPA catalysed by LPA acyltransferase (LPAT; EC 2.3.1.51) produces phosphatidic acid (PA). Following phosphate removal by phosphatidic acid phosphatase (PAP; EC 3.1.3.4), 1,2-*sn*-diacylglycerol (DAG) results and forms the precursor for TAG synthesis. The third and final acylation step is catalysed by DAG acyltransferase (DGAT; EC 2.3.1.20) that converts the DAG to TAG. In order to induce large changes in oil composition, LPAT was considered an important target enzyme because of its discrimination against saturated C16 and C18 acyl groups.[76] It has also been shown that DGATs are involved in regulating the quantity of seed TAGs and the sink size in developing seeds.[77] However, the assembly of fatty acids onto a glycerol backbone is not as straightforward as originally outlined by this pathway.

In many oilseeds, DAG represents an important branch point between the production of TAG and the synthesis of polyunsaturated-DAG compounds through phosphatidylcholine (PC) conversion. DAG may be acylated to produce TAG (via sequential acyl-CoA-dependent acylations of the Kennedy pathway as previously discussed) or converted to PC by CDP-choline:1,2-diacyl-*sn*-glycerol cholinephosphotransferase (CPT; EC 2.7.8.2). PCs are an important class of phospholipids and, with a choline head group and two fatty acid side chains, make up a major component of cell membranes and membrane-mediated cell-signaling structures. More importantly, these compounds also act as an intermediate in the biosynthesis of plant oil. In this intermediate form, additional desaturation of the acyl chains by desaturase enzymes becomes possible.[62,78] The desaturation of C18:1 to C18:2 and then to C18:3 on PC produces an abundance of polyunsaturated-PC species.[79–81] PC is thus the major site of fatty acid desaturation and to support this, evidence for a direct role for DAG or TAG as a desaturase substrate is also lacking. Therefore, with regard to TAG biosynthesis, the conversion of PC to DAG may enrich for more polyunsaturated-TAGs – as done by sequential FAD2 and FAD3 desaturases (EC 1.14.99*; to produce linoleic and linolenic fatty acids during the acyl-editing cycle, respectively). For these desaturations to occur, the reverse action of CPT is believed to make both DAG and PC interchangeable.[82] TAGs may also become enriched with polyunsaturated fatty acids by PC acyl-editing mechanisms that increase the polyunsaturated fatty acid content of the acyl-CoA pool. A third enrichment mechanism arises from the direct transfer of acyl groups from PC to DAG by the action of phospholipid:DGAT (PDAT; EC 2.3.1.158) in TAG synthesis. This means of transferring acyl groups from PC to DAG seems to be a common mode of action for the synthesis of plant TAGs.

Plants exhibit both acyltransferase and transacylase mechanisms for the acylation of DAG to TAG. In addition to the sequential acyl-CoA-dependent acylations of the Kennedy pathway, acyl chains can be incorporated into TAG, either through conversion back to DAG or by transacylase action. PDAT is a transacylase that allows the transfer of acyl groups from the *sn*-2 position of PC to DAG, producing TAG and lyso-PC products.[83] The lyso-PC product contains just one acyl chain and is presumably rapidly reacylated by a lyso-PC acyltransferase (LPCAT; EC 2.3.1.23) before/during the acyl-editing cycle. A DAG:DAG transacylase activity that produces TAG and monoacylglycerol has also been described[84] but since the contribution of monoacylglycerol to an oil body accounts for less than 2% of the endproducts, the action of this enzyme is believed negligible.[85] Hence, DAG represents an important branch point between the production of TAG and the synthesis of polyunsaturated-DAG compounds through PC conversion.

In both leaves[86] and seeds[87] the flux of nascent fatty acids into PC dominates over the synthesis of other phospholipids. Recently, it has been shown that most newly synthesised fatty acids are not directly esterified to G3P during *de novo* glycerolipid synthesis. Instead, these fatty acids are incorporated directly into a *sn*-1 or *sn*-2 lyso-PC *via* an acyl-editing mechanism, rather than through PA and DAG intermediates, as previously discussed.[88] This acyl-editing mechanism forms an integral component of eukaryotic glycerolipid synthesis. The editing process results in molecular species of PC with one newly synthesised and one previously synthesised fatty acid. It should be noted that this editing process may proceed by at least two mechanisms. The first mechanism involves CoA:PC acyl exchange producing lyso-PC and acyl-CoA (as discussed before with PDAT). A second possible mechanism for acyl editing involves hydrolysis of PC to lyso-PC by phospholipase activity, activation of the released free fatty acid, and its reuse for phospholipid synthesis from lyso-PC by LPCAT.[89] A *sn*-2-specific LPCAT could direct C18:1 toward desaturation on PC and release polyunsaturated fatty acids to the acyl-CoA pool for the Kennedy pathway.[90] Although there is now significant *in vivo* evidence for PC acyl-editing in the eukaryotic pathway in plant leaves,[88] much less *in vivo* evidence for such a cycle in oilseeds exists. To improve and extend on previous studies involved with oilseed metabolism, steps were taken to determine how fatty acids and glycerol are incorporated into membrane and storage lipids in cultured soybean embryos (as a model plant system).[85] This quantitative analysis has allowed the development of an acyl-flux model and Figure 12.3 incorporates this data into an overview of fatty acid biosynthesis and plant TAG production.

It is clear that there are a number of alternative metabolic routes for TAG synthesis. The alternatives may vary with tissue, species, and development and include the recycling of intermediates of membrane lipid biosynthesis. Different metabolic labeling experiments have been used to probe the sequence of reactions for TAG synthesis *in vivo* from a variety of oilseed species. Use of a PC-derived DAG moiety has been proposed as the major pathway for TAG synthesis in excised linseed and soybean cotyledons.[87] More recently, labeling of developing soybean embryos with [^{14}C]acetate and [^{14}C]glycerol has allowed a more quantitative and positional analysis of acyl group and glycerol backbone fluxes that comprise extraplastidic phospholipid and TAG synthesis, including acyl-editing and PC to DAG interconversion.[85] This work demonstrates for the first time that the acyl-editing of PC is a dominate flux in TAG synthesis. Here, about 60% of the newly synthesised fatty acids first enter glycerolipids through PC acyl-editing, largely at the *sn*-2 position. This flux, mostly of C18:1, was over three times the flux of nascent [^{14}C] fatty acids incorporated into the *sn*-1 and *sn*-2 positions of DAG through G3P acylation. Furthermore, the total flux for PC acyl-editing, which includes both nascent and pre-existing fatty acids, was estimated to be 1.5 to 5 times the flux of plastidial fatty acid synthesis. Thus, recycled acyl groups (commonly C16:0, C18:1, C18:2 and C18:3) in the acyl-CoA pool provide most of the acyl chains for *de novo* G3P acylation (Table 12.3).

The results of this work hence show kinetically distinct DAG pools. DAG used for TAG synthesis is mostly derived from PC, whereas *de novo* synthesised DAG is mostly used for PC synthesis. It has also been shown that newly synthesised fatty acids mix with recycled fatty acids to acylate G3P for *de novo* glycerolipid synthesis. About 25% of these fatty acids are directly incorporated into DAG through sequential acylations, and these acyl chains are often distributed equally between the *sn*-1 and *sn*-2 positions. Roughly 10% of newly synthesised fatty acids are incorporated into the *sn*-3 acylation of pre-existing DAG to generate TAG. As TAG constitutes approximately 95% of the endogenous acyl-lipid mass, eventually 25% of the nascent fatty acids must end up in the *sn*-3 position of TAG. Thus, approximately 15% of the nascent fatty acids incorporated into glycerolipids other than TAG must eventually move through into the *sn*-3 position of TAG to add to the 10% that are immediately incorporated. It is believed that this represents a second, distinctive flux of fatty acids into the *sn*-3 position of TAG providing molecular species enriched in saturated or polyunsaturated fatty acids. The relative rates of these three independent acyl transferase pathways have been incorporated into Figure 12.3.

In many plant species, oleic acid is the most common substrate that can be desaturated to linoleic and linolenic acids by the two desaturase enzymes of the endoplasmic reticulum; FAD2 and FAD3.[91,92] For desaturation to take place, oleic acid must be first incorporated into PC as this is the only substrate recognised by the desaturase enzymes.[93] As previously discussed, oleic acid can be incorporated into PC by two different mechanisms (by action of LPCAT or sequential acylation onto G3P then conversion by action of a CPT). It had been proposed that the CPT reaction is reversible and provides a means of producing polyunsaturated-DAG species for the synthesis of TAGs containing linoleic and linolenic fatty acids.[82] Very recently, a PC:DAG CPT (PDCT, EC 2.7.8.2) enzyme has been shown to provide a major reaction for the transfer of oleic acid into PC for desaturation and also for the reverse transfer of polyunsaturated fatty acids for TAG synthesis.[94] In this work, the PDCT enzyme has been shown to catalyse the transfer of the phosphocholine headgroup from PC to DAG where disruption to the gene encoding this enzyme has reduced the accumulation of polyunsaturated fatty acids in seed TAG by 40%. After analysis of the main fluxes controlling polyunsaturated fatty acid-TAG synthesis presented here, acyl-flux values relevant to the biosynthesis of *Pongamia* oil (containing 25% PUFAs) have been included in Figure 12.3.

12.5 Manipulation of *Pongamia* Seed Oil

Improvement in the physicochemical properties of *Pongamia* FAMEs for optimal diesel-engine performance will require a comprehensive understanding of fatty acid and seed oil biosynthesis. Manipulation of the fatty acid profile of *Pongamia* is predicted to be achieved by activating or repressing the expression of targeted genes involved in fatty acid biosynthesis. Activating the expression of a gene usually involves the incorporation of an extra copy of a gene of interest under the regulation of either its cognate promoter or a promoter of a different expression pattern.[95,96] However, the repression of gene expression can be accomplished by a number of methods (*e.g.*, RNAi).[97] Targeted gene repression has enabled large compositional changes to be made to the fatty acid profile of several oilseed crops.[98] Hairpin-RNA-mediated gene silencing has been used to inhibit the synthesis of linoleic acid in many plant species.[99,100] Seed-specific silencing of *FAD2* has been the most effective means of reducing seed linoleic-acid content.[98] Since the effects of unsaturated esters on the cloud point of a fuel are believed to be negligible,[50] improvement in this property can only be achieved by reducing saturated fatty acids. To maximise the characteristics of soybean seed oil for biodiesel production[101] the expression of *FAD2* and *FatB* (responsible for the production of palmitic and stearic acids) were simultaneously downregulated in a seed-specific fashion.[102] In considering the yield and composition of oil from candidate biodiesel feedstocks it is worth noting that much greater success has been found in studies targeting improvements in oil composition. Less success has been found in studies where researchers have tried to markedly improve the yield of oil. In the scenario where demands for liquid fuels are expected to increase in the near to long-term future in a climate where the supply of fossil fuels is to diminish, a significant challenge will be to meet this increased demand with concomitant increases in oil yield from biofuel feedstocks.

12.6 Propagation and Genetic Modification of *Pongamia*

To enhance the genotype and subsequent phenotypic outcomes of a plant it is widely recognised that introduction of modified gene constructs is an integral part of any plant-improvement program. This strategy can expand the number and type of modifications as well as shorten the path taken in more classical breeding programs, particularly in the case of trees. Any targeted improvement may be *via* the introduction of DNA sourced from another organism (*i.e.* transgenics) or through modification of DNA from the host itself (*i.e.* cisgenics). The production of a genetically modified plant requires an efficient technique for DNA integration into the genome and subsequent plant-regeneration methods. Therefore, the most significant phase in the improvement of any crop, including *Pongamia*, is the design and optimisation of a genetic transformation protocol that incorporates high-frequency regeneration of stably modified plants. As part of any improvement program genetic transformation is also valuable as a tool for additional molecular and physiological analyses of gene function.

While there are examples of legumes for which genetic transformation is a relatively simple process (*e.g. Lotus japonicus*),[103] many legumes are recalcitrant to current transformation technologies, limiting the possibilities for functional genetic studies and genetic improvement. In contrast to the vast majority of crop legumes that are annual herbaceous plants, *Pongamia* presents another challenge in that it is a perennial woody species. There are relatively few examples of woody and/or legume species for which a reliable transformation protocol has been developed.[104–107] Preliminary studies in our laboratory have shown that *Pongamia* is also difficult to regenerate *in vitro*. A fast, reproducible and efficient transformation procedure employing the phytopathogenic bacteria *Agrobacterium tumefaciens* is thought to be best means of genetic transformation of *Pongamia*. Development of a protocol incorporating *A. tumefaciens* with the most receptive and regenerative explant tissue will allow the stable integration of T-DNA constructs for either gene silencing or overexpression of candidate gene targets. Ultimately, the aim of such work will be the creation of well defined and characterised "elite varieties" of *Pongamia* from existing superior germplasm. The genotypes of such elite varieties will incorporate as-yet identified gene "cassettes" that will contribute important domestication traits (*e.g.*, high oleic-acid content; Table 12.1).[108]

In concert with any molecular-based improvement program is the requirement for the clonal propagation of the vast number of trees that will be needed in the foreseeable future. This is particularly important as *Pongamia* is an outcrossing species, which negates the ability to produce genetically identical plants from seed-derived propagules. It is relatively easy to generate saplings of *Pongamia* through shoot cuttings. However, labour costs and number of plants that can be produced will most likely render this approach uneconomic, at least in developed countries. What is more likely to be a successful commercial approach is through tissue culture and micropropagation. This approach has already been investigated[109,110] but questionably requires further refinement for quick and reliable commercial application. The advantage of tissue culture-based procedures is that once an appropriate genotype and explant tissue has been determined as most suitable for micropropagation it is likely that the necessary numbers of plants will be able to be generated through manipulation of the tissue culture media (*e.g.*, hormone modifications) and the consequent multiplication of plantlets.

12.7 Future Challenges and Concluding Remarks

While the energy-intensive lifestyles of the developed and emerging economies are widely acknowledged, the demand for fuel continues to increase under a "business as usual" approach by many industries and governments. In the face of diminishing fossil-fuel resources biofuels are now recognised as an emerging industry that will supply much of the future demands. More than anything else, the factor most limiting biofuel production (particularly in Australia) is the environmentally and economically sustainable supply of feedstock to biofuel industrial plants. Many plant, animal and microbial candidates have been proposed as feedstocks. However, careful consideration must be given to which feedstocks are going to be successful energy crops long into the future. In the context of warm-climate agriculture, on low agriculturally productive land, unreliable water resources, and with minimal effect on food production, *Pongamia* presents itself as a strong candidate.

Nonetheless, as a candidate bioenergy crop *Pongamia* presents many challenges in its future path to commercialisation. With little or no history of domestication the scientific community has much to do in its efforts to establish this tree legume as a reliable and significant source of feedstock for the biodiesel industry. This will require long-term research efforts supported by sizable funding from relevant government and industry groups. At the ARC Centre of Excellence for Integrative Legume Research, thanks to government and industry research partners, we have instituted a research program incorporating functional genomics, molecular breeding, plant physiology and agronomy in order to put *Pongamia* on the path to commercial success (Figure 12.4). In particular, the technology of functional genomics can provide tools that will enable the relatively rapid characterisation of the *Pongamia* genome and genetic diversity amongst currently available germplasm. These tools will also be invaluable in the genetic enhancement of current germplasm towards future defined elite varieties.

To date, almost all evaluations of *Pongamia* have been based on trees growing in the wild or urban ornamental plantation. It is anticipated that new challenges will be presented once *Pongamia* is grown in managed plantations of hundreds to thousands of hectares. For example, to date there are little or no insect or microbial pathogens that are reported to be of concern.[36] Our own anecdotal evidence indicates that there are several insect pests capable of causing minor damage to foliage without significantly affecting the overall

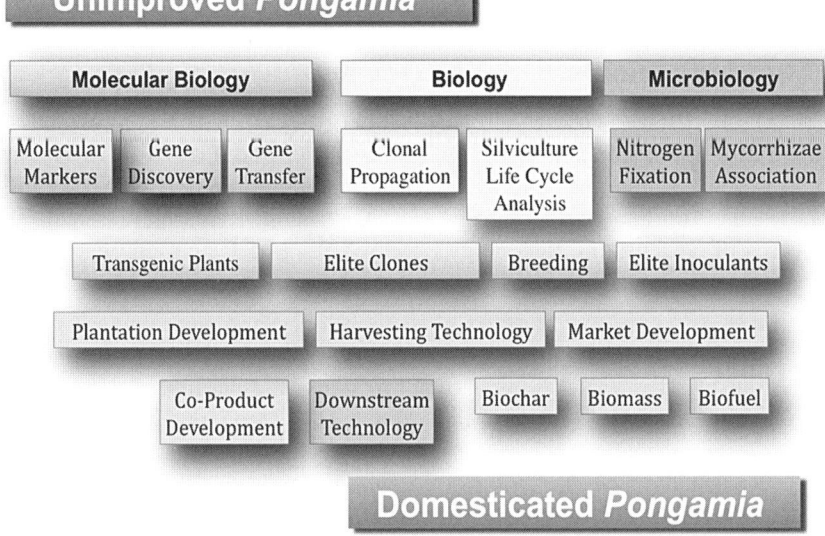

Figure 12.4 Path to *Pongamia* commercialisation: Unimproved individual *Pongamia* plants may have desirable traits. These need to be selected and combined, then propagated. Multiple approaches need to interact to facilitate domestication. What has taken 10 000 years of human intervention in corn, rice, soybean, potato and wheat, most likely needs to be accomplished in a 10- to 25-year time horizon. Accordingly modern genetic methods creating predictability and overcoming classical breeding limitations need to be employed. The understanding of gene function in *Pongamia* thus becomes a key bottleneck for future improvements.

health of the tree. Visual evidence exists for viral infections on plants. It is anticipated that once *Pongamia* is grown as a broad-acre crop there will be the opportunity for one or more pests and/or pathogens to establish themselves. In terms of nitrogen fixation, we observe that *Pongamia* nodules contain a relative limited amount of nitrogen-fixing tissue. Is this controlled by the suboptimal bacterial inoculant, or an inherent nature of the species?

Lessons should be learnt from past experiences with broad-acre monoculture crops. Oil palm for example, is propagated vegetatively through tissue culture but is troubled by genomic instabilities, most likely involving DNA methylation as well as aneuploidy, leading to abnormal flower determination and crop yield. To understand how *Pongamia* will perform under broad-acre cropping it is therefore important that many field trials be established in the near future under the proposed plantation landscapes. In addition, there are still unanswered questions as to some of the simple agronomic questions (*e.g.*, tree architecture, appropriate mechanical harvesting). The evaluation of the long-term future of any proposed biodiesel feedstock is still too early to determine. However, in considering all the pertinent environmental, agricultural and economic issues surrounding this emerging industry it is clear that *Pongamia* should be placed as one of the short-listed candidates to succeed.

Acknowledgements

We thank Drs Charles Hocart and Michael Djordjevic for their help in analysis of the fatty acid composition of *Pongamia* seeds, and Prof. Ray Rose and Dr Xin-Ding Wing for microscopy of *Pongamia* seeds. We gratefully acknowledge funding support from the Australian Research Council, the University of Queensland, BioEnergy Research Pty Ltd, Origin Energy, and the Brisbane City Council.

References

1. *International Energy Outlook 2009*. US Department of Energy, Washington DC, 2009.
2. J. Rockström, W. Steffen, K. Noone, Å. Persson, F. S. Chappin III, E. F. Lambin, T. M. Lenton, M. Scheffer, C. Folke, H. J. Schellnhuber, B. Nykvist, C. A. de Wit, T. Hughes, S. van der Leeuw, H. Rodhe, S. Sörlin, P. K. Snyder, R. Costanza, U. Svedin, M. Falkenmark, L. Karlberg, R.W. Corell, V. J. Fabry, J. Hansen, B. Walker, D. Liverman, K. Richardson, P. Crutzen and J. A. Foley, *Nature*, 2009, **461**, 472.
3. C. B. Granda, L. Zhu and M. T. Holtzapple, *Environ. Prog.*, 2007, **26**, 233.
4. R. Luque, L. Herrero-Davila, J. M. Campelo, J. H. Clark, J. M. Hidalgo, D. Luna, J. M. Marinas and A. A. Romero, *Energy Environ. Sci.*, 2008, **1**, 542.
5. L. D. Gomez, C. G. Steele-King and S. J. McQueen-Mason, *New Phytol.*, 2008, **178**, 473.
6. J. Hill, E. Nelson, D. Tilman, S. Polasky and D. Tiffany, *Proc. Natl. Acad. Sci. USA*, 2006, **103**, 11206.
7. J. Hill, *Agron. Sustain. Dev.*, 2007, **27**, 1.
8. G. Zemanek and G. Reinhardt, *Fett/Lipid*, 1999, **101**, S321.
9. P. J. Crutzen, A. R. Mosier, K. A. Smith and W. Winiwarter, *Atmos. Chem. Phys.*, 2008, **8**, 389.
10. P. B Reich, S. E. Hobbie, T. Lee, D. S. Ellsworth, J. B. West, D. Tilman, J. M. H. Knops, S. Naeem and J. Trost, *Nature*, 2006, **440**, 922.
11. E. B. Madsen, L. H. Madsen, S. Radutoiu, M. Olbryt, M. Rakwalska, K. Szczglowski, S. Sato, T. Kaneko, S. Tabata, N. Sandal and J. Stougaard, *Nature*, 2003, **425**, 637.
12. E. Limpens, C. Franken, P. Smit, J. Willemse, T. Bisseling and R. Geurts, *Science*, 2003, **302**, 630.
13. A. Indrasumunar, A. Kereszt, I. Searle, M. Miyagi, D. Li, C. D. T. Nguyen, A. Men, B. J. Carroll and P. M. Gresshoff, *Plant Cell Physiol.*, 2010, **51**, 201.
14. B. J. Ferguson, A. Indrasumunar, S. Hayashi, M.-H. Lin, Y.-H. Lin, D. Reid and P. M. Gresshoff, *J. Integ. Plant Biol.*, 2010, **52**, 61.

15. M. K. Udvardi, G. D. Price, P. M. Gresshoff and D. A. Day, *FEBS Lett.*, 1988, **231**, 36.
16. M. Kinkema, P. T. Scott and P. M. Gresshoff, *Funct. Plant Biol.*, 2006, **33**, 1.
17. S. Nontachaiyapoom, P. T. Scott, A. E. Men, M. Kinkema, P. M. Schenk and P. M. Gresshoff, *Mol. Plant Microbe Interact.*, 2007, **20**, 769.
18. S. Okamoto, E. Ohnishi, S. Sato, H. Takahashi, M. Nakazono, S. Tabata and M. Kawaguchi, *Plant Cell Physiol.*, 2009, **50**, 67.
19. K. Oelkers, N. Goffard, G. F. Weiller, P. M. Gresshoff, U. Mathesius and T. Frickey, *BMC Plant Biol.*, 2008, **8**, 1.
20. P. M. Gresshoff, A. Indrasumunar, S. Nontachaiyapoom, M. Kinkema, Y.-H. Lin, Q. Jiang, D.-X. Li, A. Miyahara, C. Nguyen, D. Buzas, B. Biswas, P. K. Chan, P. Scott, T. Hirani, M. Miyagi, M. Djordjevic, B. J. Carroll, A. Men and A. Kereszt, *Curr. Plant Sci. Biotech. Agric.*, 2008, **42**, 173.
21. I. R. Searle, A. E. Men, T. S. Laniya, D. M. Buzas, I. Iturbe-Ormaetxe, B. J. Carroll and P. M. Gresshoff, *Science*, 2003, **299**, 109.
22. L. Krusell, L. H. Madsen, S. Sato, G. Aubert, A. Geua, L. Szczyglowski, G. Duc, T. Kaneko, S. Tabata, F. de Bruijn, E. Pajuelo, N. Sandal and J. Stougaard, *Nature*, 2002, **420**, 422.
23. R. Nishimura, M. Hayashi, G. J. Wu, H. Kouchi, H. Imaizumi-Anraku, Y. Murakami, S. Kawasaki, S. Akao, M. Ohmori, M. Nagasawa, K. Harada and M. Kawaguchi, *Nature*, 2002, **420**, 426.
24. E. Schnabel, E. P. Journet, F. de Carvalho-Niebel, G. Duc and J. Frugoli, *Plant Mol. Biol.*, 2005, **58**, 809.
25. M. Kinkema and P. M. Gresshoff, *Mol. Plant Microbe Interact.*, 2008, **21**, 1337.
26. S. C. Davis, K. J. Anderson-Teixeira and E. H. DeLucia, *Trends Plant Sci.*, 2009, **14**, 140.
27. H. Halleux, S. Lassaux, R. Renzoni and A. Germain, *Int. J. LCA*, 2008, **13**, 184.
28. S. Pinzi, I. L. Garcia, F. J. Lopez-Gimenez, M. D. Luque de Castro, G. Dorado and M. P. Dorado, *Energy Fuels*, 2009, **23**, 2325.
29. M. J. Groom, E. M. Gray and P. A. Townsend, *Conserv. Biol.*, 2008, **22**, 602.
30. K. Openshaw, *Biomass Bioenergy*, 2000, **19**, 1.
31. K. Becker and H. P. S. Makkar, *Lipid Tech.*, 2008, **20**, 104.
32. W. M. J. Achten, E. Mathijs, L. Verchot, R. Aerts and B. Muys, *Biofuels Bioprod. Biorefin.*, 2007, **1**, 283.
33. W. M. J. Achten, L. Verchot, Y. J. Franken, E. Mathijs, V. P. Singh, R. Aerts and B. Muys, *Biomass Bioenergy*, 2008, **32**, 1063.
34. A. J. King, W. He, J. A. Cuevas, M. Freudenberger, D. Ramiaramanana and I. A. Graham, *J. Exp. Bot.*, 2009, **60**, 2897.
35. K. Sanderson, *Nature*, 2009, **461**, 328.
36. P. T. Scott, L. Pregelj, N. Chen, J. S. Hadler, M. A. Djordjevic and P. M. Gresshoff, *Bioenergy Res.*, 2008, **1**, 2.

37. S. K. Karmee and A. Chadha, *Bioresour Technol.*, 2005, **96**, 1425.
38. M. M. Azam, A. Waris and N. M. Nahar, *Biomass Bioenergy*, 2005, **29**, 293.
39. M. Naik, L. C. Meher, S. Naik and L. M. Das, *Biomass Bioenergy*, 2008, **32**, 354.
40. P. D. Patil and S. Deng, *Fuel*, 2009, **88**, 1302.
41. S. G. Patil, M. Hebbara and S. B. Devarnavadagi, *Ann. Arid Zone*, 1996, **35**, 57.
42. O. S. Tomar and R. K. Gupta, *Plant Soil*, 1985, **87**, 329.
43. D. Chandrasekaran, R. Kadirvel and K. Viswanathan, *Anim. Feed Sci. Technol.*, 1989, **22**, 321.
44. B. K. Konwar and G. C. Banerjee, *Indian Vet. J.*, 1987, **64**, 500.
45. R. Natanam, R. Kadirvel and K. Viswanathan, *Anim. Feed Sci. Technol.*, 1989, **27**, 89.
46. G. H. Knothe, *Fuel Process. Technol.*, 2005, **86**, 1059.
47. A. G. Raheman and A. G. Phadatare, *Biomass Bioenergy*, 2004, **27**, 393.
48. R. M. Joshi and M. J. Pegg, *Fuel*, 2006, **86**, 143.
49. A. Srivastava and R. Prasad, *Renew. Sustain. Energy Rev.*, 2000, **4**, 111.
50. H. Imahara, E. Minami and S. Saka, *Fuel*, 2006, **85**, 1666.
51. A. K. Agarwal, *Prog. Energy Combust. Sci.*, 2007, **33**, 233.
52. M. Canakci and H. Sanli, *J. Ind. Microbiol. Biotechnol.*, 2008, **35**, 431.
53. T. P. Durrett, C. Benning and J. B. Ohlrogge, *Plant J.*, 2008, **54**, 593.
54. M. J. Ramos, C. M. Fernandez, A. Casas, L. Rodriguez and A. Perez, *Bioresour. Technol.*, 2009, **100**, 261.
55. G. H. Knothe, *Energy Fuels*, 2008, **22**, 1358.
56. J. B. Ohlrogge, D. N. Kuhn and P. K. Stumpf, *Proc. Natl. Acad. Sci. USA*, 1979, **76**, 1194.
57. J. B. Ohlrogge and J. A. Browse, *Plant Cell*, 1995, **7**, 957.
58. J. Schwender, Y. Shachar-Hill and J. B. Ohlrogge, *J. Biol. Chem.*, 2006, **281**, 34040.
59. D. F. Hildebrand, R. Li and T. Hatanaka, In: *Genetics and Genomics of Soybean*, G. Stacey (ed.), Springer, New York, 2008, p. 185.
60. W. C. Plaxton and F. E. Podesta, *CRC Crit. Rev. Plant Sci.*, 2006, **25**, 159.
61. B. J. Nikolau, J. B. Ohlrogge and E. S. Wurtele, *Arch. Biochem. Biophys.*, 2003, **414**, 211.
62. J. B. Ohlrogge and J. G. Jaworski, *Annu. Rev. Plant Physiol. Plant Mol. Biol.*, 1997, **48**, 109.
63. T. A. Voelker, A. C. Worell, L. Anderson, J. Bleiaum, C. Fan, D. J. Hawkins, S. E. Radke and H. M. Davies, *Science*, 1992, **257**, 72.
64. G. Budziszewski, K. P. C. Croft and D. F. Hildebrand, *Lipids*, 1996, **31**, 557.
65. J. M. Dyer and R. T. Mullen, *Seed Sci. Res.*, 2005, **15**, 155.
66. J. J. Salas and J. B. Ohlrogge, *Arch. Biochem. Biophys.*, 2002, **403**, 25.
67. M. Pollard and J. B. Ohlrogge, *Plant Physiol.*, 1999, **121**, 1217.

68. A. J. K. Koo, J. B. Ohlrogge and M. Pollard, *J. Biol. Chem.*, 2004, **279**, 16101.
69. J. A. Schnurr, J. M. Shockey, G.-J. De Boer and J. A. Browse, *Plant Physiol.*, 2002, **129**, 1700.
70. J. M. Shockey, M. S. Fulda and J. A. Browse, *Plant Physiol.*, 2002, **129**, 1710.
71. T. McKeon, S. T. Kang, C. Turner, X. He, G. Chen, J.-T. Lin, *97th AOCS Annual Meeting & Expo*, 2006, p. 15.
72. J. L. Harwood, In: *Plant Lipid Biosynthesis Fundamentals and Agricultural Applications*, J. L. Harwood, (ed.), Cambridge University Press, New York, 1998, p. 1.
73. P. G. Roughan and C. R. Slack, *Annu. Rev. Plant Physiol. Plant Mol. Biol.*, 1982, **33**, 97.
74. S. Stymne and A. K. Stobart, in *The Biochemistry of Plants: A Comprehensive Treatise, Vol 9, Lipids: Structure and Function*, P. K. Stumpf, (ed.), Academic Press, New York, 1987, p. 175.
75. J. A. Napier, *Annu. Rev. Plant Biol.*, 2007, **58**, 295.
76. F. Bourgis, J.-C. Kader, P. Barret, M. Renard, D. Robinson, C. Robinson, M. Delseny and T. Roscoe, *Plant Physiol.*, 1999, **120**, 913.
77. C. Jako, A. Kumar, Y. D. Wei, J. T. Zou, D. L. Barton, E. M. Giblin, P. S. Covello and D. C. Taylor, *Plant Physiol.*, 2001, **126**, 861.
78. F. M. Jackson, L. Michaelson, T. C. M. Fraser, A. K. Stobart and G. Griffiths, *Microbiology*, 1998, **144**, 2639.
79. P. Sperling and E. Heinz, *Eur. J. Biochem.*, 1993, **213**, 965.
80. P. Sperling, M. Linscheid, S. Stocker, H. P. Muhlbach and E. Heinz, *J. Biol. Chem.*, 1993, **268**, 26935.
81. J. L. Harwood, *Biochim. Biophys. Acta*, 1996, **1301**, 7.
82. C. R. Slack, L. C. Campbell, J. A. Browse and P. G. Roughan, *Biochim. Biophys. Acta*, 1983, **754**, 10.
83. A. Dahlqvist, U. Stahl, M. Lenman, A. Banas, M. Lee, L. Sandager, H. Ronne and H. Stymne, *Proc. Natl. Acad. Sci. USA*, 2000, **97**, 6487.
84. K. Stobart, M. Mancha, M. Lenman, A. Dahlqvist and S. Stymne, *Planta*, 1997, **203**, 58.
85. P. D. Bates, T. P. Durrett, J. B. Ohlrogge and M. Pollard, *Plant Physiol.*, 2009, **150**, 55.
86. G. Bonaventure, X. M. Ba, J. B. Ohlrogge and M. Pollard, *Plant Physiol.*, 2004, **135**, 1269.
87. C. R. Slack, P. G. Roughan and N. Balasingham, *Biochem. J.*, 1978, **170**, 421.
88. P. D. Bates, J. B. Ohlrogge and M. Pollard, *J. Biol. Chem.*, 2007, **282**, 31206.
89. H. Shindou and T. Shimizu, *J. Biol. Chem.*, 2009, **284**, 1.
90. G. Griffiths, S. Stymne and A. K. Stobart, *Planta*, 1988, **173**, 309.
91. V. Arondel, B. Lemieux, I. Hwang, S. Gibson, H. M. Goodman and C. R. Somerville, *Science*, 1992, **258**, 1353.

92. J. Okuley, J. Lightner, K. Feldmann, N. Yadav, E. Lark and J. Browse, *Plant Cell*, 1994, **6**, 147.
93. S. Stymne and L. A. Appelqvist, *Eur. J. Biochem.*, 1978, **90**, 223.
94. C. Lu, Z. Xin, Z. Ren, M. Miquel and J. Browse, *Proc. Natl. Acad. Sci. USA*, 2009, **106**, 18837.
95. R. X. Fang, F. Nagy, S. Sivasubramania and N. H. Chua, *Plant Cell*, 1989, **8**, 141.
96. P. Hajdukiewicz, Z. Svab and P. Maliga, *Plant Mol. Biol.*, 1994, **25**, 989.
97. S. Mlotshwa, O. Voinnet, M. F. Mette, M. Matzke, H. Vaucheret, S. W. Ding, G. Pruss and V. B. Vance, *Plant Cell*, 2002, **14**, 289.
98. J. J. Thelen and J. B. Ohlrogge, *Plant Metabol. Eng.*, 2002, **4**, 12.
99. P. A. Stoutjesdijk, S. P. Singh, Q. Liu, C. J. Hurlstone, P. A. Waterhouse and A. G. Green, *Plant Physiol.*, 2002, **129**, 1723.
100. Q. Liu, S. P. Singh and A. G. Green, *Plant Physiol.*, 2002, **129**, 1732.
101. J. Duffield, H. Shapouri, M. Graboski, R. McCormick and R. Wilson, *US Biodiesel Development: New Markets for Conventional and Genetically Modified Agricultural Products*, Economic Research Service, USDA, Washington DC, 1998.
102. T. Buhr, S. Sato, F. Ebrahim, A. Xing, Y. Zhou, M. Mathiesen, B. Schweiger, A. J. Kinney, P. Staswick and T. Clemente, *Plant J.*, 2002, **30**, 155.
103. D. P. Lohar, K. Schuller, D. M. Buzas, P. M. Greshoff and J. Stiller, *J. Exp. Bot.*, 2001, **52**, 1697.
104. K.-H. Han, D. E. Keathley, J. M. Davis and M. P. Gordon, *Plant Sci.*, 1993, **88**, 149.
105. C. Franche, D. Diouf, Q. V. Le, D. Bogusz, A. N'Diaye, H. Gherbi, C. Gobe and E. Duhoux, *Plant J.*, 1997, **11**, 897.
106. T. Igasaki, T. Mohri, H. Ichikawa and K. Shinohara, *Plant Cell Rep.*, 2000, **19**, 448.
107. L. J. Cseke, S. B. Cseke and G. K. Podila, *Plant Cell Rep.*, 2007, **26**, 1529.
108. J. Gressel, *Plant Sci.*, 2008, **174**, 246.
109. T. Sugla, J. Purkayastha, S. K. Singh, S. K. Solleti and L. Sahoo, *In Vitro Cell. Dev. Biol. – Plant*, 2007, **43**, 409.
110. K. Sujatha and S. Hazra, *In Vitro Cell. Dev. Biol. – Plant*, 2007, **43**, 608.

CHAPTER 13
Willow

S. J. HANLEY

Centre for Bioenergy and Climate Change, Rothamsted Research, Harpenden, Hertfordshire, AL5 2JQ UK

13.1 Introduction

Willows have been cultivated throughout history as a source of raw materials for a range of useful products. Inherent properties such as high strength, low weight and flexibility have long been exploited in basketry and other traditional crafts such as fencing and hurdle making. Willow also provides the timber of choice for cricket bat production. In addition, willow's analgaesic properties, which result from high levels of salicin in the bark, have also been exploited by man since antiquity.

Significant interest in the use of willow biomass as a source of fuel first arose as a consequence of oil crisis of the 1970s that sparked new interest in the search for alternative and renewable energy sources. Willows are now grown commercially for bioenergy end uses in several countries including Sweden, the United Kingdom, and more recently, the United States, albeit on a somewhat limited scale at present. While energy security remains an important driver, recognition of the role that second-generation bioenergy crops could play in mitigating climate change has led to an increased level of interest in the use of willow for bioenergy purposes in recent years. Inherent properties such as rapid growth rates and low input requirements are the fundamental traits that, amongst others, make willow an attractive energy crop for this purpose. Breeding and underpinning research on biomass willows has been ongoing over the last two decades and significant advances in terms of crop improvement have already been made.

13.1.1 The Genus *Salix*

Willows (genus *Salix*), together with poplars (*Populus*), are members of the Salicaceae family. Willows present many morphological forms, ranging from large trees to shrubs, subshrubs and dwarf species. They are adapted to a wide range of habitats although, as the name *Salix* suggests (derived from Celtic – *sallis*: *sal* "near", *lis* "water"),[1] they are often found in riparian and alluvial habitats. This more reflects the requirement of moisture for seed germination due to limited seed longevity, rather than a requirement for substantial amounts of water for post-establishment survival.

The *Salix* genus is thought to have originated in the mountains of Eastern Asia and spread over considerable distances following the most recent ice age, when glaciers linking the continents of the northern hemisphere melted to form vast rivers that carried sediment across Asia, Europe and North America.[1] As the glaciers receded, *Salix* populations remained widely disseminated, mainly in temperate and arctic regions of the Northern Hemisphere where the genus remains abundant today. According to the systematic treatment of Argus,[2] 65 willow species are found in Europe, 130 species in North America and 120 and 275 species in the former Soviet Union and China, respectively. Although examples exist, few species are native to South America and Africa. Willows are not native to Antarctica or Oceania, although species introduced to the latter are thriving.

13.1.2 Classification

Willows are hugely diverse with respect to botanical characteristics and, as yet, no single satisfactory classification system has been developed. Indeed, Linnaeus described the genus *Salix* as "extremely difficult to clarify".[3] Difficulties in classification arise as a consequence of high levels of intraspecific morphological variation, dioecy and the presence of relatively few informative floral characteristics for use in systematic studies. Furthermore, natural interspecific hybridisation commonly occurs in willow, although there are limitations to this. For example, hybridisation between different subgenera does not generally occur.

Despite difficulties in classification, several authoritative treatments exist with fairly stable nomenclature.[2,4–6] Although clearly still subject to some debate, it is generally accepted that there are between 330 and 500 species within the genus and that the genus can be further divided into three main subgenera. The first of these, the subgenus *Salix*, comprises trees and large shrub species such as *Salix alba*, *S. fragilis*, *S. babylonica*, *S. pentandra*, and *S. nigra*. The second subgenus, *Vetrix*, is represented by around two thirds of the *Salix* genus and contains several small trees and shrub willow types.

Examples include *S. viminalis*, *S. caprea*, *S. cinerea*, *S. purpurea* and *S. eriocephala*. Several dwarf and creeping forms, such as *S. herbacea*, *S. retusae* and *S. myrsinites* as examples, comprise the third subgenus, *Chamaetia*. Although often treated separately, *Chosenia* is regarded by some as a smaller fourth subgenus due to the commonality of several fundamental botanical characteristics with *Salix*. However, there are some notable exceptions such as the presence of pendulous catkins and an absence of nectaries, which are both features more commonly associated with *Populus*.

13.1.3 Willow Biology

Willows are dioecious, with the vast majority of plants producing either male or female catkins exclusively. However, on occasion, hermaphrodite catkins can occur in a restricted number of species. In nature, propagation is largely *via* seed although vegetative propagation can also occur and is common in some species such as *S. fragilis*, which, as the name suggests, has brittle branches that easily break off and regenerate elsewhere. Pollination is usually insect-mediated although wind pollination may also occur. Three to eight weeks postpollination, small seeds (0.8–3 mm) are shed. These are surrounded by fine hairs that aid wind dispersal. Although the seeds contain chlorophyll, no endosperm is present, thus germination often occurs immediately upon contact with moisture.

Willows, like poplars, have a haploid chromosome number of 19. Different species range from diploid ($2n=38$), up to a maximum of dodecaploid ($12n=228$). In diploid species, estimates of DNA content range from 0.76 to 0.98 picograms per nucleus.[7] To date, no willow genomes have been fully sequenced although syntenic relationships with the *Populus trichocarpa* genome sequence[8] have been demonstrated.[9]

13.2 Willow as a Bioenergy Crop

Clearly not all willow species are suitable for biomass production. For this purpose, the vast majority of cultivars are derived from species of the subgenus *Vetrix,* although some members of *Salix* have been used in the production of useful hybrids. In Europe, the predominant species used for biomass is *S. viminalis* although other species, such as *S. dasyclados*, *S. schwerinii*, *S. triandra* and *S. caprea* have contributed in the breeding of hybrid cultivars with improved yield, disease and pest resistance. In the United States, *S. eriocephala*, *S. purpurea* and *S. miyabeana* are the major species used in biomass breeding and production at present. The reason for the use of different species in Europe and the US is largely due to the susceptibility of some European *S. viminalis*-derived cultivars to damage by the potato leaf hopper (*Empoasca fabae*).[10] Whatever the species or hybrid used, several inherent beneficial properties of willows are exploited for biomass production.

13.2.1 Fast Growth and Coppicing Ability

Perhaps of most importance in biomass production is the capacity for rapid growth, which is further promoted by cultivation as short-rotation coppice (SRC). The traditional practice of coppicing involves the removal of stems from the stool following an initial period of establishment growth, usually when the plants are dormant in the winter. This removal of apical dominance allows otherwise dormant axillary buds to grow out from the stool when dormancy breaks, resulting in the production of a greater number of harvestable proleptic shoots. Furthermore, coppicing is also believed to reinvigourate the stool and accelerate growth rates towards a theoretical maximum.[11,12]

Different willows vary in their response to coppicing and some species are more amenable to this practice than others. Generally, willows of the subgenus *Vetrix* often respond better to coppicing than do those of the *Salix* subgenus. Variation in coppicing potential has been attributed to species-specific differences in both the numbers and behaviour of buds.[13] For example, in *S. viminalis* and *S. eriocephala* the main shoot primordia develop first followed by two lateral primordia. Shoots of the lateral primordia contribute to rapid canopy development but are later suppressed and subject to self-thinning. In comparison, in *S. amygdaloides* (subgenus *Salix*) the development of the lateral primordia is inhibited, resulting in a lower resprouting potential in this species. Although coppicing ability is central to the production of willow crops for biomass, high yields may be obtained by different strategies and two distinct routes to high biomass yield have been described.[14,15] First, high yields may be obtained by plants that produce relatively few shoots, each being of large diameter. These types are also characterised by a high leaf area index (LAI) and a high specific leaf area (SLA). Alternatively, high yields can be achieved by plants with a greater number of relatively thin stems and lower LAI and SLA.

13.2.2 Low-N Input Requirements

In contrast to annual crops that are highly dependent on N inputs, willow, as a perennial, requires comparatively little in the way of fertiliser inputs to achieve impressive yields. The requirement for N application is limited to the amount lost at harvest, and is usually in the region of 60 kg N ha^{-1}. This has a significant benefit given the costs, both financial and in terms of GHG emissions that are associated with the production and use of fertilisers. Indeed, in a comparison of nitrogen intensities for a range of bioenergy crops, SRC willow ranked best with a nitrogen intensity of 90 Kg N/1000 GJ energy produced.[16]

13.2.3 Natural Adaption to Marginal Land

Given that the area of land available for agriculture is finite and that there will be an increasing need to produce greater amounts of food for a growing population, it has been suggested that bioenergy crops should be grown predominantly on subprime land that is only suitable for limited food production.

As many willow species are naturally adapted to more marginal land types, it is reasonable to expect that willow will perform well in this regard when compared to some other bioenergy crops.

13.2.4 Genetic Diversity for Breeding

The wealth of genetic diversity represented by hundreds of different species that can often easily be hybridised, and the wide geographic range of the genus are significant assets to the development of willow as a biomass crop. Furthermore, with the exception of some cultivars that have been the focus of past selections for basketry purposes, the genus remains undomesticated. As a result, there remains a vast amount of natural allelic variation for many traits that can be exploited and combined to produce improved varieties for a wide range of growing environments and end uses.

13.3 Cultivation and Agronomy

Biomass willow plantations are most commonly established by vegetative propagation of dormant stem cuttings of 15–20 cm in length. The optimum planting density has been the subject of research in several countries as it is of significant importance in maximising yields. If the density is too high, stool mortality rates increase and excessive self-thinning occurs, both of which greatly reduce yield.[17–19] Planting material is financially expensive and establishment involves the greatest single expenditure on the crop. Planting densities of 15 000–18 000 stools per hectare are most commonly used at present.

Planting occurs in early spring and growth continues until the plants become dormant at the end of the initial growing season. At this time, the plants are coppiced to promote the outgrowth of a greater number of shoots in the following Spring. Competition from weeds can be problematic during the establishment year, but this becomes of less concern later as, once established, the rapid rate of growth and canopy closure effectively suppresses potentially competitive weeds beneath. At present, harvesting most commonly occurs on a three-year cycle with stand productivity expected to be maintained at a commercially viable level for up to 25 years.

13.4 Willow Genetic Improvement

Although willow breeding for traditional end uses has been ongoing for centuries, targeted genetic improvement of willow as a bioenergy crop is a relatively recent endeavour. However, in recognition of the potential of willow as a biomass crop across a wide range of geographic areas, willow is receiving a greater level of interest than in previous years and an increasing number of crop-improvement programmes are being established in response. Underpinning research on the genetic and physiological control of complex traits is also progressing in several countries.

13.4.1 Germplasm Collections

Numerous valuable willow germplasm collections exist worldwide (reviewed by Kuzovkina[20]) that form the basis of a variety of research programmes including several that address biomass and biofuel improvement. In the UK, the most comprehensive collection of willow germplasm is the National Willow Collection (NWC) – a living collection maintained at Rothamsted Research in Harpenden, Hertfordshire. This collection was established in the early 1900s at Long Ashton Research Station (LARS) which, until its closure in 2003, was located near to the Somerset Levels – a major area for commercial production of willows for the basketry industry. The collection was originally established at LARS under Government funding in recognition of the fact that many important basketry cultivars were in danger of being lost. Furthermore, overexploitation in the production of baskets for the transportation of troop supplies and artillery during World War I resulted in a severe depletion of stocks. Throughout its history, additional species and specimens have been added to the collection and at present, the collection comprises over 1300 accessions representing more than 100 different willow species and numerous hybrids. Although historically a collection of material of most interest to the basketry industry, the NWC now contains a large number of cultivars of importance to biomass breeding programmes.

As in any crop-improvement programme, an understanding of the genetic diversity and species relationships within the available breeding material is essential for its effective deployment. However, as highlighted earlier, characterisation of *Salix* germplasm is not straightforward when based on morphological characters alone. The problem of germplasm characterisation is further confounded when diverse accessions from a wide native range are maintained as a single collection. As part of such, plants are isolated from their natural environment, thus the expression of certain discriminatory characteristics and traits may be masked. Also, as willows are often maintained as coppiced plants, some discriminatory characteristics may not be evident.

DNA-based technologies offer an alternative approach to characterisation of breeding germplasm. The earliest molecular studies used sequence-based phylogeny approaches to analyse relationships between different willow species.[21–24] However, the resolution afforded by these studies was limited. More recently, molecular markers such as microsatellites and amplified fragment length polymorphisms (AFLP) have proved more informative in the analysis of genetic diversity in *Salix*.[25–27] However, the application of microsatellites to the full range of willow species is not straightforward due to high levels of interspecific nucleotide diversity that can prevent primer annealing in PCR. Polyploidy can also confound data interpretation. AFLP analysis has the advantage that it can be applied to any sample without the requirement for knowledge of underlying sequencing information or high levels of nucleotide conservation. This approach was successfully applied in the analysis of 154 accessions of the UK National Willow Collection, encompassing 50 diverse species.[28] This study represents the most detailed molecular-marker analysis of the genus to date and indicated that molecular diversity largely agreed with current taxonomic thinking, although some new insights into species relationships were obtained. Although significant progress has now been made in the characterisation of diversity within *Salix*, further work will be required to provide a thorough understanding of the relationships between the many species within this complex genus.

13.4.2 Breeding Willows for Bioenergy End Uses

The first willow-breeding programme aimed specifically at the production of varieties suitable for biomass uses was initiated in 1987 by the Swedish plant-breeding company, Svalöf Weibull AB.[29] Early progress was significant and the first new varieties aimed specifically at biomass production were released during the early 1990s. These *S. viminalis*-derived varieties, namely Orm, Rapp, Ulv, Jorr and Jorrun, were reported to produce yields 15–20% higher than those achieved with earlier nonbred clones.[30] This rapid progress was not surprising given that the breeding of willows for biomass was a relatively novel endeavour. Later gains came from interspecific hybrids, with the varieties Bjorn and Tora also benefitting from increased rust resistance derived from *S. schwerinii*. Yield increases of around 50% were reported at this time.[31] In 1996, the European Willow Breeding Partnership was formed that brought together the existing Swedish programme and a UK-based crossing programme centred at Long Ashton Research Station. Promising hybrid varieties such as Resolution and Terra Nova emanated from this partnership. More recently, the breeding programmes have reverted to national programmes once again, with those in Sweden, the UK and more recently the US, continuing to make significant biomass yield gains.

13.4.3 Practical Breeding and Selection

Willows have several properties that make them very amenable to breeding. Sexual maturity is reached more rapidly than in many other trees with catkins normally appearing in the second year of growth. Dioecy is also helpful in that plants produce only male or female flowers and thus there is no requirement for emasculation to prevent selfing. Male and female plants are simply separated in different glasshouse compartments during the crossing process.

Crosses are usually performed in January through February. Prior to this, dormant stems bearing flower buds are collected and maintained in a cold store before being grown in water in a glasshouse. This practice offers the chance to synchronise flowering between species that would naturally flower at different times. Crosses are performed by transferring pollen to the female catkin directly or by using a small brush. If the cross is successful, seed set occurs within three to six weeks, at which time the shoots are covered with bags to collect seed and prevent dispersal. Seeds are sown immediately in trays under mist, grown to a height of around 3 cm and then transferred to individual tray compartments. Once at a height of around 10 cm, plants are transferred outside to a nursery where they are grown in trays under irrigation for the remainder of the season and selections for pest and disease resistance are often made during this time. Selections for yield traits are not reliable at this point and field assessments are required. Although different selection strategies exist in different breeding programmes, in the UK, stem cuttings from dormant nursery material are made and first planted in an initial observation trial. These trials are not replicated as the number of cuttings that can be made from nursery growth is limited. Individual genotypes that show promise in observation trials are then multiplied and planted out in a fully replicated second observation trial. Selections from within these trials are subsequently tested in full yield trials at a number of different sites.

13.4.4 Targets for Crop Improvement

13.4.4.1 Increased Biomass Yield

Limits on the area of land available for planting bioenergy crops place a strong emphasis on the need to increase harvestable yield per unit area of land. To date, breeding programmes have achieved significant yield improvements through selection of stem (diameter, height) and coppicing traits (stem numbers, vigour) and improved disease and pest resistances have greatly contributed to yield gains. While breeding programmes are still relatively young, there is still scope for further yield gains by such approaches, but it is likely that more advanced and targeted trait-improvement strategies will be required in coming years. It is important, however, that further yield improvements are achieved without increasing inputs.

Strategies for enhancing yield within these constraints have been suggested.[32] Extending the length of the growing season is one such approach. It is known that in *S. viminalis*, yield is significantly and positively influenced by an early date of bud flush[33,34] and that this trait has a greater effect on yield than a later date of autumn growth cessation. Therefore, the exploitation of natural variation in the selection for early bud flush may be beneficial, but this may be associated with an increased risk of frost damage in colder climates. Thus, improvement of cold tolerance will also be required, particularly as willow is most likely to be deployed at more northerly latitudes where other energy crops, such as *Miscanthus,* spp. may be less productive. Indeed, this requirement is already being addressed in Sweden and genetic studies that identified and mapped genetic variation in freezing tolerance have been performed.[35]

A second strategy is to optimise assimilate partitioning and increase the ratio of harvestable above-ground biomass to that which is below ground. It is known that different willows vary in this trait. For example, Sennerby-Forsse and Zsuffa[13] observed a significant difference in root/shoot ratios between *S. eriocephala* and *S. viminalis*. However, even in model plants, underlying mechanisms that lead to such variation are not yet fully understood and further research will be needed in order to provide the required information for targeted improvement of this trait. Modification is likely to be additionally complicated in willow as it is a perennial and consideration must be given to the extent of assimilate partitioning between growth and storage towards the end of the growing season. It is likely that the latter will influence both winter survival and spring regrowth, thus a carefully optimised balance should be sought.

13.4.4.2 Increased Resource-Use Efficiency

To date, biomass willow-breeding programmes have focused on the production of willow varieties for cultivation on relatively productive arable land and yield selections have historically been performed in this type of environment. However, in future, under climate-change scenarios and with a rapidly growing population, there will greater competition for agricultural land and non-food crops that are productive on subprime, resource-limited sites will be required.

Increased water-use efficiency and drought tolerance will become important targets in willow improvement. While the simultaneous growth of many shoots, rapid build up of leaf area and rapid canopy closure provides an efficient strategy for biomass accumulation in willows, high transpiration rates can result. Under future climate scenarios where water availability may be limiting, this may be of particular concern. Furthermore, higher incidences of extreme weather conditions are predicted, thus, as willow stands need to remain productive for up to 25 years, there is a requirement to breed varieties that can sustain yields despite fluctuations in water availability. Several studies have demonstrated that different willow genotypes respond differently to water stress, with the relative biomass production in stressed and unstressed plants varying significantly.[36-38] Differences in water-use efficiency have also been reported[39] indicating that there is genetic variation for use in breeding for improved water-use traits.

Similarly, nitrogen-use efficiency (NUE) in willow will play a key role in future yield gains if willows are deployed on subprime land where N is limiting and constraints regarding greenhouse-gas emissions are considered. A limited number of experiments to investigate NUE in willows have been conducted, although due to difficulties controlling conditions in field experiments, the majority of these have been pot-based. In one such study, Weih and Nordh[36] showed that willow varieties differ in response to nutrient stress, demonstrating that genetic variation for this trait exists.

13.4.4.3 Increased Resistance to Biotic Stresses

Chemical control of pests and diseases in SRC willow plantations is not viable at present for both economic and practical reasons, thus breeding for resistance to biotic stresses is an important goal. Rust diseases, caused by *Melampsora* spp. pose the most significant disease threat to commercial willow plantations with yield losses of up to 40% being reported.[40] Furthermore, severe rust infection can result in plants becoming more susceptible to secondary infections that can lead to complete crop failure. Given its importance, rust resistance was a major target in early willow breeding and remains so to date.

Although the first commercial biomass varieties were much improved in terms of yield, rust susceptibility remained a problem, largely as a consequence of the pure *Salix viminalis* pedigrees of these varieties. Although this species is a favoured species in willow breeding due to its rapid rate of biomass accumulation and desirable growth habit, it is generally susceptible to rust in field conditions, although natural variation in quantitative resistance is commonly observed.[33,41] Later varieties that were based on interspecific crosses were much improved in terms of rust resistance. For example, the variety Ashton Stott (*S. viminalis* "Bowles Hybrid"× *S. burjatica* "Korso') was initially both rust resistant and high yielding. However, this variety has since become susceptible to rust and is no longer planted commercially. The breakdown of rust resistance, which can result when strong selection pressure is encountered by the pathogen, has been observed for multiple willow varieties over the last three decades. Thus, there is a real need to breed for durable resistance. To date, the most durable source of rust resistance has come from *S. schwerinii* "L79069", a progenitor of several modern cultivars such as Tora, Sven, Olof, Tordis and Resolution, which have all remained highly resistant for several years now across multiple environments. Although it remains to be seen whether this resistance will eventually be overcome, previous experience suggests that diversification of resistance sources should continue to be of major concern within breeding programmes.

Willow yields can also be significantly compromised by insect herbivory and grazing by mammals such as deer and rabbits. In the UK, leaf-feeding willow beetles (Coleoptera: Chrysomelidae) are the most significant invertebrate pest at present. Damage can be substantial and complete defoliation of willow plantations has been observed.[42] However, numerous studies have shown that there is considerable variation in the susceptibility of different willow varieties and species to feeding damage by willow beetles[43] and selection for resistance to herbivory is common practice within breeding programmes. Additional potential pests often encountered include willow aphids (*Tuberolachnus salignus* and *Pterocomma salicis*) and galling midge species such as *Dasineura marginemtorquens*.

13.4.4.4 Composition Traits

The genetic improvement of composition traits is yet to be addressed in depth within willow-breeding programmes. Although some fundamental research is now underway, there is currently a paucity of information on composition traits, both in terms of the inherent compositional properties of willow and the extent to which such traits vary between different species and cultivars. Similarly, the effects of agronomic practices and growth conditions on final biomass composition are not yet fully understood, although research to address this issue is underway.

Similarly, research on the use of willows in the production of liquid biofuels is in its infancy although preliminary comparisons of the suitability of different feedstocks for enzymatic saccharification and subsequent fermentation to ethanol, suggest that willow may be a useful feedstock for this purpose.[44] Furthermore, in the willow-mapping family used in this study, natural variation in saccharification potential of different willows was identified and shown not to be linked to biomass yield traits. This suggests that specific selection regimes for saccharification traits may be required in willow-breeding programmes.

13.5 Genetic and Genomic Research to Underpin Breeding

Current public attitudes regarding the deployment of transgenic crops, in Europe at least, suggest that in the near term, the most likely application of biotechnology in willow improvement will come from the development of molecular markers that can be used to increase selection efficiency within conventional breeding programmes. Moreover, routine genetic transformation has not yet been achieved in *Salix*, despite increased interest in this technology as a research tool to aid the understanding of gene function.

The development of molecular markers for use in marker-assisted selection (MAS) strategies has been ongoing in willow since the mid-1990s. Several genetic maps have now been constructed for use in the identification of genetic loci governing complex traits through quantitative trait locus (QTL) analysis.[9,45,46] However, the application of genetic mapping approaches is not trivial in willow given that it is dioecious, highly heterozygous and generation times are relatively long. Despite these inherent complications, significant progress has been made in this area and QTL for several important traits, including rust resistance,[41,47] biomass yield and associated stem traits,[41,46] saccharification potential,[44] water-use efficiency,[39] phenology traits[35] and freezing tolerance[35] have now been reported. However, in order for QTL-based markers to be deployed effectively within breeding programmes, a higher QTL mapping resolution than that often achieved will be required and further marker development and validation stages will be necessary.

The K8 mapping population[41] at Rothamsted Research is currently the largest mapping population available for complex trait analysis and marker development in willow. This population of 947 full siblings is established at three contrasting sites in the UK and has been characterised for numerous traits over several harvest cycles. The large number of recombination events represented by this population offers the potential for a higher degree of mapping resolution. Well-resolved QTL for yield and disease resistance have now been identified using this family and markers derived from these loci are now being used in molecular breeding. However, as a single mapping population only represents variation that is derived from the two parental accessions, QTL mapping is limited in its ability to identify a large number of useful polymorphisms. Furthermore, variation detected in one population may not be relevant to the wider germplasm. To circumvent these limitations, additional mapping populations with different parentage can be established and used to expand the catalogue of mapped QTL, although this approach can be resource intensive. An alternative approach is provided by association mapping, also termed linkage disequilibrium (LD) mapping, which is based on natural populations that represent many more recombination events and therefore offer a greater potential to map useful polymorphisms with high resolution. Unlike QTL mapping, this approach can concurrently survey many different alleles in a population, although population history must be accounted for to avoid the identification of spurious associations. Although at an early stage at the time of writing, association mapping projects in willow are now underway.

13.6 Environmental Benefits

Energy and carbon balances have been estimated for power generation from willow in several studies (reviewed by Rowe *et al.*)[48] and although there is clearly variation in the reported ratios that arises from differences in methodology and varying suppositions regarding yield, processing and management, it is clear that SRC willow cultivation offers clear benefits in terms of carbon savings. Similar studies based on production of liquid biofuels from willow are not yet available.

Hypothetical alternative fertiliser scenarios have been analysed in a Life-Cycle Analysis based on a US willow production model[49] and potential routes for improving the performance of willow as a bioenergy crop have been identified. The addition of organic matter to the soil in the form of biosolids has the potential to further improve the carbon balance and may also lead to increased carbon sequestration. Despite concerns that such practices may lead to a build up of trace heavy metals in the soil, studies to date suggest that this is not necessarily the case in willow plantations fertilised by sewage sludge biosolids.[50,51] The potential of willows to accumulate heavy metals has been well documented, thus the use of willow in a phytoremediation role may offer an additional environmental benefit.

As for any major land-use change, increasing the area of land used to produce energy crops has been the subject of speculation regarding potential impacts. When compared to annual arable crops, as a perennial with a very different crop morphology, willow might be expected to have significant effects on biodiversity. Early studies in this area indicate that there are likely to be several beneficial impacts on biodiversity that arise as a consequence of commercial SRC willow cultivation. In terms of diversity of flora, UK willow plantations have been shown to harbour a greater richness and abundance of plant species compared to arable crop plantations.[52] Similarly for birds, numbers are greater and a more diverse range of species have been reported in willow plantations compared with arable and grassland controls in the UK,[53] and with arable or set-aside land in Sweden.[54] Furthermore, there is data to suggest that willow coppice may perform better than SRC poplar in this regard.[55] Invertebrate diversity on noncultivated willows is generally very high and it is likely that this will be reflected in SRC plantations (which are not normally treated with pesticides). Based on the knowledge gathered in recent years, this appears to be the case. For example, in a UK study,[56] 120 different species of invertebrate were recorded in the canopy of a willow plantation, with the most widespread being Chrysomelid leaf beetles. Although not compared directly, the authors suggested that as arable crops only support around 45 phytophagus invertebrate species, SRC willow plantations are likely to have a positive effect on invertebrate diversity. An additional benefit may be that willow flowers may provide an early source of nectar and pollen for bees.[57] As part the UK-based RELU-Biomass project, in which the social, economic and environmental implications of increasing rural land-use under energy crops were addressed, the most extensive survey to date of biodiversity in UK willow (and *Miscanthus*) plantations was carried out. A total of 16 willow fields representing different regions of the UK where willows are currently planted were used for assessments of weeds, plant- and ground-dwelling insects in addition to flying insects. Butterflies and bees were also counted throughout the season in headlands to assess foraging behaviour. Although the entire data sets generated in this project have not yet been analysed in their entirety, the butterfly-abundance data has been used as an indicator of biodiversity.[58] In comparison to the field margins of arable crops (cereals, maize, sugar beet, spring and winter oilseed rape), 132% more butterflies were recorded in the field margins of willow, adding further evidence of a beneficial impact of willow plantations on biodiversity.

13.7 Conclusions

Willow is poised to make a significant contribution to bioenergy production as part of a portfolio of dedicated energy crops. Although by no means limiting, perhaps the most valuable role of willow crops will be in biomass production at more northerly latitudes where sufficient yields from other crops such as *Miscanthus*, may be difficult to obtain. In contrast to some of the more novel energy crops, well-established breeding programmes are already in place and, although there is still much to be done, a substantial portfolio of fundamental research regarding willow agronomy, physiology and genetics has been generated over the past two decades. This information is now being exploited in several crop-improvement programmes that are delivering a steady flow of new varieties with increased yields that will help strengthen the feasibility of the willow crop, both in terms of economics and energy and carbon balances. Furthermore, the rate of improvement is likely to increase rapidly in the near term at least, as the natural diversity of the genus is exploited more fully and the latest molecular marker and genomics technologies are more extensively deployed in willow-improvement programmes.

Acknowledgements

The author would like to acknowledge the valuable advice and expertise provided by the current energy-crops research team at Rothamsted Research in addition to the expert willow scientists previously located at Long Ashton Research Station. The author would also like to acknowledge funding awarded to Rothamsted Research for willow research and development from the UK Biotechnology and Biological Sciences Research Council (BBSRC), the Engineering and Physical Sciences Research Council (EPSRC), the Department of Environment, Food and Rural Affairs (Defra), the Economic and Social Research Council (ESRC), and the Natural Environment Research Council (NERC).

References

1. C. Newsholme, *Willows: the genus Salix*. B.T. Batsford Ltd., London, 1992.
2. G. W. Argus, *Botanical Electronic News (BEN)*, 1999, **227**, http://www.ou.edu/cas/botanymicro/ben227.html.
3. C. Linnaeus, *Species plantarum*. Impensis Laurentii Salvii, Holmiæ, Stockholm, 1753.
4. G.W. Argus, *System. Bot. Monographs*, 1997, **52**.
5. Z. Fang, S. Zhao, A. K. Skvortsov and Salicaceae, In: *Flora of China*, Z. Wu, P. H. Raven, (ed.), Science Press-Missouri Botanical Garden Press, Beijing-St. Louis, 1999.
6. A.K. Skvortsov, *Willows of Russia and adjacent countries: taxonomical and geographical revision*. Report series vol. 39. Faculty of Mathematics and Natural Sciences, University of Joensuu, Finland, 1999.
7. J. Thibault, *Canad. J. Bot.*, 1998, **76**, 157–165.
8. G. A. Tuskan, S. DiFazio, S. Jansson, J. Bohlmann, I. Grigouriev, U. Hellsten, N. Putnam, S. Ralph, S. Rombauts and A. Salamov A, *et al.*, *Science*, 2006, **313**, 1596–1604.
9. S. Hanley, M. D. Mallott and A. Karp, *Tree Genetics Genomes*, 2006, **3**, 35–48.
10. L. B. Smart, T. A. Volk, J. Lin, R. F. Kopp, I. S. Phillips, K. D. Cameron, E. H. White and L. P. Abrahamson, *Unasylva*, 2005, **221**, 51–55.
11. T. J. Tschaplinski, *Blake Physiol. Plantarum*, 1989, **75**, 157–165.
12. L. Sennerby-Forsse, *Biomass Bioenergy*, 1995, **9**, 35–43.
13. L. Sennerby-Forsse and L. Zsuffa, *Trees – Struct. Funct.*, 1995, **9**, 224–234.
14. Stott K. In: *Ecology and Management of Forest Biomass Production Systems*, K. L. Perttu, (ed.), Uppsala, Sweden: Swedish University of Agricultural Sciences, 233–260.

15. T. A. Volk, T. Verwijst, P. J. Tharakan, L. P. Abrahamson and E. H. White, *Front. Ecology Environ.*, 2004, **2**, 411–418.
16. S. Miller, *Environ. Sci. Technol.*, 2010, **44**, 3932–3939.
17. T. Verwijst, *Biomass Bioenergy*, 1996a, **11**, 161–165.
18. T. Verwijst, *Biomass Bioenergy*, 1996b, **10**, 245–250.
19. R. F. Kopp, L. P. Abrahamson, E. H. White, K. F. Burns and C. A. Nowak, *Biomass Bioenergy*, 1997, **12**, 313–319.
20. Y. A. Kuzovkina, M. Weih and M. A. Romero *et al.*, In: *Horticultural Reviews*, J. Janick, (ed.), **Vol. 34**, 2008, Wiley, New York, pp. 447–489.
21. S. J. Brunsfeld, D. E. Soltis and P. S. Soltis, *Am. J. Bot.*, 1991, **78**, 855–869.
22. S. J. Brunsfeld, D. E. Soltis and P. S. Soltis, *System. Bot.*, 1992, **17**, 239–256.
23. E. Leskinen and Alström-Rapaport, *Plant System. Evol.*, 1999, **215**, 209–227.
24. T. Azuma, T. Kajita and J. Yokoyama, *Am. J. Bot.*, 2000, **87**, 67–75.
25. H. Beismann, J. H. A. Barker, A. Karp and T. Speck, *Molec. Ecol.*, 1997, **6**, 989–993.
26. J. H. A. Barker, M. Matthes, G. M. Arnold, K. J. Edwards, I. Åhman, S. Larsson and A. Karp, *Genome*, 1999, **42**, 173–183.
27. J. H. A. Barker, A. Pahlich, S. Trybush, K. J. Edwards and A. Karp, *Molec. Ecol. Notes*, 2003, **3**, 4–6.
28. S. Trybush, Š. Jahodová, W. Macalpine and A. Karp, *Bioenergy Res.*, 2008, **1**, 67–79.
29. I. Åhman and S. Larsson, *Norweg. J. Agr. Sci., Supplement*, 1994, **18**, 47–56.
30. S. Larsson S, In: *Energy from Crops*, Semundo Limited, Cambridge, UK, 1996.
31. S. Larsson, *Aspects Appl. Biol.*, 2001, **65**, 193–198.
32. A. Karp and I. Shield, *New Phytol.*, 2008, **179**, 15–32.
33. A. C. Rönnberg-Wästljung and U. Gullberg, *Theoret. Appl. Genet.*, 1999, **98**, 531–540.
34. M. Weih, *Tree Physiology*, 2009, **28**, 1479–1490.
35. V. Tsarouhas, U. Gullberg and U. Lagercrantz, *Theoret. Appl. Genet.*, 2004, **108**, 1335–1342.
36. M. Weih and N. E. Nordh, *Biomass Bioenergy*, 2002, **23**, 397–413.
37. L.I.G. Bonneau, *Drought resistance of willow short-rotation coppice genotypes*. PhD thesis. Cranfield Univeristy, 2004.
38. M.-L. Linderson, Z. Iritz and A. Lindroth, *Biomass Bioenergy*, 2007, **31**, 460–468.
39. A. C. Rönnberg-Wästljung, C. Glynn and M. Weih, *Theoret. Appl. Genet.*, 2005, **110**, 537–549.
40. S. R. Parker, D. J. Royle and T. Hunter, In: *Abstracts of the 6th International Congress of plant pathology*, Montreal, Canada, National Research Council Canada, Ottawa, ON, Canada, 1993, p.117.
41. S.J. Hanley, *Genetic mapping of important agronomic traits in biomass willow*. PhD thesis. University of Bristol. 2003.

42. D. A. Kendall, T. Hunter, G. M. Arnold, J. Liggitt, T. Morris and C. W. Wiltshire, *Ann. Appl. Biol.*, 1996, **129**, 379–390.
43. C. W. Wiltshire, D. A. Kendall, T. Hunter and G. M. Arnold, *Aspects Appl. Biol.*, 1997, **49**, 113–120.
44. N. J. B. Brereton, F. Pitre, S. J. Hanley, M. J. Ray, A. Karp and R. J. Murphy, *BioEnergy Res.*, 2010. DOI 10.1007/s12155-010-9077-3.
45. S. J. Hanley, J. H. A. Barker, J. W. Van Ooijen, C. Aldam, S. Harris, I. Åhman, S. Larsson and A. Karp, *Theoret. Appl. Genet.*, 2002, **105**, 1087–1096.
46. V. Tsarouhas, U. Gullberg and U. Lagercrantz, *Theoret. Appl. Genet.*, 2002, **105**, 277–288.
47. V. Tsarouhas, U. Gullberg and U. Lagercrantz, *Hereditas*, 2003, **138**, 172–178.
48. R. Rowe, N. Street and G. Taylor, *Renew. Sustain. Energy Rev.*, 2009, **13**, 271–290.
49. G. A. Keoleian and T. A. Volk, *Crit. Rev. Plant Sci.*, 2005, **24**, 385–406.
50. P. Borjesson, *Biomass Bioenergy*, 1999, **16**, 137–154.
51. M. Labrecque and T. I. Teodorescu, *Biomass Bioenergy*, 2003, **25**, 135–146.
52. M. D. Cunningham, J. D. Bishop, H. V. McKay, R. B. Sage. URN 04/961. DTI 2004.
53. R. B. Sage, M. D. Cunningham and N. Boatman, *Ibis*, 2006, **148**, 184–97.
54. A. Berg, *Agr. Ecosyst. Environ.*, 2002, **90**, 265–276.
55. R. B. Sage, P. A. Robertson, J. G. Poulson. ETSU B/W5/00277/REP, DTI, 1994.
56. R. B. Sage, K. Tucker, ETSU B/W2/00400/REP, DTI, 1998.
57. J. Redderson, *Biomass Bioenergy*, 2001, **20**, 171–179.
58. J. Haughton, A. J. Bond, A. A. Lovett, T. Dockerty, G. Sünnenberg, S. J. Clark, D. A. Bohan, R. B. Sage, M. D. Mallott, V. E. Mallott, M. D. Cunningham, A. B. Riche, I. F. Shield, J. W. Finch, M. M. Turner and A. Karp, *J. Appl. Ecol.*, 2009, **46**, 315–322.

CHAPTER 14
Poplar

S. Y. DILLEN,[a] O. EL KASMIOUI,[a] N. MARRON,[b]
C. CALFAPIETRA[c] AND R. CEULEMANS[a]

[a] University of Antwerp, Department of Biology, Research Group of Plant and Vegetation Ecology, Campus Drie Eiken, Universiteitsplein 1, BE-2610 Wilrijk, Belgium; [b] Unité Mixte de Recherche 1137, INRA-Nancy Université, Écologie et Écophysiologie Forestières, 54280 Champenoux, France; [c] Institute of Agro-Environmental and Forest Biology (IBAF), National Research Council (CNR), Via Salaria, 00016 Monterotondo Scalo, Italy

14.1 The Poplar Genus

14.1.1 Uses and Interests of the Poplar Genus

Poplars (*Populus* spp.) are versatile trees that naturally occur in a broad range of ecological habitats. Although they typically occupy river flood plains and bottomlands, some species can tolerate desert or arid upland conditions.[1] In temperate regions, they are among the highest yielding trees that are extensively cultivated for forestry purposes as well as in agricultural systems. Hence, poplar trees are used for various applications, including wood products (such as paper, veneer and lumber) and services (such as shelter, shade and protection of soil, water and livestock).[2] They can be deployed in phytoremediation of slightly polluted or degraded sites, combating desertification and reforestation, and increasingly, poplar biomass is used as a resource of bioenergy, *i.e.* renewable energy from biological sources.

In addition to the numerous applications, poplar is well recognised as a model tree in physiological and in molecular genetic research.[3] The poplar genus has

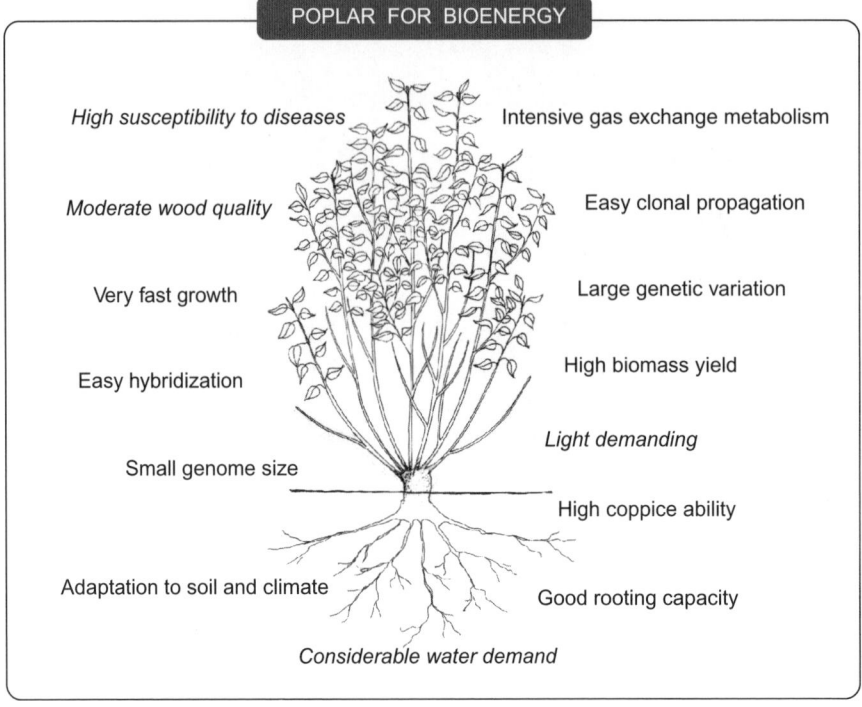

Figure 14.1 Summary overview of the principal characteristics of the poplar genus. Poplar has a considerable number of advantages for application as a bioenergy crop. Less favourable or disadvantageous characteristics have been put in italics.

indeed numerous assets justifying the interest of the scientific community: (i) most species are very easy to propagate and to grow under controlled or natural conditions; (ii) the relatively small poplar genome made it the first tree genome recently sequenced;[4] (iii) it is the first tree genus being transformed with *Agrobacterium*;[5,6] and (iv) it contains a considerable source of genetic diversity (Figures 14.1 and 14.2).[7] Nonetheless, the poplar genus also has some important weaknesses including large water requirements associated with its rapid growth, and high susceptibility to disease and insect infestations (Figure 14.1). However, there is a huge amount of variation available within the poplar genus in terms of productivity, resistance to diseases and abiotic hazards, leaf and tree morphology, *etc.* (Figure 14.2). This genetic diversity enables poplars to grow under a wide range of environmental gradients all over the world, and turns them into excellent candidates for genetic studies as well as for breeding and selection programmes.[8]

14.1.2 Subdivision of the Poplar Genus

The genus *Populus* is part of the Salicaceae family, which also includes the willows (*Salix*) (Figure 14.3). Trees within this family are dioecious, meaning

Figure 14.2 Illustration of diversity in leaf shape within the *Populus* genus for three genotypes.

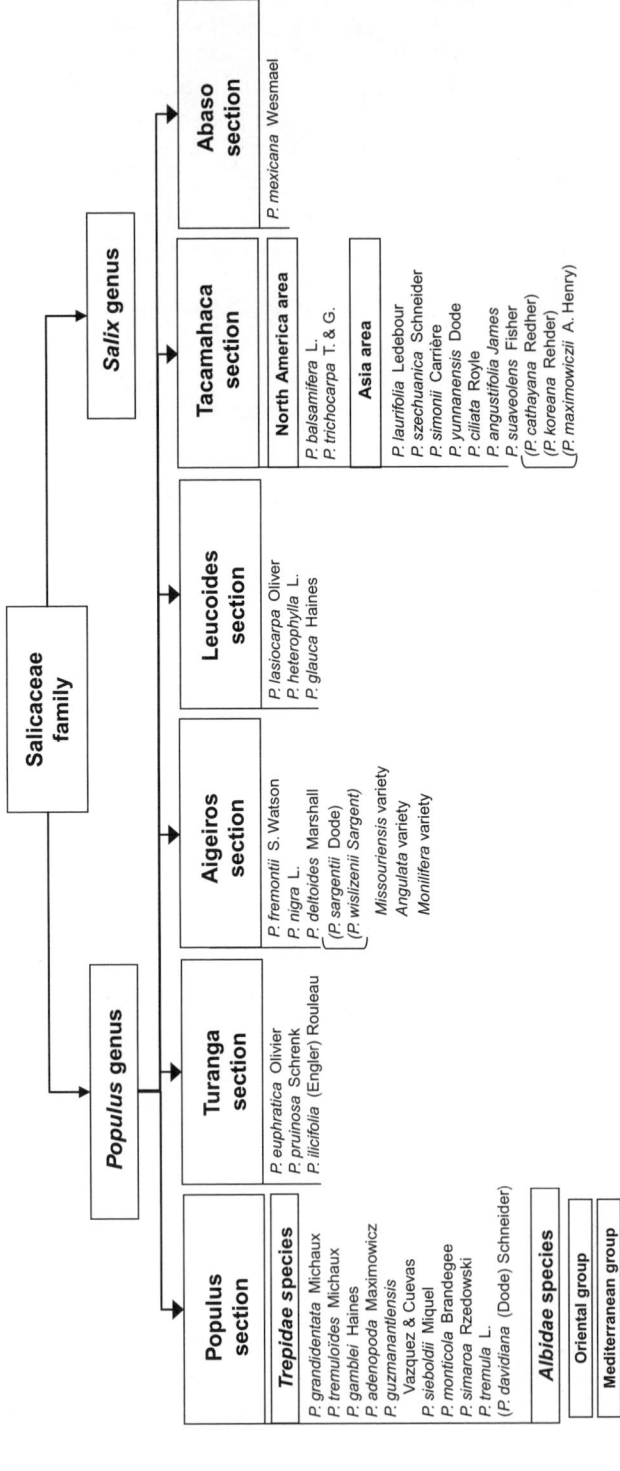

Figure 14.3 Subdivisions of the *Populus* genus; section and species within the genus have been identified. The most common species and hybrids used in poplar cultivation, including short-rotation coppice (SRC) systems involve *Populus deltoides*, *P. trichocarpa* and *P. nigra*.[10]

that they have either male or female flowers.[9] The poplar genus is complex and divided into six sections or systematic subdivisions, three of which being most employed in cultivation.[10] These three sections are (i) Aigeiros, including black poplar (*P. nigra* L.) and eastern cottonwood (*P. deltoides* Bartr.); (ii) Tacamahaca, the balsam poplars; and (iii) finally Populus (ex Leuce), which includes the aspens and white poplar (*P. alba* L.).[11] Reportedly, the majority of worldwide poplar cultivation involves species and hybrids in the section Aigeiros.[10,12] Most of these hybrids are derived from the species *P. deltoides* crossbred with the following species: *P. nigra*, *P. trichocarpa* T. & G., *P. maximowiczii* Henry, *P. ciliata* Royle, *P. szechuanica* Schneider, *P. yunnanensis* Dode, and *P. balsamifera* L. (Figure 14.4), thanks to the ease of intrasectional hybridisation and intersectional crosses with species in the Tacamahaca section. The real advantages of these hybrids over the pure *P. deltoides* parental species are the substantially improved rooting and survival from cuttings, and the higher productivity, also known as heterosis or hybrid vigour, which can be defined as the superiority of the hybrids over their parents.[13,14] The section Turanga has been gaining importance, in particular since large-scale reforestation and planting programmes have been initiated in China in 1978. The Three-North Shelterbelt Programme is a 35.6 million hectare shelterbelt project across the desert border in

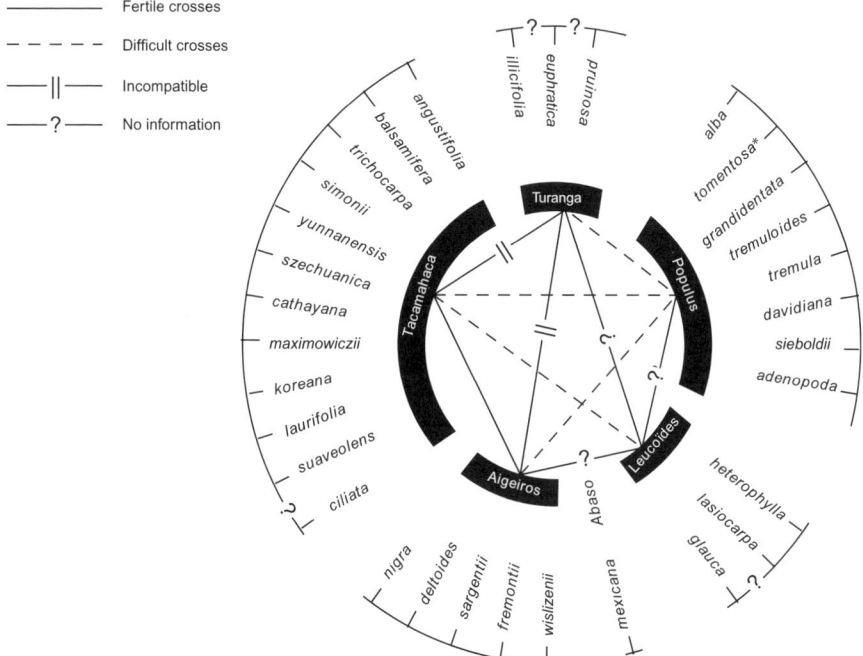

Figure 14.4 Crossability of *Populus* species.[126,127] * The *P. tomentosa* species presented here does not appear in Figure 14.3 because it has been considered as a hybrid.[11]

northern China, where poplars make up 60% of the effort, much of it *P. euphratica* and *P. simonii* Carrière × *P. nigra* hybrids.[15]

Note: In this chapter, the word "poplar" refers to the whole *Populus* genus, encompassing poplars *s. str.*, aspens, cottonwoods, and hybrids.

Worldwide, about 80 million hectares of naturally occurring poplars have been reported to the International Poplar Commission (IPC) of the United Nations.[10] Additionally, the IPC recorded globally 5.3 million hectares of poplar plantations. Countries with significant areas of poplar plantations include China (4.3 million ha), France (236 000 ha), Turkey (125 000 ha), Italy (118 500 ha) and Germany (100 000 ha). However, the share of the plantations specifically dedicated to energy purposes is very small: countries reported that only 0.9% of their total wood production (from poplar and willow) was used as fuel wood or as a resource of bioenergy.[10] Existing poplar short-rotation coppice (SRC) systems with the purpose to produce bioenergy are largely experimental. In Italy, the profitability and effects of environment, management regime and clones on biomass production are currently being evaluated for 4000 ha of poplar SRC scattered in the north and the middle of the country.[16] More experimental plantations were reported throughout the world, *e.g.* in Serbia, Romania and Argentina.[10] Although poplar has been cultivated for many years, the information on planted areas of SRC plantations remains restricted and is less focused on energy use than willow.[17]

14.2 Poplar SRC Systems

14.2.1 Synoptic History of Poplar SRC

SRC systems can be defined as high-density plantations of fast-growing perennial species with good coppice and resprout capacity. The above-ground biomass collected at the end of each rotation cycle, generally after two to five years, can be used as a resource of renewable energy, paper pulp or fibres. Management of these SRC plantations is on the interface between agricultural and forestry practices.

In temperate climates, SRC with poplar and willow is being adopted. For many decades, the vigourous growth performance and high biomass yields of the Salicaceae family have been exploited. In the 1930s, after the first artificial poplar hybrids were created, numerous poplar-breeding programmes have been established in Europe, Canada and USA.[18] The concept of SRC was initially launched in the 1960s in the USA, as the "silage" sycamore project.[19,20] SRC was stimulated in 1973 during the OPEC oil embargo, in the search for alternatives to fossil fuels, and revived in the 1980s as a result of the set-aside policy of the European Union in times of agricultural overproduction. More recently, concerns about climate change and energy security together with spectacular progress in tree genomics and biotechnology have generated renewed interest in SRC with poplar. These scientific and biotechnological advances awaken hopes that in the future more productive and more suitable poplar genotypes can be deployed in SRC. Thus far, poplar cultivation and improvement programmes have been primarily concentrated on the production of veneer and plywood.

14.2.2 SRC Principles

To achieve maximal potential yields, poplar SRC is preferably established on loam- or clay-containing soils with adequate water availability. Before planting, site preparation includes ploughing to assure good rooting and harrowing to even out the field. The field should be completely weed free prior to planting either through chemical or mechanical weeding as poplars do not tolerate shade. Given the ability of most poplar species to reproduce easily by means of asexual or vegetative propagation, SRC plantations are usually established from unrooted hardwood cuttings (Figure 14.5). These cuttings are harvested from one-year old or older stems during the dormant season and their size generally ranges from 20 to 30 cm.

A distinctive feature of SRC plantations is their high planting density in order to reach the highest possible yields per unit area. In the Swedish SRC scheme, originally set up for willow, cuttings are planted in a double-row design at an overall planting density of 10 000–15 000 cuttings per hectare.[21] A second model often used in Italy is characterised by a single-row design with 3 m between rows and 0.5–0.7 m between cuttings within a row, facilitating weed control and accommodating 6000–7000 cuttings per hectare.[16] Initial planting density determines the length of the rotation cycle and *vice versa* (Figure 14.5).

Careful inspection of the establishment of the cuttings during the first months of cultivation is needed for satisfactory yield levels. Overall, the survival rate of hardwood cuttings at the end of establishment year is rather high, about 90% for commercial poplar clones.[22–24] Poplars, being light-demanding pioneer species, require mechanical or chemical removal of weeds during the first growing season, or each year after coppice. During the following years, poplar trees attain sufficient height to prevent weed growth. Thus, establishment appears to be the most critical phase throughout the complete life span of SRC. The trees are often coppiced at the end of the first growing season to create an easily harvestable multistem coppice (Figure 14.5) and to benefit from the already existing root systems that are known to enhance growth in the subsequent rotation cycles.[25,26]

Compared to willow, poplar has a stronger apical dominance and its stools tend to produce fewer but larger shoots when coppiced.[27] This difference in coppice and resprout ability between the two genera is probably due to a different breeding strategy in the past. In Europe, poplar-breeding programmes have been focused on single-stem growth and straight stem form, but yield of multistem coppice poplar including coppice ability has been overlooked and requires more attention.[28] As a result, several commercial poplar cultivars do not support very short rotation cycles or frequent harvesting. Nevertheless, coppicing is thought to reinvigourate growth, at least in the early rotations of SRC, and obviously avoids replanting costs. In contrast to willow, poplar is

Figure 14.5 Schematic representation of the concept of short-rotation coppice (SRC) culture for the production of bioenergy using poplar. Rotation cycles range typically from 2 to 5 years. (Photo credit for picture on harvest: I. Laureysens; all other pictures: R. Ceulemans.)

availed with wider initial planting densities and longer rotations.[24,29] In a literature review, it was suggested that the optimum rotation system for poplar is about four years with densities between 2500 and 10 000 trees per hectare.[24] In denser plantations a quick succession of harvests reduces stool survival and number of resprouts per stump. However, more research on coppice regrowth traits of different clones is needed for a better insight in and management of poplar SRC.

If possible, harvests take place in winter to take advantage of the frozen soils to enter the field sites with heavy tractors and harvesters, avoiding soil compaction, and to coppice after leaf fall so that nutrients left over in senescing leaves are recycled (Figure 14.5). Out of necessity, harvests are postponed to spring in case soils are too wet in wintertime. Two different harvesting regimes can be applied: (i) the crop can be cut and chipped simultaneously *in situ*; or (ii) the crop is cut and the stems are chipped as a later operation.[30] Poplar biomass can be used as renewable energy resource either for the production of heat and electricity through (co-)combustion and gasification, or for liquid transport fuels through chemical fermentation or thermochemical conversion (Figure 14.5).[31]

The very first projections about potential above-ground dry biomass yields from poplar SRC (exceeding $20\,\text{Mg}\,\text{ha}^{-1}\,\text{yr}^{-1}$)[32,33] were too optimistic. These high yields were obtained in relatively small, experimental and carefully tended plantations and can only be achieved at a larger scale under optimal conditions in terms of climate, water and nutrient availability, *i.e.* on the best agricultural lands. More realistic and still satisfying yields range from 10 to $15\,\text{Mg}\,\text{ha}^{-1}\,\text{yr}^{-1}$.[26,34,35] The main reasons for yield failure or lower productivity are poor site quality, insufficient weed control and damage due to diseases, insects or wildlife.

14.2.3 Case Study

In 1996 an experimental SRC with 17 poplar clones was established in Boom (Antwerp, Belgium) on a former waste-disposal site, moderately polluted by heavy metals. The field site (0.5 ha) was planted with 5000 hardwood cuttings according to a double-row design. Due to the heavy loam soil, *P. deltoides* and its hybrids experienced difficult root establishment.[36] No fertilisation or irrigation was applied after the establishment of the experiment. From 1996 to date, biomass production for a large range of poplar clones has been monitored in relation to other ecophysiological traits during several rotations. This experiment is unique as only a few poplar SRC plantations have been carefully studied for more than a decade.

Coppicing stimulated productivity and biomass production, although this stimulation decreased each rotation.[26] Annual above-ground dry biomass production ranged from 1.5 to $10.8\,\text{Mg}\,\text{ha}^{-1}$.[26,34,37] During the first rotation, clones Hazendans and Hoogvorst (both *P. trichocarpa* × *P. deltoides*) were the most productive clones, confirming the enormous juvenile growth capacity of interamerican hybrids.[38] However, during the subsequent rotations, clones of the pure species *P. nigra* and *P. trichocarpa* displayed the highest yields. This study shows the need for long-term SRC trials to assess the effect of ageing, multiple coppicing and presence of pathogens and weeds on productivity in order to identify the most suitable genotypes depending on the management regime and the lifetime of the plantation.[26] Yields during the establishment year(s) are often not representative for future biomass yields as rooting problems or absence of pathogens (such as leaf rust *Melampsora larici-populina*), may give a wrong impression of which are the most suitable clones for a given SRC plantation.

Some clones in the long-term experimental plantation showed considerable potential for phytoextraction of heavy metals.[39] Wolterson (*P. nigra*) is a promising clone for phytoextraction of Al and Zn, given its high biomass production and high concentrations of both metals in the wood. Cd uptake from the soil was highest for clone Balsam Spire (*P. trichocarpa* × *P. balsamifera*). To maximize the phytoextraction potential of SRC plantations, harvests have to be advanced to late summer or early fall as metals are primarily concentrated in senescing leaves.[39] Nevertheless, poplar SRC holds the potential for the decontamination of sites moderately polluted by heavy metals.

14.3 Breeding

14.3.1 Quantitative Traits: Yield and Yield Components

The breeding of trees (*i.e.* application of genetic principles to the improvement and management of trees), is with the exception of fruit trees, a relatively recent discipline (see Section 14.2.1). The overarching objective of energy-crop-improvement programmes is the maximisation of yield on the minimum amount of land, with inclusion of environmental, social and economic aspects. Yield is, however, a complex result of numerous structural traits and physiological processes each regulated by different genes, and can be bred for either directly or indirectly through its component traits. Indirect selection for yield has proven its value in food crops, using the "ideotype" (*i.e.* a hypothetical ideal plant model) approach. Yet, the improvement of food crops usually resulted from reallocation of carbon to reproductive structures or below-ground plant parts, whereas in energy crops, yield includes the above-ground vegetative plant parts.[40,41] In spite of this apparent contradiction, identification of key yield components is essential for both food and energy crops.[42]

Many studies were dedicated to unravel the relationships between poplar yield and its component traits. Following these studies, the "SRC poplar ideotype" may consist of (i) rapid stem growth;[43] (ii) large individual leaves;[44,45] (iii) rapid canopy development;[7] (iv) early bud flush and late bud set;[46] (v) a high number of branches, particularly sylleptic branches;[47] (vi) high gas exchange rates;[48] (vii) good coppice and resprout ability;[22,49] (viii) high wood quality;[50] and (ix) low susceptibility to disease and insect pests.[51,52]

This nonexhaustive list of desirable traits for SRC poplar may vary according to planting density, rotation system, environment and endproduct, and may contain some trade-offs. For example, a high level of branchiness usually reduces the overall wood quality (see Section 14.3.2). For energy purposes, straightness and homogeneity of stems are not aimed for, and breeding programmes should therefore focus on traits more specific to short-rotation systems (2–3 years) such as growth vigour and number of resprouts after coppicing. The large genetic variation found in yield and yield component traits can be exploited in traditional selection programmes producing more suitable poplar genotypes for SRC. Beside the amount of genetic variation, the success of breeding programmes also depends on the selection method and intensity, and the nature of relationships between yield on the one hand and individual component traits on the other.[53]

14.3.2 Qualitative Traits: Wood Quality

The three main components of wood are cellulose, hemicellulose and lignin (for poplar approximately 45, 25 and 20%, respectively).[54] Poplar wood is considered a low-quality fuel wood, though largely compensated for by its rapid growth. Compared to many other tree species it is characterised by a relatively low specific gravity (0.3–0.4) [1,55] and high water content (55–60% just after coppice).[56] Major energy-related traits of biomass feedstock are: bark and moisture percentages, heating value (calorific value), cellulose/lignin ratio, specific gravity (density), amount of extractives, concentration of inorganic elements including alkali metals and ash/residue content.[57]

Compared to 3-year-old willows, poplar hybrids of the same age appeared to have significantly lower bark percentage, lower specific gravity, lower Ca concentration and higher K concentration.[57] Percentage of moisture, ash content and N, P, Mg and Na concentrations were comparable between both genera,[57] and also the calorific values of poplar and willow wood appear to fall within the same range (14–24 MJ kg^{-1} dry biomass).[58] Nevertheless, a large variability of wood traits has been observed within the poplar genus, and higher wood quality could be achieved by focusing on energy-related traits in order to create a more energy-efficient poplar biomass feedstock.[57] Generally, the presence of bark is considered adverse due to its higher moisture content and high accumulation of undesired elements (see Section 14.2.3).[39] Therefore, the biomass feedstock from poplar may be considerably improved when tree diameters exceed 4 cm and when the percentage of branches is reduced.[59] As the tree ages and tree size increases, the percentage of bark decreases together with wood and bark density. A considerable advantage of woody over herbaceous energy crops is the significantly lower amounts of pollutant byproducts of energy conversion.[59]

An important element in wood is lignin (representing 15–30% of the wood).[60] Lignin provides structural support to cell walls, enables solute conductance in the vascular system, and protects plants against pathogen attacks.

The presence of lignin is somehow conflicting and may be positive or negative depending on the type of energy conversion, *i.e.* combustion and gasification versus processing of biomass to biofuels.[61] Lignin has a higher calorific value as compared to cellulose (24 and 12 MJ kg^{-1}, respectively).[62] Hence, when poplars are cultivated for energy production through combustion or gasification, genotypes with higher lignin content are desirable. On the other hand, the presence of lignin hinders the conversion of biomass to biofuel.[63] Since the recent boost in biofuel production, lignin engineering has become a major research topic in poplar genetics and genomics, contributing to a better insight in the lignin biosynthetic pathway. In preliminary studies, transgenic poplars with modified lignin biosynthesis proved to be beneficial for conversion to biofuels, *i.e.* improved saccharification potential.[63,64] Large genetic variation has been found in the lignin content of poplar wood; so beside genetic engineering lower or higher cellulose/lignin content may also be achieved through traditional clonal selection.[65]

Tension wood, which is often present in poplar in response to stem movements caused by wind or load, is characterised by an additional secondary wall layer of cellulose microfibrils as compared to normal wood. Accordingly, it is enriched in cellulose, and deficient in lignin and hemicellulose.[66,67] As a result, wood with a high proportion of tension wood might represent a potential feedstock of lignin-poor biomass.

14.3.3 Molecular Genetics and Molecular Biological Tools

Pre-eminently, *Populus* spp. is the "guinea pig" of woody plants because of the exceptionally strong scientific basis and knowledge of the material.[3,8] Molecular genetic research on poplar has been accelerated since the sequencing of its genome in 2004.[4] To date, functions of genes or genomic regions in yield,[68] disease resistance[52] (see Section 14.3.4), phenology,[69,70] drought tolerance[71] (see Section 14.4.2) and lignin biosynthesis[63] (see Section 14.3.2) have been located on the genetic maps of several poplar species. These biotechnological advances may be of use in creating transgenic trees and in providing new breeding strategies in SRC by reducing the breeding time.[60,72] Transgenic trees with increased yield and improved quality traits have demonstrated great potential and are currently being evaluated in field trials throughout the world.[60]

14.4 Current and Future Limitations to Large-Scale Poplar SRC

Poplar is a perfect candidate for bioenergy. However, the development of large-scale poplar plantations for this purpose in the European Union and in the member countries of the International Energy Agency (IEA) is rather slow because of a variety of limitations, *e.g.* (i) the difficult societal acceptance of such plantations due to their presumed impact on the environment; (ii) the current lack of profitability of SRC plantations in many countries, mostly because of the absence of markets and value chains; and (iii) the sensitivity of poplar to diseases and pathogens that may drastically reduce biomass production during the growing season.[73] Despite these drawbacks, further improvement of the cultivation is possible through improvement of the plant material and through concrete incentives of governments (such as carbon credits, CO_2 taxes on fossil fuels and renewable-energy incentives).

14.4.1 Environmental Impacts

The large-scale development of poplar plantations for bioenergy production would require landscape-scale changes; the social and environmental impacts of such changes are not yet adequately understood or accepted. However, these impacts depend mostly on what is replaced by the bioenergy plantation. Indeed, positive effects on soil properties, biodiversity, energy balance, greenhouse-gas (GHG) mitigation, carbon footprint and visual impact are likely, when arable crops are replaced by SRC poplars. Compared to replacement of set-aside and permanent extensive grassland, benefits are less apparent.[74] Moreover, biodiversity in SRC appears to be strongly correlated with the density of the canopy (determining the abundance of the herbaceous layer), and with the rate of litter breakdown. Compared to exotic species (*Eucalyptus* and *Nothofagus*), a poplar canopy is less dense and litter turnover is fast, favouring a rich arthropod diversity.[75] In general, less is known about the consequences of large-scale deployment of the C4 grass *Miscanthus*, in comparison with SRC willow and poplar, including the effects on biodiversity and hydrology.[74]

14.4.2 Water and Nutrient Requirements

Poplar is an interesting species for bioenergy purposes because of its unequalled growth under temperate latitudes, when the rooting system is well established. However, poplar is rather demanding in terms of soil quality, notably water retention and drainage, as well as soil fertility. Productivity may be strongly limited by water availability.[76,77] For instance, in a nursery plantation with 17 000 single-stem plants per hectare, a reduction of one third of the fresh biomass and of 15% of stem height and circumference have been observed in nonirrigated versus irrigated plots, for commercial hybrids as well as for hybrids from controlled crosses.[78,79] After coppicing, a second growing season without irrigation caused a 45% reduction of fresh biomass accumulation and a 25% reduction of stem growth in height and circumference.

A partial shift from fossil-fuel derived energy toward large-scale application of highly water consuming energy-crop plantations may have strong implications for local fresh-water availability and quality.[80] In this context, more water-use-efficient genotypes are required. At the whole plant level water-use efficiency (WUE) is defined as the ratio between biomass production and water consumption. The identification of poplars combining satisfactorily high productivity and high water-use efficiency would represent a considerable advance in more sustainable poplar cultivation.

Removal of biomass after each rotation generally causes a rapid depletion of the soil fertility.[81] Harvests usually occur after leaf loss, but nutrient retranslocation of senescing leaves to the perennial organs (roots and stumps) is small in comparison with retranslocation to the bark. The bark is particularly rich in mineral elements and its proportion in the exported biomass is high for short rotations (*e.g.* 2–3 years). Still, compared to herbaceous energy species (with the exception of *Miscanthus*), poplar is characterised both by a high nutrient-use efficiency (NUE, defined as biomass produced per unit of nutrient absorbed, notably N, P and K) and by a large variation for NUE.[82–84] These two properties allow the selection of poplar genotypes maintaining high biomass production with a limited amount of fertiliser inputs.

14.4.3 Pest and Disease Sensitivity

Bioenergy plantations are dense, and the microclimate under the canopy frequently induces the development of pests and diseases. In North America and Europe, poplars are hosts to many pathogens and insect pests, including leaf rusts, stem cankers, galls, boring, sucking and defoliator insects, several of which severely impair plantation productivity.[85] A few create major limitations to economic production.[86] Leaf rust diseases are among the most devastating on poplars, notably *Melampsora larici-populina* Kleb., which is considered one of the ten most important sanitary problems threatening forests.[52] This disease can cause premature defoliation that may reduce biomass accumulation and growth vigour in severely affected plantings (Figure 14.6). Defoliation may provoke infections by secondary organisms and may also subject shoots to increased winter injury. Durable resistance to leaf rust is one of the top priorities of poplar-breeding programmes in Europe.[10,52] Once more, the wide diversity present within the poplar genus and the ease to produce new clones and hybrids is the key quality of this species, enabling the use and generation of resistant genotypes, able to compete in the poplar/rust "arms race".

14.4.4 Effects of Global Change on Poplar SRC

Fast-growing species such as poplars are usually more responsive to environmental changes than other forest species. When we mention climatic changes in experiments involving trees we often refer to one or a few factors. However, the phenomenon involves complex interactions among various factors such as atmospheric CO_2 and tropospheric O_3 concentrations, air temperature, UV-B radiation, environmental pollutants, drought events, *etc*. The impact of climate change on photosynthesis and growth of poplar SRC has been intensively studied. Although most of these experiments were carried out with poplars under elevated CO_2 over the last decades,[87,88] some studies have also examined the impacts of O_3 and warming on poplars.[89–91] Elevated CO_2 was reported to

Figure 14.6 Impact of leaf rust on a poplar plantation established in May 2003 under the framework of the European project POPYOMICs (QLK5-CT-2002-00953) near Orléans (central France) at the end of the second growing season. Two blocks were treated with fungicides while one block was not treated against leaf rust. (Photo credit: C. Bastien (aerial view), N. Marron.)

stimulate photosynthesis in many field experiments with trees by about 50%,[92,93] but fast-growing trees like poplars are usually stimulated even more by elevated CO_2 because of their indeterminate growth and the theoretical unlimited sink capacity.[94] Interestingly, in poplar, the stimulation of elevated CO_2 on photosynthesis and growth has been much higher in field experiments than in closed chambers, probably because of the limitations in sink strength in closed-chamber experiments.[87]

The stimulation by elevated CO_2 on productivity has been observed in the three largest experimental facilities involving poplar species in SRC, *i.e.* the POP-EUROFACE (Italy), the AspenFACE (USA) and the Biosphere 2 (USA) experiments with stimulations of both above- and below-ground biomass ranging from 20 to 40%.[95–97] The AspenFACE experiment evidenced a loss of productivity due to increased O_3 levels, except for the most tolerant clone.[98] The loss of productivity that O_3 causes applies to most components of growth, but root growth appears to be most severely impacted.[90] In a recent meta-analysis, the effect of O_3 on leaf gas exchange from different experiments was reported, and the genus *Populus* was listed as one of the most sensitive among those studied.[91]

14.4.5 Life-Cycle Assessment (LCA)

As opposed to agricultural crops, trees are long-lived organisms with a complex structure and considerable carbon sinks in perennial tissues. Regardless of short-rotation cycles, the long lifespan of these perennial crops might significantly influence acclimation processes.[99] In contrast to traditional forest plantations, the management of SRC is quite dynamic. The length of the rotation and the cultural practices used in management should be reconsidered to maximize the positive aspects and/or minimise the negative effects of climatic changes.[100]

There is a great potential for SRC to sequester carbon (C) in soils. Forest and agricultural soils can accrete C at average rates of $0.3\,Mg\,ha^{-1}\,year^{-1}$ with a maximum as high as $3\,Mg\,ha^{-1}\,yr^{-1}$.[101] Although an initial decline in soil organic carbon (SOC) was not found by Coleman *et al.*,[102] SOC in some cases may initially decline during the establishment of SRC plantations,[103] followed by a predicted increase after 5 years.[104] Global change and especially the increase in atmospheric CO_2 might influence the C cycle between the plantation and the atmosphere, in particular the C sequestration by soil. The elevated CO_2 concentration will probably increase the fraction of C incorporated into the soil in poplar SRC although this might be counterbalanced by the increase of O_3 levels.[105,106] It has been estimated that the long-term conversion of 20% of present arable land area of the European Union into bioenergy tree crops and natural woodlands would increase soil C stocks by about 5% over a century and would offset European C emissions by about 4%.[107] This figure is comparable with the Kyoto obligations.

Poplar SRC could be very effective in directly reducing CO_2 emissions to the atmosphere, since considerable amounts of renewable energy are produced by the SRC culture, thereby reducing the use of fossil fuels, *i.e.* a substitution effect that continues over different years. By comparison, carbon sequestration in natural forests is diminished as the forest reaches maturity. With regard to energy substitution, SRC reduces emissions with 24 to 29 $Mg\,CO_2\,ha^{-1}\,yr^{-1}$, while an oak–beech mixed forest reduces only 6 to 7 $Mg\,CO_2\,ha^{-1}\,yr^{-1}$.[108] The POP-EUROFACE experiment allowed estimating a future CO_2 reduction potential of poplar SRC between 95 and 108 $Mg\,CO_2$ equivalent $ha^{-1}\,yr^{-1}$. If expected arable area potentially available in Europe were available for production of woody biomass from SRC, by the end of 2100, 1805 up to 4212 Mton CO_2 equivalents $year^{-1}$ could be mitigated by producing energy from poplar biomass.[109]

14.5 Economic and Energetic Analysis of SRC Poplar

At present, a key constraint to the commercialisation of SRC for the production of heat, electricity and/or liquid fuels is the economic feasibility. General conditions have to be met to make SRC competitive with fossil fuel sources. First, the feedstock price and the costs of biomass conversion have to be competitive with fossil-fuel-derived energy. Secondly, the price offered by the converting facilities to the farmers has to be high enough so that energy crops are at least as profitable as conventional food crops. Accordingly, a number of measures to increase the competitiveness of energy crops in comparison with fossil fuel-derived energy and traditional agriculture appear vital in order to increase the economic viability of SRC.

14.5.1 Costs of Poplar for Bioenergy

The cultivation costs of poplar in a SRC culture involve (i) establishment costs (initial weed and pest control, soil preparation, (mechanical) planting, planting material and wildlife control); (ii) operating costs (land rent, ongoing weed and wildlife control, soil fertilisation and irrigation, harvesting, storage and transportation of the biomass product); and (iii) costs of processing and converting the poplar wood to heat, electricity and/or liquid fuels.[110] However, the latter goes beyond the scope of this chapter and will therefore not be discussed.

The estimated costs of the production of poplar biomass in SRC differ to a great extent depending on the location of the plantation. The land rent as well as the biomass yields, which are both strongly associated with the location of the plantation, account for most of the variation in estimated production costs (Table 14.1).[111] Different framework and methodological approaches make up much of the remaining variation in cost estimations. The figures presented in

Table 14.1 Overview of yields and delivered costs of hybrid poplar in SRC systems.

Location	Approach	Rotation length (years)	Yield (Mg dry biomass $ha^{-1}\,year^{-1}$)	Delivered costs[e] ($USD\,Mg^{-1}$)	References
USA	Full economic	7	6.7–20.2	40.6–94.3[a]	112
USA	Full economic	7	11.2	52.8–59.3[b]	112, 121
USA	Discounted cash flow	5–8	16	40	122
USA	Discounted cash flow	5–8	16	38.8	123
Spain	Full economic	5	13.5	27.3–35.1[c,f]	124
Italy	NA	2	10	109.8–131.8[d,f]	125

[a]Depending on the yield: the higher the yield, the lower the delivered cost.
[b]Depending on the transportation costs ranging from 5.5 to 12.0 USD per Mg dry biomass. Values for transportation costs were taken from a similar study.[112]
[c]Depending on the transportation costs ranging from 3.5 to 13.4 USD per Mg dry biomass.
[d]Depending on the size of the cultivated area: the larger the area, the lower the delivered cost.
[e]The delivered costs were calculated for the full production process from establishment of the plantation to the delivery of the biomass feedstock at the conversion plant.
[f]Costs were converted from EUR into USD using the average EUR-USD exchange rate for 2008, namely 1.46. Source: http://france.usembassy.gov/irs-euro.html.

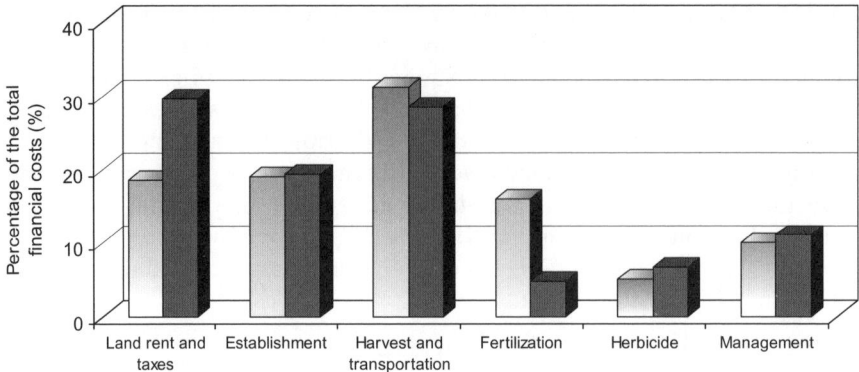

Figure 14.7 Total breakdown of total financial costs among poplar SRC activities in two studies in the USA (light grey bars: study from Strauss et al.;[113] and dark grey bars: study from Linville and Betters).[114]

Table 14.1 consider two financial models, *i.e.* full economic accounting model and discounted cash-flow model. The full economic accounting approach provides a structural *annual review* of all costs (variable and fixed cash costs, and opportunity costs of owned resources) involved in the production of biomass feedstock, from the plantation establishment to the delivery of the feedstock at the conversion plant.[112,113] The discounted cash-flow model, on the other hand, identifies the *total investments* and the related operating costs or revenues over the entire lifetime of the plantation.

A breakdown of the total financial costs among activities in two studies showed that land rent, harvest and transportation costs largely contributed to the total costs of poplar SRC (Figure 14.7).[113,114] Poplar biomass has a lower energy density than fossil fuels, which translates into relatively large amounts of biomass required to generate one energy unit. Consequently, large plantation areas distributed over large geographic areas are required, resulting in high land rent, harvest and transportation costs.[115] Irrigation was not considered in this breakdown, as several studies reported very high costs for irrigation that were not compensated by a comparable increase in biomass yield.[113,114]

14.5.2 Energy Balance of Poplar SRC

A basic requirement of any (bio)energy generation system is that the energy produced must be larger than the inputs of nonrenewable energy required for the establishment and operation of the system.[116] Thus, an analysis of the energy balance of poplar SRC in addition to the economic assessment is essential. The energy balance entirely depends on the boundaries of the considered system. In the next sections the boundary will be placed at the farm gate, unless otherwise mentioned. The transportation of the feedstock from the farm to the conversion site and conversion steps will not be considered, as this

Poplar 293

would result in widely divergent figures depending on the conversion technology used. Furthermore, solar energy and human labour will not be taken into account as energy sources.

The energy requirement (*i.e.* the ratio of total nonrenewable energy input to total energy content of biomass feedstock) for poplar production is in the order of 0.05 MJ MJ^{-1} that corresponds to an energy ratio (*i.e.* 1/energy requirement) of 20.[117] The energy ratio has to be larger than one if the system is to produce more energy than it consumes. A Belgian study assessed the energy balance of poplar SRC, considering three levels of management regimes with regard to the nature and the amount of inputs (fertilisers, pesticides, *etc*.) as well as to the scale of mechanisation.[118] The energy ratios in this study ranged from 21.5 to 25.6. When local transportation of about 30 km was included in the energy balance, the energy ratio declined to a range of 17.2 to 20. In both cases the scenario with the most intensive management regime yielded the highest positive energy ratio, even though its total energy costs were higher. Better crop management, resulting in higher biomass yields, clearly outweighs the higher energy costs.[119] The energy balance of poplar (and willow) turns out to be quite positive in comparison to other energy crops, such as alfalfa, switchgrass and reed canary grass, which all have significantly larger energy requirements than poplar.[117]

The breakdown of the total energy costs indicated that crop maintenance (fertilisation, pesticide, *etc*.), harvest and transport are the most energy intensive activities.[118] In particular, the production of fertilisers is very energy demanding, given the high energy cost of nitrogen (1 kg of nitrogen = 57.5 MJ). In this study,[118] crop maintenance represents between 32 and 39% of the total energy costs, depending on the intensity of the SRC cultivation practices. Obviously, the incidence of (local) transport on the total energy costs is subject to the distance between the farm and the conversion site. The total energy requirement of bulk road carrier transport in the UK was estimated about 1.1 ± 0.04 MJ Mg^{-1} km^{-1}.[120] The precise share of the different activities in the total energy costs depends not only on the management practice used but also on the boundaries chosen. The more activities or energy inputs considered (*e.g.* inclusion of regional transportation), the larger the total energy costs and consequently the smaller the share of the different activities will be.

The lack of comprehensive accounting methods to assess financial as well as energy costs of different activities involved in the production of SRC largely explains the differences among the results of existing studies. To provide accurate and unambiguous results, a full system analysis of poplar SRC from the planting to the energy conversion in combination with standardised calculation and accounting methods is indispensable.

14.6 Conclusions

Characterised by rapid growth rates and good coppice ability, poplars hold several trumps for bioenergy. Large genotypic differences occur in tree physiology, morphology and responses to biotic and abiotic stresses among and within poplar species and their hybrids. This rich source of genetic variation can be exploited in poplar breeding programmes aiming at improving poplar material for SRC, mainly in terms of yield and quality traits. Yet, greenhouse gas and net energy balances as well as economic analyses of a poplar SRC life cycle represent a promising area for future research.

Acknowledgements

S.Y. Dillen and O. El Kasmioui are postdoctoral research associate and research assistant, respectively, of the Flemish Science Foundation (FWO). Poplar research at the University of Antwerp is being supported by the European Research Council (ERC Adv. Grant, POPFULL), by the Research Center of Excellence ECO and by the Methusalem funding scheme of the Flemish Government.

References

1. D. I. Dickmann and J. Kuzovkina, in *Poplars and Willows of the World*, FAO, Rome, 2008.
2. J. Ball, J. Carle and A. Del Lungo, *Unasylva*, 2005, **56**, 3.
3. H. D. Bradshaw Jr, R. Ceulemans, J. Davis and R. F. Stettler, *J. Plant Growth Regul.*, 2000, **19**, 306.
4. G. A. Tuskan, S. DiFazio, S. Jansson, J. Bohlmann, I. Grigouriev, U. Hellsten, N. Putnam, S. Ralph, S. Rombauts, A. Salamov, J. Schein, L. Sterck, A. Aerts, R. R. Bhalerao, R. P. Bhalerao, D. Blaudez, W. Boerjan, A. Brun, A. Brunner, V. Busov, M. Campbell, J. Carlson, M. Chalot, J. Chapman, G.-L. Chen, D. Cooper, P. M. Coutinho, J. Couturier, S. Covert, Q. Cronk, R. Cunningham, J. Davis, S. Degroeve, A. Déjardin, C. dePamphilis, J. Detter, B. Dirks, I. Dubchak, S. Duplessis, J. Ehlting, B. Ellis, K. Gendler, D. Goodstein, M. Gribskov, J. Grimwood, A. Groover, L. Gunter, B. Hamberger, B. Heinze, Y. Helariutta, B. Henrissat, D. Holligan, R. Holt, W. Huang, N. Islam-Faridi, S. Jones, M. Jones-Rhoades, R. Jorgensen, C. Joshi, J. Kangasjärvi, J. Karlsson, C. Kelleher, R. Kirkpatrick, M. Kirst, A. Kohler, U. Kalluri, F. Larimer, J. Leebens-Mack, J.-C. Leplé, P. Locascio, Y. Lou, S. Lucas, F. Martin, B. Montanini, C. Napoli, D. R. Nelson, C. Nelson, K. Nieminen, O. Nilsson, V. Pereda, G. Peter, R. Philippe, G. Pilate, A. Poliakov, J. Razumovskaya, P. Richardson, C. Rinaldi, K. Ritland, P. Rouzé, D. Ryaboy, J. Schmutz, J. Schrader, B. Segerman, H. Shin, A. Siddiqui, F. Sterky, A. Terry, C.-J. Tsai, E. Uberbacher, P. Unneberg, J. Vahala, K. Wall, S. Wessler, G. Yang, T. Yin, C. Douglas, M. Marra, G. Sandberg, Y. Van de Peer and D. Rokhsar, *Science*, 2006, **313**, 1596.
5. J. J. Fillatti, J. Sellmer, B. McCown, B. Haissig and L. Comai, *Mol. Gen. Genet.*, 1987, **206**, 192.

6. S. Jansson and C. J. Douglas, *Annu. Rev. Plant Biol.*, 2007, **58**, 435.
7. R. Ceulemans, *Genetic Variation in Functional and Structural Productivity Determinants in Poplar*, Thesis Publishers, Amsterdam, 1990.
8. G. Taylor, *Ann. Bot.*, 2002, **90**, 681.
9. D. I. Dickmann, in *Poplar Culture in North America*, D. I. Dickmann, J. G. Isebrands, J. E. Eckenwalder J. Richardson, (ed.), NRC Research Press, Ottawa, 2001, p. 1.
10. FAO, *Synthesis of Country Progress Reports* prepared during the 23rd Session of the IPC in Beijing, FAO, Rome, 2008.
11. J. E. Eckenwalder, in *Biology of Populus and its Implications for Management and Conservation*, R. F. Stettler, H. D. Bradshaw Jr, P. E. Heilman and T. M. Hinckley, (ed.), NRC Research Press, Ottawa, 1996, p. 7.
12. B. A. Thielges, *Breeding Poplars for Disease Resistance*, FAO, Rome, 1985.
13. H. K. Hayes, in *Heterosis*, J. W. Gowen, (ed.), Iowa State University College Press, Ames, 1952, p. 49.
14. R. F. Stettler, L. Zsuffa and R. Wu, in *Biology of Populus and its Implications for Management and Conservation*, R. F. Stettler, H. D. Bradshaw Jr, P. E. Heilman and T. M. Hinckley, (ed.), NRC Research Press, Ottowa, 1996, p. 87.
15. H. Weisgerber, D. Kownatzki and M. Mussong, *Silvae Genet.*, 1995, **44**, 298.
16. R. Spinelli, C. Nati and N. Magagnotti, *Biomass Bioenerg.*, 2009, **33**, 817.
17. R. Venendaal, U. Jørgensen and C. A. Foster, *Biomass Bioenerg.*, 1997, **13**, 147.
18. D. I. Dickmann, *Biomass Bioenerg.*, 2006, **30**, 696.
19. R. G. McAlpine, C. L. Brown, A. M. Herrick and H. E. Ruark, *For. Farmer*, 1966, **26**, 6.
20. A. M. Herrick and C. L. Brown, *Agr. Sci. Rev.*, 1967, **5**, 8.
21. E. Willebrand in *Handbook on How to Grow Short Rotation Forests*, ed. S. Ledin and E. Willebrand, Swedish University of Agricultural Sciences, Uppsala, 1995, p. 2.6.1.
22. D. I. Dickmann and K.W. Stuart, *The Culture of Poplars in Eastern North America*, Michigan State University, East Lansing, 1983.
23. T. F. Strong and J. Zavitkovski, in *Intensive Plantation Culture: 12 Years's Research*, E. H. Hansen, (ed.), USDA Forest Service, St Paul, 1983, p. 54.
24. R. Ceulemans and W. Deraedt, *For. Ecol. Manage.*, 1999, **121**, 9.
25. R. E. H. Sims, T. G. Maiava and B. T. Bullock, *Biomass Bioenerg.*, 2001, **20**, 329.
26. N. Al Afas, N. Marron, S. Van Dongen, I. Laureysens and R. Ceulemans, *For. Ecol. Manage.*, 2008, **255**, 1883.
27. M. J. Aylott, E. Casella, I. Tubby, N. R. Street, P. Smith and G. Taylor, *New Phytol.*, 2008, **178**, 358.
28. J. Steenackers, M. Steenackers, V. Steenackers and M. Stevens, *Biomass Bioenerg.*, 1996, **10**, 267.

29. A. Armstrong, C. Johns and I. Tubby, *Biomass Bioenerg.*, 1999, **17**, 305.
30. I. Tubby and A. Armstrong, *Establishment and Management of Short-Rotation Coppice*, Forestry Commission, Edinburgh, 2002.
31. R. E. H. Sims, A. Hastings, B. Schlamadinger, G. Taylor and P. Smith, *Glob. Change Biol.*, 2006, **12**, 2054.
32. P. E. Heilman, G. Ekuan and D. Fogle, *Can. J. For. Res.*, 1994, **24**, 1186.
33. G. E. Scarascia-Mugnozza, R. Ceulemans, P. E. Heilman, J. G. Isebrands, R. F. Stettler and T. M. Hinckley, *Can. J. For. Res.*, 1997, **27**, 285.
34. I. Laureysens, J. Bogaert, R. Blust and R. Ceulemans, *For. Ecol. Manage.*, 2004, **187**, 295.
35. M. Labrecque and T. I. Teodorescu, *Biomass Bioenerg.*, 2005, **29**, 1.
36. I. Laureysens, W. Deraedt, T. Indeherberge and R. Ceulemans, *Biomass Bioenerg.*, 2003, **24**, 81.
37. I. Laureysens, A. Pellis, J. Willems and R. Ceulemans, *Biomass Bioenerg.*, 2005, **29**, 10.
38. P. E. Heilman and R. F. Stettler, *Can. J. For. Res.*, 1985, **15**, 384.
39. I. Laureysens, R. Blust, L. De Temmerman, C. Lemmens and R. Ceulemans, *Environ. Pollut.*, 2004, **131**, 485.
40. E. A. Hansen, *Biomass Bioenerg.*, 1991, **1**, 1.
41. J. Peng, D. E. Richards, N. M. Hartley, G. P. Murphy, K. M. Devos, J. E. Flintham, J. Beales, L. J. Fish, A. J. Worland, F. Pelica, D. Sudhakar, P. Christou, J. W. Snape, M. D. Gale and N. P. Harverd, *Nature*, 1999, **400**, 256.
42. R. F. Stettler and R. Ceulemans, in *Clonal Forestry. I. Genetics and Biotechnology*, M. R. Ahuja and W. J. Libby, (ed.), Springer, Berlin, 1993, p. 68.
43. D. I. Dickmann, M. A. Gold and J. A. Flore, *Plant Breed. Rev.*, 1994, **12**, 163.
44. C. R. Ridge, T. M. Hinckley, R. F. Stettler and E. Van Volkenburgh, *Tree Physiol.*, 1986, **1**, 209.
45. N. Marron, S. Y. Dillen and R. Ceulemans, *Environ. Exp. Bot.*, 2007, **61**, 103.
46. A. Pellis, I. Laureysens and R. Ceulemans, *Plant Biol.*, 2004, **6**, 38.
47. S. Y. Dillen, S. B. Rood and R. Ceulemans, in *Genetics and Genomics of Populus*, S. Jansson, R. P. Bhalerao and A. T. Groover, (ed.), Springer, New York, 2010, p. 39.
48. R. Ceulemans and J. G. Isebrands, in *Biology of Populus and its Implications for Management and Conservation*, R. F. Stettler, H. D. Bradshaw Jr, P. E. Heilman and T. M. Hinckley, (ed.), NRC Research Press, Ottawa, 1996, p. 355.
49. R. Ceulemans, A. J. S. McDonald and J. S. Pereira, *Biomass Bioenerg.*, 1996, **11**, 215.
50. E. Novaes, L. Osorio, D. R. Drost, B. L. Miles, C. R. D. Boaventura-Novaes, C. Benedict, C. Dervinis, Q. Yu, R. Sykes, M. Davis, T. A. Martin, G. F. Peter and M. Kirst, *New Phytol.*, 2009, **182**, 878.

51. W. J. Mattson, E. A. Hart and W. J. A. Volney, in *Poplar Culture in North America*, D. I. Dickmann, J. G. Isebrands, J. E. Eckenwalder, J. Richardson, (ed.), NRC Research Press, Ottawa, 2001, p. 219.
52. V. Jorge, A. Dowkiw, P. Faivre-Rampant and C. Bastien, *New Phytol.*, 2005, **167**, 113.
53. D. S. Falconer, *Introduction to Quantitative Genetics*, 2nd edn, Longman, London, 1981.
54. G. J. McDougall, I. M. Morrison, D. Stewart, J. D. B. Weyers and J. R. Hillman, *J. Sci. Food Agric.*, 1993, **62**, 1.
55. J. J. Balatinecz and D. E. Kretschmann, in *Poplar Culture in North America*, D. I. Dickmann, J. G. Isebrands, J. E. Eckenwalder and J. Richardson, (ed.), NRC Research Press, Ottawa, 2001, p. 277.
56. D. Kauter, I. Lewandowski and W. Claupein, *Biomass Bioenerg.*, 2003, **24**, 411.
57. P. J. Tharakan, T. A. Volk, L. P. Abrahamson and E. H. White, *Biomass Bioenerg.*, 2003, **25**, 571.
58. I. Van de Walle, N. Van Camp, L. Van de Casteele, K. Verheyen and R. Lemeur, *Biomass Bioenerg.*, 2007, **31**, 276.
59. W. Guidi, E. Piccioni, M. Ginnani and E. Bonari, *Biomass Bioenerg.*, 2008, **32**, 518.
60. W. Boerjan, *Curr. Opin. Biotech.*, 2005, **16**, 159.
61. A. Demirbas, *Energy Sources*, 2003, **25**, 309.
62. K. Raveendran and A. Ganesh, *Fuel*, 1996, **75**, 1715.
63. R. Vanholme, K. Morreel, J. Ralph and W. Boerjan, *Curr. Opin. Plant Biol.*, 2008, **11**, 278.
64. J.-C. Leplé, R. Dauwe, K. Morreel, V. Storme, C. Lapierre, B. Pollet, A. Naumann, K.-Y. Kang, H. Kim, A. Lefèbvre, J.-P. Joseleau, J. Grima-Pettenati, R. De Rycke, S. Andersson-Gunnerås, A. Erban, I. Fehrle, M. Petit-Conil, J. Kopka, A. Polle, E. Messens, B. Sundberg, S. D. Mansfield, J. Ralph, G. Pilate and W. Boerjan, *Plant Cell*, 2007, **19**, 3669.
65. R. J. Dinus, P. Payne, M. M. Sewell, V. L. Chiang and G. A. Tuskan, *Crit. Rev. Plant Sci.*, 2001, **20**, 51.
66. G. Pilate, A. Dejardin, F. Laurans and J. C. Leplé, *New Phytol.*, 2004, **164**, 63.
67. S. Andersson-Gunnerås, E. Mellerowicz, J. Love, B. Segerman, Y. Ohmiya, P. M. Coutinho, P. Nillson, B. Henrissat, T. Moritz and B. Sundberg, *Plant J.*, 2006, **45**, 144.
68. A. M. Rae, N. R. Street, K. M. Robinson, N. Harris and G. Taylor, *BMC Plant Biol.*, 2009, **9**.
69. H. D. Bradshaw Jr and R. F. Stettler, *Genetics*, 1995, **139**, 963.
70. B. E. Frewen, T. H. H. Chen, G. T. Howe, J. Davis, A. Rohde, W. Boerjan and H. D. Bradshaw Jr, *Genetics*, 2000, **154**, 837.
71. N. R. Street, O. Skogström, A. Sjödin, J. Tucker, M. Rodriguez-Acosta, P. Nilsson, S. Jansson and G. Taylor, *Plant J.*, 2006, **48**, 321.
72. D. Grattapaglia, P. Wilcox, S. McCord, B. Crane, B. H. Liu, H. Amerson, E. G. Kuhlman, S. McKeand, R. Whetten, D. O. O'Malley and

R. Sederoff, in *Proceedings of the International Wood Biotechnology Symposium*, Japan, 1994.
73. G. Alker, C. Bruton and K. Richards, *Full-Scale Implementation of SRC systems: Assessment of Technical and Non-Technical Barriers*, IEA Bioenergy Task 30, 2005.
74. R. L. Rowe, N. R. Street and G. Taylor, *Renew. Sust. Energ. Rev.*, 2009, **13**, 271.
75. P. D. Hardcastle, *A Review of the Potential Impacts of Short Rotation Forestry*, LTS International, Edinburgh, 2006.
76. T. J. Tschaplinski, G. A. Tuskan and C. A. Gunderson, *Can. J. For. Res.*, 1994, **24**, 364.
77. L. Zsuffa, E. Giordano, L. D. Pryor and R. F. Stettler, in *Biology of Populus and its Implications for Management and Conservation*, R. F. Stettler, H. D. Bradshaw Jr, P. E. Heilman and T. M. Hinckley, (ed.), NRC Research Press, Ottawa, 1996, p. 539.
78. R. Monclus, E. Dreyer, M. Villar, F. M. Delmotte, D. Delay, J. M. Petit, C. Barbaroux, D. Le Thiec, C. Bréchet and F. Brignolas, *New Phytol.*, 2006, **169**, 765.
79. R. Monclus, M. Villar, C. Barbaroux, C. Bastien, R. Fichot, F. M. Delmotte, D. Delay, J. M. Petit, C. Bréchet, E. Dreyer and F. Brignolas, *Tree Physiol.*, 2009, **29**, 1329.
80. P. B. Gerbens-Leenes, A. Y. Hoekstra and T. Van der Meer, *Ecol. Econ.*, 2009, **68**, 1052.
81. A. Berthelot, J. Ranger and D. Gelhaye, *For. Ecol. Manage.*, 2000, **128**, 167.
82. U. Jørgensen and K. Schelde, *Energy Crop Water and Nutrient Use Efficiency*, IEA Bioenergy Task 17, 2001.
83. L. S. Lodhiyal and N. Lodhiyal, *Ann. Bot.*, 1997, **79**, 517.
84. A. Jug, C. Hofmann-Schielle, F. Makeschin and K. E. Rehfuess, *For. Ecol. Manage.*, 1999, **121**, 67.
85. M. E. Ostry, L. W. Wilson, H. S. McNabb Jr and L. M. Moore, *A guide to Insect, Disease and Animal Pests of Poplars*, USDA Forest Service, Washington, 1989.
86. D. J. Royle and M. E. Ostry, *Biomass Bioenerg.*, 1995, **9**, 69.
87. B. Gielen and R. Ceulemans, *Environ. Pollut.*, 2001, **115**, 335.
88. E. A. Ainsworth and S. P. Long, *New Phytol.*, 2005, **165**, 351.
89. M. H. Turnbull, D. T. Tissue, R. Murthy, X. Wang, A. D. Sparrow and K. L. Griffin, *New Phytol.*, 2004, **161**, 819.
90. D. F. Karnosky, J. M. Skelly, K. E. Percy and A. H. Chappelka, *Environ. Pollut.*, 2007, **147**, 489.
91. V. E. Wittig, E. A. Ainsworth and S. P. Long, *Plant Cell Environ.*, 2007, **30**, 1150.
92. C. A. Gunderson and S. D. Wullschleger, *Photosynth. Res.*, 1994, **39**, 369.
93. B. E. Medlyn, F.-W. Badeck, D. G. G. De Pury, C. V. M. Barton, M. Broadmeadow, R. Ceulemans, P. De Angelis, M. Forstreuter, M. E. Jach, S. Kellomäki, E. Laitat, M. Marek, S. Philippot, A. Rey,

J. Strassemeyer, K. Laitinen, R. Liozon, B. Portier, P. Roberntz, K. Wang and P. G. Jarvis, *Plant Cell Environ.*, 1999, **22**, 1475.
94. M. J. Tjoelker, J. Oleksyn and P. B. Reich, *Glob. Change Biol.*, 1999, **5**, 679.
95. G. E. Scarascia-Mugnozza, C. Calfapietra, R. Ceulemans, B. Gielen, F. Cotrufo, P. De Angelis, D. Godbold, M. R. Hoosbeek, O. Kull, M. Lukac, M. Marek, F. Miglietta, A. Polle, C. Raines, M. Sabatti and G. Taylor, in *Managed Ecosystems and Elevated CO_2: Case Studies, Processes and Perspectives*, J. Nosberger, S. P. Long, R. J. Norby, M. Stitt, G. R. Hendrey and H. Blum, (ed.), Springer, Berlin, 2006, p. 173.
96. D. F. Karnosky and K. S. Pregitzer, in *Managed Ecosystems and Elevated CO_2: Case Studies, Processes and Perspectives*, J. Nosberger, S. P. Long, R. J. Norby, M. Stitt, G. R. Hendrey and H. Blum, (ed.), Springer, Berlin, 2006, p. 213.
97. G. A. Barron-Gafford, D. Martens, K. Grieve, J. E. T. McLain, D. Lipson and R. Murthy, *Glob. Change Biol.*, 2005, **11**, 1220.
98. D. F. Karnosky, K. S. Pregitzer, D. R. Zak, M. E. Kubiske, G. R. Hendrey, D. Weinstein, M. Nosal and K. E. Percy, *Plant Cell Environ.*, 2005, **28**, 965.
99. I. A. Janssens, M. Mousseau and R. Ceulemans, in *Climate Change and Global Crop Productivity*, K. R. Reddy and H. F. Hodges, (ed.), CAB International, Wallingford, 2000, p. 245.
100. C. Calfapietra, B. Gielen, D. Karnosky, R. Ceulemans and G. E. Scarascia-Mugnozza, *Environ. Pollut.*, 2010, **158**, 1095.
101. W. M. Post and K. C. Kwon, *Glob. Change Biol.*, 2000, **6**, 317.
102. D. C. Coleman, D. A. Crossley and P. F. Hendrix, *Fundamentals of Soil Ecology*, 2nd edn, Elsevier Academic Press, Burlington, 2004.
103. E. A. Hansen, *Biomass Bioenerg.*, 1993, **5**, 431.
104. D. F. Grigal and W. E. Berguson, *Biomass Bioenerg.*, 1998, **14**, 371.
105. W. M. Loya, K. S. Pregitzer, N. J. Karberg, J. S. King and C. P. Giardina, *Nature*, 2003, **425**, 705.
106. M. R. Hoosbeek, Y. Li and G. E. Scarascia-Mugnozza, *Plant Soil*, 2006, **281**, 247.
107. P. Smith, D. Powlson, M. Glendining and J. Smith, *Glob. Change Biol.*, 1997, **3**, 67.
108. G. Deckmyn, B. Muys, J. Garcia Quijano and R. Ceulemans, *Glob. Change Biol.*, 2004, **10**, 1482.
109. M. Liberloo, S. Luyssaert, V. Bellassen, S. Njakou Djomo, M. Lukac, C. Calfapietra, I. A. Janssens, M. R. Hoosbeek, N. Viovy, G. Churkina, G. Scarascia-Mugnozza and R. Ceulemans *Plos One*, 2010 **5**, e11648
110. R. P. Bardos, C. French, A. Lewis, A. Moffat and S. Nortcliff, *exSite Research Project Final Report*, LQM Press, Nottingham, 2001.
111. C. P. Mitchell, E. A. Stevens and M. P. Watters, *For. Ecol. Manage.*, 1999, **121**, 123.
112. M. E. Walsh, *Biomass Bioenerg.*, 1998, **14**, 341.
113. C. H. Strauss, S. C. Grado, P. R. Blankenhorn and T. W. Bowersox, *Sol. Energy*, 1989, **42**, 379.

114. W. R. Linville and D. R. Betters, *J. Sustain. Agr.*, 1997, **9**, 49.
115. D. Yemshanov and D. McKenney, *Biomass Bioenerg.*, 2008, **32**, 185.
116. R. W. Matthews, *Biomass Bioenerg.*, 2001, **21**, 1.
117. U. R. Boman and J. H. Turnbull, *Biomass Bioenerg.*, 1997, **13**, 333.
118. X. Dubuisson and I. Sintzoff, *Biomass Bioenerg.*, 1998, **15**, 379.
119. S. Nonhebel, *Biomass Bioenerg.*, 2002, **22**, 159.
120. M. A. Elsayed and N. D. Mortimer, *Carbon and Energy Modeling of Biomass systems: Conversion Plant and Data Updates*, Sheffield Hallam University, Sheffield, 2001.
121. M. E. Walsh and D. A. Becker, in *Proceedings of the Seventh National Bioenergy Conference*, Southeastern Regional Biomass Energy Program, Nashville, 1996.
122. C. H. Strauss and L. L. Wright, *Sol. Energy*, 1990, **45**, 105.
123. C. H. Strauss and S. C. Grado, *Sol. Energy*, 1992, **48**, 45.
124. C. M. Gasol, S. Martinez, M. Rigola, J. Rieradevall, A. Anton, J. Carrasco, P. Ciria and X. Gabarell, *Renew. Sust. Energ. Rev.*, 2009, **13**, 801.
125. M. Manzone, G. Airoldi and P. Balsari, *Biomass Bioenerg.*, 2009, **33**, 1258.
126. L. Zsuffa, in *Proceedings of Canadian Tree Improvement Association*, Canadian Forest Service, Ottawa, 1975, p. 107.
127. OECD (Organisation for Economic Co-operation and Development), *Safety Assessment of Transgenic Organism*, OECD, Paris, 2006.

CHAPTER 15
Developing Miscanthus for Bioenergy

JOHN CLIFTON BROWN,[a] STEVE RENVOIZE,[b] YU-CHUNG CHIANG,[c] YASUSHI IBARAGI,[d] RICHARD FLAVELL,[e] JOERG GREEF,[f] LIN HUANG,[a] TSAI WEN HSU,[g] DO-SOON KIM,[h] ASTLEY HASTINGS,[i] KAI SCHWARZ,[f] PAUL STAMPFL,[j] JOHN VALENTINE,[a] TOSHIHIKO YAMADA,[k] QINGGUO XI[l] AND IAIN DONNISON[a]

[a] Institute of Biological, Environmental and Rural Sciences (IBERS), Aberystwyth University, Gogerddan, Aberystwyth, Ceredigion, Wales, SY23 3EQ, UK; [b] The Herbarium Royal Botanic Gardens Kew, Richmond, Surrey, TW9 3AB, UK; [c] Department of Life Science, National Pingtung University of Science and Technology, 1, Shuehfu Rd., Neipu, Pingtung 91201, Taiwan; [d] Tokushima Prefectural Museum, Bunka-no-mori Park, Hachiman-cho, Tokushima city, Tokushima, 770-8070 Japan; [e] CERES, Inc., 1535 Ranchi Conejo Blvd, Thousand Oaks, CA 91320, USA; [f] Julius Kühn-Institut (JKI) Bundesforschungsinstitut für Kulturpflanzen, Messeweg 11/12, 38104 Braunschweig; [g] High Altitude Experimental Station, Endemic Species Research Institute, Council of Agriculture, Nantou County, Taiwan, Republic of China; [h] Department of Plant Science, College of Agriculture and Life Science, Seoul National University, Seoul, Korea; [i] Institute of Biological and Environmental Science, University of Aberdeen, 23 St. Machar Drive, Aberdeen, Scotland, AB24 3UU; [j] Institute of Ecology, University of Innsbruck, Sternwartestraße 15, 6020 Innsbruck, Austria; [k] Field Science Center for Northern Biosphere, Hokkaido University,

RSC Energy and Environment Series No. 3
Energy Crops
Edited by Nigel G Halford and Angela Karp
© Royal Society of Chemistry 2011
Published by the Royal Society of Chemistry, www.rsc.org

Kita 11, Nishi 10, Kita-ku, Sapporo 060-0811, Japan; [1]Agricultural Institute of Dongying, Jiaozhoulu 21, Dongying, 257091, Republic of China

15.1 Introduction

Growing international pressure for adoption of renewable sources of energy and the associated political targets have intensified the interest in using dedicated energy crops for biofuels and biopower. In the past decade European agriculture has suffered from overproduction of key food and animal feed crops and cheaper products being obtained from outside the EU. Consequently, set-aside was introduced in 1988 to reduce overproduction and this amounted to 15% of the arable land. This was reduced to 10% in 1996 and with the introduction of the single farm payment in 2005, set-aside entitlements were assessed on the number of hectares removed from agricultural production. This proportion of good-grade agricultural land therefore became available for nonfood uses. Enterprising farmers, aware that energy costs were rising, realised that set-aside land could be used for biomass production. In the early 1990s, several species were identified as being potentially suitable sources of biomass: these included willow, reed canary grass and *Miscanthus*. Of these, willow and *Miscanthus* have attracted the most attention in the UK. This review provides some of the historical development of *Miscanthus* and shows how it has become a front runner in the options for dedicated energy crops.

15.1.1 How Was *Miscanthus* Identified as a Key Candidate Bioenergy Feedstock?

Miscanthus was first introduced to Europe in 1935 by the Danish plant collector, Axel Olsen.[1] Amongst the horticultural specimens that were brought from Japan a particularly vigourous genotype was discovered. This was identified as a potential source of biomass in Europe as far back as 1960 at the Danish Institute of Landscape Plants in Hornum.[2] *Miscanthus* breeding in Germany during the 1960s was aimed at producing horticultural varieties. In the late 1980s, interest in *Miscanthus* as a potential biomass crop grew. A German scientist, Professor Wolfgang Stander, is reported to have said "I have seen the future, I can see grass growing 12 feet high that can fuel power stations". In 1988, Veba Öl, a German-based oil company, began multisite trialling with *M. x giganteus*.[3] Results from 20 field trials were mixed due mainly to considerable variation in stand establishment from rhizomes.[4] In 1992, Dr. Martin Deuter began breeding *Miscanthus* for bioenergy in Marburg, Eastern Germany. Drs. Deuter and Greef described *M. x giganteus* as a species in its own right, and proposed that it was the hybrid from a cross between *M. sacchariflorus* and *M. sinensis* since it had intermediate features.[1] European-funded research programmes during the 1990s evaluated *M. x giganteus* from Greece to Sweden.[5] As with the Veba Öl project, field trials showed both outstanding and poor performances. Some locations produced yields in excess of 25 dry tonnes per hectare per year, while others yielded less than 5 dry tonnes per hectare per year.[6]

In 1997, the European *Miscanthus* Improvement Project, funded by the European Union, recognised the importance of widening the genetic base. Field trials performed at five locations between Sweden and Portugal demonstrated that the best-performing genotype in Sweden was amongst the worst in Portugal and *vice versa*.[7] Therefore, the potential for tailoring hybrids adapted to specific climatic regions through characterisation and selection of new germplasm is key to developing *Miscanthus* as a bioenergy crop. This chapter presents a novel scheme for the classification of *Miscanthus* and includes a discussion on the phenotypic variation in traits needed to breed future varieties.

15.2 *Miscanthus*, Taxonomy and Origins

Miscanthus is a member of the grass tribe Andropogoneae, which includes maize, sorghum and sugarcane.[8] *Miscanthus* is characterised by the tough raceme rachis and bisexual paired 2-flowered spikelets. Molecular and morphological systematics places *Miscanthus* closest to sugarcane followed by sorghum.[9] There is some discussion amongst taxonomists on the number of species within the genus *Miscanthu*s. The general view is between 14 to 20 species.[10–13] Hybridisation, especially between *M. sinensis* variants, is leading to a complex range of forms that are difficult to accommodate in a practical taxonomic system.

In this review, a taxonomic scheme is presented with five sections (Figure 15.1). All are found in Eastern Asia, and it is these that are considered the most important for bioenergy. An isolated group of four species in Southern Africa, formerly placed in *Miscanthus*, are now segregated as a separate genus, *Miscanthidium*.[9,14,15]

Figure 15.2 shows the approximate geographic distributions of *Miscanthus* species throughout Eastern Asia. *M. sinensis* has the broadest distribution, stretching from Hebei province, just south of Beijing, to Hong Kong, including Korea, Taiwan and N. Japan to Ryukus. *M. sacchariflorus* has a more restricted distribution than *M. sinensis*, but it embraces a very wide range of climates from Russia, N. China, Korea and Japan (Honshu). The distribution of *M. floridulus* is confined to latitudes below 30 N, since it is relatively frost sensitive. *M. lutarioriparius*, often classified as a member of *M. sacchariflorus*, has a rather narrow distribution in the Yangtze river system. *M. transmorrisonensis*, which is morphologically rather similar to *M. sinensis*, is found at high altitudes >2500 m in Taiwan where snow is frequent. *M. longiberbis* is an endemic species of South Korea. It has been proposed that it forms a link between other members of sect. *Kariyasua* and *M. sacchariflorus* (Ibaragi and Ohashi 2004). *M. tinctorius* is similar to *M. longiberbis* but it has no awn. This grass has soft leaves and it is used as a traditional yellow colour dye in Japan.

MISCANTHUS Anderss. (1856)

- Generic synonymy

- Sclerostachya (Hack.) A. Camus (1922)
- Triarrhena (Maxim.) Nakai (1950)
- Rubimons B.S. Sun (1997)
- Diandranthus Liou (1997)

Section Diandranthus Keng (1957,1959)

Lemmas awned
Stamens 2,
Stigma apically exserted

1. **Miscanthus nudipes** (Griseb.) Hack. 1889.
- *Erianthus nudipes* Grisebach 1868:
- *Miscanthus brevipilus* Handel-Mazzetti:
- *M. eulalioides* Keng:
- *M. nudipes* ssp. *yunnanensis* A. Camus:
- *M. szechuanensis* Keng ex S. L. Zhong:
- *M. taylori* Bor:
- *M. wardii* Bor:
- *M. yunnanensis* (A. Camus) Keng.

Diandranthus breviplus (Handel-Mazetti) L. Liou:
- *D. corymbosus* L. Liou:
- *D. eulalioides* (Keng) L. Liou:
- *D. nudipes* (Griseb.) L. Liou:
- *D. szechuanensis* (Keng ex Zhong) L. Liou:
- *D. taylori* (Bor) L. Liou:
- *D. tibeticus* L. Liou:
- *D. wardii* (Bor) L. Liou:
- *D. yunnanensis* (A. Camus) L. Liou:

2. **M. nepalensis** (Trin.) Hack. (1889)
- *Eulalia nepalensis* Trin. (1833)
- *Diandranthus nepalensis* (Trin.) L. Liou.

Section Triarrhena (Max.) Honda (1930)

- Stamens 3
- Lemma awnless

1. **Miscanthus sacchariflorus** (Maxim.) Hack. (1887)
- *Imperata sacchariflora* Maxim. (1859)
- *Triarrhena sacchariflora* (Maxim.) Nakai.

2. **Miscanthus lutarioriparius** L. Liou ex Renvoize & S. L. Chen.

Section Miscanthus

- Lemmas awned
- Stamens 3
- Stigmas laterally exserted

1. **Miscanthus sinensis** Anders. (1855)
- *Miscanthus condensatus* Hack.
- *M. flavidus* Honda:
- *M. kanehirai* Honda:
- *M. purpurascens* Anders.
- *M. transmorrisonensis* Hayata.

2. **Miscanthus floridulus** (Labill.) Warburg ex K. Schum. & Lauterb. (1901)
- *Saccharum floridulum* Labill. 1824;
- *Miscanthus japonicus* Anders. (1855)

Section Sclerostachya ined.

- Lemma awnless
- Stamens 3
- 1 species

1. **M. fuscus** (Roxb.) Benth. (1881)
- *Saccharum fuscum* Roxb. (1820)
- *Sclerostachya fusca* (Roxb.) Camus (1922)

Section Kariyasua Ohwi ex Hirayoshi.
- Nishikawa & Kubono (1956)
- Lemmas awned or awnless
- Stamens 3
- Stigma exserted laterally
- 4 species

1. **M. intermedius** (Honda) Honda (1936)
- *M. longiberbis* Nakai var *intermedius* Honda. (1933)
- *M. tinctorius* (Steud.) Hack. var *intermedius* (Honda) Ohwi (1942)
- *M. oligostachyus* Stapf ssp *intermedius* (Honda) T. Koyama (1987)

2. **M. longiberbis(Hack.)** Nakai (1916)
- *M. matsumurae* Hack. var *longiberbis* Hack. (1904)
- *M. sinensis var longiberbis* (Hack.) I.C. Chung (1955)
- *M. changii* Y.N. Lee (1964)
- *M. oligostachyus var longiberbis* (Hack.) I.C. Chung (1965)

3. **M. oligostachyus** Stapf (1898)
- *M. matsumurae* Hack. (1899)

4. **M. tinctorius** (Steud.) Hack. (1889)
- *Saccharum tinctorum* Steud. (1855)
- *Miscanthus sieboldii* Honda (1930)

Section Unassigned

- **Miscanthus x giganteus** Greef & Deuter ex Hodkinson & Renvoize (2001)
- **Miscanthus paniculatus** (B.S. Sun) Renvoize & S.L. Chen comb. Nov. (2006)
- *Rubimons paniculatus* B.S. Sun (1997)
- **Miscanthus depauperatus** Merr.
- Lemmas awned
- Stamens 3

Figure 15.1 Taxonomic groups within *Miscanthus* Anderss. For Synonmy see The Online World Grass Flora – Grass Base.[16]

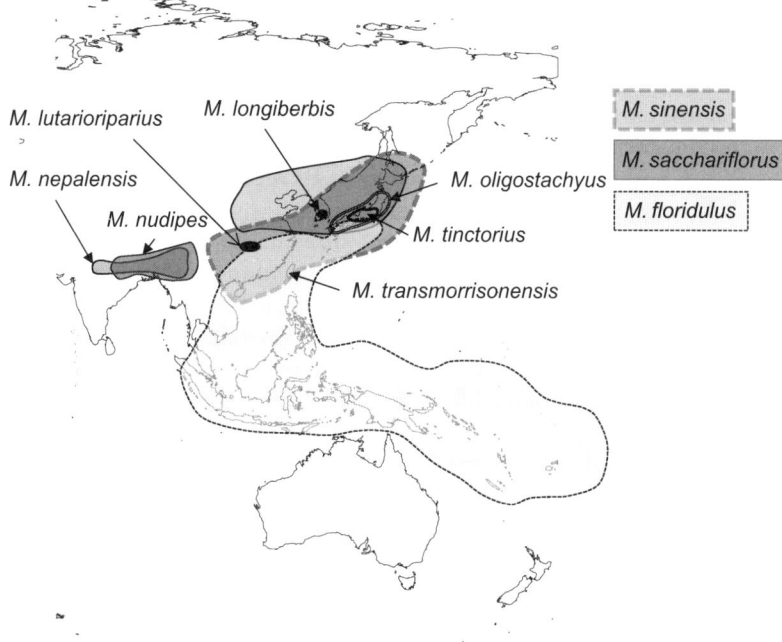

Figure 15.2 Geographical distribution of the major *Miscanthus* species. The distribution of *M. x giganteus* is not fully known but can potentially be found in regions where *M. sinensis* and *M. sacchariflorus* overlap.

M. nudipes and *M. nepalensis* have only two anthers and are sometimes treated as an independent genus *Diandranthus*.[15] They are mainly distributed around the Himalayan region discontinuously from other *Miscanthus* members.

The basic chromosome number of section *Miscanthus* is 19.[17–20] Diploid and tetraploid forms of *Miscanthus* are the most common. Hybrids between 2× and 4× forms have occurred naturally, producing vigourous triploids such as *M. x giganteus*. Cytological studies confirmed *M. x giganteus* as an interspecific hybrid between *M. sinensis* and *M. sacchariflorus*.[21] Hodkinson and Renvoize [22] renamed *M. x giganteus* with a Latin description making this name compliant with the procedures of ICBN (International Code of Botanical Nomenclature reference to be provided). In recent years, it was suspected that several *M. x giganteus* were in circulation in horticulture. Greef *et al.*[23] identified differences on AFLP; however, multilocation trails in Europe with four types showed no agronomic differences.[7,24]

Miscanthus is typically wind pollinated. Temporal separation between the maturation of the stamen and pistil combine with a genetic incompatibility system to make most *Miscanthus* accessions highly self-incompatible.[25] However, there are some reports in *M. sinensis* var. *condensatus*[26] and *M. floridulus* with reasonable levels of self-compatibility. Adati[19] reported a particular *M. sinensis* var. *condensatus* that was shown to be apomictic when progeny were found to be clones by RAPD fingerprint testing.[27]

15.3 Physiological Traits

15.3.1 C4 Photosynthesis in *Miscanthus*

M. x giganteus is exceptional among C4 species for its high productivity in temperate climates.[28,29] In C4 plants, chilling has been shown to decrease carboxylation efficiency,[30] capacity for PEP regeneration[31] and Rubisco activity.[32] Researchers in Illinois have shown that *M. x giganteus* can maintain photosynthetically active leaves at temperatures 6 °C below the minimum for maize.[33] Wang *et al.*[34] demonstrated that the activation state of Rubisco in leaf extracts did not differ in *M. x giganteus* plants that were grown under warm and cold conditions. In a different study, Wang[33] reported that pyruvate pi dikinase (PPDK) expression and activity (which regenerates PEPco) was correlated to the exceptional cold tolerance of photosynthesis in *M. x giganteus*. Interestingly, inseason photosynthesis measurements showed that under high summer temperatures in Illinois *M. x giganteus* had lower assimilation rates per unit of leaf area than maize. Indeed, in Oklahoma, where maximum summer temperatures average above 30 °C, irrigated *M. x giganteus* performed poorly in comparison to native switchgrass (Joe Bouton, personal communication).

15.3.2 Nutrient-Use Efficiency

New shoots grow from an overwintering rhizome in spring, using the rhizome as a source of both carbohydrates and nutrients. In autumn, these assimilates and nutrients are returned to the rhizome and soil before harvest. *M. x giganteus* has been evaluated for ~20 yr in plot trials in Europe. Where the crop is rainfed and the winters are frosty, responses to additional nitrogen fertilisers have rarely been found to be significant. Nutrient analysis in the harvested crop at higher latitudes revealed that offtakes at a spring harvest, after the crop has ripened and senesced over winter, are so low that soil stocks and nutrient deposition from the atmosphere are sufficient for a decade of harvests on reasonable quality soils.[35] In Ireland, *M. x giganteus* unfertilised plots >10 years old showed a gradual reduction in yield relative to fertilised plots. Elemental analysis showed that potassium offtakes were relatively high, and K levels in the soil were low. It was proposed that K fertilisation would restore crop vigour. At lower latitudes where the winters are warmer, responses to nitrogen fertilisation have been shown, particularly when the crop is irrigated and harvests have been made in the late autumn before full senescence has occurred.[36] This may be due to earlier flowering time that is often associated with the onset of senescence. Indeed, some early flowering *M. sinensis* have shown lower nutrient offtakes than *M. x giganteus*, but yields per hectare also tend to be lower.[24]

15.3.3 Water-Use Efficiency

One of the major advantages of plants with C4 photosynthesis is the higher water-use efficiencies (WUE) per unit of dry-matter produced.[37] Several studies in the field and in pots have confirmed that *M. x giganteus* has WUE as good as, and in some cases better than maize.[38] Typical WUE range from 8 to 12 g dry matter per litre water transpired. Studies in Germany during a drought period showed that *M. x giganteus* maintained high rates of photosynthesis at water potentials well below the traditional wilt point of 1.5 MPa.[39] However, as drought conditions persisted, leaves senesced to the point where there was significant reduction in the radiation interception of the canopy. When rainfall resumed, the plant had to regrow a green leaf canopy before resuming production. Interestingly, a *M. sinensis* hybrid produced in Germany had tight stomatal regulation during drought, and, unlike *M. x giganteus*, was capable of resuming full canopy photosynthesis as soon as the water status improved.[39]

15.3.4 Overwintering Frost Tolerance

Severe *M. x giganteus* stand losses have been observed during the first winter at many higher-latitude trial sites in Europe, particularly when the trials were established from tissue-cultured plantlets.[6,40] In January 1998 rhizomes from five *Miscanthus* genotypes were sampled from a field trial in Southern Germany which had been established from tissue cultured plants in May 1997. The rhizomes were subjected to a series of tightly controlled freezing temperatures using a glycol bath.[41] Curve fitting estimated the lethal temperature (LT_{50}) for *M. x giganteus* at $-3.4\,°C$. This was compared with the minimum temperatures recorded at five trials in Europe established in 1997 with exactly the same clones. The pattern of overwinter plant losses matched the minimum temperatures recorded at depths of 10 cm in the soil. This experiment explained for the first time the patterns of overwinter losses observed in many trials over period 1990 to 1999.[42] In trials in central Illinois, *M. x giganteus* established from rhizome-propagated plants survived the first winter when average air temperatures ranged between $-5\,°C$ and $-8\,°C$ in the coldest months.[43] Unfortunately, the temperature in the soil was not recorded in this experiment, but it is likely that temperatures fell below $-3.4\,°C$. This would indicate that *M. x giganteus* rhizomes can survive lower temperatures when trials are planted in environments with exceptionally good growing conditions during establishment before the onset of winter. Furthermore, we have evidence that *M. x giganteus*, given the appropriate ramping down of temperatures can acclimate to low temperatures resulting in lower LT_{50s} (Clifton-Brown and Jones, personal communication). It has often been discussed that unseasonally warm periods during winter, where soil temperatures reach $\sim 10\,°C$ (enough to stimulate growth),[44] render the rhizome particularly sensitive to sudden cold snaps. Climate changes may well increase the frequency of such events, increasing the risk of overwinter frost kill.

It is important to note that the freezing tests in Germany, showed that the rhizomes of a particular *M. sinensis* hybrid could withstand freezing to $-6.5\,°C$, $3\,°C$ lower than *M. x giganteus*.[41] The genetic potential to decrease the risk of overwinter losses is clear, and this is an important trait required for breeding hybrids suitable to regions with cold winters.

15.3.5 Radiation Capture

In field trials in Illinois, Dohleman et al.[29] reported that the total productivity of *M. x giganteus* was 59% higher than from grain maize. The yield advantage from *Miscanthus* was attributed to superior interception of incidental radiation throughout the growing season. Can this be improved? Research funded by the BBSRC in the UK is aimed at increasing the length of the growing season even further by identifying germplasm with lower thresholds for leaf emergence from the overwintering rhizomes, and with concomitant low temperature tolerance to the spring frosts. Modeling and measurements indicate that up to 20% yield improvement could be made if known variation in thermal requirements for leaf emergence and frost tolerance could be combined through breeding.[44]

15.3.6 Pests, Diseases and Weed Control

In Eastern Asia, where *Miscanthus* is indigenous, a range of insect, fungal and viral diseases have been reported. The most serious of these appear to be stem borers (*Lepidopterans*) that can cause the stems to brackle at the base (Xi, personal observations), and fungal infections in the panicle causing ergot (Clifton-Brown, personal observations). In Europe, *Miscanthus* is widely distributed as a garden ornamental. In general, the incidence of pests and diseases have been low, but there have been reports of *Fusarium*[45] and a strain of *Stagonospora*[46] and BYDV.[47] In heated glasshouses severe red spider mite and aphid infections build up, and these can damage the flowering stems. Interestingly, these are not seen outside in the field. In the UK, DEFRA released an advisory note on *Melanaphis sorini*, a large Asian aphid.[48] Importing rhizomes from Asia carries the risk of importing soil nematodes such as *Radopholus similis* (Paul Barber, personal communication, DEFRA, UK). Recently, in the United States, soil sampling has revealed plant-parasitic nematodes – two species of *Xiphinema* (*X. americanum* and *X. rivesi*) and one species of *Longidorus* (*L. breviannulatus*). Currently, nothing is known about how the feeding of these nematodes on roots directly affects bioenergy crops or how the nematodes indirectly affect these crops by transmitting viruses.[49] With the potential for increased production it is likely that more pathogens will find *Miscanthus* a suitable host. An example of this has been observed at Aberystwyth where *Stagonospora* of European origin has caused severe necrosis in several genotypes of *M. sacchariflorus* (Luis Muir, personal communication).

New plantations of *Miscanthus* establish relatively slowly and often compete badly with annual C_3 weed species in European climates. Over the past decade European developers of *M. x giganteus* developed effective protocols for weed control during establishment from rhizomes using soil residual herbicides. Ongoing research is needed because (1) many effective herbicides are being withdrawn from use due to unwanted environmental impacts, and (2) systems for seed-based propagation are in their infancy. In Illinois, Anderson[50] reported on a range of pre- (acetochlor, s-metolachlor, pendimethalin, and atrazine) and postemergence herbicides (postemergence herbicides such as PSII inhibitor (atrazine, bromoxynil)), plant-growth regulators (dicamba, 2,4-D), HPPD inhibitors (topramezone and tembotrione), and some ALS inhibitors (primisulfuron, halosulfuron). Lindsey[51] also reported potential pre- and postemergence herbicides including aminopyralid, dicamba + diflufenzopyr, and atrazine.

15.3.7 Crop Modeling

Modeling provides researchers with the tools to examine and utilise field and laboratory data in a systematic way. Models are an important part of the evaluation of the economic potential and environmental benefits. Spatial crop modeling with GIS has been used to analyse the regional yield potential in the UK,[52–54] Europe[55] and in the US.[56] Karp and Shield[57] present a list of crop models adapted and developed for use in predicting *Miscanthus* yields.

The simple growth model MISCANMOD was developed to predict peak autumn yield in *M. x giganteus* throughout Europe and has been generally validated.[58,59] MISCANFOR,[60] which built on MISCANMOD, is gradually expanding to include parameters for ecophysiological traits from new and/or hypothetical hybrids. In a recent publication, Hastings *et al.*[61] investigated the impact of genetic improvements to frost and drought tolerance on Europe-wide projected yields. Figure 15.3 shows projections for yield in 2080 based on the climate scenario A2 for *M. x giganteus* and a *M. sinensis* hybrid. The *M. sinensis* hybrid can tolerate temperatures to $-6\,°C$, $3\,°C$ below *M. x giganteus*. In addition, the *M. sinensis* hybrid has much tighter control of water loss *via* stomatal closure during periods of water deficit (detailed above). MISCANFOR projects that *M. x giganteus* will be largely unsuitable for rainfed biomass production in most of Europe. Modeling makes the challenges for breeding clear.

15.4 Towards Breeding Better *Miscanthus* for Bioenergy

15.4.1 Setting Breeding Targets

Increasing crop yield is the primary objective of any breeding programme. Crop yield is normally understood to be harvestable dry tonnes per hectare of useful product. Any statement such as "double yield" needs qualification because there are many environmental factors (*e.g.* inputs, soil quality, local climate)

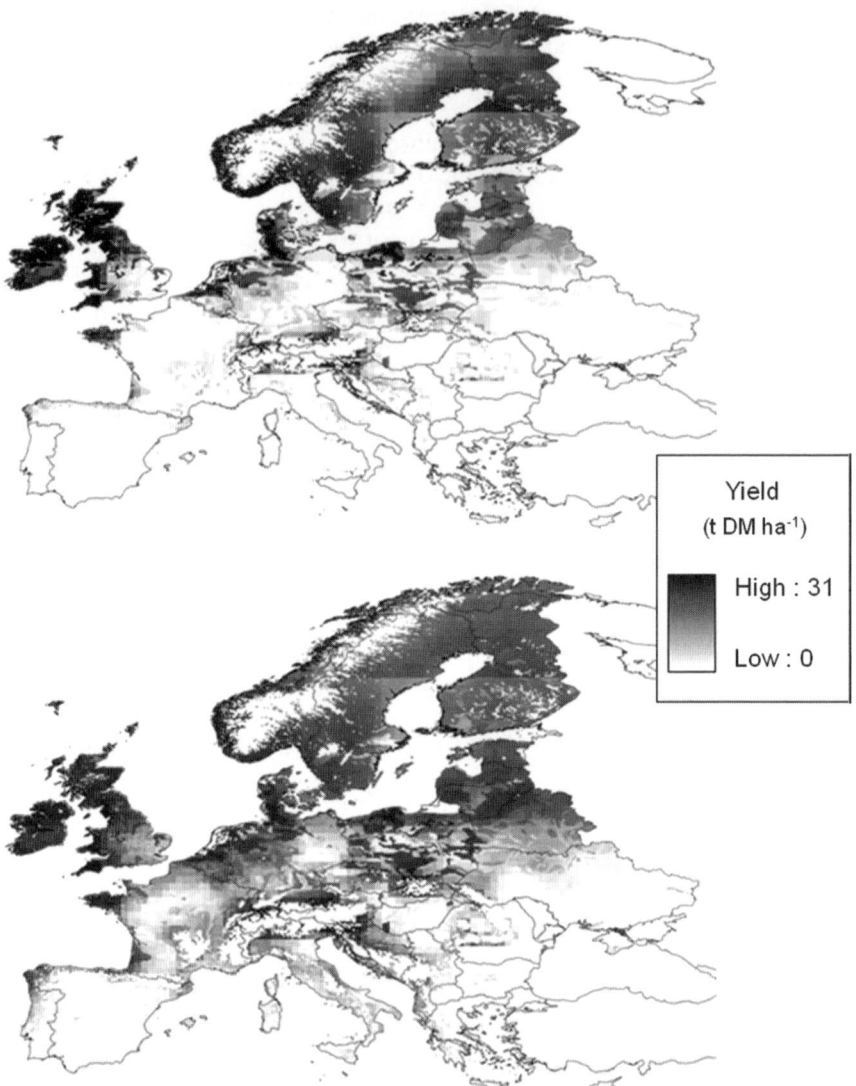

Figure 15.3 Yield projections (T DM ha^{-1} yt^{-1} at spring harvest) for *Miscanthus x giganteus* and a *Miscanthus sinensis* hybrid (Hi-tech hybrid) for the IPCC A2 scenario in 2080. White areas are predicted to be either too dry (lower latitudes) or too cold (higher and Eastern latitudes).

which adjust the yield performance of a new hybrid or variety. A more precise target is "maximise net energy yield per hectare" which is energy output minus the energy input in whatever form is needed to grow and process the crop.

Since *Miscanthus* is largely undomesticated, it is still a challenge for the breeder to identify which of many traits to incorporate into selections.[57]

One scientific approach is to use a simple model such as that proposed by Monteith.[62]

$$\text{Yield} = \text{Incident radiation} \times \text{Intercepted radiation} \times \text{radiation-use efficiency} \quad (15.1)$$

With such a model it is possible to examine the importance of observed phenotypic variation on yield. Much effort continues to be devoted to improving radiation-use efficiency by increasing photosynthesis but, in practical breeding, it is an almost impossible task to screen for photosynthetic traits because the measurements are too laborious. Traits contributing to radiation capture are easier and faster to evaluate. Once carbon is assimilated, the chemical forms in which the carbon is stored will influence downstream utilisation.

15.4.2 *Miscanthus* Germplasm Collection and Characterisation

The key to a scientific-based breeding programme is a thorough knowledge of the phenotypic and genotypic variation in wild germplasm. The general distribution of *Miscanthus* genetic resources is shown in Figure 15.2. Within the general geographic zones there is huge phenotypic variation resulting from environmental pressures (*e.g.* salinity, high altitude) and from genetic isolation leading to specific ecotypes. Descriptions of the variation of genetic resources in Taiwan,[63,64] Japan[65,66] and China[67,68] show the huge diversity of traits available for breeding.

Germplasm from a wide range of Asian locations is being evaluated in spaced plant field trials near Aberystwyth (Figure 15.4). Intensive phenotype measurements have been made over the past 3 years on a whole suite of traits including flowering time, plant architecture and morphology and lignocellulosic composition.[69] Valuable variation has been identified. Molecular genotyping has resulted in diversity maps and phylogenies of all plants that have been characterised in the field. This genotyping is based on the discovery of molecular polymorphisms between different plants at large numbers of genetic loci. Ceres has discovered many such polymorphisms and is identifying many more. Statistically significant associations are sought between particular traits in a large number of plants and markers that define versions (haplotypes) of particular chromosome segments.[70] Such chromosome segments thus become candidates for determining the traits. Many such segments can be associated with a particular trait. While such analyses are very powerful the associations can be spurious and simply be due to the genetic closeness of the lines used to make the association. Use of germplasm taken from areas distributed across Asia reduces the probability of such false associations. Other statistical tests to support the validity of the associations can be applied. The techniques of genomic selection[71,72] can also be applied to increase the efficiency of breeding when many markers are available. In this approach, markers mapping across the genome are statistically linked with traits and then used to

Figure 15.4 Characterisation of *Miscanthus* germplasm in a replicated spaced plant field trial near Aberystwyth.

select the traits instead of selecting the traits themselves. This technology is still under development in plants but may turn out to be very powerful.

15.4.3 Crossing

Hybridisation of different genotypes to create new progeny is relatively easy for day neutral *M. sinensis*. In the UK, most *M. sacchariflorus* requires photoperiod induction before flowering will occur. Synchronisation of flowering between genotypes from different latitudes is an area of intense research activity.[73] Phenotypic data and genetic diversity analyses are being used to inform novel parental combinations for crosses.

15.4.4 Selection

Once progeny have resulted from certain parental combinations, field evaluation is required and this is traditionally performed in spaced plant selection trials. In the UK, where temperatures are quite low, we are finding that it is necessary to wait 2–3 years before reliable selections can be made. Leading plants are promoted to plot trials. To date, early morphometric analysis of phenotypes following the first year proved of limited value for identifying the mature phenotype. Molecular markers will play an important role in increasing the speed and precision of the breeding programme.

From the viewpoint of breeding, vegetative propagation makes the breeding of *Miscanthus* relatively straightforward since cloning fixes the hybrid characteristics from plant to plot to commercial crop. However, cloning is expensive and a seed-propagated crop would have many benefits.[74] Synthetic varieties have been developed in other cross-pollinated crops such as perennial ryegrass.[75] A synthetic variety was defined by Allard[76] as a variety that is maintained from open-pollinated seed following its synthesis by hybridisation in all combinations among a number of selected genotypes. Allard indicated that many different procedures can be used to determine combining ability, from simple visual inspection for highly heritable characters to progeny testing. Seed of synthetics is multiplied up over several generations.

15.5 Development of the Whole Biomass Chain

It is envisaged that different hybrids (or varietal populations) will be selected for different climates and end uses. Each variety will require the development of specific agronomic knowledge to launch it commercially.

15.5.1 Propagation and Establishment

In the case of *M. x giganteus*, propagation is limited to cloning since this hybrid is sterile. Planting grants are aimed at mitigating some of the high costs associated with clonal propagation.[77] The EU has given Defra permission to raise the planting grant in England from 40% to 50% of actual costs but this will not be a sufficient incentive to make a huge difference to farmer uptake. In the medium to long term, planting grants may be unsustainable. The planting of 300 000 ha of energy crops would cost over £190 m. Improvements in the techniques to establish *M. x giganteus* include better rhizome splitting and planting machinery, and the recent advances in *in vitro* methods. Such improvements and the economies of scale might bring the costs of vegetative propagation down.

In theory, seed-based varieties could considerably reduce establishment costs, probably to around £450/ha. Harvesting and sowing seed are well-developed technologies in many species. There are a number of issues to address.

First, *Miscanthus* has a small seed and a relatively high temperature requirement for germination. Our experiments in Aberystwyth show that *M. sinensis* has considerable variation in the thermal requirements for germination, but these were well above thresholds for *Lolium perenne* varieties bred in Aberystwyth. *Miscanthus* seed germination was also improved by gibberellic acid and KNO_3, and antifungal agents such as sodium hypochlorite (domestic bleach) also improved seed germination (Do-Soon Kim, personal communication). Much research is needed to develop reliable establishment from seed in *Miscanthus*.

Secondly, since *Miscanthus* is an outbreeding species with self-incompatibility preventing selfing, new methods have to be developed to optimise heterosis and minimise inbreeding. The production of synthetics and F1 hybrids are the main options to consider. Uniformity will also have to be considered; doubtless a seed-based crop based on a number of parents will not be as uniform as a crop produced clonally.

Thirdly, we will have to ensure that in those areas where *Miscanthus* is not native, seed propagation will not cause the species to be a serious invasive weed. Any risk would be greatly reduced by making the UK crop sterile.

15.5.2 Harvest

Optimising harvest time will vary according the hybrid. *M. x giganteus* does not normally flower under UK climate conditions, and, consequently, senescence really only begins after the first frost in autumn. To ensure translocation of growth nutrients back to the rhizome and reduce nutrient offtake at harvest, *M. x giganteus* is harvested in the early spring after winter ripening. Winter ripening also reduces the moisture content from 60–80% to about 30–40%, depending on the weather conditions. During ripening, harvestable yields also decline by about 30% from the autumn peak yield.[35,78] Where novel hybrids flower and begin to senescence before the end of the growing season, earlier harvests may be possible. There is also considerable variation in stem diameter and strength. These traits will influence the machinery needed for harvest. In Europe, machinery to harvest *Miscanthus* has been developed from the existing infrastructure of grass machines. However, as the scale increases, there will be an increasing demand for more efficient and faster methods. In Ireland, a one-pass cutter and bailer has been developed (Figure 15.5).

The format in which the biomass is harvested has an impact on both storage options and transport. Recent research in Ireland evaluated storage options for round bales.[79] It would seem logical that large square bales (see http://www.hesston.agcocorp.com) appear to be the most appropriate since they stack well, are denser than round bales and allow the biomass to continue to dry out. However, large Heston-style balers require very powerful tractors, and it may be that this is prohibitively expensive in certain areas.

15.5.3 Utilisation/Fuel Matching with End Uses

For combustion, higher lignin contents improve the calorific value of the biomass. In the UK, combustion facilities using biomass range from small dedicated 5–10 kW biomass pellet burners such as RIKA central heating stoves to 1000 MW power stations employing co-combustion with coal.

The main components of dry biomass are cellulose, hemicellulose and lignin. The proportion of each of these components affect gross calorific value and the kinetics of devolatisation during combustion.[80] Lignin has a higher gross calorific value (23.3–25.6 MJ/kg) in relation to biomass combustion than

Figure 15.5 One-pass cutting and baling system developed for *Miscanthus* harvesting in Ireland (Photo: Keith Armitage).

cellulose (nearly 18.6 MJ/kg) and hemicellulose (variable but has a similar composition to cellulose), due to higher carbon and, to a lesser extent, hydrogen contents. Hemicellulose decomposes first followed by cellulose with lignin being responsible for the flat tailing in the relationship between temperature and decomposition. This affects consecutive steps in combustion and may be of importance in cofiring.

Species and genotypes with low or altered lignin are required for the generation of next-generation ethanol from lignocellulose, since cellulose and hemicellulose degrade to simple sugars that can be converted to ethanol and lignin hinders degradation. Limited work has been undertaken on differences between varieties in combustion quality. Steer *et al.*, 2008[81] showed that *M. x giganteus* had a lower ash fusion temperature index (894 °C) than *Miscanthus* cv Goliath (1034 °C). Thus, *M. x giganteus* is more likely to cause combustion issues such as slagging. For applications such as ethanol production, lower lignin is desirable because lignin makes the cell walls more difficult to degrade.

15.6 Life-Cycle Analysis and Energy Balance

As the world population is expected to rise to 9 billion in 2050, good land will be needed for food production. Bioenergy, therefore, will be confined to using lower-grade land. The recent fuel versus food debate raised many issues regarding the development of bioenergy, where the net energy produced per hectare becomes the key metric as land resources become more valuable.[82] The channeling of food crops such as maize and wheat into ethanol production because of limited long-term value since these crops require high inputs and therefore their net energy per hectare is much lower. Life-cycle analysis provides a framework for "accounting" the net benefit and is capable of identifying good and bad sources of bioenergy. *Miscanthus* is capable of achieving some of the highest energy balances of all candidate energy-crop species.[83] Further improvements to the crop need to optimise the net energy yield per hectare. Rowe *et al.*[84] comment that energy crops may offer significant benefits for the environment; the potential will only be realised if landscape-scale issues are effectively managed and the whole chain of crop growth and utilisation is placed within a regulatory framework where sustainability is the central driver.

15.7 Biodiversity and Visual Impact

Less research has been undertaken on the potential impacts of *Miscanthus* on biodiversity compared to other energy crops like short-rotation coppice (SRC). Semere and Slater[85] reported that young crops of *Miscanthus* provide a substantially improved habitat for many forms of native wildlife, due to the low intensity of the agricultural management system and the untreated headlands. *Miscanthus* fields were shown to be used as overwintering sites for invertebrates.[86] Bellamy *et al.*[87] reported that *Miscanthus* fields had a greater abundance and diversity of farmland birds than winter or summer wheat. Work undertaken in the RELU-Biomass project showed that the abundance of total butterflies was significantly greater in field margins surrounding *Miscanthus* than in field margins of arable crops.[88]

The potential for *Miscanthus* to become an invasive weed depends on many factors, including the species and the local environmental conditions. The current commercial strain of *Miscanthus* is sterile. Therefore, there are no volunteers produced from seed lost either on location or *via* transport to other places. Jorgensen and Muhs[89] considered that fertile *Miscanthus* could become a serious weed problem. However, fertile *M. sinensis* has not become a serious weed in Aberystwyth or Rothamsted. In order to become invasive, any species has to have high capacity to disperse, colonise and compete with other species. Experience to date shows that seed dispersal is limited to several metres from the maternal plant. It could potentially be eliminated by selecting genotypes that are either sterile, or do not flower under the prevailing climatic conditions.[90] Certain types of *M. sacchariflorus* have strong creeping rhizomes, which, left unmanaged have the potential to colonise large areas. The rhizomes can be controlled by ridges, similar to those used for sugarcane (Xi, personal communication).

The visual impact of *Miscanthus* has received less attention than for SRC willow, perhaps reflecting the lower crop height in comparison to SRC of 3.5 m, and the more traditional harvest cycle and appearance of *Miscanthus*.[84] Surveys of public opinion undertaken in the RELU project[91] found that more than 75% of respondents felt that *Miscanthus* would fit "very well" or "reasonably well" into the English landscape. It was concluded that it was unlikely that widescale planting of biomass crops would give rise to any substantial public concern in relation to their visual impact in the landscape.

15.8 Conclusions

The natural diversity of *Miscanthus* needs to be harnessed through breeding to produce hybrids suited to lower grade land for a wide range of climates. We have outlined the current status of taxonomic classification, phenotypic characterisation, and breeding, and identified many areas where research and development is needed to produce sustainable bioenergy from *Miscanthus* in the future. In this review we have used a model to highlight the importance of traits associated with drought and frost. Implementation of the genetic improvements will influence all parts of the life-cycle analysis. Our aim is to produce varieties that can be developed through public and private sector relationships to contribute to the global effort in reducing dependence on fossil fuels.

Acknowledgements

We are grateful for funding from the following sponsors: DEFRA, BBSRC, EPSRC (all in the UK), the Taiwan National Science Council for supporting a conference in Taiwan (November 2009) and CERES Inc. Drs. Paul Robson, Kerrie Farrar, Elaine Jensen, Gordon Allison and Ed Hodgson, are involved in the characterisation of germplasm for breeding that has been mentioned in this chapter. JCB would also like to acknowledge members of the breeding team in Aberystwyth, in particular Mrs. Sue Youell, Dr. Maurice Hinton-Jones, Dr. Charlotte Hayes, Dr. Richard Webster, Mr. Peter Roberts and Mr. Brian Jones.

References

1. J. M. Greef and M. Deuter, *Angew. Bot.*, 1993, **67**, 87–90.
2. P. N. Nielsen, *Tidsskrift Planteavl*, 1987, **91**, 275–281.
3. VEBA_OEL, Bundesministerium für Ernährung, Landwirtschaft und Forsten (BMELF), 14. Sept., Gelsenkirchen, 1993.
4. K.-U. Schwarz, J. M. Greef and E. Schnug, *Untersuchungen zur Etablierung und Biomassebildung von Miscanthus giganteus unter verschiedenen Umweltbedingungen*, Bundesforschungsanstalt für Landwirtschaft, Braunschweig-Völkenrode (FAL), Braunschweig-Völkenrode, 1995.
5. M. B. Jones and M. Walsh, (ed.), *Miscanthus-for Energy and Fibre*, James and James (Science Publishers), London, 2001.

6. J. C. Clifton-Brown, S. P. Long and U. Jørgensen, in *Miscanthus-for Energy and Fibre*, M. B. Jones and M. Walsh, (ed.), James and James (Science Publishers), London, 2001, pp. 46–67.
7. J. C. Clifton-Brown, I. Lewandowski, B. Andersson, G. Basch, D. G. Christian, J. Bonderup-Kjeldsen, U. Jørgensen, J. Mortensen, A. B. Riche, K.-U. Schwarz, K. Tayebi and F. Teixeira, *Agron. J.*, 2001, **93**, 1013–1019.
8. J. Daniels and B. T. Roach, *Sugar cane Spring Supplement*, 1987, 16–19.
9. T. R. Hodkinson, M. W. Chase, M. D. Lledo, N. Salamin and S. A. Renvoize, *J. Plant Res.*, 2002, **115**, 381–392.
10. Y. N. Lee, *J. Jpn. Bot.*, 1964, **39**, 196–205.
11. Y. N. Lee, *J. Jpn. Bot.*, 1964, **39**, 257–265.
12. Y. N. Lee, *J. Jpn. Bot.*, 1964, **39**, 289–298.
13. W. D. Clayton and S. A. Renvoize, *Genera graminum : Grasses of the world.*, Her Majesty's Stationery Office, London, 1986.
14. T. Hoshino and G. Davidse, *Ann. Missouri Botanical Garden*, 1988, **75**, 866–873.
15. Y. Ibaragi, *Acta Phytotaxonomica et Geobotanica*, 2003, **54**, 109–125.
16. W. D. Clayton, K. T. Harman and H. Williamson, Royal Botanic Gardens, Kew, 2002. http://www.kew.org/data/grasses-syn/
17. I. Hirayoshi, K. Nishikawa, R. Kato and M. Kitagawa, *Jpn. J. Breed.*, 1955, **5**, 49–50 (in Japanese).
18. I. Hirayoshi and S. Mitsuishi, *Bull. Fac. Agric. Mie Univ.*, 1956, **12**, 1–10 (in Japanese).
19. S. Adati, *Bull. Fac. Agric. Mie University*, 1958, **17**, 1–112.
20. S. Adati and I. Shiotani, *Bull. Fac. Agric. Mie University*, 1962, **25**, 1–24.
21. I. B. Linde-Laursen, *Hereditas*, 1993, **119**, 297–300.
22. T. R. Hodkinson and S. Renvoize, *Kew Bulletin*, 2001, **56**, 759–760.
23. J. M. Greef, M. Deuter, C. Jung and J. Schondelmaier, *Genet. Resources Crop Evol.*, 1997, **44**, 185–195.
24. I. Lewandowski, J. C. Clifton-Brown, B. Andersson, G. Basch, D. G. Christian, U. Jørgensen, M. B. Jones, A. B. Riche, K.-U. Schwarz, K. Tayebi and F. Teixeira, *Agron. J.*, 2003, **95**, 1274–1280.
25. I. Hirayoshi, K. Nishikawa and R. Kato, *Jpn. J. Breed.*, 1955, **5**, 19–22.
26. C. H. Chou, Y. C. Chiang and T. Y. Chiang, *Canad. J. Bot.*, 2000, **78**, 1262–1268.
27. C.-H. Chou, Y.-C. Chiang and T.-Y. Chiang, *Can. J. Bot.*, 2000, **78**, 1262–1268.
28. C. V. Beale and S. P. Long, *Plant, Cell Environ.*, 1995, **18**, 641–650.
29. F. G. Dohleman and S. P. Long, *Plant Physiol.*, 2009, **150**, 2104–2115.
30. A. H. Kingston-Smith, J. Harbinson and C. H. Foyer, *Plant Cell Environ.*, 1999, **22**, 1071–1083.
31. Y. C. Du, A. Nose and K. Wasano, *Plant Cell Physiol.*, 1999, **40**, 298–304.
32. D. S. Kubien and R. F. Sage, *Plant Cell Environ.*, 2004, **27**, 1424–1435.

33. D. F. Wang, A. R. Portis, S. P. Moose and S. P. Long, *Plant Physiol.*, 2008, **148**, 557–567.
34. D. Wang, S. L. Naidu, A. R. Portis, S. P. Moose and S. P. Long, *J. Exp. Bot.*, 2008, **59**, 1779–1787.
35. J. C. Clifton-Brown, J. Breuer and M. B. Jones, *Glob. Change Biol.*, 2007, **13**, 2296–2307.
36. S. L. Cosentino, C. Patane, E. Sanzone, V. Copani and F. Salvatore, *Indust. Crops Prod.*, 2007, **25**, 75–88.
37. S. P. Long, *Plant, Cell Environ.*, 1983, **6**, 345–363.
38. C. V. Beale, J. I. L. Morison and S. P. Long, *Agr. For. Meteorol.*, 1999, **96**, 103–115.
39. J. C. Clifton-Brown, I. Lewandowski, F. Bangerth and M. B. Jones, *New Phytol.*, 2002, **154**, 335–345.
40. D. G. Christian and E. Haase, in *Miscanthus - for Energy and Fibre*, M. B. Jones and M. Walsh, (ed.), James and James (Science Publishers), London, Editor edn, 2001, pp. 21–45.
41. J. C. Clifton-Brown and I. Lewandowski, *New Phytol.*, 2000, **148**, 287–294.
42. U. Jørgensen and K.-U. Schwarz, *New Phytol.*, 2000, **149**, 190–193.
43. E. A. Heaton, F. G. Dohleman and S. P. Long, *Glob. Change Biol.*, 2008, **14**, 2000–2014.
44. A. D. Farrell, J. C. Clifton-Brown, I. Lewandowski and M. B. Jones, *Ann. Appl. Biol.*, 2006, **149**, 337–345.
45. K. Thinggaard, *Acta Agr. Scand.*, 1997, **47**, 238–241.
46. N. R. O'Neill and D. F. Farr, *Plant Disease*, 1996, **80**, 980–987.
47. D. G. Christian, J. N. L. Lamptey, S. M. D. Forde and R. T. Plumb, *Eur. J. Plant Pathol.*, 1994, **100**, 167–170.
48. R. Hammon, S. Reid, L. Matthews and D. Eyre, *DEFRA plant pest notice*, 2006, **44**.
49. T. Mekete, M. E. Gray and T. L. Niblack, *Glob. Change Biol. Bioenergy*, 2009, **1**, 257–266.
50. E. Anderson, A. Hager, T. Voigt, G. Bollero, *Proceeding of 6th Annual Bioenergy Feedstocks Symposium*, Illinois, 2009.
51. A. Lindsey, W. Everman, C. Glaspie, *Proceeding of 2009 North Central Weed Science Society Conference*, 2009.
52. L. Price, M. Bullard, H. Lyons, S. Anthony and P. Nixon, *Biomass Bioenergy*, 2003, **26**, 3–13.
53. A. A. Lovett, G. M. Sunnenberg, G. M. Richter, A. G. Dailey, A. B. Riche and A. Karp, *BioEnergy Res.*, 2009, **2**, 17–28.
54. G. M. Richter, A. B. Riche, A. G. Dailey, S. A. Gezan and D. S. Powlson, *Soil Use Manage.*, 2008, **24**, 235–245.
55. P. Stampfl, J. C. Clifton-Brown and M. B. Jones, *Glob. Change Biol.*, 2007, **13**, 2283–2295.
56. M. Khanna, B. Dhungana and J. Clifton-Brown, *Biomass Bioenergy*, 2008, **32**, 482–493.
57. A. Karp and I. Shield, *New Phytol.*, 2008, **179**, 15–32.

58. J. C. Clifton-Brown, B. M. Neilson, I. Lewandowski and M. B. Jones, *Indust. Crops Prod.*, 2000, **12**, 97–109.
59. J. C. Clifton-Brown, P. Stampfl and M. B. Jones, *Glob. Change Biol.*, 2004, **10**, 509–518.
60. A. Hastings, J. C. Clifton-Brown, M. Wattenbach, C. P. Mitchell and P. Smith, *Glob. Change Biol. Bioenergy*, 2009, **1**, 154–170.
61. A. Hastings, J. Clifton-Brown, M. Wattenbach, C. P. Mitchell, P. Stampfl and P. Smith, *Glob. Change Biol. Bioenergy*, 2009, **1**, 180–196.
62. J. L. Monteith, *Philos. Trans. Roy. Soc.*, 1977, **B281**, 277–294.
63. C. H. Chou, S. Huang, S. H. Chen, C. S. Kuoh, T. Y. Chiang and Y. C. Chiang, *National Science Council Monthly*, 1999, **27**, 1158–1169.
64. C. H. Chou, *Renew. Energy*, 2009, **34**, 1908–1912.
65. J. Ryan Stewart, Y. Toma, F. G. Fernandez, A. Nishiwaki, T. Yamada and G. Bollero, *Glob. Change Biol. Bioenergy*, 2009, **1**, 126–153.
66. H. Iwata, T. Kamijo and Y. Tsumara, *Conserv. Genet.*, 2005.
67. S. L. Chen and S. A. Renvoize, *Flora China*, 2006, **22**, 581–583.
68. Q. Xi and S. Jezowkski, *Plant Breed. Seed Sci.*, 2004, **49**, 63–77.
69. J. Clifton-Brown, P. Robson, G. Allison, S. Lister, R. Sanderson, C. Morris, E. Hodgson, K. Farrar, S. Hawkins, E. Jensen, S. Jones, L. Huang, P. Roberts, S. Youell, B. Jones, A. Wright, J. Valentine and I. Donnison, *Aspects Appl. Biol.*, 2008, **90**, 199–206.
70. F. Breseghello and M. Sorrells, *Crop Sci.*, 2006, **46**, 1323–1330.
71. E. Heffner, A. Lorenz, J. Jannink and M. Sorrells, *Crop Sci.*, 2010, in press.
72. T. Meuwissen, B. Hayes and M. Goddard, *Genetics*, 2001, **157**, 1819–1829.
73. E. Jensen, S. Thomas-Jones, K. Farrar, J. Clifton-Brown and I. Donnison, *Compar. Biochem. Physiol., Part A*, 2008, **150**, S181–S181.
74. C. J. Atkinson, *Biomass Bioenergy*, 2009, **33**, 752–759.
75. P. W. Wilkins and M. O. Humphreys, *J. Agr. Sci.*, 2003, **140**, 129–150.
76. R. W. Allard, *Principles of Plant Breeding*, Wiley International Edition John Wiley & Sons, Inc., New York, London, 1960.
77. Defra, *The Energy Crops Scheme Business Case*, 2006.
78. S. A. Gezan and A. B. Riche, *Aspects Appl. Biol.*, 2008, 219–223.
79. A. Nolan, K. M. C. Donnell, M. M. C. Siurtain, J. P. Carroll, J. Finnan and B. Rice, *Biosyst. Eng.*, 2009, **104**, 345–352.
80. A. Gani and I. Naruse, *Renew. Energy*, 2007, **32**, 649–661.
81. J. M. Steer, M. Hinton-Jones, J. Valentine, Y. Liu, C. K. Tan, J. Ward and S. J. Wilcox, The characterisation of the physical and chemical properties of biomass for power generation, *Energy from Biomass and Waste Project*, CU Report No. 3161, 45pp.
82. J. Vidal, *The Guardian (UK Newspaper)*, 2007, guardian.co.uk/environment.
83. I. Lewandowski and U. Schmidt, *Agr. Ecosyst. Environ.*, 2006, **112**, 335–346.
84. R. L. Rowe, N. R. Street and G. Taylor, *Renew. Sustain. Energy Rev.*, 2009, **13**, 260–279.
85. T. Semere and F. M. Slater, *Biomass Bioenergy*, 2007, **31**, 20–29.
86. T. Semere and F. M. Slater, *Biomass Bioenergy*, 2007, **31**, 30–39.

87. P. E. Bellamy, P. J. Croxton, M. S. Heard, S. A. Hinsley, L. Hulmes, S. Hulmes, P. Nuttall, R. F. Pywell and P. Rothery, *Biomass Bioenergy*, 2009, **33**, 191–199.
88. A. J. Haughton, A. J. Bond, A. A. Lovett, T. Dockerty, G. Sunnenberg, S. J. Clark, D. A. Bohan, R. B. Sage, M. D. Mallott, V. E. Mallott, M. D. Cunningham, A. B. Riche, I. F. Shield, J. W. Finch, M. M. Turner and A. Karp, *J. Appl. Ecol.*, 2009, **46**, 315–322.
89. U. Jørgensen and H.-J. Muhs, in *Miscanthus - for Energy and Fibre*, M. B. Jones and M. Walsh, (ed.), James and James (Science Publishers), London, Editor edn, 2001, pp. 68–85.
90. K. Jakob, F. S. Zhou and A. Paterson, in *Vitro Cell. Develop. Biol.-Plant*, 2009, **45**, 291–305.
91. A. Karp, *NNFCC Conference on "The Impact of Land Use Changes – Myth or Reality"*, January, 2010, London.

CHAPTER 16
Subtropical and Tropical Reeds for Biomass

MIHÁLY CZAKÓ AND LÁSZLÓ MÁRTON

University of South Carolina, Department of Biological Sciences, 700 Sumter St., Columbia, SC 29208, USA

16.1 Introduction

Biomass crops have attracted increasing interest since as far back as the 1970s because they may satisfy a significant part of the energy demand and at the same time reduce carbon-dioxide emission (CO_2).[1–6] Their use reduces anthropogenic CO_2 emissions because during their combustion less CO_2 is released than was fixed into biomass by photosynthesis while the crop was growing.[7–9] Furthermore, for perennial rhizomatous crops soil management (planting and related tillage) needs are limited, thus reducing the risk of soil erosion[10] and increasing the potential for enhancing soil carbon as well as biodiversity.[11] Moreover, since perennial rhizomatous grasses, including giant reed, recycle nutrients by their rhizome systems,[12] they have a low demand for nutrient inputs. They also have few natural pests, especially when cultivated far away from their natural origin, they may also be produced without pesticide use.[13]

Various large perennial rhizomatous grasses have been considered for energy crops because they produce lignocellulosic biomass that is ideal for fuel and because they also display a good adaptability to a wide range of environments and climates from tropical and subtropical to Mediterranean and mild continental. As there is abundant literature on certain well-established economic reeds/canes like sorghum and biomass sugarcanes, and *Miscanthus* spp., and since the tropical Napier grass (*Pennisetum purpureum* Schumach.) is not a typical reed but a leafy crop, this review focuses on *Arundo donax* L., and its congeneric and more distant relatives such as broom grass (*Thysanolaena latifolia* Honda) and caña brava (*Gynerium sagittatum* Beauv.).

16.2 Giant Reed – *Arundo*: Botany and Nonbiomass Uses

Arundo is a genus of the grass family (Poaceae) represented by four species distributed over a range from the Mediterranean Europe to China and Taiwan.[14–16] *Arundo donax* L., a perennial reed, is by far the most robust species growing up to 10 m.[14] Unlike its relatives, *A. donax* does not set fruit because developmental steps are missing in seed production. It has been shown that the microspores do not become pollen grains, and that while some ovaries occasionally simulate the form of fruit, the female gametophyte development ceases with the occasional appearance of archesporial cells. Egg cells are never produced.[17,18] It can be propagated only vegetatively, especially by rhizomes[19] but the more cost-efficient, large-scale propagation method is by *in vitro* culture.[20,21]

Arundo donax is known as giant reed, Spanish cane, or wild cane in English and is native to areas from Mediterranean Europe to India and the foothills of the Himalayas in Nepal.[22–24] It is believed to have originated from the giant grass thickets of middle Asia and India[24,25] and was domesticated in lands from the Mediterranean Region to Caucasia and in the valleys of the Tiger and Euphrates Rivers.[26] Giant reed has been introduced into most countries around the world with warmer tropical and temperate climates and it is often found as an escaped ornamental[27,28] even in northern Europe and Canada, where it adapted very well to very cold winters and short growing seasons (with decreased yield).[29]

It was present in Europe in the 1600s but when it was introduced in the USA is not certain, maybe as early as the 1700s.[30] Records indicate that it was introduced into the USA in California in the early 1800s[31,32] from where it was distributed throughout the country, mostly intentionally. It has naturalised in California, Texas, Arkansas, Missouri and Virginia and in between and southward.[33] It was introduced into Australia over 150 years ago.[34]

Giant reed can grow on a wide range of soil types and is drought, salinity and inundation tolerant, but grows most vigorously in moist, well-drained substrates along lakes, ditches, canals, and rivers.[28] In the southeastern US, giant reed plantings are common and persistent along roadside ditches, but it does not spread aggressively.[33,35] Giant reed has naturalised in coastal areas of the several Hawaiian Islands to form thickets[36] but so far it is not highly invasive in these areas.[33] It thrives in some areas of Florida, such as near Panama City and Cape Canaveral.[33,37] It has become invasive along muddy banks of waterways in warmer areas of the Southwest such as the coastal rivers of southern

California[38,39] and along the lower Rio Grande in the Big Bend region of Texas.[40]

Giant reed forms clumps of culms. It has stout and sometimes globular rhizomes but not long, fast-creeping stolons such as produced by the common reed (*Phragmites communis*) or certain bamboo species. The colony grows slowly by the extension of rhizomes[5] and by layering from lodged stems in moist areas. Spreading is facilitated by moving water when a colony is undermined by water or otherwise dislodged, *e.g.* by dredging, and rhizomic rafts float away on water or are carried away attached to vehicles. The well-publicised spread of giant reed in California happens during floods; the rhizomic rafts lodge on banks downstream where they take root[41] indicating the need of a buffer zone between the Arundo stands and moving water.

Giant reed can outcompete native vegetation by virtue of its rapid growth rate and ability to recover after fires (because of its relatively high moisture 55–70% it is much less of a fire hazard than other perennial grasses that dry out completely by the end of the growing season). The resulting extensive climax stands are considered to have much lower wildlife value[41,42] and to be a flood-management problem in southern California coastal rivers where this originally man-made problem is attempted to be remedied by expensive mechanical and chemical controls.[38,43] Considerable effort has been invested into the much less effective biological control by imported arthropod pests that were actually released in the USA[44,45] but in the light of contemplated application of giant reed in the USA[46] the pests are bound to cause overall economic damage.

Giant reed has been the best source of reeds for woodwind instruments. It is cultivated for this purpose in France,[28] Australia[47] and California, USA.[33] Giant reed has been in cultivation for its fibre in Europe, Eurasia and North and South America to make paper, textiles, and construction boards.[28,33,48–51] As much as 6000 hectares had been once in cultivation as a raw material for rayon in Torviscosa, Italy, in the late 1930s.[52]

In India, giant reed supplemented with salt was adequate for maintenance of cattle that relish the young shoots (also in the USA[33,53]) while older shoots are of relatively low palatability.[54] Young, tender shoots, especially of the variegated are used as fodder for cattle, as well as for horses, pigs and poultry[55] in Australia. Cattle are reported to eat Arundo in China.[56] Sheep also relish giant reed in the Imperial Valley of California.[33] Giant reed and the related, smaller, *Arundo formosana*,[14] are used in the feed of exotic mammals (African elephant, white rhinoceros, giraffe and antelope, at Disney's Animal Kingdom Theme Park in Orlando, Florida, USA but neither species is reported to have naturalised there.[33]

Giant reed hedges are planted for water- and wind-erosion control in Eurasia, Australia, and the US[28,55,57] but shading produced by the tall reeds to crops in its proximity must be taken into account.[58] Stream-bank stabilisation was found to be easier with giant reed compared to certain willows.[59] Superior varieties of giant reed for erosion control plantings were sought but not found among various accessions from the southeastern US.[60] Giant reed was also suitable for sand-dune stabilisation in Texas[54] and Australia, where it survived in areas with areas with as low as 10 inch of annual rainfall.[55] In Australia, the more stout and leafy variegated variety was found to be superior to the wild type for dune stabilisation.[55]

Giant reed is planted as an ornamental hedge or screen, perennial border or specimen plant.[61,62] It is also used for landscaping in parks, golf courses and zoos.[62] Few horticultural varieties exist. Striped giant reed is a variety that is shorter (2–4 m), *Arundo donax var. versicolour* or *A. donax* "Variegata", has white-striped foliage,[14,61,62] and is reportedly noninvasive.[63]

Giant reed produces a diversity of chemicals,[43] some of which have been researched for their use as medicines[23,64] and insect feeding-deterrents.[65] The presence of a large variety of secondary metabolites and alkaloids explains why no devastating insect pest has been identified and not even most of the herbivorous mammals chose this plant as food. White tail deer in the USA and kangaroos in South Australia hardly damage the fresh growth of giant reed even in the absence of other green plants[57] (Márton and Czakó, South Carolina, personal observation, Heading in Australia, personal observation).

Arundo as a pioneer species has been shown to be effective on manganese mine tailings and slurry ponds in Guangxi China.[66,67] In addition, Arundo is able to tolerate and take up heavy metals.[66,68–77] By virtue of its exceptionally high biomass yield giant reed has a great potential for removing unwanted excess nutrients and contaminants from soil and water. Phytoremediation uses have been numerous albeit not extensive so far.[78–82]

Stems of giant reed are employed in a wide variety of light construction work, such as for making thatch roofs, fences, baskets, matting, arrows and fishing poles, Giant reed has also been used in folk remedies for external and internal problems.[64]

16.2.1 Giant Reed – *Arundo* for Biomass

16.2.1.1 Energy Input and Output Considerations

The most important value of giant reed is its exceptional high biomass yield. From an economic point of view, energy crops should ideally give high output of useable energy and be produced with minimum energy input. Perennial biomass crops have an advantage over annual crops because their planting can be distributed over the cultivation period and this gives a more favourable energy output/input ratio.[83] Perennial grasses have high yields in high-resource environments (high inputs).[84,85] For giant reed the high productive potential has been demonstrated in several Mediterranean environments[13,84,86,87] but better definition of the best management system for producing plant material with the highest yield and the lowest production costs.

Application of fertilisers at high rates is required for high biomass production in crops, but this requires high energy input.[88] Furthermore, there is concern with nitrate contamination of ground water and this requires an analysis of giant reed yield response to nitrogen fertilisation. As in any biomass species, the yield of giant reed is influenced by harvest time and delaying harvest time into winter in temperate climates may increase biomass yield and increase dry matter content making harvesting and storing easier. Planting density is another agronomic aspect that has been considered because of its effects on crop establishment and duration, yield and production costs.[83] Vegetative propagation by rhizomes has traditionally been the chief way to plant giant reed and the difficulty of mechanisation of the rhizome propagation is the cause of high planting costs.[89] Recently, a high-efficiency micro-propagation technology was developed utilising sustained embryogenic cultures (Figures 16.1A–F). The cultures regenerate homogenous high-quality plant material even after more than 10 years in culture *in vitro*.[20,21] The efficiency could be further improved such that megafarm-scale plantations can now be established by combining the above technology with the Fitobioreaktor/artificial plant ovary (APO) technology (patent pending).[90]

16.2.1.2 Natural Biomass Accumulation Potential

An equation for estimating giant reed shoot dry mass from shoot length has been developed[91] using natural stands of giant reed from northern California. The equation provides above-ground biomass estimates from stem counts and heights more rapidly than harvest methods and it provided good biomass estimates also for giant reed populations in Texas and Mississippi. The predicted *vs.* actual harvested yields were in the range of 38–256 odt ha^{-1} yr^{-1} *vs.* 68–302 odt ha^{-1} yr^{-1} in California, 108–389 odt ha^{-1} yr^{-1} *vs.* 113–400 odt ha^{-1} yr^{-1} in Mississippi, and 25–156 odt ha^{-1} yr^{-1} *vs.* "not determined", the average for all these selected sites being 156 *vs.* 171 odt ha^{-1} yr^{-1} and the maximum measured yield being 400 odt ha^{-1} yr^{-1}.[91] Although the equation was developed from and validated on data from natural stands it is neutral to the source of data and can be applied to cultivated stands.

There is not much published data with which to compare the above results. A natural stand on the University of South Carolina campus (Columbia, SC, USA) was found to have biomass densities ranging from to 30–120 odt ha^{-1} (Figures 16.2 and 16.3) with a calorific value of 19 MJ kg^{-1}.[47] These figures from US natural stands encompass a range wider than that of the above-ground biomass densities between 36–167 kg m^{-2} found in Rajashtan, India[92] or the 97 kg m^{-2} previously reported from the US.[28]

16.2.1.3 Sustainable Harvested Yields

The above values for undisturbed natural stands represent growth over several years and are not directly comparable to the amount of dry matter that can be sustainably harvested yearly but[28] reported annual yields of 7–83 kg odt ha^{-1} yr^{-1} of natural stands of giant reed in the USA. Annual yields in Alabama on test plots were 3.1, 19.7, 28.9, 32.1, 32.7 and 44.2 odt ha^{-1} yr^{-1} from 1999 to 2004.[93] According to the "Handbook of Energy Crops" annual biomass productivity

Figure 16.1 Mass propagation of giant reed (*Arundo donax* L.).

ranges from 10 to 59 odt ha^{-1} yr^{-1} that can reach 75 odt ha^{-1} yr^{-1} in tropical regions.[23] Over 100 t ha^{-1} (44.6 tons per acre) was reported from Turkey.[94]

In Australia, giant reed has been promoted by the South Australian Research and Development Institute (SARDI) as a promising biofuel.[13] Annual yields with irrigation with saline wastewater and fertilisation were in the range of 45.2–51.0 odt ha^{-1} yr^{-1}.[47]

Figure 16.2 Multiyear biomass accumulation of giant reed (*Arundo donax* L.) on campus, University of South Carolina, Columbia, SC, USA, November 2009, *ca.* 120 tonnes ha^{-1}.

Figure 16.3 New growth of *Arundo donax* in May 2004 and overwintered stems with the authors, László Márton and Mihály Czakó, on the campus of the University of South Carolina, Columbia, SC, USA.

Systematic research on giant reed and its appropriate cultivation techniques started in the 1980s and continued in 1997, in the 45-month European project "Giant Reed Network" with the ultimate objective of introduction of giant reed, a high-yielding, nonfood plant, into EU agriculture for energy and/or pulp production.[29] Efforts to establish giant germplasm collection have been made and the yields and various other crop parameters have been determined.[7,95] In a study in France, biomass yield reached 22.5 odt ha^{-1} yr^{-1} by year three with irrigation and fertilisation.[5] In a Spanish study, the average yield was 45.9 odt ha^{-1} yr^{-1} ranging from 29.6 to 63.1 odt ha^{-1} yr^{-1} by the end of year three.[87] In a study in Greece, the dry matter yields of several out of 40 wild populations were more than 25 odt ha^{-1} yr^{-1}, the maximum being 30.9, by the end of the second growing period.[95] In a study comparing some 200 wild populations in the south of the European Union, biomass yields by the 2nd year were odt ha^{-1} yr^{-1} in Greece, 3–34 odt ha^{-1} yr^{-1} in Italy, 5–23 odt ha^{-1} yr^{-1} in France (1st year), and 8–37 odt ha^{-1} yr^{-1} in Spain.[29,96,97] By comparison, yields were lower in northern Europe: 15–20 odt ha^{-1} yr^{-1} in Germany and 8 odt ha^{-1} yr^{-1} in the UK, but it is remarkable that giant reed had no problem adapting to the northern climate and hard winters.[29] By year three the averaged from 40 giant reed populations were 15, 20, 30 and 39 odt ha^{-1} yr^{-1}, for the first, second third and fourth growing periods, respectively, on irrigated plots.[13] Furthermore, the effects of irrigation and nitrogen fertilisation on growth and biomass productivity were assessed. Irrigation significantly affected both, especially the yield. Without irrigation the biomass productivity reached 18.8 odt ha^{-1} yr^{-1} by the second year and dropped to 11.5 odt ha^{-1} yr^{-1} by year 4. With high irrigation, the yield reached 29.9 odt ha^{-1} yr^{-1} by year three and stayed as high as 24.2 odt ha^{-1} yr^{-1}. Annual applications of nitrogen (40 and 120 kg ha^{-1}) did not influence the yields, indicating a very efficient nitrogen fixation by the root-colonising bacterial consortium, allowing the omission of expensive N fertilisation.[98]

A study on giant reed's ability to adapt to low-input cultivation methods was carried out between 1996 and 2001 in central Italy where giant reed grows spontaneously and abundantly all over the country.[83]

Response of giant reed crop yield to fertilisation (200-80-200 kg ha^{-1} N-P-K), harvest time (autumn and winter) and plant density (20 000 and 40 000 plants per ha) was evaluated. A 50% in crop yield was observed from the establishment period to the 2nd year of growth when it reached the highest dry matter yield. The mature crop represented an average yield of 30 odt ha^{-1} yr^{-1} over six years with maximum values obtained from winter harvest in fertilised plots. Fertilisation mainly increased dry matter yield in the initial period (+7 odt ha^{-1} yr^{-1}). The biomass moisture content decreased by about 10% from autumn to winter. With respect to plant density, higher dry matter yields were obtained with 20 000 plants per ha (+3 odt ha^{-1} yr^{-1}) than with 40 000 plants per ha.

Giant reed biomass calorific mean value determined by combustion in an adiabatic system was about 17 MJ kg^{-1} dry matter (7309 Btu lb^{-1}) and it was not affected by fertilisation, or by plant density or harvest time. Fertilisation enhanced crop biomass yield from 23 to 27 odt ha^{-1} year^{-1} (years 1–6 mean value), however, this 15% increase was at the expense of an energy consumption of 70% of the overall energy cost. Maximum energy yield was 496 GJ ha^{-1} obtained with 20 000 plants per ha and fertilisation. From the establishment period to the 2nd–6th year of growth the energy-production efficiency (defined as the ratio of energy output to input per ha) and the net energy yield (defined as the difference between the energy output and input per ha) increased due to the low biomass yield and the high energy costs for planting in the first year. Furthermore, both fertilisation and planting density affected energy production efficiency and net energy yield. In the mature crop the energy efficiency was highest without fertilisation both with 20 000 (131 GJ ha^{-1}) and 40 000 plants per ha (119 GJ ha^{-1}).[83]

This study was expanded into a 12-year long-term comparison of giant reed and *Miscanthus*,[99] with respect to the above-ground biomass production and the energy balance. The crops were cultivated from 1992 to 2003 in the temperate climate of Central Italy with 20 000 plants ha^{-1}, 100–100–100 kg N, P$_2$O$_5$, K$_2$O per hectare, autumn harvest and without irrigation. Each year, biometric characteristics, productive parameters, and energy analysis of biomass production was carried out to determine energy output, energy input, energy efficiency (output/input) and net energy yield (output–input). Results showed high above-ground biomass yields that were higher in giant reed than in *miscanhus* (37.7 vs. 28.7 odt ha^{-1} yr^{-1} averaged from 2 to 12 years of growth). The yields positively correlated to the number of stalks in *Miscanthus*, and to plant height and stalk diameter in giant reed. Moreover, these perennial species were characterised by a favourable energy balance with a net energy yield of 467 and 637 GJ ha^{-1} (1–12 yr average) for *Miscanthus*, and giant reed, respectively.[99]

In another recent 5-year study conducted in the semiarid hilly interior area of Sicily, compared a local clone of *Arundo donax* L. to *Miscanthus x giganteus* with respect to crop yield, net energy yield and energy ratio as affected by two different irrigation treatments (75 and 25% of ETm restoration) and two nitrogen fertilisation levels (100 and 50 kg ha^{-1}).[100] Neither fertiliser nor irrigation was used in years 4 and 5. Above-ground dry matter yield increased over all studied factors in the first three years to 38.8 odt ha^{-1} year^{-1} in giant reed and 26.9 odt ha^{-1} yr^{-1} in *Miscanthus*, and still remained high in years 4 and 5 in contrast to a dicot crop, *Cynara cardunculus* where the yield by year 5 was as low as 1 odt ha^{-1} yr^{-1}.

Irrigation had a significant effect on both species and on all parameters, while only giant reed was affected by nitrogen fertilisation. Maximum yield of giant reed was achieved in the third year, which was 8.9 up from 31.3 odt ha^{-1} yr^{-1} at lower nitrogen fertilisation and lower irrigation level due to increased fertilisation and further increased by 2.8 odt ha^{-1} yr^{-1} by the increased irrigation to 43.0 odt ha^{-1} yr^{-1} at higher nitrogen fertilisation and irrigation level. Maximum yield of *Miscanthus*, was in the third year, which was 0.6 up from 23.5 odt ha^{-1} yr^{-1} at lower nitrogen fertilisation and lower irrigation level due to increased fertilisation and further increased by 6.5 odt ha^{-1} yr^{-1} by the increased irrigation and N fertilisation to 30.6 odt ha^{-1} yr^{-1}. At the higher nitrogen fertilisation level, giant reed benefited much less from increased irrigation, while *Miscanthus*, was affected strongly at both nitrogen fertilisation levels.

Net energy yield was low or negative in the establishment year in giant reed *A. donax* and *Miscanthus*, also due to irrigation. In the subsequent two years, net energy yield of giant read was much higher (487.2 and 611.5 GJ ha^{-1}, respectively) than in *Miscanthus*, (232.2 and 425.9 GJ ha^{-1}, respectively) but *Miscanthus*, reached its highest net energy yield in the fourth year (447.2 GJ ha^{-1}).[100]

16.2.1.4 Water-Use Efficiency

The efficiency of the water used was calculated both for dry biomass and for the output energy. In terms of the efficiency of water used (rainfall + irrigation), giant reed always outperformed the other two species, and except for the first year, the values for this crop ranged between 4.08 g dry matter l^{-1} (third year) and 7.63 g dry matter l^{-1} (fifth year). At the same level of water used (rainfall + irrigation), the higher water-use efficiency of giant reed compared to *Miscanthus*, was due its higher yields.[100] Giant reed sprouted earlier and produced new stems continuously until November, while *Miscanthus*, stops its growth at the end of August according to photoperiodic control.[101] It is remarkable that the C_3 species giant reed maintains such high productivity at high temperatures and outperforms the C_4 species *Miscanthus*, under both irrigated and even semiarid conditions. It is due to its exceptional ability to maintain high rates of photosynthesis.[102] Giant reed tolerate high fluctuations of water levels[103] and can evapourate as much as 2000 l of water per metre of standing crop in a year, but it prefers well-drained soil and does not like to grow in standing water.[104]

Another advantage of *A. donax* is that in dormant, winterised stage, most of the moisture is retained in the standing cane, thus greatly decreasing fire hazard, as well as allowing gradual harvest from fall to spring. *Miscanthus*, and many other biomass grasses completely dry out at the end of the growing season, generating very serious fire hazard, and for this reason they need to be harvested and stored.

16.2.1.5 Harvesting

Planting and harvesting can be done with existing or modified equipment such as a pine planter or a sugarcane planter and a sugarcane harvester, respectively. The hard stumps may pose a puncture hazard for the tires of the harvester. Modifications take into consideration the strength and the splintering of the base of the cane and the relatively long stems of the crop.[105]

16.2.2 Other Subtropical and Tropical Reeds

There is very little information on the cultivation of tropical relatives of giant reed. *Arundo formosana* Hack[14] is very similar in appearance except it is much

Figure 16.4 Broomgrass (*Thysanolaena latifolia* Honda) in Disney's Animal Kingdom, Orlando, Florida, USA, June 2006 (photo by M. Czakó).

Figure 16.5 Sandbar covered with caña brava (*Gynerium sagittatum* Beauv.) near the left bank of the Amazon River upstream of the confluence with Rio Napo, Peru (photo by M. Czakó, August 2007).

Figure 16.6 Caña brava (*Gynerium sagittatum* Beauv.) on the right bank of the Amazon River, Peru (photo by M. Czakó, August 2007).

Figure 16.7 Robust canes of caña brava (*Gynerium sagittatum* Beauv.) harvested on the right bank of the Amazon River, Peru (photo by M. Czakó, August 2007).

shorter (2–2.5 m) and thinner. Tiger grass or broom grass (*Thysanolaena latifolia* Honda) is similar in texture to Arundo but the leaf blades are wider and the canes are thinner (Figure 16.4).[22,106] It is up to 3–4 m tall, its main use being brooms but it has been evaluated for biomass yield in India 5.4–10.4 odt ha^{-1} yr^{-1}.[107,108] A tropical cane that clearly has a high biomass yield potential is caña brava (*Gynerium sagittatum* Beauv.).[109–111] While natural stands must have very high biomass accumulation, the reported sustainable yield is about odt ha^{-1} yr^{-1},[112] in Iquitos, Peru (Figures 16.5–16.7). Availability of an *in vitro* propagation system was cited on the company website but no details are available. Land availability for plantation is limited and the logistics of harvest from wild stands are complicated by the transportation system and complexities of land-use rights in Peruvian Amazonia.

References

1. R. Venendaal, U. Jorgensen and C. A. Foster, *Biomass Bioenergy*, 1997, **13**, 147–185.
2. R. Venendaal and H. Stassen, European Energy Crops Overview. Country Report for Italy. EU FAIR Contract n. FAIR1-CT95-0512 (http://www.nf-2000.org/secure/Fair/S331.htm).
3. M. Hanegraaf, C. Biewinga and E. Gert van der Bijl, *Biomass Bioenergy*, 1998, **15**, 345–355.
4. C. Dalianis, C. Sooter and M. Christou, in *Biomass for energy, environment, agriculture and industry. Proceedings of the 8th EU Biomass Conference, Vol. 1*, Chartier, *et al.* (ed.), Pergamon Press, UK, 1995, pp. 575–582.
5. M. Arnoux, A. Davenel and M. Long, in *Energy from Biomass Vol 5, Proceedings of the Workshop and EC Contractors" Meeting held in Capri, 7–8 June 1983*, W. Palz and D. Pirrwitz, D. Reidel, (ed.), Publishing Company, Dordrecht, 1983, pp. 74–81.
6. G. Trebbi, *Bioresource Technol.*, 1993, **46**, 23–29.
7. S. L. Cosentino, V. Copani, G. M. D'Agosta, E. Sanzone and M. Mantineo, *Ind. Crops Prod.*, 2005, **23**, 212–222.
8. I. Lewandowski and A. Kicherer, *Eur. J. Agron.*, 1997, **6**, 163–177.
9. I. Lewandowski, A. Kicherer and P. Vonier, *Biomass Bioenergy*, 1995, **8**, 81–90.
10. E. Heaton, T. Voigt and S. P. Long, *Biomass Bioenergy*, 2004, **27**, 21–30.
11. I. Lewandowski and U. Schmidt, *Agr. Ecosys. Environ.*, 2006, **112**, 335–346.
12. K. P. Sharma and S. P. Kushwaha, *Int. J. Ecol. Environ. Sci.*, 1999, **25**, 1–20.
13. I. Lewandowski, J. M. O. Scurlock, E. Lindvall and M. Christou, *Biomass Bioenergy*, 2003, **25**, 335–361.
14. H. J. Conert, *Die Systematik and Anatomie der Arundineae*, J. Cramer, (ed.), Weinheim, Germany, 1961.
15. G. C. Tucker, *J. Arnold Arbor. Harv. Univ.*, 1990, **71**, 145–177.
16. A. Danin, *Willdenowia*, 2004, **34**, 361–369.

17. R. K. Bhanwra, S. P. Choda and S. Kumar, *Proc. Indian Natn. Sci. Acad.*, 1982, **B48**, 152–162.
18. E. Balogh, J. M. Herr Jr., M. Czako and L. Marton, *2010 Annual Meeting of the Association of Southeastern Biologists*, Asheville, North Carolina, April 7–10, 2010.
19. M. Christou, M. Mardikis and E. Alexopoulou, in *Biomass for energy and industry: Proceeding of the First World Conference*, Sevilla, Spain, June 5–9, 2000, 2000, pp. 1622–1628.
20. L. Márton and M. Czakó, *Sustained totipotent culture of selected monocot genera*. United States Patent 6,821,782 B2, 2004.
21. L. Márton and M. Czakó, *Sustained totipotent culture of selected monocot genera*. United States Patent 7303916, 2007.
22. C. D. K. Cook, *Aquatic and Wetland Plants of India : A Reference Book and Identification Manual for the Vascular Plants Found in Permanent or Seasonal Fresh Water in the Subcontinent of India South of the Himalayas*, Oxford University Press, Oxford, 1998.
23. J. A. Duke, *New Crop Resource Online Program, Center for New Crops and Plant Products*, Department of Horticulture and Landscape Architecture, Purdue University, West Lafayette, Indiana. (http://www.hort.purdue.edu/newcrop/duke_energy/Arundo_donax.html), 1984.
24. O. Polunin and A. Huxley, *Flowers of the Mediterranean*, Hogarth Press, London, 1987.
25. O. L. Krzyhanovskii, *The Composition and Origin of the Terrestrial Fauna of Middle Asia*, Smithsonian Institution and the National Science Foundation, Washington, DC, 1965, pp. 390.
26. A. C. Zeven and J. M. J. de Wel, *Dictionary of Cultivated Plants and their Regions of Diversity*, Centre for Agricultural Publishing and Documentation, Wageningen, 1982.
27. H. Ogawa, K. Yoda and T. Kira, in *Nature and Life in Southeast Asia. Vol. I*, T. Kira and U. Tadao, (ed.), Fauna and Flora Research Society, Kyoto, 1961, pp. 21–157.
28. R. E. Perdue Jr, *Econom. Bot.*, 1958, **12**, 368–404.
29. M. Christou, M. Mardikis and E. Alexopoulou, in *1st World Conference on Biomass for Energy and Industry*, Sevilla, Spain, 5–9 June, 2000. Vols. 1–2., ed. S. Kyritsis, A. A. C. M. Beenackers, P. Helm, A. Grassi and D. Chiaramonti, James & James (Science Publishers) Ltd., London, 2001a, pp. 1803–1806.
30. J. M. DiTomaso, in *Arundo and Salt Cedar: The Deadly Duo, A Workshop on Combating the Threat from Arundo and Salt Cedar*, C. E. Bell, (ed.), University of California Cooperative Extension, Imperial County, Ontario, CA, 1998, pp. 1–5.
31. W. W. Robbins, M. K. Bello and W. S. Ball, *Weeds of California. Sacramento*, California Department of Agriculture, Sacramento, CA, 1951.
32. M. E. Iverson, in *Arundo donax workshop proceedings,* Nov. 19, 1993, N. E. Jackson, P. Frandsen, S. Douthit, (ed.), Riverside County Parks Department, Ontario, CA, 1993, pp. 19–26.

33. J. L. Tracy and C. J. DeLoach, in *Arundo and Salt Cedar: The Deadly Duo, A Workshop on Combating the Threat from Arundo and Salt Cedar*, C. E. Bell, (ed.), University of California Cooperative Extension, Imperial County, Ontario, CA, 1998, pp. 73–110.
34. J. Jessop, G. R. M. Dashorst and F. M. James, *Grasses of South Australia*, Wakefield Press, Adelaide, 2006.
35. J. T. Morris, *Personal Communication* University of South Carolina, Columbia, SC, USA, 2005.
36. W. L. Wagner, D. R. Herbst and S. H. Sohmer, *Manual of the Flower Plants of Hawai'i, Volume 2*, University of Hawaii Press & Bishop Museum, Honolulu, Hawaii, 1991.
37. M. Bodle, *Wildland Weeds*, 1998, **1**, 11–13.
38. N. E. Jackson, in *Arundo donax Workshop Proceedings, 19 November 1993. Ontario, California. Team Arundo and California Exotic Pest Plant Council, Pismo Beach, California*, N. E. Jackson, P. Frandsen and S. Dulhoit, (ed.), 1994, pp. 27–32.
39. T. Dudley and B. Collins, *Biological invasions in California wetlands: the impacts and control of non-indigenous species in natural areas*, Pacific Institute for SIDES, Oakland, CA, 1995.
40. B. G. Hughes and K. J. Mickey, *A Reanalysis of recreational and livestock trespass impacts on riparian zone of the Rio Grande, Big Bend National Park, Texas. USDA National Park Service, Southwest Region, Santa Fe, New Mexico, Final Report*, Department of Biology, Sul Ross Stale University, Alpine, Texas, 1995.
41. G. Bell, in *Arundo donax workshop proceedings, Nov. 19, 1993*, N. E. Jackson, P. Frandsen and S. Douthit, (ed.), Riverside County Parks Department, Ontario, CA, 1994, pp. 1–6.
42. J. P. Rieger and D. A. Kreager, in *Protection, Management and Restoration for the 1990s. Proceedings of the California Riparian Systems Conference. 22–24 September, 1988, Davis. California. General Technical Report PSW-110* USDA Forest Service. Pacific Southwest Forest and Range Experiment Station, Berkeley, CA, 1989, pp. 222–225.
43. G. Bell, in *Plant Invasions: Studies from North America and Europe*, Blackouts Publishers, Leiden, The Netherlands, 1997, pp. 103–113.
44. S. Wager-Page, Field Release of the Arundo Wasp, *Tetramesa romana* (Hymenoptera: Eurytomidae), an Insect for Biol. Contr. of *Arundo donax* (Poaceae), in the Continental United States Environmental Assessment, http://www.aphis.usda.gov/plant_health/ea/downloads/Tetramesa-romana-ea.pdf, 2009.
45. J. A. Goolsby and P. Moran, *Biol. Contr.*, 2009, **49**, 160–168.
46. D. Allen, *N. Am. Clean Energy*, 2010, **4**, 97–97.
47. C. M. J. Williams, T. K. Biswas, I. D. Black, L. Marton, M. Czako, P. L. Harris, R. Pollock, S. Heading and J. G. Virtue, *Acta Horticult.*, 2009, **806**, 595–602.
48. A. A. Shatalov and H. Pereira, *Indust. Crops Prod.*, 2002, **15**, 77–83.

49. A. A. Shatalov and H. Pereira, in *1st World Conference on Biomass for Energy and Industry, Sevilla, Spain, 5–9 June, 2000. Vols. 1–2*, ed. S. Kyritsis, A. A. C. M. Beenackers, P. Helm, A. Grassi and D. Chiaramonti, James & James (Science Publishers) Ltd., London, 2001, pp. 1183–1186.
50. L. Paavilainen, in *1st World Conference on Biomass for Energy and Industry, Sevilla, Spain, 5–9 June, 2000. Vols. 1–2.*, ed. S. Kyritsis, A. A. C. M. Beenackers, P. Helm, A. Grassi and D. Chiaramonti, James & James (Science Publishers) Ltd., London, 2001, pp. 2100–2102.
51. M. Lewis and M. Jackson, in *Trends in New Crops and New Uses*, J. Janick and A. Whipkey, (ed.), ASHS Press, Alexandria, VA, 2002, pp. 371–376.
52. J. Taylor, *A Future Woven in Rayon*, Museo Territoriale Bassa Friulana, Torviscosa, Italy, 2008.
53. R. B. Singh, G. C. Banerjee and B. N. Gupta, *Ind. J. Animal Nutr.*, 1988, **5**, 144–146.
54. F. L. Wynd, G. P. Steinbauer and N. R. Diaz, *Lloydia*, 1948, **11**, 181–184.
55. W. J. Spafford, *S. Austral. Dept. Agric. J.*, 1941, **45**, 75–83.
56. C. H. Chung, in *Ecological Engineering: An Introduction to Ecotechnology*, W. J. Mitsch and S. E. Jorgensen, (ed.), Wiley and Sons, New York, 1989, pp. 255–289.
57. J. S. Horton, *Trees and Shrubs for Erosion Control of Southern California Mountains*, USDA Forest Service. California [Pacific Southwest] Forest and Range Experiment Station, California Department of Natural Resources, Division of Forestry, Berkeley, CA, 1949.
58. M. Lane and Douglas, J., *USDA/NRCS Jamie L Whitten Plant Materials Center-Technical Note* (Coffeeville, Mississippi), 1986, **12**, 1–7.
59. J. Snider, *USDA/NRCS Jamie L. Whitten Plant Materials Center-Technical Note* (Coffeeville, Mississippi), 1996, **12**, 1–8.
60. J. A. Wolfe and B. B. Billingsley, *Advanced evaluations of giant reed IV. Comparison of a Coffeeville PMC selection with five accessions from Brooksville (1987–1989). Coffeeville Plant Materials Center Technical Notes 6. USDA/NRCS Jamie L. Whitten Plant Material Center.* Coffeeville, Mississippi, 1990.
61. J. Greenlee, *The Encyclopedia of Ornamental Grasses*, Rodale Press, Emmaus, PA, 1992.
62. A. J. Oakes, *Ornamental Grasses and Grasslike Plants*, Krieger Publishing Company, Malabar, FL, 1993.
63. Heronswood Nursery, *Plant List Grasses*, Heronswood Nurseries, Ltd., Kingston, WA, 1998.
64. S. Cheatham and M. C. Johnston, *The Useful Wild Plants of Texas, the Southeastern and Southwestern United States, the Southern Plains and Northern Mexico, Volume I*, Useful Wild Plants, Inc., Austin, TX, 1995.
65. D. H. Miles, K. Tunsuwan, V. Chittawong, P. A. Hedin, U. Kokpol, C. Z. Ni and J. Clardy, *J. Nat. Prod.*, 1993, **56**, 1590–1593.

66. S.-Q. Zhou and S. H. Wu, *Trop. Geog.*, 2003, **23**, 226–230.
67. W.-J. Tang and M.-S. Li, *J. Agro-Environ. Sci.*, 2008, **27**, 741–745.
68. C. G. van der Merwe, H. J. Schoonbee and J. Pretorius, *Water SA*, 1990, **16**, 119–124.
69. G. Mavrogianopoulos, D. Bilais and G. Kyritsis, in *1st World Conference on Biomass for Energy and Industry,* Sevilla, Spain, 5–9 June, 2000. Vols. 1–2., ed. S. Kyritsis, A. A. C. M. Beenackers, P. Helm, A. Grassi and D. Chiaramonti, James & James (Science Publishers) Ltd., London, 2001, pp. 1578–1581.
70. G. Mavrogianopoulos and S. Kyritsis, *Optimizing of Plants by Metabolic Engineering*, Meeting at Milano (Italy), 15–16 December 2000 "Hopes and limits of metabolic engineering in phytoremediation", 2000.
71. E. G. Papazoglou, *Desalination*, 2007, **211**, 304–313.
72. E. G. Papazoglou, G. A. Karantounias, S. N. Vemmos and D. L. Bouranis, *Environ. Int.*, 2005, **31**, 243–249.
73. E. G. Papazoglou, K. G. Serelis and D. L. Bouranis, *Eur. J. Soil Biol.*, 2007, **43**, 207–215.
74. Z. Han, *Techn. Equip. Environ. Pollut. Contr.*, 2005, **6**, 31–33.
75. Z. Han, *Ying Yong Sheng Tai Xue Bao/The J. Appl. Ecol.*, 2005, **16**.
76. Z. Han, *Environ. Sci. Technol.*, 2006, **29**, 106–108.
77. Z. Han and C. Wang, *Shengtai Haunjing/Ecol. Environ.*, 2007, 16.
78. M. Czakó, X. Feng, Y. He, S. Gollapudi and L. Márton, in *Focus on Biotechnology Vol. 9A: Phytoremediation/Rhizoremediation*, M. Macková, D. Dowling and T. Macek, (ed.), Springer ISBN 1-4020-4952-8, Berlin and New York, 2006, pp. 217–225.
79. M. Vecchiet, *Workshop "The Reuse Of Urban Waste Water In Agriculture"*, 10th May 2000 Jolly Hotel, Bologna, 2000.
80. M. Vecchiet and R. Jodice, in *1st World Conference on Biomass for Energy and Industry,* Sevilla, Spain, 5–9 June, 2000. Vols. 1–2, S. Kyritsis, A. A. C. M. Beenackers, P. Helm, A. Grassi and D. Chiaramonti, (ed.), James & James (Science Publishers) Ltd., London, 2001, pp. 418–421.
81. G. Mavrogianopoulos, V. Vogli and S. Kyritsis, *Bioresource Technol.*, 2002, **82**, 103–107.
82. M. M. Karpiscak, C. P. Gerba, P. M. Watt, K. E. Foster and J. A. Falabi, *Water Sci. Technol.*, 1996, **33**, 231–236.
83. L. G. Angelini, L. Ceccarini and E. Bonari, *Eur. J. Agron.*, 2005, **22**, 375–389.
84. M. Vecchiet, R. Jodice, E. Smedile and F. Parrini, *Quaderni Agricolture e Innovazione*, 1994, **30/31**, 78–85.
85. M. Vecchiet, R. Jodice and G. Schenone, in *Proceedings of the 9th European Bioenergy Conference, Copenhagen,* 24–27 June, 1996, pp. 644–648.
86. S. Foti and S. L. Cosentino, *Rivista di Agronomia*, 2001, **35**, 200–215.
87. M. Hidalgo and J. Fernández, in *1st World Conference on Biomass for Energy and Industry,* Sevilla, Spain, 5–9 June, 2000. Vols. 1–2, S. Kyritsis, A. A. C. M. Beenackers, P. Helm, A. Grassi and D. Chiaramonti, (ed.), James & James (Science Publishers) Ltd., London, 2001, pp. 1881–1884.

88. W. Lockeretz, in *Handbook of Energy Utilisation in Agriculture*, D. Pimentel, (ed.), CRC Press, Boca Raton, FL, 1980, pp. 23–26.
89. L. Pari, in *Proceedings of the International Conference on Biomass for Energy and Industry. Wurzburg,* Germany, 8–11 June, H. Kopetz, T. Weber, W. Pals, P. Chartier and G. L. Ferrero, (ed.), CARMEN, Rimpar, Germany, 1999, pp. 824–826.
90. M. G. Fári, T. Kertesz, M. Czakó and L. Márton, *Syn-Plant cultures raised in artificial plant ovary (APO)*, Abstract, 2010 In Vitro Biology Meeting, June 6–11 2010, St. Louis, Missouri, USA, 2010.
91. D. F. Spencer, P.-S. Liow, W. K. Chan, G. G. Ksander and K. D. Getsinger, *Aqua. Bot.*, 2006, **84**, 272–276.
92. K. P. Sharma, S. P. S. Kushwaha and B. Gopal, *Trop. Ecol.*, 1998, **39**, 3–14.
93. D. I. Bransby, H. Gu, S. R. Duke, G. A. V and H. T. Cullinan, in *Association for the Advancement of Industrial Crops/New Uses Council Joint Annual Meeting: Industrial Crops and Uses to Diversify Agriculture, Minneapolis, MN, September 19–22,* 2004.
94. K. Günes and Ö. Saygin, *Fresenius Env. Bull.*, 1996, **5**, 756–761.
95. M. Mardikis, M. Christou and E. Alexopoulou, in *1st World Conference on Biomass for Energy and Industry,* Sevilla, Spain, 5–9 June, 2000, Vols. 1–2, S. Kyritsis, A. A. C. M. Beenackers, P. Helm, A. Grassi and D. Chiaramonti, (ed.), James & James (Science Publishers) Ltd., London, 2001, pp. 1626–1629.
96. M. Christou, M. Mardikis, S. Kyritsis, S. Cosentino, R. Jodice, M. Vecchiet and G. Gosse, in *1st World Conference on Biomass for Energy and Industry,* Sevilla, Spain, 5–9 June, 2000. Vols. 1–2, S. Kyritsis, A. A. C. M. Beenackers, P. Helm, A. Grassi and D. Chiaramonti, (ed.), James & James (Science Publishers) Ltd., London, 2001, pp. 2048–2051.
97. M. Christou, in *1st World Conference on Biomass for Energy and Industry,* Sevilla, Spain, 5–9 June, 2000. Vols. 1–2, S. Kyritsis, A. A. C. M. Beenackers, P. Helm, A. Grassi and D. Chiaramonti, (ed.), James & James (Science Publishers) Ltd., London, 2001, pp. 2089–2091.
98. M. Christou, M. Mardikis and E. Alexopoulou, *Aspects Appl. Biol., Biomass Energy Crops II*, 2001, **65**, 47–55.
99. L. G. Angelini, L. Ceccarini, N. N. O. Di Nassa and E. Bonari, *Biomass Bioenergy*, 2009, **33**, 635–643.
100. M. Mantineo, G. M. D'Agosta, V. Copani, C. Patane and S. L. Cosentino, *Field Crops Res.*, 2009, **114**, 204–213.
101. S. L. Cosentino, C. Patane, E. Sanzone, V. Copani and S. Foti, *Ind. Crops Prod.*, 2007, **25**, 75–88.
102. B. Rossa, A. V. Tuffers, G. Naidoo and D. J. von Willert, *Bot. Acta*, 1998, **111**, 216–221.
103. M. R. Rezk and T. Y. Edany, *Egypt. J. Bot.*, 1979, **22**, 157–172.
104. M. Hochovsky, *Element stewardship abstract for Arundo donax-giant reed*, The Nature Conservancy, Arlington, VA, 1986.

105. P. Barbucci, P. Andreucetti, G. Frati, P. Bacchiet, D. Vannucci and L. Pari, in *Biomass for energy and industry. 7th E.C. Conference,* Florence, Italy, 5–9 October 1992, D. O. Hall, G. Grassi and H. Scheer, (ed.), Ponte Press, Bochum, Germany, 1994.
106. R. Darke and M. Griffiths, *Manual of Grasses*, Timber Press, Portland, OR, 1994.
107. D. B. Basnet and P. Lama, *Environ. Ecol.*, 1998.
108. N. S. Bisht and S. P. Ahlawat, *SFRI, Information Bulletin No. 6 State Forest Research Institute, Department of Environment & Forests Government of Arunachal Pradesh, Itanagar, India (http://www.sfri.org/images/pdf/broom_grass.pdf)*, 1998.
109. C. Hsiao, S. W. L. Jacobs, N. P. Barker and N. J. Chatterton, *Aust. Syst. Bot.*, 1998, **11**, 41–52.
110. M. E. Barkworth, K. M. Capels, S. Long and M. B. Piep, (ed.), *Flora of North America north of Mexico. Vol. 25. Magnoliophyta: Commelinidae (in part): Poaceae, part 2*, Oxford University Press, New York, 2003.
111. J. K. Francis, *Gynerium sagittatum* (Aubl.) Beauv. wild cane (POACEAE) http://www.fs.fed.us/global/iitf/pdf/shrubs/Gynerium%20sagittatum.pdf, International Institute of Tropical Forestry, Puerto Rico, 2003.
112. J. van den Berg and L. Rademakers, in BIOPACT, http://news.mongabay.com/bioenergy/2007/02/closer-look-at-gynerium-sagittatum-new.html, 2007.

CHAPTER 17
Switchgrass

KENNETH P. VOGEL, GAUTAM SARATH, AARON J. SAATHOFF AND ROBERT B. MITCHELL

Grain, Forage, and Bioenergy Research Unit, Agricultural Research Service, U. S. Department of Agriculture, Keim Hall, Rm 317, P.O. Box 830937, University of Nebraska, Lincoln, NE 68583, USA

17.1 Introduction

Switchgrass (*Panicum virgatum* L.) is a warm-season perennial grass that is native to North America that is being developed into a biomass energy crop. It has been used in pastures and for conservation purposes in the Great Plains and the Midwest, USA, for over 70 years.[1] The research supporting its use as a pasture and conservation species was largely conducted by US Department of Agriculture (USDA) research programs, most notably the Agricultural Research Service (ARS) project located at the University of Nebraska-Lincoln and USDA Plant Materials Centers that are located throughout the United States. In this report, its development as a biomass energy crop will be emphasised.

Beginning in 1984, the US Department of Energy (DOE) *via* the Oak Ridge National Laboratory (ORNL) funded multiyear screening of about 34 herbaceous species in the main crop-producing areas of the USA for their suitability for biomass energy production.[2] The screening took place at 31 different sites in seven states and was largely conducted by universities. Switchgrass was among the top 2 or 3 species in the majority of the trials.[2] Based on these studies, switchgrass was selected as a model species in 1991 by DOE. In addition to its high yields, it also was a widely adapted native species and had significant conservation attributes. It could be propagated by seed, there was an existing seed industry, and it could be grown and harvested with available hay equipment. Because of the previous USDA work, there were adapted switchgrass cultivars available for most areas of the USA where the tests were held. Preliminary results indicated that it had the capability to fix significant amounts of soil carbon. From 1992 to 2002, DOE funded switchgrass production and breeding research *via* the Biofuels Feedstock Development Program that was managed by the ORNL.[2,3] This work was conducted by several USA universities and by the USDA-ARS project at Lincoln, NE. The emphasis on switchgrass in the USA led to its evaluation in multitrials in Europe.[4-6] In 2002, the comprehensive DOE Feedstock Development Program was discontinued although significant research progress had been made.[3,7,8] When the program was terminated, DOE was placing considerable emphasis on using crop residues such as corn stover for biomass energy because of their assumed availability and projected low cost.

In 2002, the USDA-ARS expanded its funding on biomass energy that included some funding for switchgrass genetics, breeding, and management as a biomass energy crop. This funding enabled the ARS project at Lincoln, NE led by coauthor Vogel to continue a five-year, large-scale onfarm production study with switchgrass which had been initiated with DOE funding in 2000 (Figure 17.1). In 2005 and 2006, prepublication results from this study[9-11] and from a study that demonstrated that the use of corn stover for bioenergy had major sustainability issues[12] were presented at a series of DOE, DOE and British Petroleum, and USDA planning session. As a result of these planning sessions, new research programs on perennial dedicated energy crops including switchgrass were initiated beginning in 2006 in the USA. We estimate that over $1 billion has been allocated in the USA to biomass energy research since 2006 by both government and commercial companies, with much of the emphasis on perennial grass energy crops such as switchgrass. USDA research with switchgrass has focused on its potential use as a biomass energy crop that would be grown on marginal croplands similar to land that is currently being held out of production in the Conservation Reserve Program (CRP). In this chapter we provide information on switchgrass biology, its potential use in agriculture for biomass energy, and its compositional properties as they affect its conversion to energy.

17.2 Adaptation and Distribution

Switchgrass's native habitat originally included the prairies, open woods, brackish marches, and pine-woods (Pinus spp.) of most of North America except for the areas west of the Rocky Mountains and north of 55° N. lat.[13,14] Its native range includes USDA Plant Hardiness Zones 3 to 9, which range from southern Canada to Baja California, Florida, and Central Mexico. In most areas of its original range, especially in the prairie regions, the land has been converted to agricultural uses and less than 1% of the original ecosystems exist intact today. Remnant areas do exist in most of its original range and

Figure 17.1 An upland cultivar of switchgrass, Cave-in-Rock, at flowering in eastern Nebraska, USA. Plants are being evaluated for maturity stage by USDA-ARS agronomists Rob Mitchell (left) and Marty Schmer.

these areas serve as *in situ* gene banks for switchgrass and other native species. Ecotypes of switchgrass can tolerate a wide range of abiotic conditions including flooding, drought, extreme heat and cold, and as a species it has relatively few major insect or disease pests that have significant negative impacts on its productivity. Switchgrass has two major ecotypes, lowland and upland.[15,16] Lowland ecotypes are found on flood plains and other areas subject to inundation, while upland ecotypes are found in upland areas that are not subject to flooding. Lowland ecotypes are typically taller, more coarse, generally more rust (*Puccinia* spp.) resistant, have a more bunch-type growth and may be more rapid growing than upland types. Switchgrass tolerates soil with pH values ranging from 3.9 to 7.6.[17–19]

Switchgrass is continentally adapted because it has regionally adapted sub-ecotypes within each of the two major ecotypes, as demonstrated in early common garden studies.[20–23] Moisture and thermal zones or ecoregions characterize conditions for plant growth in a geographical area.[24,25] Latitude affects plant adaptation within and across ecoregions by its effect on day length, length of the growing season, and temperature during both the growing and non-growing or dormant seasons. These parameters, which also are influenced by geographical features such as large bodies of water, are used to develop Plant Hardiness Zones such as those for the USA.[26] Plant Adaptation Regions (PAR) for perennial plants such as switchgrass were developed by Vogel *et al.*[27]

for the USA by overlaying Bailey's ecoregion map[28] with the USDA Plant Hardiness Zone Map.[26] The use of Plant Adaptation regions for classifying switchgrass cultivars and germplasm for adaption in the USA was validated by Casler et al.[29,30] in large multistate tests. Based on these and earlier adaptation research, a switchgrass cultivar should be able to be grown in the PAR of the origin of its parent germplasm and areas of adjacent PARs. It may be grown in other PARs if proven to be adapted in multiyear trials. Switchgrass cultivar adaptation areas outside the USA can be determined by identifying analogous Plant Adaptation Regions. An ecoregion-associated factor that determines specific adaptation is response to precipitation and the associated humidity. Plant materials from the more arid Great Plains states in the USA may be more susceptible to foliar diseases when grown in the more humid eastern USA. Cultivars developed from eastern germplasm may not be as well adapted to drought stress as those based on western germplasm.

17.3 Morphology and Taxonomy

Switchgrass is a member of the Paniceae tribe of grasses. Depending on genotype, ecotype and location, plants grow 0.5 to 3 m tall and although most genotypes are caespitose in appearance, some are rhizomatous. The caespitose genotypes have short rhizomes and can form a sod over time. The inflorescence is a diffuse panicle 15 to 55 cm long with spikelets toward the end of long branches.[13,31] Spikelets disarticulate below the glumes and are two-flowered with the upper floret perfect and the lower floret either empty or staminate. Spikelets are 3 to 5 mm long and florets are glabrous and awnless. The lemma and palea of the fertile floret are smooth and shiny. Leaves have rounded sheaths and firm flat blades that vary from 10 to 60 cm in length. The number of leaves per culm will vary depending on genotype and environment.[32] The ligule is a fringed membrane 1.5 to 3.5 mm long and consists mostly of hairs. Switchgrass reproduces by seeds, tillers, and rhizomes. It has the Panicoid type of seedling root.[33,34] Roots of established plants may reach depths of 3 m.[35] Seed consists of the indurate and smooth lemma and palea that hold tightly to the caryopsis. The margins of the lemma are enrolled over the margin of the palea. Glumes are almost entirely removed by combining and cleaning. On the average, there are approximately 850 seeds g^{-1}.[36] Seed weight is affected by environmental conditions and even among cultivars, seed weight can vary by almost 200%.[37] In addition to its two major ecotypes, switchgrass also has two main ploidy levels, tetraploid and octaploid[38–41] (see section on genetics and breeding for additional details).

17.4 Physiology and Growth

17.4.1 Abiotic Factors

Switchgrass has a C_4 photosynthetic pathway[42] and it has the anatomical and physiological characteristics of C_4 grasses. Since it is a warm-season grass, the germination and growth of switchgrass seedlings are reduced at soil temperatures less than 20 °C.[43,44] The recommended seeding dates for switchgrass in a region correspond to those for maize (*Zea mays* L.). Its seedlings have the Panicoid seedling morphology. Switchgrass seedlings emerge by elongation of the mesocotyl or the subcoleoptile internode that pushes the crown node and the coleoptile, which stays short, to the soil surface.[33,34,45] When the coleoptile reaches the soil surface, light induces the mesocotyl to stop elongating. Adventitious roots that are necessary for seedling and plant survival arise from the crown node at the base of the coleoptile near the soil surface. Planting seed deeper than 1 cm can adversely affect field establishment because more seedling reserves are needed for mesocotyl elongation. Dry soil conditions at the soil surface can prevent seedlings from developing adventitious roots to ensure survival, therefore planting dates should be targeted for periods when the probability of rain is high and soil temperatures are warm enough to germinate the seed.[46] Planting too late in the summer will result in stand failures because seedlings will not have adequate time to become established and develop the root reserves necessary to become perennial. Within 6 weeks of emergence several tillers may be produced. Growth of switchgrass in the establishment year depends upon soil moisture, fertility, and competition from weeds and other plants. Under optimum conditions, switchgrass will produce seed during the establishment year but flowering occurs several weeks later than in following years. Switchgrass growth the establishment year is slow relative to annual grasses because plant resources are being devoted to development of an extensive root system.

In the spring, new growth is initiated from auxiliary buds on the stem, crown, or rhizomes.[47–49] The amount of new growth from each type of bud varies with genotype, ecotype and strain. Bunch types produce new tillers from both crown buds and rhizomes[48,49] but sod-forming genotypes produce new tillers primarily from rhizomes.[47] Depending upon the physiological stage and environmental conditions, new growth may be initiated after biomass harvest from all three types of buds. Genotypes with short rhizomes produce bunch-type plants that can be pushed above the soil line by roots while sod-forming genotypes have longer rhizomes.[47] The growth and development of a switchgrass plant depends upon its genotype and the location where it is evaluated. The development of switchgrass is location dependent because flowering depends on photoperiod as discussed previously but also growing-degree-days (GDD) that measure accumulated heat or photosynthesis energy.

Switchgrass plants are photoperiod sensitive and require short days to induce flowering.[50] Their photoperiod requirement is based on the latitude where they evolved. In nature, flowering is induced by decreasing day length during early summer. In North America, moving northern ecotypes south provides them a shorter than normal daylength during summer months and they flower early. The opposite occurs when southern ecotypes are moved north. They remain vegetative longer and produce more forage than northern strains moved south.[51] When grown in the central Great Plains, switchgrass plants from the Dakotas (northern ecotypes) flower and mature early and are short in stature while those from Texas and Oklahoma (southern ecotypes) flower late and are tall.[20–23] The photoperiod response is also associated with winter survival. Southern types moved too far north will not survive winters because they stay vegetative too late in the fall. At this time, the genetic regulation mechanisms of these latitude-associated traits is unknown.

A maturity staging system can be used to quantify the physiological development of switchgrass.[52] Switchgrass maturity is highly correlated to day-of-the-year (DOY) and GDD in temperate climates such as the Great Plains of the USA.[53–55] In the Central Great Plains, photoperiod as measured by DOY was more predictive of physiological development than GDD, indicating the photoperiod is the primary determinant of switchgrass development but photosynthesis or heat units can modify the developmental response.[54] Plants in a population of switchgrass will have tillers at different stages of development.[54] Genetically broad-based populations will have some plants at anthesis over a 3-week period.[56] Florets in an individual panicle will be undergoing anthesis for up to 12 d.[57] Peak pollen-shedding periods are from 1000 to 1200 h or from 1200 to 1500 h depending upon environmental conditions.[57] Heading dates for cultivars are typically population means. Because flowering is variable, the development of ripe seed is also variable within a population or cultivar.

Stem bases, roots, and rhizomes are the primary sites of nonstructural carbohydrate storage in switchgrass. Starch is the primary nonstructural carbohydrate in switchgrass stem bases and rhizomes.[58] Nonreducing sugars, primarily sucrose, are secondary in importance to starch and fluctuate in a similar manner during the growing season. Total nonstructural carbohydrates (TNC) concentrations in the stem bases of unharvested plants are greatest at the beginning and end of the growing season. Stem-base TNC concentrations reach the lowest levels at the time of tiller elongation or when regrowth is initiated following harvest.[58] A management study in which N concentration of biomass was monitored indicates that switchgrass may actively transport N and nonstructural carbohydrates from above-ground biomass to stem bases and roots after anthesis but before a killing frost.[59]

Switchgrass has deep roots that extend to two metres in depth[35] that contribute to its drought tolerance. Water-use efficiency (WUE) of switchgrass ranged from 1.8 to 3.6 mg biomass g^{-1} water in the eastern USA,[60] while in the Great Plains WUE of switchgrass ranged from 3.5 to 5.0 mg biomass g^{-1} water.[61] The WUE of switchgrass is similar to the WUE of maize when grown for biomass production but is over 5× that of adapted C_3 grasses.[60,61] WUE of switchgrass can be improved by N fertilisation because fertilisation increases biomass yields

17.4.2 Biotic Factors

Insects that feed on herbage such as grasshoppers (family Acidiidae) are the primary insect pests affecting the biomass productivity of switchgrass. They are normally not present in densities that require insecticide applications.

Individual switchgrass plants can be susceptible to an array of diseases[62] but a range of resistance is found in most populations or cultivar so that chemical control of diseases to date has not been practiced. Principal diseases and causal agent (in parenthesis) include rusts (*Puccinia emaculata* Schwein., *Puccinia graminis* Pers., and *Uromyces gramincola* Burrill), smuts (*Tilletia maclaganii* (Berk.) G.P. Clinton), anthracnose (*Colletotrichum graminicola* (CES) G.W. Wils); Elsinoë leaf spot (*Elsinoë panici* Tiffany and Mathra), Helmintosporium spot blotch (*Bipolaris sorokiniana* (Sacc. ex Sorok.) Shoem = *Helminthosporium sativum* Pam., King, & Brakke), Phoma leaf spot (*Phoma* sp.), Fusarium root rot (*Fusarium* spp.).[62-68] Severe outbreaks of rusts are not common if adapted germplasm is used but rusts can reduce seed yields in seed production fields. Smuts can have a significant impact on both seed yields and biomass production.[64,65] In Iowa, switchgrass fields with smut had no seed yields and half of the expected biomass yield.[64] These fields had been in the Conservation Reserve Program in which biomass was not harvested but allowed to accumulate, which produced conditions favouring heavy disease infestation. The cultivar "Cave-in-Rock" was the main susceptible cultivar susceptible to seed smut. Heminthosporium spot blotch can be a serious pathogen of switchgrass in Pennsylvania and other eastern states.[68] Genetic variability exists among and within cultivars and populations for resistance to Helminosporium[68] and other pathogens. Panicum mosaic virus (PMV) infestation can cause death of tillers and plants of switchgrass.[69-71] All cultivars that have been evaluated have some susceptible plants but most plants appear to be resistant or do not exhibit symptoms. Because diseases could become more prevalent if extensive use is made of switchgrass for bioenergy, additional research is needed on switchgrass diseases and their control. Tolerant or resistant plants offer the most economical means for reducing losses.

Switchgrass requires the establishment of a symbiotic relationship with arbuscular mycorrhizal fungi (AMF) in its roots to become well established and persist.[72] Rhizosphere microflora from numerous native prairies and old seeded stands of switchgrass were all effective in enhancing seedling growth of switchgrass in greenhouse trials.[72] A field study on two different soils demonstrated that indigenous AMF in cultivated fields of the central Great Plains establish a symbiotic relationship with switchgrass and that inoculation offers little potential to increase switchgrass production unless the soils have been severely degraded.[73] Mycorrhizal inoculation may be needed for switchgrass only in extremely disturbed sites such as mine land-reclamation sites.

17.5 Genetics and Breeding

17.5.1 Genetics

Although it is a polyploid, switchgrass has a basic chromosome number of $x = 9$.[31] A wide range of chromosome numbers has been reported in the literature including somatic counts of 18, 36, 54, 72, 90, and 108 chromosomes.[74,75]

Switchgrass has small chromosomes that are difficult to count. Recent studies aided by the use of flow cytometry indicate that most switchgrass cultivars are either tetraploid ($2n = 4x = 36$) or octaploid ($2n = 8x = 72$).[38,41] The tetraploids and octaploids average 3.1 and 6.1 pg 2C^{-1} DNA.[41] The 2C ("C" stands for "constant") value is the DNA content of a diploid somatic nucleus expressed in pg (picogram or 10^{-12} g) and can be converted to daltons or nucleotide pairs using the formulas: 1 nucleotide pair = 660 Da; 1 pg = 0.965×10^9 nucleotide pairs.[76] All lowland ecotypes are tetraploids ($2n = 4x = 36$), while upland ecotypes can be either tetraploids or octaploids ($2n = 8x = 72$). Tetraploid and octaploid plants were found occurring together in over half of the remnant prairies that were evaluated by Hultquist et al.[40] Other than the diploid counts reported by Nielsen,[75] no diploid plants have been identified in recent analyses using flow cytometry. Several studies attempted to relate ploidy levels to morphological traits and geographical distribution but with inconclusive results.[23,74,75,77] Normal bivalent pairing has been reported for tetraploid and octaploid switchgrass plants.[78,79] Frequencies of aneuploid variants and multivalent chromosome associations are more frequent at higher ploidy levels.[15,74] Segregation and linkage relationships of molecular markers have been used in studies to determine if switchgrass is an auto- or allopolyploid.[80,81] The most comprehensive results to date identified 18 linkage groups, suggesting an allotetraploid genome with disomic inheritance.[80]

Switchgrass is a crosspollinated species that is enforced by a gametophytic self-compatibility system that is similar to the S-Z incompatibility system found in other Poaceae.[82] Pollen is dispersed by wind. Percentages of self-compatibility as measured by seed set from bagged panicles is typically less than 1%.[82,83] A postfertilisation incompatibility system also exists that inhibits intermatings among octaploid and tetraploid plants.[82] The postfertilisation incompatibility system between ploidy levels in switchgrass appears to be similar to the endosperm balance number system found in other species. The postfertilisation incompatibility system is probably responsible for the lack of hexaploid plants in native prairies. The tetraploid and octaploids plants in native prairies may exist as separate and distinct breeding populations.

Switchgrass has two cytoplasm types, "L" and "U" based on a chloroplast DNA (cpDNA) polymorphisms that are associated with the lowland and upland ecotypes, respectively.[39,84] The "L" cytoplasm types are tetraploids while the "U" types can be either tetraploids or octaploids.[39] The chloroplast DNA of switchgrass is maternally inherited as determined using molecular markers in reciprocal crosses.[78] The lowland and upland ecotypes and associated cytoplasm types of switchgrass are completely crossfertile at the tetraploid level and there is a high degree of similarity among their nuclear genomes as indicated by normal bivalent pairing during meiosis.[78] Molecular-marker analyses of nuclear DNA of upland and lowland cultivars separated the cultivars into upland and lowland groups.[85] Although the tetraploids are crossfertile, the lowland and upland ecotypes are genetically distinct.

Many regions where switchgrass was originally found such as the tallgrass prairie ecoregion are highly fragmented with only a small percentage of the

Switchgrass 349

original ecosystem remaining intact. However, in the tallgrass prairie ecoregion, a vast array of genetic variability has been preserved both within and among prairie remnant sites.[86] Molecular-marker analyses of plants collected from prairie remnants from the Dakotas to New York suggest that these isolated prairie remnants continue to act as one large remnant population that contains many subpopulations each capable of representing much of the variability present within the population as a whole.[87] Molecular-marker and adaptation data for switchgrass plant materials collected from a broad geographical area suggests that switchgrass germplasm for plant-breeding purposes can be classified into relatively few functional gene pools for the eastern 2/3 of the USA.[29,87] The switchgrass effective gene pools represent four Plant Hardiness zone groups and two main biomes, Prairie and Eastern Forest biomes. A molecular map for switchgrass has not been developed to date nor has the genetic relationship of switchgrass to other Panicum species been determined. A very large array of expressed sequence tags (ESTs) has been developed for switchgrass[88,89] and the switchgrass genome is currently being sequenced.

17.5.2 Breeding

Much of the initial breeding work in switchgrass involved collecting a large array of native accessions (ecotypes or strains) from a specific geographic region, screening them in a common nursery for various agronomic traits, selecting one or more of these accessions for testing in additional environments, and based on these tests, directly increasing the most desirable accession for release as a cultivar. This procedure was used by state experiment stations and the Plant Material Centers (PMC) of the Natural Resource Conservation (NRCS), USDA, in developing the initial switchgrass varieties for different geographical regions of the USA (Table 17.1). Several switchgrass cultivars including "Blackwell" and "Nebraska 28" were developed by this procedure. Nebraska 28 was the first switchgrass cultivar for which certified seed was produced. In trials in which endemic strain or accessions of switchgrass have been evaluated in their collection region, significant genetic variation has been found among strains.[86,90,91] The basic breeding procedure used to develop the initial switchgrass cultivars capitalised on this between strain or population genetic variability. The superior strains identified by this procedure form the germplasm base for further switchgrass improvement by breeding.

Genetic studies in Nebraska, North Carolina, Oklahoma, Wisconsin and other locations have demonstrated that there is also significant genetic variability within strains or accessions for most traits that have been evaluated including: forage or biomass yield, *in vitro* dry-matter digestibility (IVDMD), cell-wall composition, protein concentration, plant height, seed yield, seedling tiller number, rust resistance, maturity, and biotic and abiotic stress tolerance.[19,83,90,92–102] In these studies, heritability estimates, genetic correlations, and responses to selection were often determined. Heritability estimates

Table 17.1 Switchgrass cultivars developed by ecotype selection and recurrent breeding programs. The principal traits evaluated in ecotype selection were vigour and excellent adaptation in the region in which it was collected and biomass yield.

Cultivar	Ecotype	Ploidy	Year	Origin or Principal traits selected during cultivar development	Adapted to USDA Hardiness Zones
Cultivars developed by direct increase of a selected accession					
Alamo	Lowland	4×	1978	Southern Texas	6, 7, 8, 9
Kanlow	Lowland	4×	1963	Northern Oklahoma	6, 7
Miami	U/L[1]	4×	1996	Southern Florida	9, 10
Pangburn	Lowland	4×	NA‡	Arkansas	6, 7
Dacotah	Upland	4×	1989	Southern North Dakota	2, 3, 4a
Falcon	Upland	4×	1963	New Mexico	4, 5, 6
Grenville	Upland	NA[2]	1940	Northeastern New Mexico	4, 5, 6
High Tide	Intermediate[3]	NA	2007	Northeastern Maryland	5, 6, 7
KY1625	Upland	4×	1987	Southern West Virginia	5, 6, 7
Blackwell	Upland	8×	1944	Northern Oklahoma	5, 6, 7
Caddo	Upland	8×	1955	Central Oklahoma	6, 7
Carthage	Upland	NA	2006	North Carolina	5, 6, 7
Cave-in-Rock	Intermediate[3]	8×	1973	Southern Illinois	5, 6, 7
Forestburg	Upland	8×	1987	Eastern South Dakota	3, 4
Nebraska 28	Upland	8×	1949	Northeast Nebraska	3, 4
Shelter	Upland	8×	1986	Central West Virginia	4, 5, 6
Cultivars developed by recurrent breeding programs					
Pathfinder	Upland	8×	1967	Biomass yield and vigour	4, 5
Shawnee	Upland	8×	1996	IVDMD,[4] biomass yield	5, 6, 7
Sunburst	Upland	8×	1998	Heavy seeds	3, 4, 5
Trailblazer	Upland	8×	1984	IVDMD, biomass yield	4, 5
Summer	Upland	4×	1963	Earliness, rust resistance	4, 5
BoMaster	Lowland	4×	2006	IVDMD, biomass yield	6, 7, 8
Performer	Lowland	4×	2006	IVDMD, biomass yield	6, 7, 8

[1]Classified as upland based on chloroplast DNA markers and lowland based on nuclear DNA markers (Hulquist; Gunter et al.,[85]).
[2]NA = information not available.
[3]Intermediate-type, based on a mixture of phenotypic traits of both upland and lowland ecotypes. Unequivocal classification is not possible at this time.
[4]IVDMD = *in vitro* dry matter digestibility.

quantify the proportion of phenotypic plant-to-plant variation for a trait that is due to genetic differences among plants. Genetic correlations express the genetic relationship among traits. Heritability estimates and genetic correlations are used by plant breeders to estimate the gains that can be achieved by breeding to improve one or more traits. These studies determined that it should be possible to exploit the within-strain genetic variability of switchgrass as well as the between-strain variability to develop improved switchgrass cultivars for almost all traits that have been evaluated to date.

Since switchgrass is a crosspollinated species, breeding procedures that have been developed for other crosspollinated crops are used.[103–106] Because switchgrass has two main ploidy levels, tetraploid and octaploid, that are largely cross incompatible,[82] plants in a breeding population must have the same ploidy level. Controlled matings for genetic studies can be made using the procedure described by Martinez-Reyna and Vogel[107] or similar methods. Population improvement breeding procedures are the primary breeding methods that have been used to develop new cultivars because the primary breeding methods do not require tedious hand emasculations. Selected plants can be transplanted into field or greenhouse isolations and simply allowed to intermate using natural crosspollination. All released switchgrass cultivars that have been developed to date (Table 17.1) are improved populations or synthetic cultivars. Released cultivars have been developed for use in most areas of North America where switchgrass is adapted (Table 17.1).

The primary objectives of switchgrass breeding programs have been improving establishment capability, forage or biomass yield and quality, and insect and disease resistance. Breeding for persistence has been achieved by the use of adapted germplasm in a breeding program. Breeding work on improving establishment has targeted increased seed size[108] and seedling growth and development *per se*,[109,110] or both. Sunburst switchgrass is a cultivar with improved establishment because of its large seed.[111] Because switchgrass is a very good seed producer, little emphasis has been placed on breeding for improved seed yield. Hopkins and Taliaferro[19] demonstrated that while switchgrass seedlings have good tolerance to acid soils, there was little genetic variation in the germplasm they evaluated for this trait. Breeding for disease resistance has occurred by eliminating diseased strains and plants from breeding programs. Currently, there is an emphasis in several programs for breeding switchgrass for improved conversion efficiency in a biorefinery. Significant improvements have been made in breeding for improved forage digestibility for ruminants that has genetically altered cell-wall composition of switchgrass,[101,102,112,113] demonstrating that switchgrass cell walls can be genetically manipulated.

Emphasis has been and is being placed on breeding for increased forage or biomass yield. Released cultivars typically are higher yielding than germplasm accessions originating from the same geographical area or ecoregion.[86] These gains have been achieved by selecting and intermating the highest yielding or most vigourous plants from the highest yielding and vigourous accessions. Gains in yield can be made by selecting strains originating up to 300 to 500 km south of the intended area of use and then releasing them for use in areas north of their origin. Additional breeding for increased yield within a maturity group will be required if switchgrass yields are to be further increased. Breeders used upland germplasm to develop forage cultivars for use in pastures because they lack the thick, coarse stems of lowland switchgrasses. Lowland ecotypes typically are higher yielding than upland ecotypes adapted to the same region especially in southern USDA Plant Hardiness Zones in the USA and will be highly utilised by breeders in developing cultivars for biomass energy.[114,115] In general, switchgrass cultivars are stable for yield and quality traits over years and locations in the geographical region where their base germplasm originated, but multiyear evaluation trials are still needed to test experimental strains prior to release as cultivars.[29,30,112,116,117] Detailed information on conventional breeding procedures and systems that can be used to improve switchgrass can be found elsewhere.[103,105,106] Research to develop comprehensive molecular markers and rapid, inexpensive genotyping tools for switchgrass and many other potential bioenergy crops is currently in progress and when developed could accelerate breeding progress. Most current efforts are focused on development of expressed sequence tag (EST) markers or EST-SSR (simple sequence repeat) markers.[89,118,119]

There is potential to produce switchgrass hybrid cultivars using the species genetic self-incompatibility. Hybrid progeny of upland × lowland crosses have demonstrated an average of 19% midparent heterosis for biomass yield of spaced plants, while upland × upland and lowland × lowland hybrids of similar genetic origin showed no heterosis for biomass yield.[120] In sward plots, upland – lowland hybrids averaged 30 to 38% high-parent heterosis.[121] The upland and lowland populations used to produce the hybrids apparently are equivalent to the heterotic groups used to produce maize hybrids. Seed of switchgrass F1 hybrids could be produced at a commercial scale by growing alternate rows of the clonal pieces or ramets of the parent genotypes in seed production fields. If proper isolations distances from other switchgrass fields were maintained, virtually all seed produced would be F1 seed because of sexual self-incompatitibility. Since switchgrass is a perennial, a transplanted field could be in production for ten or more years. Methods for regenerating switchgrass plants from *in vitro* cultured cells and tissues have been developed[122–124] including a method for regenerating switchgrass plants from cells in suspension culture[125] but additional improvements will be needed. Thousands of plants can be transplanted into alternating rows of two parental clones using existing transplanting technologies for horticultural crops.

Genetic improvements of switchgrass for biomass energy production can also be made using molecular breeding technology.[126] Efficient and repeatable methods for regenerating switchgrass plants from *in vitro* cultured cells and tissues have been developed.[122–125,127–129] Deployment of transgenes in switchgrass cultivars will likely require the use of a hybrid system that prevents the introduction of transgenes into wild or natural switchgrass populations. Such a system could be based on the hybrid seed-production scheme of Martinez-Reyna and Vogel[120] in which two parental clones are increased by somatic embryogenesis and transplanted into alternating rows. Using this system, the transgenic parent could be utilised as the female and the non-transgenic parent utilised as the male. In addition, the female transgenic parent would need to have a sterility system, such that the parental clone itself and all of its progeny are male sterile, incapable of releasing transgenic pollen into the wild.

17.6 Management as a Biomass Energy Crop

17.6.1 Establishment

Switchgrass is slower to establish than annual grasses or many cultivated cereals largely because establishment-year development is oriented toward extensive root and crown development, often resulting in intense above-ground competition between switchgrass shoots and annual weeds. Both pre- and postemergence herbicides are valuable tools that aid in the establishment of switchgrass, shortening the time required to reach successful establishment and maximum biomass yields. Establishment costs and duration can significantly affect the economics of switchgrass production for biomass energy.[10] It is feasible for switchgrass stands to be fully established in Year 1 with postfrost harvestable yields equivalent to half that of full production yields and to be at full production in Year 2.

Recommended seeding rates for switchgrass are 200 to 400 pure live seeds (PLS) m^{-2}.[130] Because there is significant variation in seed weight of switchgrass seed, the number of pure live seeds needs to be determined when developing field planting rates which is from 3 to 5 kg ha^{-1}.[1,37] Establishment-year stands with 20 or more plants m^{-2} will produce harvestable forage in the year of establishment if weeds are controlled by herbicides and can be in full production the year after establishment.[130-132] Establishment-year stands of 10 plants m^{-2} are adequate but will require one or more year to achieve full production yields. Stands of less than 10 plants m^{-2} may need to be overseeded or reseeded. The minimum germination temperature for switchgrass is about 10 °C, while temperatures of 19 to 30 °C give near maximum germination with the optimal temperature between 27 and 30 °C.[43,133] Optimum germination temperatures for switchgrass may be lower than those for seedling development.[134] Seedling growth of switchgrass at 20 °C is much slower than at 25 or 30 °C.[44] As a general guideline, switchgrass can be planted during the period of 3 weeks before or after the time maize is typically planted in a region.[134,135] This general guideline for time of planting would be suitable in most areas where switchgrass is adapted. In some areas "dormant plantings" are made very late in the fall, late enough that the seed will not germinate. The seed will overwinter and the cool moist spring conditions results in a natural cold stratification and they will germinate as the weather warms. Dormant plantings are not recommended for biomass production fields because of the potential for stand failure. Switchgrass should not be planted in late summer because it does not have time to develop sufficiently before winter and it can winterkill.

Stand failures for switchgrass are typically due to planting the seed too deeply, poor quality seed, and weeds. Switchgrass seed should be planted about 1 to 2 cm deep so the seedbed needs to be firm to prevent a drill from placing the seed too deeply. No-till seeding into crop residues or chemically killed sods is effective and the most economical method of seed-bed preparation.[10,136] Corrective applications of P or K should be made before seeding but N applications are generally not made until the grass is established because it will stimulate excessive weed growth during the seeding year. High-quality, certified seeds less than 3 years old with germination percentages greater than 75% should be used. Old seed can have good laboratory germination but may have poor seedling vigour and fail to produce acceptable stands under field conditions. Physiological seed dormancy of some cultivars and seed lots of switchgrass can result in seeding failure. Although alive, dormant seed will not germinate under normally suitable conditions. Simple dormancy will be broken if the seed is aged long enough or if it is given cold treatments or cold stratified to break dormancy.[137] Producers should conduct a germination test without chilling if they suspect dormant seed and want to determine the percentage of seed that will germinate when planted. Variation exists among and within cultivars for seed size but seed size *per se* is not an indicator of establishment capacity of a seedlot. Smart and Moser[138] graded switchgrass seed into lots differing in seed weight and evaluated the seed lots in field plantings. Seedlings from the heavy seed had greater germination, earlier shoot and adventitious root growth than seedlings from light seed but growth and development were similar 8 to 10 weeks after emergence. Seed quality is more important than seed size.

Weed competition is one of the major reasons for stand failure of switchgrass. Seedlings do not develop rapidly until conditions are warm, which is when annual weeds germinate and develop. Most dicot weeds can be controlled with 2,4-D (2,4-dichlorophenoxyacteic acid).[139] Generally, 2,4-D should applied after switchgrass seedlings have approximately four to five leaves. Atrazine [6-chloro-N-ethyl-N'-(1-methylethyl)-1,3,5-triazine-2,4-diamine] has been used to improve establishment of switchgrass by controlling broadleaf weeds and C_3 weedy grasses.[140,141] Switchgrass can metabolize atrazine.[142] Acceptable stands of switchgrass could be established at a reduced seeding rate of 100 to 200 pure live seed m^{-2} when weed interference was reduced following atrazine application at time of planting.[130] Imazethapyr {2-[4,5-dihydro-4-methyl-4-(1–methylethyl)-5-oxo-1H-imidazol-2-yl]-5-ethyl-3-pyridine carboxylic acid}, applied at 70 g active ingredient (ai) ha^{-1} before the grass seedlings emerged, provided excellent weed control and enabled excellent stands of switchgrass to be obtained within 1 yr of planting.[143] Recent research conducted in the central and northern Great Plains of the USA demonstrated that switchgrass establishment was improved following application of atrazine at 1.12 kg ai ha^{-1} and quinclorac (3,7, dicholo-8-quinolinecarboxylic acid) at 1.1 kg ai ha^{-1}.[144] This herbicide treatment controlled broad-leaf weeds and weedy grasses and resulted in acceptable stands and high biomass yields. Application of imazapic {2-[4,5-dihydro-4-methyl-4-(1–methylethyl)-5-oxo-1H-imidazol-2-yl]-5-

methyl-3-pyridine carboxylic acid}on switchgrass, although effective in some trials, has resulted in significant stand reductions in other tests. Maize has been successfully used as a cover crop for switchgrass.[145] Atrazine is applied for weed control after both crops are planted. Corn is harvested for grain and is the primary crop the year of establishment. In addition to herbicides that can be used during establishment, other herbicides are available for use on established stands of switchgrass. Established switchgrass stands are not affected by metolachlor [2-chloro-N-(2-ethyl-6-methylphenyl)-N-(2-methoxy-1-methylethyl)acetamide] at rates needed to control annual weedy grasses.[143] Commercial products containing both atrazine and metolachlor are labeled for use in seed production in some regions. Herbicides should be used only in geographical regions and applications for which they are labeled.

17.6.2 Fertility Management

Switchgrass can tolerate low-fertility conditions in native stands and conservation plantings but it responds to fertiliser when grown for agricultural purposes including biomass energy.[146,147] Fertilisation with N results in significant increases in forage and biomass yield.[59,115,147–155] Nitrogen-use efficiency is greater for switchgrass than for perennial cool-season grasses.[156] Recommended N-fertilisation rates vary with location and are primarily dependent upon precipitation, cultivar, and harvest management. When switchgrass is managed for optimal biomass production, approximately 10 to $12 \, \text{kg} \, \text{ha}^{-1}$ N needs to be applied for each oven dry metric ton per hectare ($\text{odt} \, \text{ha}^{-1} = \text{Mg} \, \text{ha}^{-1}$) of biomass yield if harvested at flowering.[59,149,150] At fertility rates above this level, nitrates can accumulate in the soil profile. Harvesting a switchgrass field at flowering that produces $11 \, \text{odt} \, \text{ha}^{-1}$ of biomass with a N concentration of 1.2% will remove about $130 \, \text{kg}$ of $\text{N} \, \text{ha}^{-1}$, whereas material harvested after a killing frost may remove only half of that amount of $\text{N} \, \text{ha}^{-1}$. The N concentration of the biomass decreases as plants senescence due to N translocation to the crown and roots. If switchgrass is harvested after a killing frost, biomass yields will be 80 to 90% of biomass yields at flowering but N-fertilisation requirements will be 30 to 40% lower. Fertiliser rates should be based on soil tests and potential biomass yields. Soil samples for switchgrass production should be taken to a depth of 1.5 to 2 m due to the deep-rooting capability of switchgrass.[59] Other factors that can affect N-fertiliser requirements are the soil mineralisation potential of some soils, atmospheric N deposition, and residual soil N from previous crops that may be distributed deep in the soil profile. Fertiliser-application rates for switchgrass should be based on the difference between the requirements of the crop and available soil N. Switchgrass and other C_4 grasses should be fertilised in late spring when they are initiating growth. Early-spring fertilisation will stimulate invasion by C_3 grasses and forbs.[147]

Switchgrass may respond to P fertilisation if the availability of P in the soil is low.[147,154] On a strongly acid (pH 4.3 to 4.9), low-P soil, unfertilised switchgrass and big bluestem (*Andropogon gerardii* Vitman) produced 50% as much forage as that receiving a low level of nutrients.[146] When P declined from 35 to 5 mg kg^{-1}, switchgrass yields declined 12% compared to C_3 grasses which declined 35%.[134] Additional fertiliser research is needed for switchgrass in long-term biomass production fields to establish minimum P, K, and other nutrient values for switchgrass. Until research-based guidelines are developed, the P, K and other nutrient requirements for switchgrass can be assumed to be 60 to 75% of those for adapted cool-season grasses.

17.6.3 Harvest Management

Several trials have been conducted in the USA and other countries on optimum management practices including time of harvest. In general, a single harvest when switchgrass is fully headed gives the highest yields.[4,59,115,151,157] Biomass yield of switchgrass increases up to anthesis, after which biomass yield decreases by 10 to 20% until killed by frost.[59] Switchgrass plants harvested after a killing frost are able to mobilise N and other nutrients into roots for storage during winter and use then in new growth the following spring. In some locations with long growing seasons, two harvests provides higher biomass yields than one harvest, but the extra fossil fuels required to conduct two harvests may not warrant a two-harvest management system.[59,149] Optimal harvest management for switchgrass used for combustion may require delaying harvest until spring when most of the minerals have leached from the plant.[158] Biomass yield reductions during the winter averaged 40% in Pennsylvania.[158] Depending on location and cultivar, biomass yields of the best adapted cultivars ranged from 10 to more than 20 odt ha^{-1}.

The most efficient method to harvest switchgrass is with a self-propelled swather that will cut or mow the switchgrass and lay the cut biomass into windrows (continuous rows of loosely stacked forage or biomass) where it can field dry. The cutting height of switchgrass should be approximately 10 to 15 cm. This cutting height allows the windrows to be elevated above the soil surface, facilitating drying. Switchgrass also stores carbohydrates in stem bases and shorter cutting heights could affect stands. The biomass in the windrows should be field dried to less than 20% moisture prior to baling. Switchgrass can be baled for storage and transportation to biorefineries using conventional forage balers that produce either large rectangular or round bales. Biomass can lose weight and quality if improperly stored. Biomass dry-matter loss can range from 1 to 5% in indoor storage to over 13% in bales stored outside in a year.[159] In high-precipitation areas, losses in outdoor, uncovered storage can be even higher. Improper storage also can affect biomass quality. Unprotected switchgrass bales stored outside can lose significant amounts of ethanol extractables due to spoilage.[160] Improved methods of optimizing switchgrass harvest, storage, and transportation are needed to improve the fiscal and net energy economics of switchgrass production. Optimal harvest and postharvest management practices will likely vary with the type of bioenergy conversion method that is utilised.

17.6.4 Seed-Production Management

Seed production of switchgrass is based on practices initially recommended by Cornelius[161] for the Great Plains, subsequent research in other areas of the USA, and on anecdotal results of seed producers. Cultivated seed-production fields produce more and higher-quality seed from native prairies; row plantings produce more seed than solid stands; fertilisation and weed control are necessary for good seed production; and spring burning of seed fields usually improves seed yields.[161] In the central Great Plains where most of the commercially available switchgrass seed is produced, the seed fields are usually planted in rows spaced about 1 m apart, and are fertilised each spring with 50 to 110 kg ha^{-1} N after the fields are burned and cultivated to maintain the grass in rows. Herbicides are usually applied for weed control (see herbicides in establishment section). In Iowa, Cave-in-Rock had higher yields when grown in narrow rows spaced 20 cm apart than in wider rows spaced 1 m apart.[162] In contrast, the cultivars Blackwell and Pathfinder had higher seed yields in wide rows. Nitrogen fertiliser significantly increased seed yields in Iowa.[163] Phosphorus should be applied when soil tests indicate available soil P is low. Some seed producers irrigate, but many seed fields in the eastern Great Plains are not irrigated. Irrigation should be used when optimal, stable, seed production is need. Switchgrass seed, in contrast to seed of many native grasses, is heavy and smooth and is easily combined and cleaned with conventional combines and cleaning equipment.[36,161] Seed is usually harvested by direct combining. Grazing switchgrass seed fields early in the season can significantly reduce seed yields and is not recommended.[163,164] Seed yields in an Iowa study ranged from 200 to 1000 kg ha^{-1}.[162] while in Missouri trials, seed yields ranged from 460 to 700 kg ha^{-1}.[164] Addition research is needed on switchgrass seed production to develop improved harvesting equipment, fertiliser recommendations, and disease and insect control.

17.7 Composition and Conversion

17.7.1 Composition

Various technology platforms exist for converting switchgrass or other lignocellulosic materials into liquid fuels and chemicals. In general, these platforms fall into either thermochemical conversion or biochemical conversion strategies. Thermochemical conversion encompasses three different technologies: gasification, pyrolysis, and liquefaction, while biochemical conversion includes two different technologies: digestion and fermentation.[165] All of these platforms represent a type of biorefinery. Although there are clear differences in conversion platforms, biochemical conversion may offer the advantage of being scale neutral and less capital intensive than thermochemical processes.[166] Biochemical conversion consists of several process steps, which can change based on technologies employed and advances in research or technological evolution. In general, the main processes are pretreatment, hydrolysis, fermentation, and product recovery.[167,168] Although cost has been reduced by around a factor of four since the 1980s, research and development continues on all of these process steps as greater efficiencies, lower energy use, and overall reduced cost are sought in order to make biochemical conversion more competitive.[169] A cost analysis indicated that operating costs, listed in order of magnitude, were driven by feedstock supply, pretreatment, enzymatic hydrolysis, and enzyme production.[170] This same analysis also indicated that installed equipment costs were dominated by electrical generation (from waste fuels) and pretreatment equipment. Pretreatment, because of its overall importance to downstream processing steps as well as its significant capital and operating cost requirements, will be further discussed in a separate section (Section 17.7.2).

Biomass quality affecting these processes will be different; for example high lignin content might favour or at least not impede gasification or pyrolysis processes.[8,171–173] Whereas, lower lignin switchgrass biomass will be more suitable for generation of fuels such as ethanol and butanol, resulting from fermentative processes.[174,175] Lignin negatively influences release of free sugars from cell walls and is considered to be the major barrier for efficient enzymatic extraction of cell-wall sugars.[166,174] Lignin content can affect plant fitness.[176] Switchgrass experimental populations with genetically reduced lignin content may exhibit poor winter hardiness.[102,112] Genetic and management manipulation of lignin content of switchgrass biomass is feasible but the end use and the effects on yields and associated economic costs and plant fitness need to be considered.

At harvest, the major energy-rich structural components in switchgrass biomass available for downstream processing are cell-wall glucans, lignin and esterified and etherified phenolic acids, p-coumaric acid and ferulic acid. Grass cell-wall glucans consist of hemicellulose (a polymer of xylose substituted with a variety of 5- and 6-carbon sugars) and cellulose (a linear polymer of glucose).[177] Lignin is a heterogeneous polymer ultimately derived from differentially substituted hydroxycinnamic acid derivatives.[178] Within cell walls, these polymers and phenolic acids are enmeshed and covalently linked to each other and are not readily separable into individual components.[177] Nonstructural carbohydrates in switchgrass biomass will vary with the time of harvest; early season and regrowth harvests will contain more sugars and starch as compared to late season and postfrost harvests.[174,179]

Plant biomass is a mixture of different tissue types and composition, and the overall quality of plant biomass will be determined by the ratios of the individual tissues at the time of harvest. Switchgrass biomass at harvest consists of tillers, leaves and reproductive structures.[180] Each of these individual components has a distinct quality signature, tissue density and energy content. As an example, leaves will contain lower lignin:cellulose ratios but will also be less dense and have less total energy. In contrast, stems will possess a greater lignin:cellulose ratio but will have greater tissue density and energy content. Tissue density affects the compressibility, shattering and particle size that will be obtained postmilling. Particle size is an important parameter that influences recovery of sugars from cell-wall polysaccharides.[181–183] There is little data on management and genetic effects on tissue shattering, particle size and density for switchgrass. Based on their structure and composition, different plant parts will likely exhibit differences in these characteristics and will be capable of being modified by management and genetics.

Over an annual growing season, there is an inverse relationship between cell-wall materials and cell soluble materials that is a function of plant growth and senescence. In many perennial grasses, including switchgrass, soluble components are translocated from the aerial tissues to the below-ground tissues once seed set is completed. Albeit critical from both a sustainability and management perspective, the loss of solubles combined with leaf and potentially stem senescence often leads to diminished yields and quality of biomass as the season progresses and declines further if the crop is left standing over the winter.[158,174]

Switchgrass bales can show significant deterioration in quality when stored uncovered in the fields, although apparent dry-matter losses can be minimised by storing bales dry and under cover.[159,184] A recent report by Monti et al.[185] suggests that there was minimal loss in quality for field harvests in Italy. It is likely that environmental conditions will have a strong impact on bale storage and quality. However, considering the tonnage of switchgrass needed for a moderate sized (20–50 million gallons ethanol per year) biorefinery,[184] it is critical that storage and handling of harvested biomass results in little if any loss in quality.

The relative amounts of stems to leaves will change over a growing season. Early in the growing season, leaves constitute a significant portion of the above ground biomass. Once flowering is initiated, leaf production stops, whereas internode elongation and stem dry-matter accumulation is still proceeding.[186,187] Leaves present in the more basal nodes can begin to senesce as seed set progresses leading to further losses in leaves as a per cent of total biomass. Secondary cell-wall accretion appears to take place in all internodes of a tiller after seed fill.[187] Increase in stem dry matter as the growing season progresses leads to stems becoming the major portion of harvested biomass. In terms of quality, the maturity at harvest has a distinct effect (at least for ethanol potential on fuel yield). There appears to be an optimal harvest date from a quality perspective generally coinciding with the onset of flowering.[174] These data suggest that switchgrass biomass harvested at boot-stage contains a good mix of leaves and stems, and the enhanced conversion efficiency is partially due to increased levels of nonstructural carbohydrates, and lowered levels of lignification, particularly in the stems.[174,187,188] However, late-season, postfrost harvests are likely to provide biomass with lower levels of nutrients, especially N, P, K and minerals,[188,189] which is expected to lead to greater sustainability of production.

Lignin is the key molecule that negatively affects conversion of cell-wall glucans to sugars and presents a good target for breeding.[166,180,190] It has been known for some time that improving forage digestibility by herbivores leads to plants including switchgrass with lowered lignin levels.[191–193] Recent data from sorghum indicates that biomass from genotypes with lowered lignin through the incorporation of two independent brown midrib (bmr) alleles show significantly better conversion to ethanol.[194] These, and a host of studies in woody and other model species,[166,195,196] confirm the utility of lowering lignin in bioenergy species. As discussed earlier, lowering lignin is not without negative consequences in perennial grasses.[112,176] Nevertheless, it is possible to select for both greater digestibility as well as winter survival in perennial grasses.[102]

In general, forages bred for improved forage digestibility in ruminants such as cattle by utilising an artificial rumen procedure (*in vitro* dry-matter digestibility or IVDMD[197,198]) also exhibit a greater potential for conversion to ethanol.[194,199] In a detailed analysis of cell-wall accessibility of biomass obtained from genetically different switchgrass plants developed by divergent breeding for IVDMD[102,200] it was evident that there were multiple factors that impacted cell-wall deconstruction.[201] The plants differed in both structure and composition with plants with high IVDMD having lower lignin amounts. The ratio of *p*-coumaric acid: ferulic acid was lower for high-IVDMD–low-lignin plants and greater in the low-IVDMD–high-lignin plants. Total guaiacyl + syringyl-lignins were consistently higher in low IVDMD-high lignin genotypes as compared to high-IVDMD–low-lignin genotypes. When switchgrass biomass from these plants was incubated with esterases and/or cellulases, release of *p*-coumaric acid was greater than ferulic acid in all genotypes, indicating that most ferulate derivatives were probably etherified and therefore not amenable for hydrolysis by exogenously supplied esterases. Cellulase digestion of biomass samples indicated that more lignified, low-IVDMD plants exhibited lower loss of cell-wall materials, suggesting that cell-wall architecture was differentially modified.[201] This research demonstrated that significant, exploitable plasticity for cell-wall composition and content exists in switchgrass. Results from another recent study has shown that switchgrass germplasm with improved ethanol yields can be identified through a combination of pretreatment and fermentation strategies.[202] These studies suggest that although lowered lignin plays an important role in improving release of sugars from switchgrass cell walls, other components in the biomass also influence overall ethanol yields.[203] One of these components could be ferulic acid. Wall-bound ferulic acid is often negatively associated with plant biomass digestibility and could also limit the release of sugars from biomass by exogenous glucanases.[193,204,205] Overexpression of a fungal ferulic acid esterase in tall fescue led to greater digestibility of cell-wall polysaccharides, suggesting that efficient removal of ferulic acid and p-coumaric acid esters improved accessibility of both cellulose and hemicellulose.[204] In contrast to these previous findings, Shen *et al*.[186] have reported a positive correlation between ferulic acid ester content and apparent saccharification of switchgrass biomass. Since ferulic acid is both ester and ether linked to other cell-wall components, increasing ester-linked ferulic acid content and decreasing ether-linked ferulic acid could be an effective strategy to improve saccharification. Other aspects of cell walls, including architecture, crosslinking between the various cell-wall polymers, cell shape and size and distribution of different tissues, such as fibres, vascular bundles and parenchyma cells will also impact deconstruction. Fortunately, many of these quality traits can be manipulated through both conventional and transgenic breeding strategies.[126,186,203,206]

A search for genotypes exhibiting improved quality traits is going to require both high-throughput and laboratory-scale quality-evaluation tools for the continued improvement of switchgrass and other bioenergy species.[180] In addition to well-established methods such as near-infrared reflectance spectroscopy (NIRS) and fibre analyses protocols,[180,198,207] several other protocols have been published recently.[171,208–211] Widescale validation of these methods could lead to better analytical evaluation of the "true" quality of switchgrass and other bioenergy germplasm. There is significant potential for improving the biomass quality of switchgrass biomass for bioenergy. The increasing availability of molecular resources for switchgrass[88,126] combined with bioinformatic analyses for key genes[212–214] should accelerate this process.

17.7.2 Switchgrass Pretreatment

Pretreatment is currently necessary for biochemical conversion of switchgrass. This necessity is driven by the basic structural characteristics and components of switchgrass cell walls, as discussed previously.[177,215] The goals of pretreatment include size reduction, lignin removal, hemicellulose removal, increased material porosity for enzyme accessibility, and reduction of cellulose crystallinity.[166,216] Additional requirements were listed by Sun and Cheng[216] as improving sugar formation or the ability to form sugars, little or no carbohydrate degradation, no production of inhibitory byproducts, and cost effectiveness. Pretreatment methods include physical, chemical, physical–chemical combinations, and biologically based approaches.

Physical pretreatment encompasses chipping, grinding, or milling that reduces cellulose crystallinity and increases the available surface area for hydrolytic enzyme attack in downstream processing steps. Biomass particle size has been shown to be important in reducing both heat[217] and mass[218]-transfer limitations for hydrolysis reactions. Various types of equipment are available for comminution of biomass, including hammer, knife, pin, chain, and ball mills. However, particle-size reduction takes a considerable amount of process energy and thus it is desirable to limit the need for excessively small particles in biochemical conversion.[219] Also, the choice of equipment is not trivial: energy requirements were shown to vary substantially depending on mill and biomass type,[220,221] which indicated a need to match biomass type to the appropriate technology. Switchgrass, like other grasses, was found to have higher tensile than shear stress, with mean tensile stress nearly doubling as moisture content decreased from 60–10%.[222] Baled switchgrass will be mature and it will have a low moisture content (and therefore high tensile stress). Milling equipment that employs shear failure for particle-size reduction such as pinch-points, shear bars, and knives was suggested to be more energy efficient in ref. 222. Research using a knife grid as an early processing step showed that energy requirements depended on knife-grid spacing and moisture content of switchgrass stems, with more energy required on a mass basis for smaller grid spacing[223,224] and for stems with high moisture content.[224] Although shear stress of individual switchgrass stems was relatively independent of moisture content,[222] the packed bed used by Igathinathane *et al.*[224] represented a different testing methodology, which likely explains the differing results. Total specific energy usage also was reported to increase as a function of knife mill speed, with switchgrass having the highest optimum feed rate of 7.6 kg m^{-1} and lowest specific energy requirement (7.6 kW h Mg^{-1}) when compared to wheat straw and corn stover.[223] Results from a hammer-mill study indicated that energy consumption more than doubled for switchgrass with 8% moisture content (from 23.8 kW h t^{-1} to 62.6 kW h t^{-1}) when screen opening was decreased from 3.2 to 0.8 mm.[182] Additionally, switchgrass required more specific energy than any of the other biomass types that included wheat and barley straw and corn stover. Other research has indicated that hammer-mill speed and hammer orientation can significantly affect energy requirements for grinding switchgrass.[225] Ball-mill studies in which different fraction sizes (<90 μm and 90–600 μm) were generated, demonstrated size fractions had differing properties; cellulose, hemicellulose, and lignin preferentially accumulated in the larger-sized particles.[226] These results indicated a need to avoid excessively small particle sizes so carbohydrate losses are minimised for biochemical conversion platforms.

Chemical pretreatments that are low cost and relatively fast-acting pretreatments are needed.[219] The most promising chemical pretreatments are dilute sulfuric acid, sulfur dioxide, controlled pH, ammonia fibre expansion (AFEX), ammonia recycle percolation (ARP), lime, flow through acid, uncatalysed steam explosion, liquid hot water, and ionic liquids.[219,227,228] Dilute acid pretreatment consists of adding acid to the pretreatment reactor either directly followed by heating or through direct steam injection. Sulfuric acid appeared to be the most common choice, although Mosier *et al.*[227] noted other acids have been used. Acid hydrolysis reactions release monomers and oligosaccharides from the cell wall and are capable of releasing large amounts of hemicellulose, while only small amounts of lignin are dissolved.[219] However, a cost estimate using dilute acid for ethanol production from wood chips indicated that significant fixed capital costs (18.5%) were required due to the corrosive nature of this chemical.[229] Dilute acid pretreatment also requires neutralisation and creates byproducts, such as furfural and acetic acid, that need to be removed in order to obtain high ethanol yields from fermentation processes.[219,230]

An early comparison of switchgrass and other biomass types using dilute acid pretreatment and SSF found that switchgrass had lower ethanol yields than corn stover and wheat straw, particularly at intermediate cellulase and β-glucosidase loadings.[231] Pretreatment conditions of 0.5% (v/v) sulfuric acid, 30 min., and 160 °C yielded over 80% digestibility compared to around 70% at 140 °C for switchgrass biomass.[230] Another study looked at the impacts of acid concentration (0.6–1.2%) and temperature (140–180 °C) on xylose yield and found that temperature was the dominant factor in xylose release from switchgrass, poplar, and corn stover.[232] Later work from the same laboratory showed that a 0.5 min, 180 °C treatment with 1.2% sulfuric acid resulted in 80–90% cellulose conversion, although long incubation times (~120 h or greater) were needed, especially for the lower cellulase loading levels of 25 FPU.[233]

The impacts of plant maturity and switchgrass germplasm differences have also been investigated. Dien et al.[174] examined switchgrass at various maturities and found that glucose recovery efficiency was inversely correlated with plant maturity, although total yields were higher with more mature samples due to their higher initial glucose content. In contrast, plant maturity did not impact conversion efficiency of nonglucose sugars. However, higher-temperature pretreatment with dilute acid was found to reduce nonglucose recovery and yields, while at the same time having a positive effect on glucose yields.[174] This study demonstrated the difficulty in optimizing pretreatment conditions for glucose and nonglucose sugar recovery. Other work that utilised dilute acid pretreatment on switchgrass compared three different germplasms and suggested that germplasm had an influence on glucan conversion and subsequent ethanol yields.[202] Also, the addition of xylanase did not generally result in higher glucan conversion levels in this study.

In addition to acid, another commonly used chemical for pretreatment has been ammonia. Unlike acid, ammonia pretreatment removes a significant amount of the lignin, but solubilizes significantly lower amounts of the hemicellulose component of cell walls.[219] Removal of lignin is advantageous based on work that demonstrated that lignin content was associated with reduced cell-wall degradation and lower biomass reactivity.[234,235] Also, sorghum cultivars with differing lignin content were compared for ethanol conversion efficiency, and it was shown that higher lignin levels led to reduced glucose recovery and lower ethanol yields.[194] At high temperatures, ammonia depolymerizes lignin and cleaves lignin–carbohydrate bonds, while leaving carbohydrates mostly intact.[219,227] There are various pretreatment technologies that use ammonia, including ARP, aqueous ammonia soaking (SAA), and AFEX. ARP is basically a flow-through method that flushes biomass with a 5–15% ammonia solution at elevated temperatures (80–180 °C) and pressures (325 psi) with relatively short residence times ranging from minutes to hours.[227,236] SAA was developed to avoid the xylan loss and energy penalties associated with high temperatures and pressures, but it required long residence times (on the order of days) for it to be effective on corn stover.[237]

Pretreatment research on switchgrass using ARP found that conditions of 170 °C and 10% ammonia resulted in hemicellulose and glucose removal of 64% and 14%, respectively, with about 80% lignin removal.[238] Additionally, the resulting biomass had 87% enzymatic digestibility. Iyer et al.[238] also demonstrated that the great majority of lignin and hemicellulose removal occurs within the first 30 min of ARP pretreatment of switchgrass. The same laboratory later used ammonia in conjunction with hydrogen peroxide to better fractionate biomass by improving lignin removal and hemicellulose recovery.[239] This modification resulted in increasing the lignin removal to over 90% and hemicellulose recovery to about 80% for switchgrass biomass, although the impact of hydrogen peroxide use on process economics was not addressed.

ARP has also been combined with dilute (0.0784%) sulfuric acid pretreatment in a two-stage reaction process.[240] When ARP was followed by dilute acid, 83% of the lignin was removed as was all of the hemicellulose, which was recovered with 96% efficiency. Glucan was also shown to be higher in biomass pretreated with ARP and dilute acid than with dilute acid alone.[240] However, the impact of this pretreatment strategy on process economics was again unknown.

Aqueous ammonia soaking has also been used to disrupt switchgrass cell walls. In one case, switchgrass was soaked in 29.5% ammonia hydroxide at loading rates of $5\,\text{mL}\,\text{g}^{-1}$ and $10\,\text{mL}\,\text{g}^{-1}$ for 5 or 10 days.[211] Approximately 36–47% of the lignin was removed depending on ammonia amount and reaction time; higher loading amounts and longer reaction times resulted in better lignin removal. Also, approximately 45–50% of the hemicellulose was left unrecovered, although washing steps may have contributed to these losses. Simultaneous saccharification and fermentation SSF results showed that ammonia soaking time and amount were important determinants for ethanol yields at moderate (26–38.5 $\text{FPU}\,\text{g}\,\text{cellulose}^{-1}$) cellulase loading rates, and the process could be further optimised.[211] SAA was also used for pretreatment in pilot-scale experiments that used 50 L and 350 L fermentations to produce ethanol from switchgrass.[241] Here, 73% theoretical ethanol yield was achieved in the 50 L reactor, while a 52% ethanol yield was found for the 350 L reactor. Lactic acid buildup in the larger vessel due to bacterial contamination may have contributed to this lower yield.[241] Thus, it may be possible to obtain reasonable ethanol yields from SAA pretreatment. However, hemicellulose losses and long pretreatment times may hinder the commercial viability of this approach.

Another pretreatment approach has used lime (calcium hydroxide) to create a low-pH environment to disrupt cell-wall structure. Lime offers advantages in terms of cost and has fewer handling concerns than stronger, more expensive bases such as NaOH.[242] The effectiveness of lime for switchgrass pretreatment was demonstrated using a loading rate of 0.1:1 g lime to g biomass at 120 °C for two hours.[243] Under these conditions, 5 FPU cellulase and 28.4 CbU cellobiase (per g biomass basis) were able to release approximately 60% of the available glucose and xylose during a 72 h hydrolysis period. Like other bases such as NaOH, KOH, and LiOH, lime is able to deacetylate cell-wall polymers, which enhances hydrolysis.[244,245] This same laboratory later used lime pretreatment on switchgrass biomass followed by SSF, and reported that an ethanol yield of approximately 70% was achieved within 6 days.[246] An alternative use of lime for corn-stover pretreatment used lower temperatures over longer time periods in order to reduce energy consumption of the pretreatment process.[244] A temperature of 55 °C over four weeks provided the optimal pretreatment conditions using lime; these conditions resulted in over 90% glucose hydrolysis and about 50% xylose hydrolysis upon treatment with cellulase. Due to the four-week time period needed for low-temperature lime pretreatment, the use of large biomass piles was suggested as a possible way to implement this approach.[244] The use of large, relatively unprotected biomass piles for pretreatment has not yet been evaluated for technical or economic feasibility.

Also, there are currently no reports showing the efficacy of low-temperature lime pretreatment on switchgrass.

Other chemical pretreatments are possible that do not rely on acids or bases. For instance, ionic liquids (IL), which are salts at room temperatures and have good thermal stability and low volatility, have been shown to be capable of dissolving cellulose at up to 25% (wt%) solubility due mainly to their high hydrogen bonding capability with cellulose.[247] Using the model substrate Avicel, an ionic liquid (1-n-butyl-3-methylimidazolium chloride) was used to create a 5% cellulose solution.[248] After precipitation and treatment with cellulase at 50–60 FPU g^{-1} glucan, cellulose conversion to glucose was markedly higher in the pretreated cellulose vs. untreated Avicel. Other ionic liquids, such as 1-ethyl-3-methylimidazolium acetate [EMIM][OAc], also demonstrated the ability to enhance sugar release from Avicel and may represent ionic liquids with more desirable properties.[249] This IL, [EMIM][OAc], was used to treat switchgrass cell walls in order to examine the effects of ILs on herbaceous biomass.[228] Using confocal fluorescence microscopy, these investigators demonstrated that [EMIM][OAc] was able to dissolve both cellulose and lignin from switchgrass cell walls. More importantly, cellulose was precipitated from the IL, while lignin remained in solution, which thus removed a major barrier for cellulose hydrolysis; subsequent hydrolysis with cellulase reached maximum theoretical glucose levels after 30 h.[228] These studies demonstrated the potential of ILs for pretreatment uses, although significant process development remains and economic viability, for now, is an open question.

Physical–chemical pretreatments have consisted of exposing biomass to a solvent at elevated temperatures and pressures, followed by rapid pressure release. The two main forms of this pretreatment are steam explosion and AFEX. Steam explosion has been generally operated at temperatures of 160–260 °C and pressures of 0.69–4.83 MPa.[216] Although steam explosion is relatively low cost, requires less energy to achieve particle-size reduction, and causes lignin transformation, it degrades hemicellulose and creates inhibitory byproducts to downstream fermentation.[216] No reports using this pretreatment on switchgrass were found. AFEX uses liquid ammonia at elevated pressures (~1.5 MPa) and moderate temperatures (~100 °C or less) for 15–30 min exposure periods.[227,250,251] Pressure is then rapidly released, causing the ammonia to flash and disrupt the biomass structure through cellulose decrystallisation, lignin alteration, and hemicellulose prehydrolysis.[219,251] AFEX pretreatment has been used on a wide variety of biomass types and resulted in high cellulose and hemicellulose conversion at low enzyme loadings.[219] However, AFEX does not appear to be suitable for wood chips, due to their high lignin content.[216] For switchgrass, AFEX treatment using a 2:1 ammonia:biomass ratio at 90 °C was shown to result in 546 mg g biomass^{-1} sugar yield; subsequent ethanol yield was 83% based on the available glucose and xylose residues, although xylose was not fermented effectively.[252] Later work focused on optimisation of AFEX for switchgrass and found that a 1 kg ammonia:1 kg biomass (dry weight) ratio processed at a temperature of 100 °C for 5 min resulted in over 90% glucan and 70% xylan conversion, compared to less than

20% and 5% glucan and xylan conversion, respectively, for untreated controls.[253] Combining AFEX pretreatment with protein extraction from switchgrass was also proposed.[254] However, this study used switchgrass harvested during the early stages of the growing season (May) that appeared to be sharply at odds with the suggested management practices of maximizing dry-matter production, N translocation to the roots, and a harvest time during anthesis or after a killing frost, as discussed previously.

Alternative pretreatment approaches are possible. For example, NaOH levels were varied with microwave heating and 99% sugar release was achieved when 2 h presoak in 0.1 g:1 g ratio of NaOH to biomass (50 g L^{-1} solids loading) followed by a 30-min microwave treatment that kept the temperature at 190 °C.[255] In another study, microwaves were used to heat and disrupt switchgrass cell walls that were soaked in a 3% NaOH solution for 10 min.[256] This approach proved superior to acid soaking with microwave irradiation due to higher sugar release upon hydrolysis. However, it is unknown how well microwave heating would work when scaled up in a biorefinery setting. RF heating has also been used to heat switchgrass in the presence of NaOH, although the pretreatment temperature was only capable of reaching 90 °C with the equipment used in this study, which likely was a factor in limiting maximum sugar yields to 80%.[257] However, RF pretreatment was effective at solid loadings of up to 50% when hydrolysis results were compared, and RF pretreatment also appeared effective on 0.25–2 mm particle sizes. The ability to work with larger particle sizes and at high solids loading is a useful feature, but RF pretreatment strategies are at an early stage of technical development, and process economics are currently unknown. Also, the use of NaOH imposes a high cost penalty when making liquid fuels from biomass,[219] so successful implementation will likely require the use of cheaper chemical. Other approaches using different chemicals have been used as well. These have included CO_2 explosion, ozonolysis, organosolv (use of organic solvents such as methanol), and even biological methods, but such pretreatments have generally suffered from technical, economic, or a combination of technical and economic drawbacks.[171,216,219] No reports of using these methods for switchgrass pretreatment were found.

The ideal pretreatment for switchgrass, or lignocellulosic biomass in general, does not yet exist for integration into biochemical conversion strategies. Selection of a given pretreatment strategy must be made in the context of all other unit operations in the biorefinery as this choice will have an economic impact on these other stages.[258] Fortunately, process improvements continue to be reported. For instance, development in the AFEX ammonia recovery process reduced the predicted minimum ethanol selling price (MESP) from $1.41 to $1.03 per gallon; further cost reductions were expected if advances in bioprocessing are realised.[259] Large-scale biorefinery development is currently in its nascent phase, with first-generation plants currently being built. These plants will provide valuable insight into the logistical and technical problems that occur in full-scale operations and provide guidance for further research and technical development.

17.8 Economics and Net Energy

A biomass energy crop needs to be both economically feasible and have positive net energy value. There have been several studies on the economic feasibility of using switchgrass as an energy crop since the mid-1990s.[260–267] Except for the study by Monti et al.,[264] which was conducted on a field in Italy, many of these studies were based on results from small research plots and extrapolated to the field scale and their production cost estimates ranged from $25 to $150 Mg^{-1}. Results from a recent large, regional field-scale trial in the USA provides some of the best available production cost information.[10] The study was conducted in fields on 10 farms over 5 years in Nebraska, South Dakota, and North Dakota. Only two of the ten farmers had previously grown switchgrass. Average production costs were $65 Mg^{-1}. Five producers had average production costs of $53 Mg^{-1} over the five-year period. When projected to a ten-year production cycle, the low-cost producers had an average cost of $46 Mg^{-1}. With experience, it was assumed that other farmers could achieve switchgrass production costs of 40 to $55 Mg^{-1}. Assuming a switchgrass farm-gate cost of $40 to $55 Mg^{-1} and conversion of 0.329 liters of ethanol per kg of switchgrass, the farm-gate feedstock cost would range from $0.12 to $0.16 per liter.[10] Ethanol from maize feedstock costs were $0.13 L^{-1} at a maize price of $80 Mg^{-1}, or $0.26 L^{-1} at a maize price of $160 Mg^{-1} during 2006 and 2007 in the USA.[10] These results indicate that if economical biomass-to-ethanol conversion technology can be developed, switchgrass can be an economically viable biomass fuel crop. Land costs were a significant part of the production costs and these vary significantly from region to region so production costs will be regionally biased.

Net energy efficiency of ethanol produced from grains and cellulosics has been evaluated using net energy value (NEV), net energy yield (NEY), and the ratio of the biofuel output to petroleum input (petroleum energy ratio or PER).[11] Net energy of cellulosic biomass energy crops such as switchgrass had been modeled in several studies using yield information from small plot trials.[268,269] Only one study to date has used field production and input information to model NEV, NEY, and PER for switchgrass. The input and yield information from the large regional switchgrass economic study[10] was used to determine the NEV, NEY, and PER for switchgrass.[11] An estimated conversion rate of 0.38 L ethanol kg^{-1} biomass[268] was used in the net energy analyses. Switchgrass fields on the ten farms produced 540% more renewable fuel than nonrenewable fuel consumed over a 5-year period.[11] The estimated onfarm NEY for switchgrass was 60 GJ ha^{-1} y^{-1} and switchgrass produced an estimated average of 13.1 MJ ethanol for every MJ of petroleum input (PER). This study demonstrated that switchgrass can be net energy positive by the three primary methods used to measure bioenergy efficiency (Figure 17.2). The results of this study represent a baseline because the cultivars used in the study were developed for use in pastures. Improved biomass cultivars and improved management practices should result in improvements in NEV, NEY, and PER.

Figure 17.2 An illustration of the net energy of switchgrass based on Schmer et al.[11] This big round bale of switchgrass weighs 0.6 Mg on a dry matter (DM) basis. At an ethanol conversion rate of 300 L per Mg of DM, this bale will produce 180 L of ethanol (fuel barrel). The total input energy required to produce the ethanol in the fuel barrel is equivalent to the energy in 30 L of ethanol (2 small fuel containers).

17.9 Sustainability and Ecological Services

Several methods can be used to measure the sustainability of a bioenergy crop including net greenhouse-gas emissions (GHG) and associated soil carbon sequestration and ecological services. Net GHG from switchgrass grown for biomass energy have been modeled in several studies.[11,268] The model results to date indicate that switchgrass is either greenhouse-gas neutral or slightly positive. Sources of variation in models include differing assumptions on conversion efficiency, biomass yield, and the amount of soil C sequestered. All results to date indicate that switchgrass sequesters substantial amounts of C in the soil because of its extensive root system.[9,270–273] Switchgrass grown in Conservation Reserve Program land in South Dakota stored soil organic carbon at a rate of 2.4 to 4.0 Mg ha^{-1} yr^{-1} at the 0 to 90 cm depth.[272] In a 5-year, 10-field study in Nebraska, South Dakota, and North Dakota discussed previously,[10,11] SOC increased significantly at 0 to 30 cm and 0 to 120 cm soil depths, with an average increase in SOC of 1.1 and 2.9 Mg C ha^{-1} year^{-1}, respectively.[9] These C sequestration rates are higher than the rates used in some of the GHG models. Switchgrass has been used for conservation purposes for over 50 years in the Great Plains and Midwest of the USA for preventing wind and water erosion.[1] In comparison to grain crop production fields, switchgrass grown for biomass energy may provide additional ecological services including increasing wildlife habitat and increasing landscape and biological diversity, and improving water quality by its combined use in filter strips.[274–279]

References

1. K. P. Vogel, in *Warm-season (C4) grasses*, L. E. Moser, L. Sollenberger and B. Burson, (ed.), ASA-CSSA-SSSA, Madison, WI, 2004, pp. 561–588.
2. L. Wright, *Historical perspective on how and why switchgrass was selected as a "model" high-potential energy crop* ORNL/TM-2007/109, US Department of Energy, Oak Ridge National Laboratory, Environmental Sciences Division, Oak Ridge, TN, 2007.
3. S. B. McLaughlin and L. Adams Kszos, *Biomass Bioenergy*, 2005, **28**, 515–535.
4. D. G. Christian, A. B. Riche and N. E. Yates, *Bioresour. Technol.*, 2002, **83**, 115–124.
5. H. W. Elbersen, D. G. Christian, N. El Bassem, W. Bacher, G. Sauerbeck, E. Alexopoulou, N. Sharma, I. Piscioneri, P. de Visser and D. van den Berg, *Switchgrass (Panicum virgatum L.) as an alternative energy crop in Europe* FAIR 5-CT97-3701, 2003.
6. A. Monti, P. Venturi and H. W. Elbersen, *Soil Tillage Res.*, 2001, **63**, 75–83.
7. D. J. Parrish and J. H. Fike, *Crit. Rev. Plant Sci.*, 2005, **24**, 423–459.
8. M. A. Sanderson, P. R. Adler, A. A. Boateng, M. D. Casler and G. Sarath, *Can. J. Plant Sci.*, 2006, **86**, 1315–1325.
9. M. Liebig, M. Schmer, K. Vogel and R. Mitchell, *Bioener. Res.*, 2008, **1**, 215–222.
10. R. Perrin, K. Vogel, M. Schmer and R. Mitchell, *Bioener. Res.*, 2008, **1**, 91–97.
11. M. R. Schmer, K. P. Vogel, R. B. Mitchell and R. K. Perrin, *Proc. Nat. Acad. Sci. USA*, 2008, **105**, 464–469.
12. G. E. Varvel, K. P. Vogel, R. B. Mitchell, R. F. Follett and J. M. Kimble, *Biomass Bioenergy*, 2008, **32**, 18–21.
13. A. S. Hitchcock, *Manual of the grasses of the U.S.*, USDA Misc. Pub. 200, Washington, D.C., 1951.
14. J. L. Stubbendieck, S. L. Hatch and C. H. Butterfield, *North American Range Plants*, 5th edn. University of Nebraska Press, Lincoln, NE, 1997.
15. J. N. Brunken and J. R. Estes, *Southwest. Nat.*, 1975, **19**, 379–385.
16. C. L. Porter, *Ecology*, 1966, **47**, 980–992.
17. J. A. Duke, in *Crop Tolerance to Suboptimal Land Conditions: Proceedings of a Symposium*, G. A. Jung, (ed.), American Society of Agronomy, Madison, WI, 1978, pp. 1–61.
18. J. D. Hansen and H. A. Johnson, *Seed Technol.*, 2005, **27**, 203–210.

19. A. A. Hopkins and C. M. Taliaferro, *Crop Sci.*, 1997, **37**, 1719–1722.
20. D. R. Cornelius and C. O. Johnston, *J. Am. Soc. Agron.*, 1941, **33**, 115–124.
21. C. McMillian, *Ecol. Monogr.*, 1959, **29**, 285–308.
22. C. McMillian, *Am. J. Bot.*, 1965, **52**, 55–65.
23. C. McMillian and J. Weiler, *Am. J. Bot.*, 1959, **46**, 590–593.
24. R. G. Bailey, *Description of the ecoregions of the United States* Misc. Pub. No. 1391, US Forest Service, Washington, D.C., 1995.
25. R. G. Bailey and L. Ropes, *Ecoregions: The Ecosystem Geography of the Oceans and Continents*, Springer-Verlag, Inc., New York, NY, 1998.
26. H. M. Cathey, *USDA plant hardiness zone map*, US National Arboretum, USDA-ARS, Washington, D.C., 1990.
27. K. P. Vogel, M. R. Schmer and R. B. Mitchell, *Rangeland Ecol. Manage.*, 2005, **58**, 315–319.
28. R. G. Bailey, *Ecoregions of the United States (scale 1:7,500,000)*, USDA Forest Service, 1995.
29. M. D. Casler, K. P. Vogel, C. M. Taliaferro, N. J. Ehlke, J. D. Berdahl, E. C. Brummer, R. L. Kallenbach, C. P. West and R. B. Mitchell, *Crop Sci.*, 2007, **47**, 2249–2260.
30. M. D. Casler, K. P. Vogel, C. M. Taliaferro and R. L. Wynia, *Crop Sci.*, 2004, **44**, 293–303.
31. F. W. Gould, *The Grasses of Texas*, Texas A&M University Press, College Station, TX, 1975.
32. D. D. Redfearn, K. J. Moore, K. P. Vogel, S. S. Waller and R. B. Mitchell, *Agron. J.*, 1997, **89**, 262–269.
33. P. R. Newman and L. E. Moser, *Agron. J.*, 1988, **80**, 383–387.
34. C. R. Tischler and P. W. Voigt, *J. Range Manage.*, 1993, **46**, 436–439.
35. J. E. Weaver, *North American Prairie*, Johnsen Publishing Company, Lincoln, NE, 1954.
36. W. A. Wheeler and D. D. Hill, *Grassland Seeds*, Van Nostrand Company, Inc., Princeton, NJ, 1957.
37. K. P. Vogel, *Seed Technol.*, 2002, **24**, 9–15.
38. A. A. Hopkins, C. M. Taliaferro, C. D. Murphy and D. Christian, *Crop Sci.*, 1996, **36**, 1192–1195.
39. S. J. Hultquist, K. P. Vogel, D. J. Lee, K. Arumuganathan and S. Kaeppler, *Crop Sci.*, 1996, **36**, 1049–1052.
40. S. J. Hultquist, K. P. Vogel, D. J. Lee, K. Arumuganathan and S. Kaeppler, *Crop Sci.*, 1997, **37**, 595–598.
41. K. Lu, S. M. Kaeppler, K. P. Vogel, K. Arumuganathan and D. J. Lee, *Great Plains Res.*, 1998, **8**, 269–280.
42. S. S. Waller and J. K. Lewis, *J. Range Manage.*, 1979, **32**, 12–28.
43. F. H. Hsu, C. J. Nelson and A. G. Matches, *Crop Sci.*, 1985, **25**, 215–220.
44. F. H. Hsu, C. J. Nelson and A. G. Matches, *Crop Sci.*, 1985, **25**, 249–255.
45. K. Hoshikawa, *Bot. Gaz.*, 1969, **130**, 192–203.
46. A. J. Smart and L. E. Moser, *Agron. J.*, 1997, **89**, 958–962.

47. E. R. Beaty, J. L. Engel and J. D. Powell, *J. Range Manage.*, 1978, **31**, 361–365.
48. G. S. Heidmann and G. E. Van Riper, *J. Range Manage.*, 1967, **20**, 236–241.
49. P. L. Sims, L. A. Ayuko and D. N. Hyder, *J. Range Manage.*, 1971, **24**, 357–360.
50. H. M. Benedict, *J. Agric. Res.*, 1941, **61**, 661–672.
51. L. C. Newell, *Crop Sci.*, 1968, **8**, 205–210.
52. K. J. Moore, L. E. Moser, K. P. Vogel, S. S. Waller, B. E. Johnson and J. F. Pedersen, *Agron. J.*, 1991, **83**, 1073–1077.
53. R. Mitchell, J. Fritz, K. Moore, L. Moser, K. Vogel, D. Redfearn and D. Wester, *Agron. J.*, 2001, **93**, 118–124.
54. R. B. Mitchell, K. J. Moore, L. E. Moser, J. O. Fritz and D. D. Redfearn, *Agron. J.*, 1997, **89**, 827–832.
55. M. A. Sanderson and D. D. Wolf, *Agron. J.*, 1995, **87**, 908–915.
56. M. D. Jones and J. G. Brown, *Agron. J.*, 1951, **43**, 218–222.
57. M. D. Jones and L. C. Newell, *Pollination cycles and pollen dispersal in relation to grass improvement*, Lincoln, NE, 1946.
58. D. Smith, *J. Range Manage.*, 1975, **28**, 389–391.
59. K. P. Vogel, J. J. Brejda, D. T. Walters and D. R. Buxton, *Agron. J.*, 2002, **94**, 413–420.
60. W. L. Stout, *Soil Sci. Soc. Am. J.*, 1992, **56**, 897–902.
61. J. R. Kiniry, L. R. Lynd, N. Greene, M. V. V. Johnson, M. D. Casler and M. S. Laser, in *New Research on Biofuels*, J. H. Wright and D. A. Evans, (ed.), Nova Science Publishers, Hauppauge, NY, 2008, p. 136.
62. R. Sprague, *Diseases of Cereals and Grasses in North America (Fungi, Except Smuts and Rusts)*, Ronald Press Company, New York, NY, 1950.
63. D. F. Farr, G. F. Bills, G. P. Chamuris and A. Y. Rossman, *Fungi on Plants and Plant Products in the United States*, APS Press, St. Paul, MN, 1989.
64. C. E. Gravert and G. P. Munkvold, *J. Iowa Acad. Sci.*, 2002, **109**, 30–34.
65. C. E. Gravert, L. H. Tiffany and G. P. Munkvold, *Plant Dis.*, 2000, **84**, 596–596.
66. W. W. Ray, *Plant Dis. Rep.*, 1954, **38**, 583–587.
67. L. H. Tiffany and J. H. Mathre, *Mycologia*, 1961, **53**, 600–604.
68. K. E. Zeiders, *Plant Dis.*, 1984, **68**, 120–122.
69. M. R. McLaughlin, R. C. Larsen, L. E. Trevathan, C. E. Eastman and A. D. Hewings, in *Pasture and Forage Crop Pathology*, S. Charkraborty, K. T. Leath, R. A. Skipp, G. A. Pederson, R. A. Bray, G. C. M. Latch and F. W. Nutter, (ed.), American Society of Agronomy, Madinson, WI, Edition edn, 2002, pp. 323–361.
70. C. L. Niblett and A. Q. Paulsen, *Phytopathology*, 1975, **65**, 1157–1160.
71. W. H. Sill, Jr. and R. C. Pickett, *Plant Dis. Rep.*, 1957, **41**, 241–249.
72. J. J. Brejda, L. E. Moser and K. P. Vogel, *Agron. J.*, 1998, **90**, 753–758.
73. J. J. Brejda, *Evaluation of arbuscular mycorrihiza populations for enhancing switchgrass yield and nutrient uptake*, Ph.D. diss., Univ. of Nebraska-Lincoln, NE USA, 1996.

74. F. L. Barnett and R. F. Carver, *Crop Sci.*, 1967, **7**, 301–304.
75. E. L. Nielsen, *J. Agric. Res.*, 1944, **69**, 327–353.
76. M. D. Bennett and J. B. Smith, *Philos. Trans. R. Soc. Lond., Ser. B: Biol. Sci*, 1991, **334**, 309–345.
77. E. L. Nielsen, *J. Am. Soc. Agron.*, 1947, **39**, 822–827.
78. J. M. Martinez-Reyna, K. P. Vogel, C. Caha and D. J. Lee, *Crop Sci.*, 2001, **41**, 1579–1583.
79. R. D. Riley and K. P. Vogel, *Crop Sci.*, 1982, **22**, 1082–1083.
80. X. F. Ma, J. Bouton, M. Saha, B. Narasimhamoorthy, T. Russell, A. Hernandez, L. Anusauskiene, R. Zapata, I. Zheveleva, S. Brover and T. J. Swaller, *Plant & Animal Genomes XVI Conference*, San Diego, CA, 2008.
81. A. M. Missaoui, A. H. Paterson and J. H. Bouton, *Theor. Appl. Genet.*, 2005, **110**, 1372–1383.
82. J. M. Martinez-Reyna and K. P. Vogel, *Crop Sci.*, 2002, **42**, 1800–1805.
83. L. E. Talbert, D. H. Timothy, J. C. Burns, J. O. Rawlings and R. H. Moll, *Crop Sci.*, 1983, **23**, 725–728.
84. A. M. Missaoui, A. H. Paterson and J. H. Bouton, *Genet. Resour. Crop Evol.*, 2006, **53**, 1291–1302.
85. L. E. Gunter, G. A. Tuskan and S. D. Wullschleger, *Crop Sci.*, 1996, **36**, 1017–1022.
86. A. A. Hopkins, K. P. Vogel, K. J. Moore, K. D. Johnson and I. T. Carlson, *Crop Sci.*, 1995, **35**, 565–571.
87. M. D. Casler, C. A. Stendal, L. Kapich and K. P. Vogel, *Crop Sci.*, 2007, **47**, 2261–2273.
88. C. M. Tobias, G. Sarath, P. Twigg, E. Lindquist, J. Pangilinan, B. W. Penning, K. Barry, M. C. McCann, N. C. Carpita and G. R. Lazo, *Plant Genome*, 2008, **1**, 111–124.
89. C. M. Tobias, P. Twigg, D. M. Hayden, K. P. Vogel, R. M. Mitchell, G. R. Lazo, E. K. Chow and G. Sarath, *Theor. Appl. Genet.*, 2005, **111**, 956–964.
90. M. D. Casler, *Crop Sci.*, 2005, **45**, 388–398.
91. S. A. Eberhardt and L. C. Newell, *Agron. J.*, 1959, **51**, 613–616.
92. A. Boe and M. D. Casler, *Crop Sci.*, 2005, **45**, 2465–2472.
93. M. K. Das, R. G. Fuentes and C. M. Taliaferro, *Crop Sci.*, 2003, **44**, 443–448.
94. E. B. Godshalk, D. H. Timothy and J. C. Burns, *Crop Sci.*, 1988, **28**, 825–830.
95. D. M. Gustafson, A. Boe and Y. Jin, *Crop Sci.*, 2003, **43**, 755–759.
96. A. A. Hopkins, K. P. Vogel and K. J. Moore, *Crop Sci.*, 1993, **33**, 253–258.
97. A. Missaoui, V. Fasoula and J. Bouton, *Euphytica*, 2005, **142**, 1–12.
98. L. C. Newell and S. A. Eberhart, *Crop Sci.*, 1961, **1**, 117–121.
99. L. W. Rose, M. K. Das, R. G. Fuentes and C. M. Taliaferro, *Euphytica*, 2007, **156**, 407–415.

100. G. A. Van Esbroeck, M. A. Hussey and M. A. Sanderson, *Crop Sci.*, 1998, **38**, 342–346.
101. K. P. Vogel, F. A. Haskins and H. J. Gorz, *Crop Sci.*, 1981, **21**, 39–41.
102. K. P. Vogel, A. A. Hopkins, K. J. Moore, K. D. Johnson and I. T. Carlson, *Crop Sci.*, 2002, **42**, 1857–1862.
103. M. D. Casler and E. C. Brummer, *Crop Sci.*, 2008, **48**, 890–902.
104. K. P. Vogel, in *Native Warm-season grasses: research trends and issues*, K. J. Moore and B. Anderson, (ed.), Crop Sci. Society of America, Madison, WI, 2000.
105. K. P. Vogel and B. L. Burson, in *Warm-Season (C4) Grasses*, L. E. Moser, L. Sollenberger and B. Burson, (ed.), ASA-CSSA-SSSA, Madison, WI, 2004, pp. 83–106.
106. K. P. Vogel and J. F. Pedersen, *Plant Breed. Rev.*, 1993, **11**, 251–274.
107. J. M. Martinez-Reyna and K. P. Vogel, *Crop Sci.*, 1998, **38**, 876–878.
108. A. Boe and P. O. Johnson, *Crop Sci.*, 1987, **27**, 147–148.
109. A. J. Smart, L. E. Moser and K. P. Vogel, *Crop Sci.*, 2003, **43**, 1434–1440.
110. A. J. Smart, K. P. Vogel, L. E. Moser and W. W. Stroup, *Crop Sci.*, 2003, **43**, 1427–1433.
111. A. Boe and J. G. Ross, *Crop Sci.*, 1998, **38**, 540–540.
112. M. D. Casler, D. R. Buxton and K. P. Vogel, *Theor. Appl. Genet.*, 2002, **104**, 127–131.
113. K. P. Vogel, *J. Soil Water Conserv.*, 1996, **51**, 137–139.
114. J. H. Fike, D. J. Parrish, D. D. Wolf, J. A. Balasko, J. T. Green, M. Rasnake and J. H. Reynolds, *Biomass Bioenergy*, 2006, **30**, 207–213.
115. M. A. Sanderson, J. C. Read and R. L. Reed, *Agron. J.*, 1999, **91**, 5–10.
116. M. D. Casler and A. R. Boe, *Crop Sci.*, 2003, **43**, 2226–2233.
117. A. A. Hopkins, K. P. Vogel, K. J. Moore, K. D. Johnson and I. T. Carlson, *Crop Sci.*, 1995, **35**, 125–132.
118. B. Narasimhamoorthy, M. Saha, T. Swaller and J. Bouton, *Bioener. Res.*, 2008, **1**, 136–146.
119. C. M. Tobias, D. M. Hayden, P. Twigg and G. Sarath, *Mol. Ecol. Notes*, 2006, **6**, 185–187.
120. J. M. Martinez-Reyna and K. P. Vogel, *Crop Sci.*, 2008, **48**, 1312–1320.
121. K. P. Vogel and K. B. Mitchell, *Crop Sci.*, 2008, **48**, 2159–2164.
122. K. S. Alexandrova, P. D. Denchev and B. V. Conger, *Crop Sci.*, 1996, **36**, 175–178.
123. K. S. Alexandrova, P. D. Denchev and B. V. Conger, *Crop Sci.*, 1996, **36**, 1709–1711.
124. P. D. Denchev and B. V. Conger, *Crop Sci.*, 1994, **34**, 1623–1627.
125. S. D. Gupta and B. V. Conger, *Crop Sci.*, 1999, **39**, 243–247.
126. J. H. Bouton, *Curr. Opin. Genet. Dev.*, 2007, **17**, 553–558.
127. M. Mazarei, H. Al-Ahmad, M. R. Rudis and C. N. Stewart, Jr., *Biotechnol. J.*, 2008, **3**, 354–359.
128. H. A. Richards, V. A. Rudas, H. Sun, J. K. McDaniel, Z. Tomaszewski and B. V. Conger, *Plant Cell Rep.*, 2001, **20**, 48–54.

129. M. N. Somleva, Z. Tomaszewski and B. V. Conger, *Crop Sci.*, 2002, **42**, 2080–2087.
130. K. P. Vogel, *Agron. J.*, 1987, **79**, 509–512.
131. M. R. Schmer, K. P. Vogel, R. B. Mitchell, L. E. Moser, K. M. Eskridge and R. K. Perrin, *Crop Sci.*, 2006, **46**, 157–161.
132. K. P. Vogel and R. A. Masters, *J. Range Manage.*, 2001, **54**, 653–655.
133. B. L. Dierberger, *Switchgrass germination as influenced by temperature, chilling, cultivar, and seed lot*, M. S. Thesis, Univ. of Nebraska, Lincoln, NE, USA, 1991.
134. M. T. Panciera and G. A. Jung, *J. Soil Water Conserv.*, 1984, **39**, 68–70.
135. T. L. Vassey, J. R. George and R. E. Mullen, *Agron. J.*, 1985, **77**, 253–257.
136. J. F. Samson and L. E. Moser, *Agron. J.*, 1982, **74**, 1055–1060.
137. Z. X. Shen, D. J. Parrish, D. D. Wolf and G. E. Welbaum, *Crop Sci.*, 2001, **41**, 1546–1551.
138. A. J. Smart and L. E. Moser, *Agron. J.*, 1999, **91**, 335–338.
139. Anonymous, *Guide for weed management in Nebraska*, University of Nebraska, Lincoln, NE, 2009.
140. C. C. Bahler, K. P. Vogel and L. E. Moser, *Agron. J.*, 1984, **76**, 891–895.
141. A. R. Martin, R. S. Moomaw and K. P. Vogel, *Agron. J.*, 1982, **74**, 916–920.
142. M. R. Weimer, B. A. Swisher and K. P. Vogel, *Weed Sci.*, 1988, **36**, 436–440.
143. R. A. Masters, S. J. Nissen, R. E. Gaussoin, D. D. Beran and R. N. Stougaard, *Weed Technol.*, 1996, **10**, 392–403.
144. R. B. Mitchell, K. P. Vogel, J. Berdahl and R. A. Masters, *BioEner. Res.*, Online 31 March 2010, DOI 10.1007/s12155-010-9084-4.
145. R. L. Hintz, K. R. Harmoney, K. J. Moore, J. R. George and E. C. Brummer, *Agron. J.*, 1998, **90**, 591–596.
146. G. A. Jung, J. A. Shaffer and W. L. Stout, *Agron. J.*, 1988, **80**, 669–676.
147. G. W. Rehm, R. C. Sorensen and W. J. Moline, *Agron. J.*, 1976, **68**, 759–764.
148. K. E. Hall, J. R. George and R. R. Riedl, *Agron. J.*, 1982, **74**, 47–51.
149. M. Haque, F. M. Epplin and C. M. Taliaferro, *Agron. J.*, 2009, **101**, 1463–1469.
150. A. H. Heggenstaller, K. J. Moore, M. Liebman and R. P. Anex, *Agron. J.*, 2009, **101**, 1363–1371.
151. I. C. Madakadze, K. A. Stewart, P. R. Peterson, B. E. Coulman and D. L. Smith, *Crop Sci.*, 1999, **39**, 552–557.
152. W. E. McMurphy, C. E. Demman and B. B. Tucker, *Agron. J.*, 1975, **67**, 233–236.
153. L. J. Perry, Jr. and D. Baltensperger, *Agron. J.*, 1979, **71**, 355–358.
154. G. W. Rehm, *Agron. J.*, 1984, **76**, 731–734.
155. G. W. Rehm, R. C. Sorensen and W. J. Moline, *Agron. J.*, 1977, **69**, 955–961.
156. T. E. Staley, W. L. Stout and G. A. Jung, *Agron. J.*, 1991, **83**, 732–738.

157. I. C. Madakadze, K. Stewart, P. R. Peterson, B. E. Coulman and D. L. Smith, *Agron. J.*, 1999, **91**, 696–701.
158. P. R. Adler, M. A. Sanderson, A. A. Boateng, P. I. Weimer and H. J. G. Jung, *Agron. J.*, 2006, **98**, 1518–1525.
159. M. A. Sanderson, R. P. Egg and A. E. Wiselogel, *Biomass Bioenergy*, 1997, **12**, 107–114.
160. A. E. Wiselogel, F. A. Agblevor, D. K. Johnson, S. Deutch, J. A. Fennell and M. A. Sanderson, *Bioresour. Technol.*, 1996, **56**, 103–109.
161. D. R. Cornelius, *Ecol. Monogr.*, 1950, **20**, 1–27.
162. P. C. Kassel, R. E. Mullen and T. B. Bailey, *Agron. J.*, 1985, **77**, 214–218.
163. J. R. George, G. S. Reigh, R. E. Mullen and J. J. Hunczak, *Crop Sci.*, 1990, **30**, 845–849.
164. J. J. Brejda, J. R. Brown, G. W. Wyman and W. K. Schumacher, *J. Range Manage.*, 1994, **47**, 22–27.
165. A. Faaij, *Mitigation and Adaption Strategies for Global Change*, 2006, **11**, 343–375.
166. A. Carroll and C. Somerville, *Annu. Rev. Plant Biol.*, 2009, **60**, 165–182.
167. L. R. Lynd, J. H. Cushman, R. J. Nichols and C. E. Wyman, *Science*, 1991, **251**, 1318–1323.
168. C. E. Wyman, *Bioresour. Technol.*, 1994, **50**, 3–16.
169. C. E. Wyman and B. Yang, *Calif. Agric.*, 2009, **63**, 185–190.
170. R. Wooley, M. Ruth, D. Glassner and J. Sheehan, *Biotechnol. Prog.*, 1999, **15**, 794–803.
171. D. R. Keshwani and J. J. Cheng, *Bioresour. Technol.*, 2009, **100**, 1515–1523.
172. A. A. Boateng, K. B. Hicks and K. P. Vogel, *J. Anal. Appl. Pyrolysis*, 2006, **75**, 55–64.
173. A. A. Boateng, P. J. Weimer, H. G. Jung and J. F. S. Lamb, *Energy Fuels*, 2008, **22**, 2810–2815.
174. B. S. Dien, H. J. G. Jung, K. P. Vogel, M. D. Casler, J. F. S. Lamb, L. Iten, R. B. Mitchell and G. Sarath, *Biomass Bioenergy*, 2006, **30**, 880–891.
175. N. Qureshi and T. C. Ezeji, *Biofuels, Bioprod. Biorefin.*, 2008, **2**, 319–330.
176. J. F. Pedersen, D. L. Funnell and K. P. Vogel, *Crop Sci.*, 2005, **45**, 812–819.
177. J. Vogel, *Curr. Opin. Plant Biol.*, 2008, **11**, 301–307.
178. W. Boerjan, J. Ralph and M. Baucher, *Annu. Rev. Plant Biol.*, 2003, **54**, 519–546.
179. D. M. Burner, T. L. Tew, J. J. Harvey and D. P. Belesky, *Biomass Bioenergy*, 2009, **33**, 610–619.
180. G. Sarath, R. B. Mitchell, S. E. Sattler, D. Funnell, J. F. Pedersen, R. A. Graybosch and K. P. Vogel, *J. Ind. Microbiol. Biotechnol.*, 2008, **35**, 343–354.
181. S. P. Chundawat, B. Venkatesh and B. E. Dale, *Biotechnol. Bioeng.*, 2007, **96**, 219–231.

182. S. Mani, L. G. Tabil and S. Sokhansanj, *Biomass Bioenergy*, 2004, **27**, 339–352.
183. S. C. Yat, A. Berger and D. R. Shonnard, *Bioresource Technol.*, 2008, **99**, 3855–3863.
184. R. Mitchell, K. P. Vogel and G. Sarath, *Biofuels, Bioprod. Biorefin.*, 2008, **2**, 530–539.
185. A. Monti, S. Fazio and G. Venturi, *Biomass Bioenergy*, 2009, **33**, 841–847.
186. H. Shen, C. Fu, X. Xiao, T. Ray, Y. Tang, Z. Wang and F. Chen, *Bioener. Res.*, 2009, **2**, 233–245.
187. G. Sarath, L. M. Baird, K. P. Vogel and R. B. Mitchell, *Bioresource Technol.*, 2007, **98**, 2985–2992.
188. J. Yang, E. Worley, M. Wang, B. Lahner, D. E. Salt, M. Saha and M. Udvardi, *Bioener. Res.*, 2009, **2**, 257–266.
189. H. M. El-Nashaar, G. M. Banowetz, S. M. Griffith, M. D. Casler and K. P. Vogel, *Bioresource Technol.*, 2009, **100**, 1809–1814.
190. K. P. Vogel and H. J. G. Jung, *Crit. Rev. Plant Sci.*, 2001, **20**, 15–49.
191. M. D. Casler and K. P. Vogel, *Crop Sci.*, 1999, **39**, 12–20.
192. R. B. Mitchell, B. E. Anderson, R. A. Masters, K. P. Vogel and T. J. Klopfenstein, *Crop Sci.*, 2005, **45**, 2288–2292.
193. M. D. Casler and H. J. G. Jung, *Anim. Feed Sci. Technol.*, 2006, **125**, 151–161.
194. B. Dien, G. Sarath, J. Pedersen, S. Sattler, H. Chen, D. Funnell-Harris, N. Nichols and M. Cotta, *Bioener. Res.*, 2009, **2**, 153–164.
195. F. Chen and R. A. Dixon, *Nature Biotechnol.*, 2007, **25**, 759–761.
196. A. M. Anterola and N. G. Lewis, *Phytochemistry*, 2002, **61**, 221–294.
197. J. M. A. Tilley and R. A. Terry, *J. Braz. Grass. Soc.*, 1963, **18**, 104–111.
198. K. P. Vogel, J. F. Pedersen, S. D. Masterson and J. J. Toy, *Crop Sci.*, 1999, **39**, 276–279.
199. W. F. Anderson, B. S. Dien, S. K. Brandon and J. D. Peterson, *Appl. Biochem. Biotechnol.*, 2008, **145**, 13–21.
200. K. P. Vogel, G. Sarath and R. Mitchell, in *XX International Grassland Congress: Offered Papers*, F. P. O'Mara, R. J. Wilkins, L. 't Mannetje, D. K. Lovett, P. A. M. Rogers and T. M. Boland, (ed.), Wageningen Academic Publishers, Dublin, Ireland, 2005, p. 975.
201. G. Sarath, D. E. Akin, R. B. Mitchell and K. P. Vogel, *Appl. Biochem. Biotechnol.*, 2008, **150**, 1–14.
202. Y. Yang, R. Sharma-Shivappa, J. C. Burns and J. J. Cheng, *Energy Fuels*, 2009, **23**, 3759–3766.
203. M. C. McCann and N. C. Carpita, *Curr. Opin. Plant Biol.*, 2008, **11**, 314–320.
204. M. M. Buanafina, T. Langdon, B. Hauck, S. Dalton and P. Morris, *Plant Biotechnol. J.*, 2008, **6**, 264–280.
205. J. H. Grabber, J. Ralph and R. D. Hatfield, *J. Agr. Food Chem.*, 1998, **46**, 2609–2614.

206. G. Sarath, K. P. Vogel, R. B. Mitchell and L. M. Baird, in *XX International Grassland Congress: Offered Papers*, F. P. O'Mara, R. J. Wilkins, L. 't Mannetje, D. K. Lovett, P. A. M. Rogers and T. M. Boland, (ed.), Wageningen Academic Publishers, Dublin, Ireland, 2005, p. 975.
207. T. Hooks, J. F. Pedersen, D. B. Marx and K. P. Vogel, *Crop Sci.*, 2006, **46**, 751–757.
208. B. C. King, M. K. Donnelly, G. C. Bergstrom, L. P. Walker and D. M. Gibson, *Biotechnol Bioeng*, 2009, **102**, 1033–1044.
209. J. F. Li, E. Park, A. G. von Arnim and A. Nebenfuhr, *Plant Methods*, 2009, **5**, 6.
210. D. Mann, N. Labbé, R. Sykes, K. Gracom, L. Kline, I. Swamidoss, J. Burris, M. Davis and C. Stewart, *Bioener. Res.*, 2009, **2**, 246–256.
211. A. Isci, J. Himmelsbach, A. Pometto, D. Raman and R. Anex, *Appl. Biochem. Biotechnol.*, 2008, **144**, 69–77.
212. L. L. Escamilla-Trevino, H. Shen, S. R. Uppalapati, T. Ray, Y. Tang, T. Hernandez, Y. Yin, Y. Xu and R. A. Dixon, *New Phytol.*, 2009.
213. C. N. Stewart, L. G. Abercrombie, H. Baxter, J. Burris, F. Chen, R. A. Dixon, R. Equi, M. Halter, H. Hisano, S. Hisano, Z. King, P. Lafayette, D. Mann, M. Mazarei, R. Nandakumer, R. S. Nelson, W. A. Parrott, C. Poovaiah, H. Ramanna, M. Rudis, Z. Y. Wang, J. Zale, J. S. Hawkins, R. Percifield and J. L. Bennetzen, *In Vitro Cell. Dev. Biol. Anim.*, 2009, **45**, S6–S7.
214. H. Shen, Y. Yin, F. Chen, Y. Xu and R. Dixon, *Bioener. Res.*, 2009, **2**, 217–232.
215. M. Pauly and K. Keegstra, *Plant J.*, 2008, **54**, 559–568.
216. Y. Sun and J. Y. Cheng, *Bioresour. Technol.*, 2002, **83**, 1–11.
217. L. Tillman, A. Abaseed, Y. Lee and R. Torget, *Appl. Biochem. Biotechnol.*, 1989, **20–21**, 107–117.
218. L. Tillman, Y. Lee and R. Torget, *Appl. Biochem. Biotechnol.*, 1990, **24–25**, 103–113.
219. B. Yang and C. E. Wyman, *Biofuels, Bioprod. Biorefin.*, 2008, **2**, 26–40.
220. L. Cadoche and G. D. López, *Biol. Wastes*, 1989, **30**, 153–157.
221. D. Schell and C. Harwood, *Appl. Biochem. Biotechnol.*, 1994, **45–46**, 159–168.
222. M. Yu, A. R. Womac, C. Igathinathane, P. D. Ayers and M. J. Buschermohle, *Biomass Bioenergy*, 2006, **30**, 214–219.
223. V. S. P. Bitra, A. R. Womac, C. Igathinathane, P. I. Miu, Y. C. T. Yang, D. R. Smith, N. Chevanan and S. Sokhansanj, *Bioresour. Technol.*, 2009, **100**, 6578–6585.
224. C. Igathinathane, A. R. Womac, S. Sokhansanj and S. Narayan, *Bioresour. Technol.*, 2008, **99**, 2254–2264.
225. V. S. R. Bitra, A. R. Womac, N. Chevanan, P. I. Miu, C. Igathinathane, S. Sokhansanj and D. R. Smith, *Powder Technol.*, 2009, **193**, 32–45.
226. T. G. Bridgeman, L. I. Darvell, J. M. Jones, P. T. Williams, R. Fahmi, A. V. Bridgwater, T. Barraclough, I. Shield, N. Yates, S. C. Thain and I. S. Donnison, *Fuel*, 2007, **86**, 60–72.

227. N. Mosier, C. Wyman, B. Dale, R. Elander, Y. Y. Lee, M. Holtzapple and M. Ladisch, *Bioresour. Technol.*, 2005, **96**, 673–686.
228. S. Singh, B. A. Simmons and K. P. Vogel, *Biotechnol. Bioeng.*, 2009, **104**, 68–75.
229. N. Hinman, D. Schell, J. Riley, P. Bergeron and P. Walter, *Appl. Biochem. Biotechnol.*, 1992, **34–35**, 639–649.
230. R. Torget, P. Werdene, M. Himmel and K. Grohmann, *Appl. Biochem. Biotechnol.*, 1990, **24–25**, 115–126.
231. C. E. Wyman, D. D. Spindler and K. Grohmann, *Biomass Bioenergy*, 1992, **3**, 301–307.
232. A. Esteghlalian, A. G. Hashimoto, J. J. Fenske and M. H. Penner, *Bioresour. Technol.*, 1997, **59**, 129–136.
233. Y. C. Chung, A. Bakalinsky and M. H. Penner, *Appl. Biochem. Biotechnol.*, 2005, **121**, 947–961.
234. V. S. Chang and M. T. Holtzapple, *Appl. Biochem. Biotechnol.*, 2000, **84–6**, 5–37.
235. J. H. Grabber, D. R. Mertens, H. Kim, C. Funk, F. C. Lu and J. Ralph, *J. Sci. Food Agric.*, 2009, **89**, 122–129.
236. H. Yoon, Z. Wu and Y. Lee, *Appl. Biochem. Biotechnol.*, 1995, **51–52**, 5–19.
237. T. Kim and Y. Lee, *Appl. Biochem. Biotechnol.*, 2005, **124**, 1119–1131.
238. P. Iyer, Z. -W. Wu, S. Kim and Y. Lee, *Appl. Biochem. Biotechnol.*, 1996, **57–58**, 121–132.
239. S. Kim and Y. Lee, *Appl. Biochem. Biotechnol.*, 1996, **57–58**, 147–156.
240. Z. Wu and Y. Lee, *Appl. Biochem. Biotechnol.*, 1997, **63–65**, 21–34.
241. A. Isci, J. N. Himmelsbach, J. Strohl, A. L. Pometto, D. R. Raman and R. P. Anex, *Appl. Biochem. Biotechnol.*, 2009, **157**, 453–462.
242. M. Winugroho, M. N. M. Ibrahim and G. R. Pearce, *Agr. Wastes*, 1984, **9**, 87–99.
243. V. Chang, B. Burr and M. Holtzapple, *Appl. Biochem. Biotechnol.*, 1997, **63–65**, 3–19.
244. S. Kim and M. T. Holtzapple, *Bioresour. Technol.*, 2005, **96**, 1994–2006.
245. F. Kong, C. Engler and E. Soltes, *Appl. Biochem. Biotechnol.*, 1992, **34–35**, 23–35.
246. V. S. Chang, W. E. Kaar, B. Burr and M. T. Holtzapple, *Biotechnol. Lett.*, 2001, **23**, 1327–1333.
247. R. P. Swatloski, S. K. Spear, J. D. Holbrey and R. D. Rogers, *J. Am. Chem. Soc.*, 2002, **124**, 4974–4975.
248. A. P. Dadi, S. Varanasi and C. A. Schall, *Biotechnol. Bioeng.*, 2006, **95**, 904–910.
249. H. Zhao, C. L. Jones, G. A. Baker, S. Xia, O. Olubajo and V. N. Person, *J. Biotechnol.*, 2009, **139**, 47–54.
250. B. E. Dale, C. K. Leong, T. K. Pham, V. M. Esquivel, I. Rios and V. M. Latimer, *Bioresour. Technol.*, 1996, **56**, 111–116.

251. M. Holtzapple, J.-H. Jun, G. Ashok, S. Patibandla and B. Dale, *Appl. Biochem. Biotechnol.*, 1991, **28–29**, 59–74.
252. S. Reshamwala, B. Shawky and B. Dale, *Appl. Biochem. Biotechnol.*, 1995, **51–52**, 43–55.
253. H. Alizadeh, F. Teymouri, T. Gilbert and B. Dale, *Appl. Biochem. Biotechnol.*, 2005, **124**, 1133–1141.
254. B. Bals, L. Teachworth, B. Dale and V. Balan, *Appl. Biochem. Biotechnol.*, 2007, **143**, 187–198.
255. Z. Hu and Z. Wen, *Biochem. Eng. J.*, 2008, **38**, 369–378.
256. D. R. Keshwani, J. J. Cheng, J. C. Burns, L. Li and V. Chiang, *2007 ASABE Annual International Meeting*, Paper Number 077127, Minneapolis, MN, 2007.
257. Z. Hu, Y. Wang and Z. Wen, *Appl. Biochem. Biotechnol.*, 2008, **148**, 71–81.
258. L. d. C. Sousa, S. P. S. Chundawat, V. Balan and B. E. Dale, *Curr. Opin. Biotechnol.*, 2009, **20**, 339–347.
259. E. Sendich, M. Laser, S. Kim, H. Alizadeh, L. Laureano-Perez, B. Dale and L. Lynd, *Bioresour. Technol.*, 2008, **99**, 8429–8435.
260. M. Duffy and V. Nanhou, in *Trends in New Crops and New Uses*, J. Janick and A. Whipkey, (ed.), ASHS Press, Alexandria, VA, Edition edn, 2002, pp. 267–274.
261. F. M. Epplin, *Biomass Bioenergy*, 1996, **11**, 459–467.
262. A. Hallam, I. C. Anderson and D. R. Buxton, *Biomass Bioenergy*, 2001, **21**, 407–424.
263. M. Khanna, B. Dhungana and J. Clifton-Brown, *Biomass Bioenergy*, 2008, **32**, 482–493.
264. A. Monti, S. Fazio, V. Lychnaras, P. Soldatos and G. Venturi, *Biomass Bioenergy*, 2007, **31**, 177–185.
265. P. Vadas, K. Barnett and D. Undersander, *Bioener. Res.*, 2008, **1**, 44–55.
266. M. E. Walsh, *Biomass Bioenergy*, 1998, **14**, 341–350.
267. M. E. Walsh, D. G. D. Ugarte, H. Shapouri and S. P. Slinsky, *Environ. Resour. Econ.*, 2003, **24**, 313–333.
268. A. E. Farrell, R. J. Plevin, B. T. Turner, A. D. Jones, M. O'Hare and D. M. Kammen, *Science*, 2006, **311**, 506–508.
269. D. Pimentel and T. W. Patzek, *Nature Resour. Res.*, 2005, **14**, 65–76.
270. A. B. Frank, J. D. Berdahl, J. D. Hanson, M. A. Liebig and H. A. Johnson, *Crop Sci.*, 2004, **44**, 1391–1396.
271. C. T. Garten and S. D. Wullschleger, *J. Environ. Qual.*, 2000, **29**, 645–653.
272. D. K. Lee, V. N. Owens and J. J. Doolittle, *Agron. J.*, 2007, **99**, 462–468.
273. M. A. Liebig, H. A. Johnson, J. D. Hanson and A. B. Frank, *Biomass Bioenergy*, 2005, **28**, 347–354.
274. B. Eghball, J. E. Gilley, L. A. Kramer and T. B. Moorman, *J. Soil Water Conserv.*, 2000, **55**, 172–176.
275. J. E. Gilley, B. Eghball, L. A. Kramer and T. B. Moorman, *J. Soil Water Conserv.*, 2000, **55**, 190–196.

276. S. B. McLaughlin, D. G. D. L. Ugarte, C. T. Garten, L. R. Lynd, M. A. Sanderson, V. R. Tolbert and D. D. Wolf, *Environ. Sci. Technol.*, 2002, **36**, 2122–2129.
277. S. B. McLaughlin and M. E. Walsh, *Biomass Bioenergy*, 1998, **14**, 317–324.
278. A. M. Roth, D. W. Sample, C. A. Ribic, L. Paine, D. J. Undersander and G. A. Bartelt, *Biomass Bioenergy*, 2005, **28**, 490–498.
279. M. A. Sanderson, R. L. Reed, S. B. McLaughlin, S. D. Wullschleger, B. V. Conger, D. J. Parrish, D. D. Wolf, C. Taliaferro, A. A. Hopkins, W. R. Ocumpaugh, M. A. Hussey, J. C. Read and C. R. Tischler, *Bioresour. Technol.*, 1996, **56**, 83–93.

CHAPTER 18
Algae

IRA A. LEVINE

University of Southern Maine, 51 Westminster Street, Lewiston, Maine 04240, USA

18.1 Introduction

Alga; Algae (plural): an assemblage of predominantly aquatic, photosynthetic, chlorophyll-containing, eukaryotic organisms, either unicellular or multicellular, lacking true roots, stems, and leaves. The approximately 30 000 described species are segregated by photosynthetic pigment content, carbohydrate food reserve, cell wall components, and flagella construction and orientation. This eclectic group has occupied niches on earth for more than 450 million years and will be described in this chapter as macroalgae (seaweeds) and microalgae (phytoplankton) as a source of biofuel feedstock.

"Vilor Alga" (translated as more vile or worthless than algae), wrote Virgil, the Latin Poet. However, the importance and value of algae had been realised long before the poems of Virgil were composed in 30 BC. The use of algae dates back to the time of Shen Nung,[1] the father of husbandry and medicine, approximately 3000 BC.[2] Algae have been reported to be utilised in Iceland (960 BC), the Chinese Book of Poetry (800–600 BC) praised housewives for cooking with algae. The Chinese Materia Medica (600 BC) refers to algae as; "Some algae are a delicacy fit for the most honorable of guest, even for the King himself".[1,3] Algae have been and are used as a source of food, industrial and fine chemicals, pharmaceuticals, nutraceuticals, cosmaceuticals, animal fodder, energy source, waste water and aquaculture effluent bioremediation, and fertiliser.

RSC Energy and Environment Series No. 3
Energy Crops
Edited by Nigel G Halford and Angela Karp
© Royal Society of Chemistry 2011
Published by the Royal Society of Chemistry, www.rsc.org

Phycologists (one who studies algae) have enjoyed their lives' work in relative obscurity until the explosion in government funding of research and development, private equity investments and the nearly 200 corporate startups for the singular purpose of developing algal-based biofuels. Working with algae is now "cool" (personal communication, author's daughter) and enjoys the focus and funding to move our discipline significantly forward in: cultivar development and enhancement, photobioreactor design, harvesting, processing, and new product development.

Algal utilisation as a feedstock for biofuel development is not a new concept. Indications of burning beach-dried algae have been passed down in Pacific Island nation oral histories. Algae to biofuel efforts of various scale and duration have been initiated over the past forty years. The significant differences in the present global effort to develop algal-based biofuels concerns the growing awareness of global climate change, tracking petroleum reserves and the value of a barrel of oil, governmental mandates of energy independence, the marketplace acceptance of the first-generation biofuels, and regulations to develop second-generation biofuels.

The following chapter describes the various technologies and opportunities for macro- and microalgae to be a significant biofuel feedstock. Utilisation of algae as a potential biofuel feedstock, reducing global reliance on petroleum-based transportation fuels, has attracted significant governmental, industrial, academic, and capital market attention due to: the ability to consume greenhouse gases, significantly higher areal productivities as compared to terrestrial-based plant feedstocks (Table 18.1), use of marginal lands and variable water supplies, and avoidance of competition with the human agriculture supply chain.

Algal-based biofuels represent many different forms of liquid transportation fuel or energy sources, including: alkanes, biobutanol, biodiesel, bioethanol, biohydrogen, biomethane, gasoline, green diesel, and jet fuel. These products represent three distinct pathways within the process chain: algal cell direct biosynthesis of biofuel molecule, whole algal cell processed into biofuel (anaerobic digestion, gasification, and pyrolysis), and the extraction of macromolecules from algal cell and subsequent processing into biofuels. The biofuel

Table 18.1 Biodiesel yields by source[4]

Crop	Oil yield	
	(L/Hectare)	(Gall/Acre)
Corn	172	18
Soybean	446	48
Canola	1190	127
Jatropha	1892	202
Coconut	2689	288
Oil Palm	5950	636
Microalgae[a]	58 700	6278

[a]cell lipid concentration calculated at 30% by weight

Algae

products are designed to reduce the dependence on petroleum-based transportation fuels and their chemical nature allows for the direct entry into existing transportation fuel infrastructure and distribution systems. Biofuel regulatory specifications
and certification processes are in place, minimizing challenges to market penetration. Yet to be completed and distributed is a carbon and energy life-cycle analysis; reports of private corporate efforts abound but efforts should continue in developing a public analysis, supporting the development of this nascent industrial effort.

The United States (Energy Independence and Security Act 2007) and the European Union (Directives on Renewable Energy) have each mandated the development of biofuels to supplant petroleum-based transportation fuels. The US law mandates the development of 36 billion gallons of renewable fuels with at least 21 billion gallons of advanced biofuel (noncorn ethanol) by 2022 and the EU has established a baseline 10% substitution of petroleum-based fuels with biofuels by 2020. The first generation of biofuels, corn- and cane-based bioethanol and terrestrial plant oil-based biodiesel have experienced market and political backlash as the competition with the human food and animal-feed market sectors have resulted in elevated agricultural product prices. Shifts in crop plantings and competition for quality farm acreage, fertiliser, pesticides, and water resulted in the US Department of Energy freezing the target production totals of the first-generation bioethanol at 15 billion gallons, while mandating the development of an equal number of gallons of second-generation cellulosic bioethanol and an additional six billion gallons of biofuels from alternative feedstocks including algae.

The commercial viability of algal-based biofuels has been shown to be inextricably tied to the value of petroleum-based transportation fuels. Chisti developed a mathematical relationship between the cost of a barrel of crude oil and that of algal oil.[4] The value index assumes the following: lipid content is 30% by weight; algal oil contains ~80% of the energy content of crude petroleum; value of a United States crude oil is 52% barrel price, 20% taxes, 19% refining, and 9% marketing and distribution. The following equation develops the cost of a liter of algal oil relative to current crude prices.

$$\text{Cost Algal Oil (\$/L)} = (0.0069)(\text{Cost Barrel of Crude Oil})$$

E.g. oil at \$80/barrel = \$0.552/liter of algal oil or \$2.10/gallon of algal oil

Calculations and estimates illustrating the replacement of the annual US thirst for petroleum-based transportation fuels of 210 billion gallons indicate the need for a land mass nearly the size of Florida, producing 2.5 billion dry tons of algae, sequestering over 5 billion tons of the greenhouse gas, carbon dioxide. Algae will play a significant role in the development of an alternative transportation fuel, but one must keep the discussion within the realm of reality to maintain credibility for the field and emerging enterprise, a reality evolving into a commercial opportunity (Table 18.2).

18.2 Macroalgae

Macroalgae or seaweeds have been collected or farmed as a source of food, feeds, medicine, dyes, soil conditioner, fertiliser, hydrocolloids, cosmaceuticals, nutraceuticals, and industrial chemicals for thousands of years. Macroalgae can be sourced from wild standing crops, traditional near-shore farming efforts, and developing offshore mariculture efforts. Seaweeds represent an alternative to microalgal cultivation as a source of algal-based biofuels, exhibiting several advantages: harvesting ease, long-established family, corporate and farming cooperative farming efforts, highly regulated permitted mariculture sites and the primary producer in multitrophic integrated aquaculture systems.

The annual yields of macroalgae have been reported at 16.2 million wet tons (~one million dry tons).[5] As the wild harvests continue to decline, 93% of the global harvest is farmed seaweeds, 99% of which are from Asian mariculture efforts. Historically, the United States' wild harvest for industrial utilisation reached its zenith in 1919 at ~400 000 tons, decreasing to the current levels of 50 000 wet tons.[6] Productivity as compared with total yields has been reported for wild standing stocks from 3.3 to 11.3 kg dry t/m^2/yr, which is equivalent to terrestrial bioenergy feedstocks of 35 to 129 dry t/ha/yr. Algal mariculture productivities vary by species and location; *Laminaria* and *Ulva* productivities range from 20–60 and 2–45 dry t/ha/yr, respectively.

Considering macroalgae as an energy source; the energy heat content of *Macrocystis pyrifera* ranges between 9.5 and 11.4 MJ/kg dry weight representing 50% of the heat content of switchgrass or poplar. One disadvantage of using seaweeds as an energy feedstock is the large but varied (11–40%) ash content; the ash content is somewhat offset by the limited amounts of cellulose, hemicellulose, and lignin. The potential use of seaweeds as a direct combustion source is limited; internal moisture contents range from 80–90%, high ash content, and the costs to dry and handle would eliminate its consideration.

The productivity and standing crop of seaweeds are enough to warrant efforts to develop them as a feedstock for biofuels. Additionally, the relative ease of harvesting is a significant advantage not realised by microalgal cultivation efforts. Global efforts for the development of microalgal-based biofuels include: Japanese-based; Ocean Sunrise Project whose goals include the establishment of a 4.47 million km^2 Exclusive Economic Zone to farm and harvest 150 tons of *Sargassum horneri* yielding 1.2 million gallons of bioethanol;[7] Tokyo Gas Company's effort to produce biomethane from a 1 ton/day digestion facility utilising *Laminaria* and *Ulva*; Scotland's Energy Minister, Jim Mather, initiated the BioMara research project in collaboration with the Scottish Association for Marine Science by stating "Effectively, seaweed harvested off a beach in the Outer Isles could be heating a crofter's kettle for their cup of tea the next morning": this project represents a 6 million euro effort to determine the feasibility of utilising macroalgae to produce biofuels; Biolsystems Co. Ltd of South Korea collaborating with Bohol Province, Philippines to establish up to 100 000 hectares of *Eucheuma cottonii* mariculture for the production of bioethanol. Danish National Environmental Research Institute and

Table 18.2 Commercial algal cultivation efforts.

Company	Location	Culture Strategy	Cultivar Enhancement	Open Pond	Raceway	Closed Bioreactor	Non Biofuel
A2BE	CO, USA	Horizontal panel			x		
Advanced Biofuel Technologies (KwikPower)	UK		x				
Algaenesis	Israel	Closed reactor	Light Delivery				
Alganol	FL, USA	Open ponds – ethanol direct					
Algatech	Israel	Closed tubular reactors				x	
Algae Link	Spain	Horizontal tubes				x	
Algae Wheel	IN, USA	Open pond – Wasterwater		x			x
Algenol Biofuel	CO, USA	Pond – Direct to Ethanol		x			
Algetech	Norway	Greenhouse vertical tubular				x	
Algoil	India	Tubular bioreactors				x	
AquaCarotene	Australia	Open Ponds		x			x
Aquaflow Bionomic	New Zealand	Open Ponds – wastewater		x			x
Aurora Biofuels	CA, USA	Open Ponds	GMO	x			
Betatene	Australia	Open Ponds		x			
BFS Biopetroleo	Spain	Annular Column					
Bionavitas	WA, USA	Closed Reactor	Light Delivery			x	
BioReal	Sweden	Closed Cylindrical reactor				x	x
Blue Marbel Energy	WA, USA	Wild Harvest					
Bodega Algae	MA, USA	Closed Reactor	Light Delivery			x	
C3	UK	Tubular Reactor					
Carbon Capture Corp	CA, USA	Open Pond		x			
Cellana	HI, USA	Duel closed:open	GMO	x	x		

Table 18.2 (continued)

Company	Location	Culture Strategy	Cultivar Enhancement	Open Pond	Raceway	Closed Bioreactor	Non Biofuel
Cequesta	Israel	Vertical Flat Panels				x	x
Cognis Nutritional & Health	Australia	Open Pond		x			x
Columbia Energy	WA, USA	Open Pond		x			
Culturing Solutions	FL, USA	Vertical Biofence				x	
Cyanotech	HI, USA	Open raceways		x			x
Desert Sweet Biofuels	AZ, USA	Covered raceways			x		
Diversified Energy	AZ, USA	Ground Tubes			x		
Eastern Algae	LO, USA	Open Ponds		x			
Earthrise Nutritionals	CA, USA	Open Ponds		x			x
Enhanced Biofuels Tech	UK	Closed G.F. old reactor				x	
Eni Division R&M	Italy	Duel closed:open		x		x	
E.ON	Germany	Vertical tube				x	
Fraunhofer Institute	Germany	Tubular reactor				x	
Fuji Health Science	Hawaii	Enclosed hemispherical				x	
General Atomics	CA, USA	Open Pond	x	x			
Global Green Solutions (Valcent)	CA, USA	Vertical reactor				x	x
Global Seawater	Mexico	Open Pond		x			
Greenfuel Technologies Corp, Inc	MA, USA	Vertical Fabric				x	
Greenshift/GS Cleantech	NY/GA, USA	Fabric Panels				x	
Green Gold Algae & Seaweed Sciences	Israel	Seaweeds					x
Greenstar	South Africa	Covered raceways			x		
HR BioPetroleum	HI, USA	Hybrid system			x	x	
IGV Institut Getreideverarbeitung	Germany	Vertical Tubular bioreactor				x	x

Chapter 18

Company	Location	Type	Col1	Col2	Col3	Col4
Infinifuel	NV, USA	Geothermal Ponds		x		
Inner Mongolia Biological Eng	China	Open Raceway		x		x
International Energy	DC, USA	Closed Reactor			x	
Inventure Chemicals	WA, USA	Postharvest processing				x
Kent SeaTech	CA, USA	Open Pond – Polyculture		x		
LiveFuels	CA, USA	Open Pond		x		
Martek Biosciences	MD, USA	Heterotrophic				x
Mera Pharmaceuticals	HI, USA	Mera Growth Module				x
Nature Beta Technologies	Israel	Open Raceway		x		x
Neptune Industry	FL, USA	Floating socks – Polyculture				x
OriginOil	CA, USA	Helix bioreactor			x	
Parry Agro Industries	India	Open Raceway		x		x
Parry Nutraceuticals	Australia	Open Pond		x		x
Petroalgae	FL, USA	Open Raceway	x	x		
PetroSun	AZ, USA	Saltwater Ponds		x		
Phycal	OH, USA	Open Pond	GMO			
Reed Mariculture	CA, USA	Open Pond		x		x
Sapphire Energy	CA, USA	Open Pond	GMO			
Seambiotic	Israel	Open Pond		x		x
Solazyme	CA, USA	Heterotrophic	GMO		x	
Solix Biofuels, Inc	CO, USA	Vertical Panel			x	x
Subitec Gmbh	Germany	Vertical Tube reactor			x	x
Taiwan Chlorella Manufacturing	Taiwan	Open Pond		x		
Targeted Growth	WA, USA	Equipment providers	x			
Texas Clean Fuels	TX, USA	Open Raceways			x	
Tianjin Lantai Biotechnology	China			x		x
	MI, USA	Northern Climates			x	

Table 18.2 (continued)

Company	Location	Culture Strategy	Cultivar Enhancement	Open Pond	Raceway	Closed Bioreactor	Non Biofuel
The Power Alternative (TPA)							
Trident Exploration	Canada	Light Stream, CO$_2$ concentrate				x	
Valcent	TX, USA	Vertical Panel				x	
Western Biotechnology	Australia	Open Ponds		x			x
Phyco Biosciences	AZ, USA	Ground Trough			x		

the University of Aarhus are examining the potential of *Ulva lactuca* mariculture with estimates of 200–500 wet tons of *Ulva*/hectare with a total yield of 80–100 000 tons producing bioethanol.

The bioethanol value of macroalgae is focused on the free sugar and carbohydrate content of the selected cultivar but the commercial value of the seaweed must include the consideration of the protein ($\sim$10–50%) and lipid ($\sim$1–10%) contents. Postfermentation algal mash has potential value as a feed and/or soil supplement at $\sim$\$150–200/ton.

Macroalgae are globally sourced by manual or mechanised harvesting or *via* highly organised marine agronomy. Presently, maricultured macrophytes (Figure 18.1) represent 94% of the annual global seaweed yield of 15 million wet tons valued at >\$7.2 billion;[5] *Laminaria* and other kelps represent more than 50% of the total macroalgal landings.

China dominates global mariculture efforts, originally focused on kelp production, China transformed the *Porphyra* farming (Figure 18.2) industry with its entry in the early 1990s. China is followed by the Philippines, Indonesia and all other nations by producing 10 800, 1300, 900, and 2000 tons, respectively.[5]

Numerous offshore mariculture programs are envisioned in conjunction with wind farms, oil rigs, and finfish aquaculture farms. The future offshore efforts will need to concentrate on aquacultural infrastructure engineering to

Figure 18.1 Kerala State, India *Kappaphycus* farming operation.

(a)

(b)

Figure 18.2 (a) Coastal Plantations International, Maine USA *Porphyra* farm, (b) Aerial of CPI 75 net *Porphyra* farm.

withstand the rigours of open water cultivation, *i.e.* storm surge, wind, and waves. The Japanese are designing 40-km^2 farming modules to produce 1 million wet tons of macroalgae for deployment just 5 miles off their coast to take advantage of the nutrient-laden runoff enriching the culture grounds.

Algae 391

Wild harvests, which have significantly decreased in the past 30 years, are centreed in China, Chile, Norway, Japan, France, and Ireland, with estimated landings of 323, 305, 145, 113, 75, and 29 thousands of wet tons, respectively.

A third, underutilised source of algal macrophytes, are drift algae. The world witnessed the potential productivity of the algae to absorb excessive near shore nutrient enrichment in Qingdao, China (Figures 18.3 and 18.4) during 2008's Beijing summer Olympics; almost resulting in the cancellation of the sailing events due to the overwhelming standing stocks of drift *Ulva*. French coastal drift *Ulva* estimates of 60 000 tons represent just one European coastal enrichment experience.

With species selection and method of acquisition determined (wild stocks *vs*. marine agronomy), the remaining challenges include algal harvest, preprocessing, and energy production. Unlike microalgae, harvesting is not a critical choke point in the value chain. Harvesting is either manual or mechanical. Manual harvesting of both wild (Figure 18.5) or maricultured standing stocks continues presently. *Ascophyllum, Eucheuma, Fucus, Gracilaria, Kappaphycus, Laminaria,* and *Porphyra* are still manually harvested in Asia, Africa, N. & S. America, Europe, and Southwest Pacific. Boats, animal- or motor-driven vehicles are utilised to access the harvest grounds or mudflat-based algal farms. Mechanical harvesters (Figures 18.6 and 18.7) have developed from floating barge-bound rotary blades to modified troop-landing vehicles, trawlers equipped with rotating hooks, and belt-driven harvesting collectors.

Postharvesting pretreatment includes the removal of foreign material, including invertebrates, floating debris, and inert materials. Sand, plastic, and snails are removed by soaking, washing, and/or manual manipulation.

Figure 18.3 Qingdao, China harbour overwhelmed by floating *Ulva* bloom, seen here as the lighter colour in the bay.

Figure 18.4 Ulva bloom hand harvest, Qingdao, China.

Figure 18.5 Hand harvesting of Ascophyllum Maine, USA.

Algae 393

Figure 18.6 *Ascophyllum* harvester.

Figure 18.7 Mechanised harvest of giant Kelp, California, USA.

Dewatering may follow pretreatment, as a function of biofuel processing requirements, typically using solar energy, air drying, fan-induced drying, generated power-assisted drying, sponges, and pressing the ~90% moisture content down to ~20% water concentration.

The digestion of seaweeds does not require further drying as compared to microalgal-based biodiesel refining. Energy inputs to remove 88% of the cellular moisture would overwhelm the potential energy balance of the seaweed-based bioethanol proformas.

Seaweed biofuel production and utilisation has a checkered past with pilot efforts dating back to the 1980s in Florida, USA, where washed-up algae was collected and digested to heat a hospital in St. Petersburg, Florida. Additional biofuel generation efforts include: SOPEX, a Morocco–France collaboration, and the Tokyo Gas company.

The Japanese effort to develop industrial biogas (methane) production from algae is best illustrated by the efforts of the Energy and Industrial Technology Development of Organization of Japan and the Tokyo Gas Co., Ltd (Figures 18.8 and 18.9).[8]

The effort included the harvest of *Laminaria* and *Ulva* (Table 18.3), transport to production facility, pretreatment and fermentation stages. Pretreatment consisted of the removal of shells and sand *via* soaking (1–5% solids) and the chopping of the algal biomass. Biogas fermentation is separated into a two phase process: prefermentation phase – organic acid production and the methane-fermentation phase – biogas production. Prefermentation retention times of two to three days, maintaining temperatures of 25 to 35 °C, results in

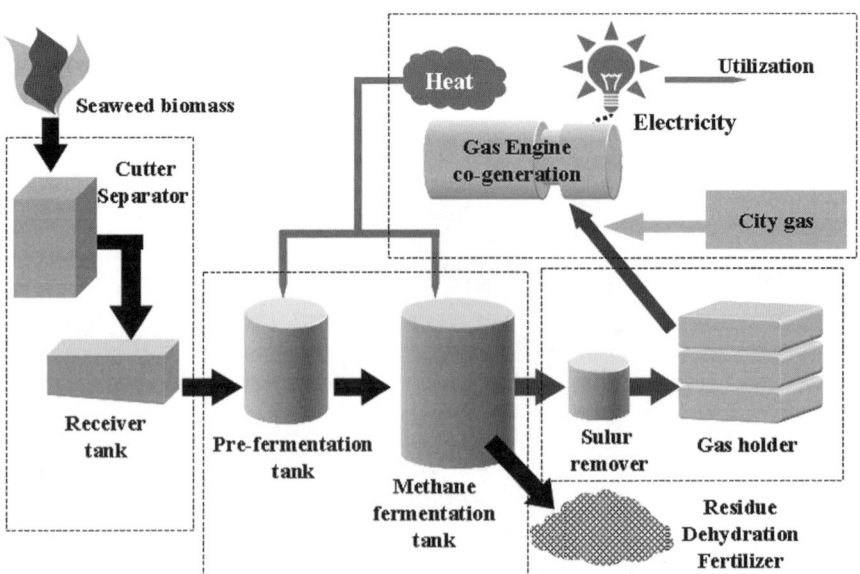

Figure 18.8 Schematic of macroalgae biomethane fermentation facility.[8]

Figure 18.9 Tokyo Gas Company methane-processing facility.[8]

Table 18.3 Composition of Tokyo Gas Company source material[8]

(wt%)	Ulva sp.	Laminaria sp.
Lipid	1.6	1.9
Protein	27.9	15.2
Carbohydrate	54.3	45.6
Ash	16.3	37.3

the production of organic acids, predominantly acetic, lactic and butylic acids. Acid production increases concentrations 5- and 3-times utilising *Laminaria* and *Ulva*, respectively. The subsequent methane-fermentation phase, lasting 15 to 25 days, maintaining a temperature of 55 °C, and consuming the organic acids, yields 60% methane and 40% carbon dioxide. One ton of *Laminaria* and *Ulva* produce 22 and 15–17 cubic metres of methane gas, respectively. Hydrogen sulfide is refined through the use of iron oxide, reducing its effect on methane quality. Pilot-run duration time varies in *Laminaria* and *Ulva*, lasting 150 and 70 days, respectively. Temperature, reaction duration, and pH remain critical production parameters with maintenance of pH > 7.5 necessary for maximal methane yields. Desulfured biogas, when blended with traditional natural gas supplies, provides near identical thermal efficiency and power output.

Seaweed-based bioethanol production is simplified by limited amounts of cellulose and lack of hemicellulose and lignin, but is complicated by the free sugar and carbohydrate composition; many of the marine algae polysaccharides, *e.g.* alginate, cellulose, and agar, are traditionally nonfermentable. The development of enhanced microbes, *via* genetic transformation, is a constant research focus but its success represents a permitting challenge that might not be overcome within the next ten years. The enhancements of these natural processes *via* the development of unique micro-organisms, thermal pretreatments, and enzymatic and/or catalytic assistance, are underway in the USA, Norway, China, Ireland, and Japan. Bioprospecting has identified a yeast (*Pichia angoophorea*) and bacterium (*Zymobacter*) capable of converting mannitol and laminaran to ethanol, although challenges remain in utilising these organisms in a commercial marine process.[9] The author has attempted the fermentation and distillation of an algal-based alcoholic beverage and with the exception of an algal-flavored ale, has been largely unsuccessful. The potential of macroalgal-based bioethanol energy content as a function of marine agronomic productivity is estimated between 250–300 GJ/ha of macroalgal farming area.

All developing corporate proformas dealing with the commercialisation of algal-based biofuels include coproduct value, which is essential to the sustainability of any industrial venture. All profit and loss statements include animal feed and/or soil conditioners and fertilisers. The algal mash or post-digestion sludge contains proteins, biosolids, lipids, minerals and salts. The development of animal feeds needs further review in the United States by its Departments of Drug Administration and Agriculture. Proximate analyses, total protein and amino acid speciation, feed compatibility, digestibility, nutrient absorptive and meat-flavor profiles will all have to be determined. The value of feed mash ranges from $0.15–0.26/kg that does not include the cost of shipping and handling of the large amounts of biomass.

The development of commercial, algal-based biofuels assumes direct and unlimited access to biofuel refineries and market service points. Algal-based bioethanol, biogas, and/or biodiesel are similar to other plant-based biofuels; whereas additional formulation modifications may be required to reduce the oxygen concentration that affects both the physical and chemical properties of the biofuel as compared to the fossil-fuel-based transportation fuels. Biogas has been used in modified personal vehicles in Mexico and Southwest USA for several decades. Specialised private fleets have been modified but national USA and European Union standards have not been developed to date.

18.3 Microalgae

Microalgae or phytoplankton (Greek – wanderer) have rapidly gained acceptance as a significant second-generation biofuel feedstock. Several advantages of microalgae over terrestrial crops include:

> 1. large areal productivities and yields (estimates 5000–7000 gallons/acre/year);

2. short life cycle;
3. greenhouse-gas reductions;
4. second-generation biofuel carbon credits;
5. wastewater and heavy-metal bioremediation;
6. valuable postfuel molecule-extraction coproducts;
7. variable water supply usage (fresh, brackish, marine, wastewater, effluents);
8. use of marginal, nonarable land;
9. avoidance of feedstock *vs.* food conflict;
10. bioremediation of human and animal wastewater effluents.

As significant progress is made in the algae biofuels value chain, the resultant menu of available cultivars, standardised commercial raceways, and photo-bioreactors offer the "new-generation farmer" predictable and stable cultivation, harvesting, extraction, and conversion, with risks reduced to a manageable level. The challenge will then shift to the sourcing of land and water, training the next generation of algal farmers, and the development of the institutional and extension experience to transform this grand experiment into a traditional agriculture/aquaculture enterprise.

18.4 Microalgae Biofuel Value Chain (Table 18.4, Figure 18.10)

18.4.1 Stage 1: Project Planning

The Algal biofuel value chain includes farm scale, site identification and preparation, environmental inputs, and nutritional supply determination. This stage is quite similar to those activities experienced by any new agricultural effort, therefore will not be further discussed in this chapter.

Corporate business plans and press releases have indicated algal-based biofuel cultivation efforts would be sized at 500–25 000 hectares. As a former owner of a 50-hectare algal farm, the author can clearly state the lack of farm experience and the fundamental underestimating of farming's functional challenges that may represent the largest barrier to the success of algal-based biofuel development. Challenges include: inexperienced aquaculturists, carbon-dioxide supply, nutrient application, climate and environmental variability, herbivory, culture contamination, pathogens, storms, regulations, land- and water-use conflicts, permitting, availability and consistency of operational capital, electrical supply, vandalism, waste removal and disposal, market variability and competition. These factors and realities need to be incorporated into the corporate psyche before algal-based biofuels can be a reality.

18.4.2 Stage 2: Cultivar Selection and Enhancement

The microalgae *Chlorella, Dunaliella, Haematococcus,* and *Spirulina* are presently commercially cultivated in open systems along with *Chaetoceros,*

Table 18.4 Organisational chart identifying and summarizing key topic areas and issues for the overall algal biofuels.

Systems Integration & Interdependences

Siting & Inputs	Algal Biology	Cultivation Systems	Harvesting Dewatering	Extraction Fractionation	Conversion processes Biofuels, Coproducts	Policy Regulatory
Geo location/ elevation	Species selection & matching to growth conditions	Photoautotrophic Open Ponds lined unlined raceway wastewater treatment	Filtration Flocculation/ settling Airlift flocculation Centrifuge Drying	Extract Processes Solvent Acoustic EM Other	Conversion processes Biochemical Thermochemical Digestion Hydrotreatment/ refine	Taxes Incentives Permitting Environmental Impact
Land characteristics	Characterization					Health and Safety
Climate	Performance					Algae control and regulation
Solar insolation			Biological harvest brine shrimp fish Other	Separation/ fractionation Membrane Distillation Centrifuge Other		other
Water source/issues brackish wastewater desal concentrate marine fresh losses, reuse salt build up	Strain improvement Biomass growth and oil content optimization	Closed PBRs horizontal tube vertical tube vertical planar flexible panels emersed fabrics			Fuels biodiesel green diesel aviation gasoline-like ETOH biogas/methane other	
CO_2 Source power plants cement plants fermentation/ other	Photoautotrophic organism operation Heterotrophic organism operation	Hybrid system combo covered ponds pond/PBR mix		Intermediate products TAG Oil Other lipids Polar Neutral Carbohydrates Proteins Other compounds Water	Coproducts feed fertiliser chemicals	
Chemicals/materials Energy/power Infrastructure	Algae pathogens predators, and mitigation	Heterotrophic industrial bioreactors wastewater facilities Operations monitoring & maintenance		Direct secretion of ETOH or hydrocarbon fuel precursors into growth medium, avoiding harvest and dewatering steps	Services carbon capture water treatment	

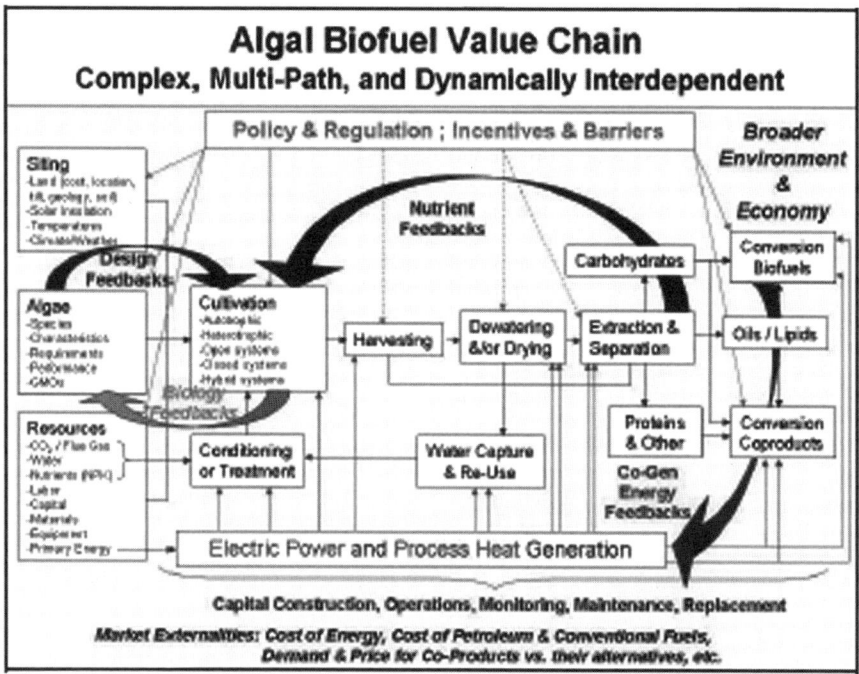

Figure 18.10 Illustration of the broad systems analysis perspective needed to address the dynamic coupling and interdependencies across the overall algal biofuels and coproducts value chain.[18]

Isochrysis, Nannochloropsis, and *Skeletonema* cultured to support larval stages of fin and shellfish aquaculture. Historically, cultivar selection was dictated by proximate analyses, indigestibility, and digestibility of aquaculture feeds normally grown in culture rooms or greenhouse batch cultures. Open-system cultivation of microalgae has used partial extremophiles, reducing potential contamination, disease, and herbivory for the production of nutraceutical, cosmaceuticals, and dyes.

Biofuel development efforts attempt to balance the culture productivity versus the yield and speciation of desired macromolecules. Biodiesel efforts focus on lipid content (Table 18.5), whereas bioethanol requires fermentable carbohydrates and free sugars.

Cyanobacteria, previously termed blue-green algae, are photosynthetic prokaryotes and are the focus of R&D efforts to develop them as a potential biofuel feedstock. Cyanobacteria advantages include: ability to fix atmospheric nitrogen, variable growth rates, culture hardiness, short life history cycles, production of long-chain hydrocarbons (*i.e.* alkanes, *e.g.* heptadecane (C17)), and ease of genetic manipulation (transgenesis and mutation). Despite these advantages, cyanobacteria are not considered primary candidates for commercial biofuel-based efforts due to natural cultivar's limited lipid production (see Table 18.6) and realised growth rates. The prioritisation will change as a

Table 18.5 Triglyceride costs per gallon from various sources.[18]

Source	Scenario	Reactor type	Lipid yield wt% dry mass	Areal yield	~Cost per gallon
Benemann	per hectare basis	open pond	50%	30	1
Benemann	per hectare basis	open pond max	50%	60	1.5
NREL	current case	open pond	25%	20	10.5
NREL	aggressive case	open pond	50%	40	4
NREL	maximum case	open pond	60%	60	3
NMSU	current yield	open pond	35%	35	25–38
NMSU	highest yield	open pond	60%	58	10–14
Solix	current	hybrid	16–47%	0–24.5	32
Solix	anticipated	hybrid	16–47%	30–40	1.5
Seambiotic	best yield	open raceway	35%	20	25
Sandia	raceway & pbr	both	35%	30	15–35
Bayer Tech Services	Germany & Texas	pbr	33%	52–110	15
General Atomics	100 acres	open/hybrid	?	?	20–33
Cal Poly	100 hectares	wastewater treatment & digester	25%	20	17
Tapie & bernard	10 hectares	T–pbr	35%	20	40–42

Adapted from National Algal Biofuels Technology Roadmap (draft) 2009.

function of predicted biotechnological advancements, *e.g.* direct production of sucrose, hydrogen, and/or ethanol. Recent claims by Algenol Biofuels indicate near commercialisation of cyanobacterial direct secretion of ethanol, annually producing 6000 gallons of ethanol per acre. Algenol's development of the "Direct to Ethanol" system represents a unique enclosed raceway design (Figure 18.11) facilitating the collection of the excreted, evapourated and condensed ethanol without compromising the culture's health and productivity. The elimination of harvesting, dewatering, extraction and conversion significantly reduces the operational and capital expenses, offering a real opportunity to be immediately competitive with other bioethanol feedstocks. In a recent collaboration with Dow Chemicals, Algenol plans to divert some of their ethanol to the manufacture of bioplastics by Dow, further reducing the reliance on petroleum-based products.

Cultivars can be obtained from public, nonprofit culture collections, *e.g.* The Culture Collection of Algae at the University of Texas at Austin, Texas (UTEX) and the Provasoli-Guillard National Center for Culture of Marine

Table 18.6 Microalgal oil content.[4,19]

Microalga	Oil Content (% dry wt)
Anabaena cylindrical	4–7
Botryococcus braunii	25–75
Chlamydomonas rheinhardii	21
Chlorella pyrenoidosa	2
Chlorella sp.	28–32
Chlorella vulgaris	14–22
Crypthecodinium cohnii	20
Cylindrotheca sp.	16–37
Dunaliella bioculata	8
Dunaliella primolecta	23
Dunaliella salina	6
Euglena gracilis	14–20
Isochrysis sp.	25–33
Monallanthus salina	>20
Nannochloris sp.	20–35
Nannochloropsis sp.	31–68
Neochloris oleoabundans	35–54
Nitzschia sp.	45–47
Phaeodactylum tricornutum	20–30
Porphyridium cruentum	9–14
Prymnesium parvum	22–38
Scenedesmus obliquus	12–14
Scenedesmus quadricauda	1.9
Scenedesmus dimorphus	16–40
Schizochytrium sp.	50–77
Spirogyra sp.	11–21
Spirulina maxima	6–7
Spirulina platensis	4–9
Synechoccus sp.	11
Tetraselmis maculata	3
Tetraselmis sueica	15–23

Phytoplankton at the Bigelow Laboratory for Ocean Sciences in West Boothbay Harbor, Maine (CCMP). The collections are the cultivar source for many academic and commercial interests. Numerous private microalgal collections exist but access is limited and at the discretion of the principle investigator. Culture collections, although convenient, represent organisms that may have lost physiological vigour after years in suboptimal growth environments. Alternatively, global field collections continue and isolation, identification, and evaluation are either by classical methodologies or modern high-throughput cell-sorting technologies.[10,11] As with acquired or collected cultivars the determination of cultivar hardiness is essential; determinations include stability, consistency, resilience to competition, environmental variation, herbivory, and disease.

Physiological ecology or the intersection of productivity, primary metabolic production, density tolerance, and environmental parameter sensitivities are all involved with cultivar selection. The ability to maintain stable, long-term

Figure 18.11 Algenol's direct synthesis of ethanol covered-raceway photobioreactor.

culture operations is a function of cultivar resistance to infection, herbivory, and competitive autotrophs.

Numerous, well-funded corporate and academic efforts have focused on the genetic manipulation (transgenic, site-directed mutagenesis, and protoplast fusion) of the algae to accomplish the following: increase photosynthetic efficiencies and yields, decrease light saturation, decrease photoinhibition, expand thermotolerance, regulate osmotic mechanisms, decrease photooxidation, increase fatty acid metabolism, modify fatty acid synthesis endpoints, and stimulate direct biofuel molecular synthesis. The successful genetic enhancement of commercial algal cultivars represents a potential strategic breakthrough for the developing industry. Cultivar biology represents one of seven gate-checks in the development of the value chain. GMO algae also represent a tremendous challenge, especially for its intended use in open-raceway systems. Local, State, and Federal regulatory bodies may impose their will upon the permitting process. Several states within the United States, e.g. Hawaii, and the European Union expressly forbid the introduction of genetically modified algae. The use of mutagenesis and/or protoplasmic fusion may not violate the GMO regulations that make these techniques more desirable.

18.4.3 Stage 3: Cultivation System

Bioreactor design has been the focus of over 100 academic, corporate, and research institute's efforts to date. There is no single, best design when considering areal productivity, footprint, capital cost per gram dry algae produced, operational expenses, lipid production rates and profiles, culture stability, and

Algae

maintenance of local environmental parameters. The initial strategic decision in choosing a cultivation system is light versus organic carbon. Autophototrophic or heterotrophic culture systems represent fundamentally divergent operational and investment strategies. The development of heterotrophic systems has been limited to a few commercial operations. Solazyme (California, USA) has worked for five years on the development of heterotrophic algal oil production. Martek Biosciences (Maryland, USA), a pioneer in heterotrophic algal cultivation producer high-value omega 3 fatty acids for infant formula and animal-feed markets, has recently indicated a shift in corporate strategies, utilising their cultivation infrastructure to harvest oils for biofuel market segments.

Heterotrophic advantages include:

1. controlled, sterile environment eliminating culture contamination, herbivory, and pathogens;
2. elevated, sustained growth rates without the energy limitations of photoautotrophic systems;
3. increased cell-culture densities;
4. direct release of biofuel molecules (ethanol and alkanes);
5. avoidance of genetically modified algae permitting challenges;
6. enhanced algal lipid content ($> 75\%$).

Heterotrophic challenges include:

1. reliance on external sources of organic carbon;
2. zero reduction of greenhouse gases;
3. no carbon credits;
4. extremely high capital expense.

The fundamental choice between photoautotrophic open pond or raceway versus a closed or hybrid photobioreactor has not been concluded either by the academic or commercial communities (Table 18.7). The marketplace will ultimately decide the fate of each production system. The large capital expenses of closed photobioreactor systems (up to 15× the cost of raceway construction and assembly) must be overwhelmed by an increase in culture yields or designated biomolecules. The final analyses will determine which system can reliably deliver algal biomass or oil at a sustainable price.

Regardless of the chosen production system, either form requires a culture inoculation system. Seambiotic of Israel (Figure 18.12a) utilises an open raceway, sequential, scalable system. The open scalable inoculum systems experience similar biological and physical challenges to that of the open raceway commercial production raceways. Alternatively, closed inoculum systems have been developed, initially to support the shellfish aquaculture industry including columnar (Figure 18.12b), flat (Figure 18.12c) and "u"-shaped soft hanging panels (Figure 18.12d). The closed inoculum photobioreactors minimise the threats to the young diluted cultures, supporting the delivery of healthy monocultures to the commercial-scale production system.

Table 18.7 Comparative utilisation parameters for closed and open photo-bioreactors. Modified from 20

Feature	Open System	Closed System
Area-to-volume ratio	large (4–10× of closed pbr)	small
Algal species	restricted	flexible
Species selection criteria	competition, herbivory, disease	shear-resistance
Population density	low	high
Harvesting efficiency	low	high
Cultivation period	limited	extended
Contamination	probable	unlikely
Evapourative water loss	probable	prevented
Light-utilisation efficiency	poor/fair	fair/excellent
Gas transfer	poor	fair/high
Temperature control	none	excellent
Costly parameter	mixing	O_2 and temp. control
Biofouling	fair	high

Open ponds and lakes have traditionally been used for the culture and/or wild harvest of microalgae. Wild populations of *Arthrospira* (*Spirulina*) are harvested from four lakes in Myanmar by the Myanmar Pharmaceutical Factory.[12] Additional efforts to grow microalgae in unmixed ponds include Mexican efforts by Sosa Texcoco Co. utilising 40 hectares of the caracol to produce 300 tons of *Arthrospira* (*Spirulina*), realizing productivities of 10 g dry matter/m^2/day, and Australia's duel efforts by the Western Biotechnology Ltd utilising 250-ha ponds for *Dunaliella*, producing six tons of β carotene per year, and Betatene Ltd utilising 460 ha to produce 7–10 tons of β carotene per year, exhibiting productivities of 1 g dry matter/m^2/day.[13–15] Although valuable, none of these efforts are appropriate for biofuel production as densities are too low and harvesting too expensive for a commodity endproduct. Therefore, open ponds and lakes will not be further considered in this review.

Constructed open-raceway culture ponds have developed over the past fifty years and utilised to culture microalgae producing food, nutraceutical, cosmaceuticals and dye products. Maximum commercial areal productivities have reached 20–25 g/m^2/day, whereas long-term productivities average 12–13 g/m^2/d.[16] Commercial estimates for open-pond culture have exceeded 60 g/m^2/d but no long-term data have been presented to support these claims. Raceway designs include the classic shallow oval, paddlewheel driven (Figure 18.13) structures utilised throughout the aquaculture world. An alternative design of Phyco Biosciences, Inc (formerly XL Renewables), an Arizona, USA company, utilised standard agronomic experience and equipment to develop the "Super Trough" system (Figures 18.14 and 18.15), a "V"-shaped, inlaid liner, into an agronomic trough. High-efficiency pumps limit electricity to 0.5 horsepower per acre, limiting energy expenses for the control of circulation, turbulence and the delivery of carbon dioxide, which is developed through the digestion of dairy-cow-farming effluents. An optional, removable cover minimises evapourative losses, dust deposition, pathogen and invasive species

Algae

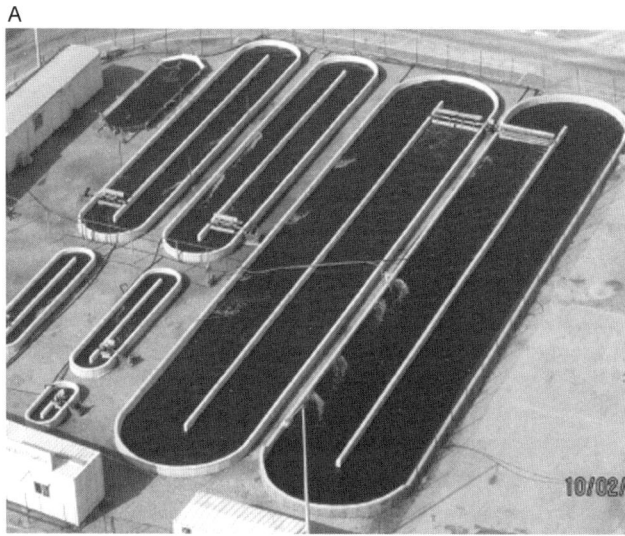

Figure 18.12 A. Seambiotic Ltd. Ashkelon, Israel. Classic open-pond sequential scaleup from microraceway to production scale unit. B. Greenhouse 500-l columnar inoculum culture system. C. Arizona State University's glass flat-panel inoculum culture system. D. GreenFuels Technologies inoculum bags.

C

D

Figure 18.12 Continued.

contamination and extends the cultivation system by providing a semblance of temperature control. XL's pilot operation is producing dried pellets of *Nannochloropsis* sp. (Figure 18.16). Recent efforts include the incorporation of a heterotrophic segment to their system, increasing yields.

Open-raceway systems are vulnerable to cultivar, bacterial, and fungal contamination, protistan herbivory, fluctuations in temperature, nutrient loads,

Algae 407

Figure 18.13 Cyanotech Corporation's Kona, Hawaii, USA algae raceway farm.

solar quality and quantity, culture densities, salinity, pH, waste concentrations, excreted metabolites, catabolic compounds, and damage by hurricanes, tornados, wind storms, dust and debris. These challenges represent significant hurdles yet to be either completely understood or overcome. Research and development on raceway design continues, *e.g.* recent collaboration between the Israeli company, Seambiotic, and the US National Aeronautical and Space Administration focused on modeling and engineering enhancements.

Closed photobioreactors (pbr) technologies take on many physical forms; the diversity of phototrophic pbr designs include tubular, fixed flat panel, flexible panels, and immersed fabric designs. A unique third-generation closed photobioreactor design by Solix Biofuels of Colorado, USA, offers up culture security and environmental parameter control (Figure 18.17). Vertically oriented, flexible, oblong panels are suspended within a water-filled retention pond. Future designs transfer the oblong panels from constructed retention facility to ponds, lakes and rivers. Electronically monitored and controlled environmental parameters maintain optimal culture conditions throughout the year, even through the harsh Colorado winter. Solix has reported continuous culture data for 30 months, perhaps the longest continuous production history for a second-generation algal system. The company reports production values > 2500 gallons algal oil/acre/year from their Fort Collins pilot facility. Recent efforts include the development of a five-acre demonstration AGS^{TM} Technology system in Coyote Gulch, Colorado, on the Southern Ute Alternative Energy Fund's coal-bed methane-production facility in southwest Colorado.

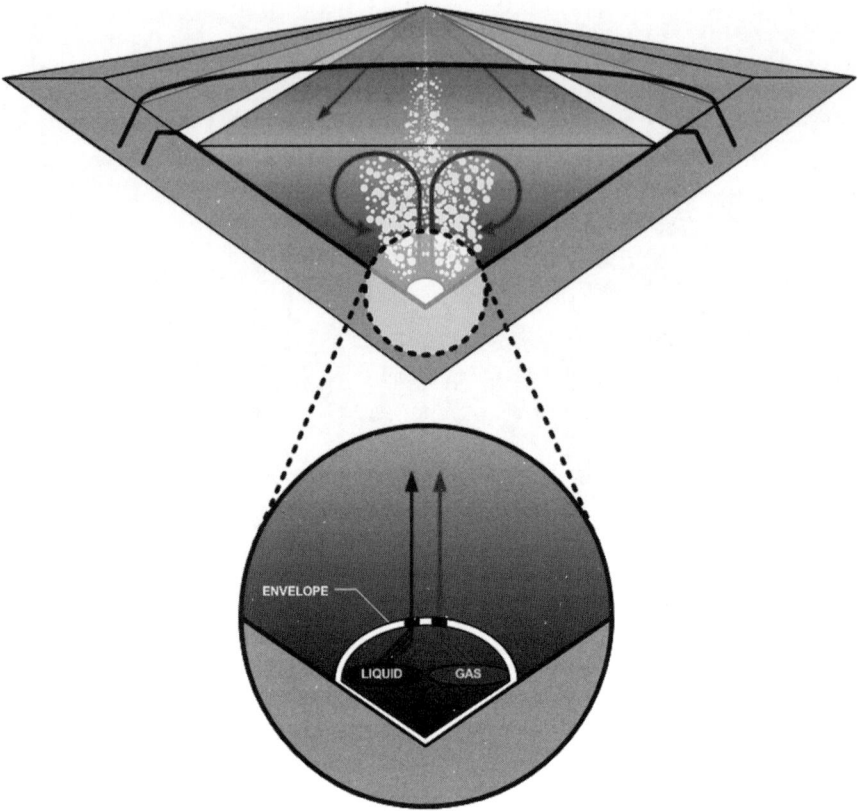

Figure 18.14 Phyco Biosciences, Inc. (formerly XL Renewables) Super Trough™ System.

Incremental improvements and mass cultivation experience have increased their production estimates to ∼ 3500 gall/acre/yr (Figure 18.18).

18.4.4 Stage 4: Harvesting

The development of microalgal-based biofuels requires the harvest, dewatering, and concentration of the culture slurry from 0.1–0.15% to 15–20% solids (Table 18.8). Final culture concentration percentages are a function of extraction requirements of the ultimate desired biofuel product sector, *e.g.* biodiesel or bioethanol. Harvest and dewatering can be accomplished in a single phase, potentially preceded by a pretreatment phase. Traditional harvesting methodologies, *i.e.* centrifugation and spray drying, require high energy inputs, representing 25–60% of the total energy content of the (W h/g) algal cells. The concentration of cell slurries must rely on either a combination of existing lower-energy-input technologies or the development of a new method

Algae

Figure 18.15 Phyco Biosciences, Inc., Arizona, USA demonstration Super Trough facility adjacent to dairy farm (upper left).

in which energy inputs are reduced from the current 35% to 13% of the total cost to produce a gallon of algal biodiesel.[17]

18.4.5 Stage 5: Extraction

Extraction fundamentals revert back to the initial product development strategies of: direct biosynthesis and excretion of targeted fuel precursors (extraction unwarranted), direct processing of whole algal cell (extraction unwarranted), and the conversion of macromolecules into biofuels (extraction warranted). Regardless of the targeted molecules, fermentable polysaccharides and sugars or lipids and fatty acids, an industrial/scalable process, as opposed to the myriad of laboratory extraction technologies, must be developed to overcome the biological challenges (water content, cell wall dynamics, cell proportionality and size) that microalgae present.

Experimental, lab- or pilot-scale extraction practices each exhibit advantages and challenges in the extraction and fractionation of transportation fuel precursors. The ultimate question remains: is it scalable, energy sustainable, product and coproduct conservable and process chain compatible? Traditionally, hexane-extraction systems, an industrial process for plant-seed-oil extraction, modified to "fit" the algal model, are suggested as the extraction system of

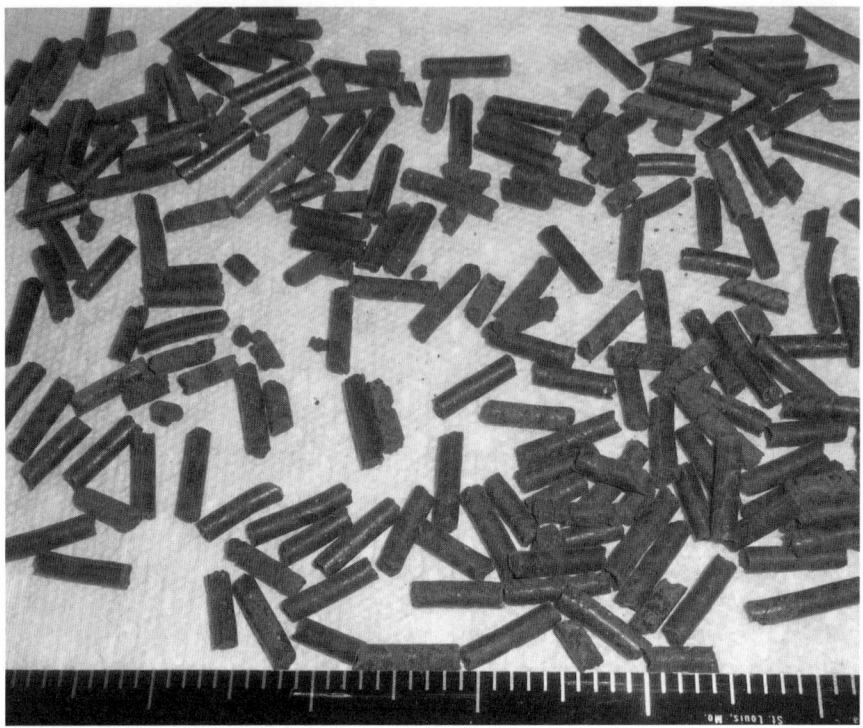

Figure 18.16 Phyco Biosciences" *Nannochloropsis* pellets.

choice. Additional proposed extraction systems include two solvent systems, solvent milking, subcritical water, supercritical alcohol or carbon dioxide, physical disruption, and accelerated solvent extractions. Challenges in part include: scalability, the need for completely dried algal harvest, solubility quotient of target molecules, compatibility of solvent fronts, solvent separation and recycling, elevated temperatures and pressures.

The US National Algal Biofuels Technology Roadmap describes an ambitious goal set for commercial extraction technologies including:[18] consumption of a maximum of 10% of the endproduct energy value; 95% conversion efficiency of extracted lipids, proteins, and carbohydrates; and recycling of water with zero environmental discharge. The evolving technologies need to borrow from the mature bioethanol processing experience while developing a greater understanding of cell wall biochemistry, lipid biosynthesis, and their interactions with solvents at different temperatures and pressures.

18.4.6 Stage 6: Conversion

The downstream utilisation of directly synthesised or extracted fuel precursors along with the use of the entire cell as a fuel source must result in a fully compatible, market acceptable fuel. The ultimate conversion pathway will be a

Algae 411

Figure 18.17 Solix Biofuels Colorado, USA Pilot Facility. Vertically orientated flexible panels submerged in water tank.

function of net energy realised per unit capital input, including primary and coproduct values. Figures 18.19 and 18.20 review the potential conversion pathways, representing similar conversion systems utilised by terrestrial plant feedstock programs.

Figure 18.18 Solix Biofuels, Coyote Gulch Production Facility (lower) flanked by the coal-bed methane-production facility.

Table 18.8 Harvesting technology effectiveness.[18]

Harvesting Technologies	~ Final Cell Concentration
Chemical or Biological Flocculation	0.5–1.0%
Filter Press	2%
Centrifugation	10–15%
Differential or Membrane Filtration	15–20%
Drying (solar or thermal activated)	80%

18.4.7 Stage 7: Policy and Regulatory Review

The lack of clarity and uncertainty adds a level of risk to the nascent industry. Risk increases the cost and access to capital, a chokehold on the development of the algal-based biofuel value chain. The regulatory bodies and industrial partners need to determine what existing laws and regulations pertain to this new commercial effort and what additional regulations need to be promulgated. Areas of review include: organisms, nutrient effluents, land use, water and air discharge, CO_2 discharge dispersion, product formulation, and safety.

A critical area, ripe for regulatory review, concerns the very nature of the species, variety or cultivar. What comprises a native indigenous species? Should cultivars acquired from culture collections be regulated as nonindigenous

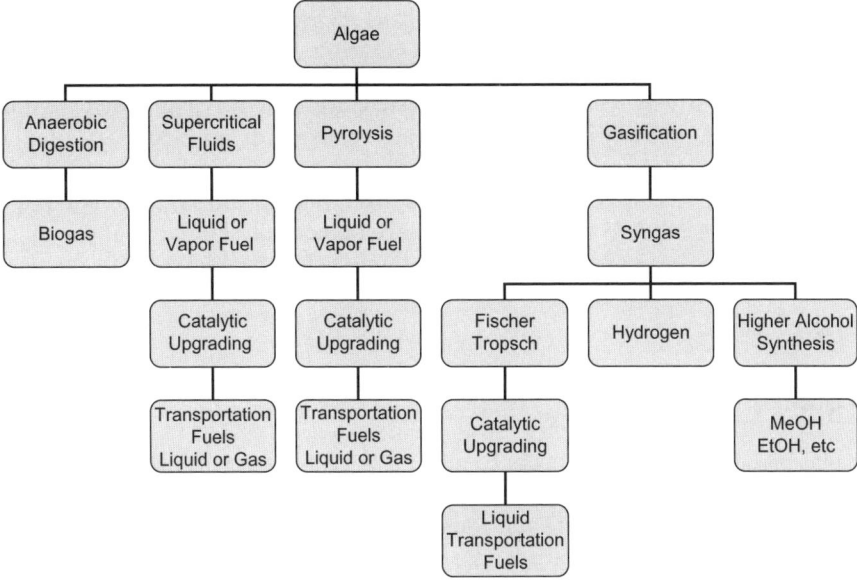

Figure 18.19 Schematic of the potential conversion routes for whole algae into biofuels.[18]

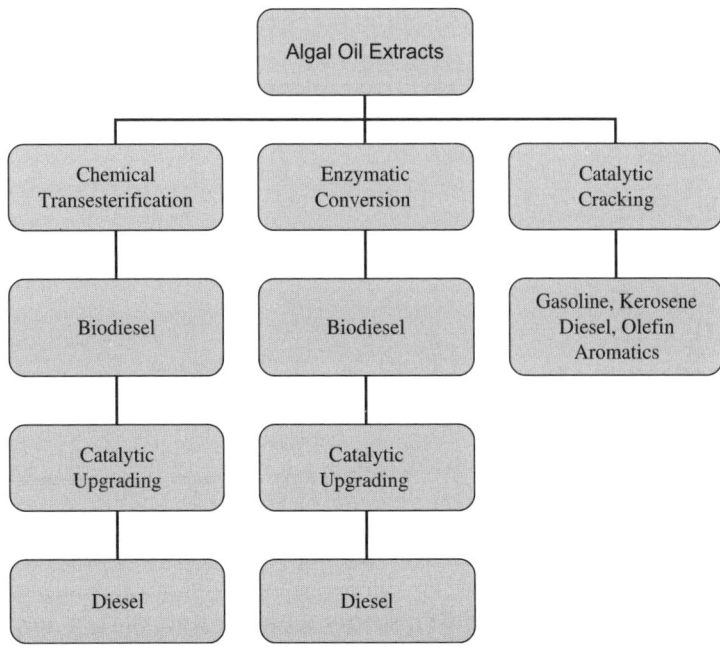

Figure 18.20 Schematic of the various conversion strategies of algal extracts into biofuels.[18]

introductions? What is the role of genetically altered, modified or transformed cultivars developed utilising molecular genetics, transformation or directed mutational technologies? Are heterotrophic or photoautotrophic systems to be treated differently by regulatory bodies? Are culture security issues divergent in closed, hybrid, open systems? What, if any, safeguards are necessary for the permitting of GMO cultivars? Which local, state, and federal regulatory body should be responsible for developing the algal value chain regulations? What data sets are needed to assist with the development of standards and regulations? All these are important questions that need to be addressed in the near future to position the industry and reduce the uncertainty associated with a nascent industry.

18.5 Conclusion

Algal-based biofuel stands ready to take its place amongst the evolving second-generation biofuel feedstocks. All stages of the emerging commercial effort have been identified and are the focus of intense, well-financed research and development efforts by academic, corporate, and research institutions. Advancements in cultivar physiological ecology, cultivation systems, harvesting, extraction, conversion, and product formulation have made tremendous strides this century. Incremental biological and technological enhancements along with the development of new methodologies will drive the commercialisation of this nascent industry to realize its fullest potential: a significant source of biologically based transportation fuels.

References

1. C. G. Wood, *J. Chem. Ed.*, 1974, **51**(7), 449–452.
2. M. S. Doty, *Proc. Int. Seaweed Symp.*, 1979, **9**, 35–58.
3. W. M. Porterfield, *Bull. Torrey Bot. Club*, 1922, **49**, 297–300.
4. Y. Chisti, *Biotech. Adv.*, 2007, **25**, 294–306.
5. FAO, *Yearbooks of Fishery Statistics*, UN, 2006, Rome.
6. G. Roesijadi, A. E. Copping, M. H. Huessmann, J. Forster and J. R. Benemann, *Report PNWD-3139*, Battelle Pacific Northwest Division, 2008.
7. M. Aizawa, K. Asaoka, M. Atsumi and T. Sakou, *Oceans*, 2007, 1–5.
8. T. Matsui, T. Amano, Y. Koike, A. Saiganji and H. Saito. *Am. Inst. Chem. Eng. Conference* – Session 412, 2006.
9. S. J. Horn, I. M. Aasen and K. Ostgaard, *J. Indus. Microbiol. Biotechnol.*, 2000, **25**, 249–254.
10. M. Sieracki, N. Poulton and N. Crosbie, in *Algal Culturing Techniques*, ed. R. Anderson, Elsevier Academic Press, Amsterdam, 2005, p. 101–116.
11. R. Andersen and M. Kawachi, in *Algal Culturing Techniques*, ed. R. Anderson, Elsevier Academic Press, Amsterdam, 2005, p. 83–100.

12. M. Thein, *4th Symp. Microalgae and Seaweed Products in Agriculture*, 2008, Mosonmagyarovar Hungary.
13. E. Becker Microalgae, Cambridge University Press, Cambridge, 1994.
14. Y. K. Lee, *J. Appl. Phycol.*, 1997, **9**, 403–411.
15. A. Ben-Amotz, in *Handbook of Microalgal Culture, Biotechnology and Applied Phycology*, ed. A. Richmond, Blackwell Science, Oxford, UK, 2004, p. 273.
16. M. Tredici, in *Handbook of Microalgal Culture, Biotechnology and Applied Phycology*, ed. A. Richmond, Blackwell Science, Oxford, UK, 2004, p. 178–214.
17. B. Wilson, *11th International Conf. Applied Phycology*, Galway, 2008, Ireland.
18. US DOE, *National Algal Biofuels Technology Roadmap*, 2009 draft.
19. T. Bruton, H. Lyons, Y. Lerat, M. Stanley, M. Rasmussen, *Sustainable Energy Ireland*, 2009.
20. A. Carvalho, L. Meireles and F. Malcata, *Biotechnol. Prog*, 2006, **22**, 1490–1506.

Subject Index

Page references to *figures* and *tables* are shown in *italics*.

"A Perfect Storm" 2
acid detergent fibre (ADF) 73
Acidothermus cellulolyticus 47
advanced generation biofuel 7–8
agriculture
  energy use by 2
  rise of 1–3
Agrobacterium tumefaciens mediated genetic engineering
  brassicas 117
  Pongamia pinnata 251–2
  poplars 276
  soybeans 155–7
  sugarcane 83
'alcopops' from sugar beet 111
algae 381–414
  background and use 10, 381–3
  biodiesel production from *382, 383,* 394, 408–9
  bioethanol from 382–4, 394–404, 408–10
  biomethane from 382
  costs 383
  macroalgae 384–96
  microalgae 396–414
Alopecurus myosuroides 136
alternative power generation systems 2
ammonia fibre explosion (AFEX) treatment 48
amylopectin in cornstarch 34–6
amylose in cornstarch 34–6

anaerobic digestion (AD) 5–6
Anthracnose spp., in sorghum 63–4
aphid control
  in oilseed rape 136
  in sorghum 64
  in willow 268
Arabidopsis 42–4, 157, 217
arabinogalactan peptide (AGP), wheat endosperm cell walls 17
arabinoxylan (AX)
  biosynthesis in wheat 18–21, *22*
  pathway *19*
  wheat endosperm cell walls 14–15, 17
Arundo donax see giant reed
Ascophyllum
  manual harvesting of 391, *392*
  mechanical harvesting of *393*
Assessment Report of the Intergovernmental Panel on Climate Change, IPCC (2006) 2
Assured Combinable Crops Scheme (ACCS), UK 124

bagasse
  ethanol yield from *38*
  from sorghum 73–4
  from sugarcane 80, 82, *90,* 91
Beddington, Prof. Sir John 2
Belgium, Boom short-rotation coppice experiment 283–4
Beta vulgaris see sugar beet

betaine from sugar beet 106
biochar from sunflower 178–9
biodiesel production 4, 6–7, 9
   from algae *382*, 383, 394, 408–9
   cold-flow properties *vs.*
      petroleum *159*
   crops 4, 9
   European and American Biodiesel
      Standards *172*
   European Union
      production 117–20
   global production and trade *124,*
      169–71, *190, 198, 382*
   from *Jatropha curcas* 207–11
   from oilseed rape 121–4, 127–31
   from *Pongamia pinnata* 236–41
   from soybeans 149, 150–3,
      157–61
   from sunflower 169–76, 181–2
bioenergy crops 3–4
   biological conversion 5–6
   chemical conversion 6–7
   classification 8–10
   direct combustion 4
   ethical debates 10–11
   first, second and advance
      generation technologies 4, 7–8
   thermal conversion 5
bioethanol
   from algae 382–4, 394–404,
      408–10
   corn-ethanol 27–9, 36–9
   crops 4
   from *Jatropha curcas* 198
   from *Miscanthus* spp. 315–16
   from palm oil 191, 195
   production 6
   sorghum starch-based ethanol 56
   from soybean 150
   from sugar beet 105, 109–11
   from sugarcane 6, 78–82, 90–5,
      97–8
   from switchgrass 356–68
biofuel 4
   crops 4
   'frenzy' 169–70

biological control of pests,
   sugarcane 84–6
biological nitrogen fixation (BNF)
   in cyanobacteria 399
   in giant reed 329
   in *Pongamia pinnata* 234–6,
      237
   in sugarcane production
      87
biomass crops 4
   dedicated 9–10
   Miscanthus spp. 314–15
   poplars 280–4, 286–90
   subtropical and tropical
      reeds 322–34
   switchgrass 353–69
   willow 261–3
biomethane
   from algae 382
   from *Jatropha curcas* 221
   from sorghum 59
   from stillage digestion 23
   from sugar beet 110–11
   from sunflower 176–8
   in thermal conversion 5
bioprospecting 396
biorefinery concept 3
biorenewables 3
black grass 136
Bradyrhizobium japonicum 234,
   237
Brassica napus and other spp. *see*
   oilseed rape
Brazil
   agricultural productivity,
      sugarcane 82, 86–8
   Agroecological Zoning for
      Sugarcane 96–8
   internal energy sources *80, 81*
   National Alcohol Program
      (Proálcool) 6, 78–80
broom grass *(Thysanolaena
   latifolia) 332,* 334

C_4 plants 56, 81–2, 306
   switchgrass 344–5

cake
  from *Jatropha curcas* 213
  from sugarcane 91
  from sunflower 178–9
caña brava *(Gynerium sagittatum)* 332–3, 334
canker in oilseed rape 135
Canola *see* oilseed rape
cell wall composition
  maize 39–47
  wheat 14–18
cellulose synthesis, maize 40–7
  see also lignocelluloses and their residues
cellulosic ethanol 47–9
  from sugarcane 91
cetane number (diesel combustion) 129, 218
Charlock *(Sinapis arvensis)* 136
China, Three-North Shelterbelt Programme 279–80
Claviceps afriana (Ergot), in sorghum 63
Clean Energy Myth, Time magazine cover, 2008 28–9
cleavers 136
club root in oilseed rape 135–6
Colza *see* oilseed rape
combined heat and power (CHP) systems 4
corn *see* maize
corn-ethanol 27–9
cornstarch 34–6
Cotesia flavipes 84
Cundiff's formula for TSSj 69
cyanobacteria *see* microalgae

Dasineura marginemtorquens 268
deforestation by sugarcane production 95–6
diacylglycerol acyltransferase (DGAT) 160–1
Diatraea spp. 84–5
'Direct to Ethanol' covered-raceway photobioreactor 400, *402*

diseases *see* pests, diseases and weed control
distillation 6
dry distillers grains (DDGs) 37

Economics of Climate Change. The Stern Review. (2006) 2
ecosystem damage and protection, sugarcane production 95–9
Elaeis guineensis see palm oil
elephant grass (European) *see Miscanthus* spp.
emissions from transport 2
Empoasca fabae, willow damage by 261
energy and carbon balances
  giant reed 325–6
  Jatropha curcas cultivation 218–23
  oilseed rape cultivation 138–40
  poplars 292–3
  pre21st C *vs.* 21st C *3*
  sugarcane ethanol production, Brazil 92–5
  sunflower cultivation and processing 179–81
  switchgrass 367, *368*
  US energy supply from bioethanol 37–8
  willow cultivation 270–1
energy-crops *see* bioenergy crops
Energy Independence and Security Act, 2007 (USA) 27–8
Energy Policy Act, 1978 (USA) 27–9
enzymes and products from oils *242*
ERA hybrids, maize 29–30
Erianthus spp. 82
Eucheuma cottonii, bioethanol production 384–9
Eucheuma, manual harvesting of 391
European Biodiesel Board 120
European *Miscanthus* Improvement Project 302–3

fatty acid methyl esters (FAME) 6, 117, 150, 151, 241–3

Subject Index

fatty acids
  biosynthesis in *Pongamia* 244–5, 246–50
  biosynthesis in soybean seeds *154*
  Jatropha curcas 208
  sunflower 173–4
fermentation and distillation 4–8
  Jatropha curcas 213
  maize 36–7, 47–9
  poplar 283
  sugarcane 90–5
  sweet sorghum juice 59–60, 68–74
  willow 269
feruloylation of AX and GAX 15, 18, 19, 21
filter cake *see* cake
fire, first use 1
first generation biofuel 4, 7
fossil-fuels 1–3
Fucus, manual harvesting of 391
fungal diseases in sorghum 62–4
Fusarium spp. in sorghum 63

Galium aparine 136
gasification 5
genes and genetic loci
  BAHD genes (barley, rice, wheat) 20–1, *22*
  bm genes (brown midrib, maize) 44–5
  CesA genes (glycosyl-transferase, maize) 41, 43
  COMT gene (maize) 45–6
  Csl genes (cellulose synthase-like) 43–4
  du1 (sucrose synthase, maize) 35–6
  FAD genes (soybean) 153–4, 156–7
  Fat B (soybean) 153–4, 157
  GT genes (barley, wheat) 19–21
  IRX (xylan synthesis, barley, rice, wheat) 20
  Kor gene (cellulose synthesis) 42
  sh1 (sucrose synthase, maize) 35–6

genetically modified (GM)
  brassicas 117
  Pongamia 251–2
  poplars 276
  soybeans 155–7
  sugarcane 83
genome
  switchgrass 347–9
  Zea mays B73 32
giant reed 323–31
  biology 323–4
  energy and carbon balances 325–6
  grazing resistance 325
  harvesting 331
  heavy metal uptake 325
  nitrogen fixation in 329
  non-biomass uses 324–5
  vs. Miscanthus spp. 330–1
  water-use efficiency 331
  yields 326–31
β-glucan, wheat endosperm cell walls 15
glucuronoarabinoxylans (GAXs) 18, 43–4
glycerol production from oilseed rape 127–8
Glycine spp. *see* soybeans
glycosyl-transferase subunits 41
Gracilaria, manual harvesting of 391
grain and seed crops 8
greenhouse gas (GHG) emissions
  balance in energy-crop production 10–11
  reviews of 2
Gynerium sagittatum (caña brava) *332–3*, 334
gypsum in sugarcane production 86–7

harvest index (HI) 39
Hedges, Isaac 57
Helianthus annuus see sunflower
hemicellulose synthesis, maize 43–4
high-oleic soybean oil 157–9
husk, sunflower 179

hydrodeoxygenation (HDO) in biodiesel production 150–1

India
  diesel production 196–7
  Kappaphycus farming, Kerela *389*
  Planning Commission, *Jatropha* cultivation 197
indirect land-use change (ILUC) 28–9, 95–9, 140
inositol from sugar beet 106
Intergovernmental Panel on Climate Change (IPCC) 2
internal energy sources
  Brazil *80, 81*
  world *81*
irrigated water usage, corn-ethanol production 29

Jatropha curcas 196–227
  beliefs, claims and facts *215*
  biochemical composition 202–3, *204, 205*
  bioethanol from 198
  biofuels from 207–11
  biomethane from *221*
  botanical description 198–200
  crop improvement 216–18
  cultivation 201–2
  distribution and ecological requirements 200–1
  economic balance 223–5, *226*
  energy and environmental considerations 218–23
  fatty acid composition *208*
  fermentation 213
  global cultivation and production 196–8
  greenhouse-gas balance 223
  land reclamation and hedging 211–13, 214
  meal (cake) 213
  medicinal and veterinary uses 213–14
  methanol from 209–11
  oil composition and characteristics *209*
  pests and diseases 214, *215*
  soap manufacture 213
  toxicity 203–7
  transesterification *210*
  utilisation summary *212*
  vernacular names for *199*
'just-in-time' harvesting, sugar beet 112

Kappaphycus manual harvesting, Kerela *389*, 391
kelp *see Laminaria*

Laminaria
  biofuel production 384, 394–6
  global harvest 389
  manual harvesting of 391
  mechanical harvesting of *393*
land efficiency of energy crops 197
land-use changes, sugarcane production, Brazil 95–9
leaf rust, poplar 288, *289*
light leaf spot in oilseed rape 135
light refractometers, sugar content measurement *67, 68*
lignin synthesis, maize 44–6
lignocelluloses and their residues
  from maize 37–49
  from sunflower 178–9
  switchgrass 357–61
lignocellulosic crops 4

macroalgae 384–96
  biofuel production 391–6
  biomethane fermentation *394–5*
  bioprospecting 396
  chemical composition 394–6
  commercial algal cultivation *385–8*
  energy crop production 384–96
  food production 384
  postharvesting pretreatment 391–4
Macrocystis pyrifera, energy content 384

Mahanarva spp. 84–5
maize 27–50
  B73 genome 32
  cell wall composition and synthesis 39–47
  celluloses and lignin 37–49
  corn-ethanol 27–9, 36–9
    irrigated water usage 29
  corn stover 37–40
    chemical composition 46–7
    processing 47–9
    removal 49–50
  cornstarch 34–6
  critical growth stages 31
  fermentation and distillation 36–7, 47–9
  grain processing 36–7
  grain-yield traits 31–2
  heterosis (hybrid vigour) 32–4
  nitrogen-use efficiency (NUE) 32–3
  yield improvement 29–32
    and acreage *30*
Malaysian Palm Oil Council (MPOB) 193
Melampsora larici-populina 288, *289*
Metarhizium anisopliae 84–5
methanol production 5
  from *Jatropha curcas* 209–11
methyl tertiary butyl ether (MTBE), substitution of 28
microalgae
  advantages as feedstock 396–7
  Biofuel Value Chain *398, 399*
  closed *vs.* open photobioreactors *404–12*
  conversion routes and strategies 410–11, *413*
  cultivar selection and enhancement 397–402
  cultivation systems 402–8
  'Direct to Ethanol' covered-raceway photobioreactor 400, *402*
  extraction 409–10
  global biodiesel production *382*
  harvesting 408–9, *412*
  oil content of various spp. *401*
  policy and regulation 412–14
  triglyceride costs per gallon *400*
Miscanthus spp. 301–17
  biodiversity impact 316
  bioethanol from 315–16
  biomass utilisation 314–15
  breeding targets 309–11
  C_4 pathway 306
  energy balance 315–16
  European *Miscanthus* Improvement Project 302–3
  frost tolerance 307–8
  germplasm collection 311–13
  harvesting 314
  interspecific hybrids 82–3
  life-cycle analysis 315–16
  modelling yield 309
  nutrient requirements 306
  origins, taxonomy and biology 303–5
  pests, diseases and weed control 308–9
  propagation 313–14
  radiation capture efficiency 308, 311
  RELU-Biomass project 271
  visual impact 316–17
  vs. giant reed 330–1
  water requirements 306–7
MixAlco from sorghum biomass 59
modelling growth and yield
  Miscanthus spp. 309
  poplars 288–90
molasses
  from beet *90*.107, 109–10
  from cane *90*
moonshine (illegal whiskey) production from sorghum 58–9
mosaic viruses, sugarcane 83

Narenga spp. 82
neutral detergent fibre (NDF) 73
nitrogen fixation *see* biological nitrogen fixation (BNF)

nitrogen-use efficiency (NUE),
    maize 32–3
NO$_X$ emissions, soybean
    biodiesel 152

oil crisis, first (1970s) 1–2, 57–8
oil crops 9
oilseed rape 116–42
    biodiesel production and
        properties 121–4, 127–31
    cooking oils from 121
    current and potential
        production *120*
    disease, pest and weed
        control 135–6
    double-low varieties 117
    energy economy of 138–40
    fatty acid composition 129,
        130–1
    fertiliser use 137
    financial considerations 125–7
    future developments 141–2
    global production and
        yields 116–20, 123–4
    glycerol production 127–8
    land-use change 140
    oil composition 129–30
    rapeseed meal (RSM) 127–8
    straw 128
    yield improvement 131–7
orange rust disease, sugarcane 85

palm oil 187–95
    bioethanol from 191, 195
    economics 192
    global production 187–9
    issues and challenges 192–4
    livestock crop integration
        (LCI) 193
    palm oil *vs.* palm kernel oil
        187
    potential uses 189–92
    research and development
        194–5
    yield *vs.* other crops *190*
Panicum virgatum see switchgrass

pectins from sugar beet 106
pests, diseases and weed control
    Jatropha curcas 214, *215*
    Miscanthus spp. 308–9
    oilseed rape 135–6
    poplars 288, *289*
    sorghum 62–4
    sugarcane borers 84–6
    willow 261, 267–8
phenylpropanoid pathway 44–5
phorbol esters in *Jatropha* 205–7,
    238–9
physic nut *see Jatropha curcas*
phytoplankton *see* microalgae
Pichia stipitis 48–9
polysaccharides in cell walls
    digestion of 18
    marine algae 396
    switchgrass 358
    wheat *16,* 17–18
Pongamia pinnata 233–54
    background 233–4
    biodiesel production 236–41
    botanical description 238–40
    domestication *239*
    environments conditions 236,
        239–41
    future developments 252–4
    nitrogen fixation 234–6, *237*
    propagation 251–2
    seed oil properties and
        processing 241–50
    sustainability 233–4, 236
POP-EUROFACE experiment,
    poplars 290
poplar
    fermentation 283
poplars 275–94
    characteristics and uses
        275–6
    classification and
        hybridisation 276–80
    economic balance 290–2
    effects of global climate
        change 288–90
    energy balance 292–3

Subject Index

environmental impact 287
genetic modification 276
life-cycle assessment (LCA) 290
modelling growth and
  yield 288–90
pests and diseases 288, *289*
POP-EUROFACE
  experiment 290
short-rotation coppice 280–4,
  286–90
water and nutrient
  requirements 287–8
wood quality 285–6
yield 284–5, 291
Populus spp. *see* poplars
Porphyra
  biofuel production 389
  farming *390*
  manual harvesting of 391
potato leaf hopper *(Empoasca fabae)*,
  willow damage by 261
press-cake *see* cake
purging nut *see Jatropha curcas*
Pyrenopeziza brassicae 135
pyrolysis 5
pyruvate orthophosphate dikinase
  (PPDK) 35–6

radiation capture efficiency,
  Miscanthus spp. 308, 311
rape methyl ester (RME) 117
rapeseed meal (RSM) 127–8
Raphanus raphanistrum 136
reeds, subtropical and
  tropical 322–34
  background 322–3
  broom grass *(Thysanolaena
    latifolia) 332*, 334
  caña brava *(Gynerium
    sagittatum) 332–3*, 334
  giant reed *(Arundo donax)*
    323–31
RELU-Biomass project, UK 271
Renewable Transport Fuel Obligation
  (UK) 111
Rhizobium leguminosarum 234–6

Rothamsted, England, Broadbalk
  experiment (wheat) 23
runch *(Raphanus raphanistrum)* 136

Saccharomyces cerevisiae 6, 48–9
Saccharum spp. *see* sugarcane
Salix spp. *see* willow
Saltend, England, biofuel facility 13
Sargassum horneri, bioethanol
  production 384
Sclerostachya spp. 82
Sclerotinia sclerotiorum 135
Shen Nung 381
short-rotation coppice
  poplars 280–4
  principles and practice 281–4
  willow 262
simultaneous saccharification and
  fermentation (SSF), corn-ethanol
  production 36–7
Sinapis arvensis 136
sorghum (sweet sorghum, sorgo)
  56–74
  biofuel from 57–8
  biomass uses 59
  biomethane from 59
  byproduct utilisation 73–4
  C_4 pathway 56
  cultivars 60–2
  cultivation history 56–7
  diseases 62–4
  fermentation 68–72
  illegal whiskey (moonshine)
    production 58–9
  integrated pest management 64
  juice
    fermentation and
      distillation 59–60, 68–74
    sucrose *vs.* fructose 70–1
    sugar content 66–7
    sugar extraction methods 60
  milling (traditional) 58, *59*
  planting density 64–5
  seed head removal 67–8, *69*
  soil fertility and fertiliser
    regimes 65–6

sorghum (sweet sorghum, sorgo) (*continued*)
   Sorghum bicolour 56
   Sorghum halepense 56
Sorgo or the Northern Sugar Plant (1863) 57
soybeans 148–61
   biodiesel production and properties 149, 150–3, 157–61
   bioethanol from 150
   Bradyrhizobium japonicum nodules 234, *237*
   cold-flow properties of biodiesel *vs.* petroleum *159*
   cultivation history 148–9
   fatty acid biosynthesis *154*
   global production 149
   high-oleic oil 157–9
   NO_X emissions 149
   oil quality and volume improvement 151–5, 159–61
   seed composition 149–51
Spanish cane *see* giant reed
spittlebug 84–5
starch biosynthesis 35–6
stem rot in oilseed rape 135
Stern, Nicholas 2
straight vegetable oil (SVO) fuel 183
sucrose synthase isoforms 35–6
sugar and ethanol production phases *90*
sugar beet 104–14
   biofuels from 109–11
   cane *vs.* beet 107, *108*
   cultivation history 104–5
   future developments 111–14
   growth 106–7
   production efficiency 108–9
   products and byproducts 105–6
sugar crops 8–9
sugarcane 77–99
   bioethanol from 6, 78–82, 90–5, 97–8
   cane *vs.* beet 107, *108*
   cultivation history 77–80
   fermentation 90–5

land-use change 95–9
mechanical harvesting 88–9
National Alcohol Program (Proálcool), Brazil 6, 78–80, 92–5
nutrient requirements 86–8, 89
pests and diseases 84–6
production volumes in recent times *78*
products and coproducts 90–2
residue management 88–9
Saccharum Complex breeding and engineering 81–4
varieties 81–4
water use 89–90
sugarcane borers 84–6
sunflower 165–84
   biodiesel production and trade 169–76, 181–2
   biogas production 176–8
   cultivated area and yield *166, 167*
   energy balance 179–81
   fatty acid composition 173–4
   global cultivation and production 165–9
   issues and future perspectives 182–4
   lignocellulosic residues 178–9
   oil quality 171–6
      genetic and environmental factors 173–6
      predicting, simulation models 173–6
sustainability
   Jatropha curcas 196
   palm oil 124, 189
   sugarcane 91
   switchgrass 368–9
   wheat production for biofuel 23
sweet sorghum (sorgo) *see* sorghum
switchgrass 341–69
   background 341–2
   bioethanol from 356–68
   breeding programmes 349–53
   C_4 pathway 344–5
   conversion pretreatment 361–6

distribution 342–4
energy balance 367, *368*
fertility 355–6
genetics 347–9
growth 344–6
harvesting 356
lignocellulosic
  composition 357–61
morphology 344
mycorrhizal fungi 347
seed production 357
seed sowing and
  establishment 353–5
sustainability 368–9
taxonomy 344
weed control 354–5
sycamore, 'silage' project 280

Telchin licus 85
theoretical ethanol potential
  (TEP) 38–9
theoretical ethanol yields (TEY) 39
Three-North Shelterbelt Programme,
  China 279–80
Thysanolaena latifolia (broom
  grass) *332, 334*
tiger grass *see* broom grass
  (*Thysanolaena latifolia*)
tobacco cellulose 48
total plant above ground dry matter
  (TDM) 39
total soluble sugars in juice (TSSj)
  69
triacylglycerols (TAG)
  biosynthesis in *Pongamia* 244–5,
    246–50
  in *Pongamia* 243–50
  in soybean oil 151, 160–1
Trichoderma reesei 47
trimethylglycine from sugar beet 106
Triticum spp. *see* wheat
turnip yellows virus 136

Ulva
  biofuel production 384, 389, 394–6
  bloom *391*

global harvest 391
manual harvesting of *392*
USA
  bioethanol production 6
  United States Department of
    Agriculture (USDA), corn-
    ethanol policies 28–9

vinasse
  from sorghum 73–4
  from sugarcane 90–1
vinegar, preventing production in
  fermentation 68, *69*

water demand for crop production 2
  giant reed 331
  maize, corn-ethanol 29
  Miscanthus spp. 306–7
  poplars 287–8
  sugarcane 89–90
weed control *see* pests, diseases and
  weed control
wet distillers grains with solubles
  (WDGS) 37
wheat 13–23
  aleurone composition 17
  annual grain harvest from 13–14
  arabinoxylan biosynthesis
    in 18–21, *22*
  cell wall
    composition 14–18
    structure and digestibility 18
  endosperm composition 14–17
  histological comparison of
    tissue *15*
  polysaccharide composition of cell
    walls *16*
  starch content 13
  sustainability of production for
    biofuel 23
whole stillage 37
willow 259–72
  background 259–60
  basketry industry 264
  biology 261
  biomass production 261–3

willow (*continued*)
   classification 260–1
   cultivation 263
   energy and carbon balances 270–1
   environmental benefits 270–1
   fermentation 269
   genetics and breeding 263–6, 269–70
   nutrient requirements 262, 267
   pests and diseases 261, 267–8
   short-rotation coppice 262
   yield improvement 266–7

Wilton, England, biofuel facility 13
World Wildlife Foundation (WWF), reports on sugarcane production 96–7

Xanthomonas albilineans 83
xylose fermentation 48–9

Zea mays see maize
Ziegler, Jean 29